Advances in Intelligent Systems and Computing

182

Editor-in-Chief

Prof. Janusz Kacprzyk
Systems Research Institute
Polish Academy of Sciences
ul. Newelska 6
01-447 Warsaw
Poland
E-mail: kacprzyk@ibspan.waw.pl

For further volumes:
http://www.springer.com/series/11156

Ajith Abraham and Sabu M. Thampi (Eds.)

Intelligent Informatics

Proceedings of the International Symposium on Intelligent Informatics ISI'12 Held at August 4–5 2012, Chennai, India

Editors
Dr. Ajith Abraham
Machine Intelligence Research Labs
(MIR Labs)
Scientific Network for Innovation and
Research Excellence
Auburn
Washington
USA

Dr. Sabu M. Thampi
Indian Institute of Information Technology
and Management - Kerala (IIITM-K)
Technopark Campus
Trivandrum
Kerala
India

ISSN 2194-5357
e-ISSN 2194-5365
ISBN 978-3-642-32062-0
e-ISBN 978-3-642-32063-7
DOI 10.1007/978-3-642-32063-7
Springer Heidelberg New York Dordrecht London

Library of Congress Control Number: 2012942843

Printed on acid-free paper

Springer is part of Springer Science+Business Media (www.springer.com)

Preface

This book contains a selection of refereed and revised papers originally presented at the first International Symposium on Intelligent Informatics (ISI'12), August 4–5, 2012, Chennai, India. ISI'12 provided an international forum for sharing original research results and practical development experiences among experts in the emerging areas of Intelligent Informatics. ISI'12 was co-located with International Conference on Advances in Computing, Communications and Informatics (ICACCI-2012).

Credit for the quality of the conference proceedings goes first and foremost to the authors. They contributed a great deal of effort and creativity to produce this work, and we are very thankful that they chose ISI'12 as the place to present it. All of the authors who submitted papers, both accepted and rejected, are responsible for keeping the ISI papers program vital. The total of 165 papers coming from 17 countries and touching a wide spectrum of topics related to both theory and applications were submitted to ISI'12. Out of them 54 papers were selected for regular presentations.

An event like this can only succeed as a team effort. We would like to acknowledge the contribution of the program committee members and thank the reviewers for their efforts. Many thanks to the honorary chair Lotfi Asker Zadeh, the general chair Axel Sikora as well as the program chairs Adel M. Alimi, Juan Manuel Corchado and Michal Wozniak. Their involvement and support have added greatly to the quality of the symposium.

We wish to express our thanks to Thomas Ditzinger, Senior Editor, Engineering/Applied Sciences Springer-Verlag for his help and cooperation.

August 2012

Ajith Abraham
Sabu M. Thampi

ISI'12 Conference Committee

Honorary Chair

Prof. Lotfi Asker Zadeh	Founder of Fuzzy Logic, University of California Berkeley, USA

General Chair

Axel Sikora	University of Applied Sciences Offenburg, Germany

Program Chairs

Adel M. Alimi	University of Sfax, Tunisia
Juan Manuel Corchado Rodriguez	University of Salamanca, Spain
Michal Wozniak	Wroclaw University of Technology, Poland

Publication Chairs

Ajith Abraham	Machine Intelligence Research Labs (MIR Labs), USA
Sabu M. Thampi	Indian Institute of Information Technology and Management - Kerala, India

TPC Members

Abdelmajid Khelil	TU Darmstadt, Germany
Aboul Ella Hassanien	University of Cairo, Egypt
Agostino Bruzzone	University of Genoa, Italy
Ajay Singh	Multimedia University, Malaysia
Ajay Jangra	UIET Kurukshetra University, Kurukshetra, India
Algirdas Pakaitas	London Metropolitan University - North Campus, United Kingdom

Amer Dawoud	University of Southern Mississippi, Canada
Anirban Kundu	Kuang-Chi Institute of Advanced Technology, P.R. China
Anjana Gosain	Indraprastha University, India
Ash Mohammad Abbas	Aligarh Muslim University, India
Asrul Adam	Universiti Teknologi Malaysia, Malaysia
Athanasios Pantelous	University of Liverpool, United Kingdom
Atul Negi	University of Hyderabad, India
Avinash Keskar	V.N.I.T., India
Azian Azamimi Abdullah	Universiti Malaysia Perlis, Malaysia
Belal Abuhaija	University of Tabuk, Saudi Arabia
Brajesh Kumar Kaushik	Indian Institute of Technology, Roorkee, India
Cecilia Nugraheni	Parahyangan Catholic University, Indonesia
Crina Grosan	Norwegian University of Science and Technology, Norway
Dayang Jawawi	Universiti Teknologi Malaysia, Malaysia
Dhiya Al-Jumeily	Liverpool John Moores University, United Kingdom
Edara Reddy	Nagarjuna University, India
Edward Williams	PMcorp, USA
Emilio Jimanez Macaas	University of La Rioja, Spain
Eri Shimokawara	Tokyo Metropolitan University, Japan
Farrah Wong	Universiti Malaysia Sabah, Malaysia
Fatos Xhafa	UPC, Barcelona Tech, Spain
G. Ganesan	Adikavi Nannaya University, India
Gancho Vachkov	Yamaguchi University, Japan
Georgi Dimirovski	Dogus University of Istanbul, Turkey
Ghulam Abbas	Liverpool Hope University, United Kingdom
Gregorio Romero	Universidad Politecnica de Madrid, Spain
Gurvinder-Singh Baicher	University of Wales Newport, United Kingdom
Habib Kammoun	University of Sfax, Tunisia
Hanif Ibrahim	Universiti Teknologi Malaysia, Malaysia
Ida Giriantari	Udayana University, Bali, Indonesia
Imran Bajwa	The Islamia University of Bahawalpur, Pakistan
Issam Kouatli	Lebanese American University, Lebanon
J. Mailen Kootsey	Simulation Resources, Inc., USA
Jasmy Yunus	University of Technology Malaysia, Malaysia
Javier Bajo	University of Salamanca, Spain
Jayashree Padmanabhan	Anna University, India
Jeng-Shyang Pan	National Kaohsiung University of Applied Sciences, Taiwan
Jiri Dvorsky	Technical University of Ostrava, Czech Republic
Josip Lorincz	University of Split, Croatia
K. Thangavel	Periyar University, Salem, India
Kambiz Badie	Iran Telecom Research Center, Iran

Kenneth Nwizege	University of SWANSEA, United Kingdom
Kumar R.	SRM University, India
Kumar Rajamani	GE Global Research, India
Lee Tian Soon Lt	Multi Media University, Malaysia
Mahendra Dixit	SDMCET, India
Manjunath Aradhya	Dayananda Sagar College of Engineering, India
Manu Sood	Himachal Pradesh University, India
Mario Kappen	Kyushu Institute of Technology, Japan
Martin Tunnicliffe	Kingston University, United Kingdom
Mikulas Alexik	University of Zilina, Slovakia
Mohamad Noh Ahmad	Universiti Teknologi Malaysia, Malaysia
Mohamed Baqer	University of Bahrain, Bahrain
Mohamed Dahmane	University of Montreal, Canada
Mohand Lagha	Saad Dahlab University of Blida - Blida - Algeria, Algeria
Mohsen Askari	University of Technology, Sydney, Iran
Mokhtar Beldjehem	Sainte-Anne's University, Canada
Monica Chis	Siemens PSE Romania, Romania
Muhammad Nadzir Marsono	Universiti Teknologi Malaysia, Malaysia
Narendra Bawane	RTM, Nagpur University, India
Nico Saputro	Southern Illinois University Carbondale, USA
Nor Hisham Khamis	Universiti Teknologi Malaysia, Malaysia
Obinna Anya	Liverpool Hope University, United Kingdom
Otavio Teixeira	Centro Universitajrio do Estado do Pari (CESUPA), Brazil
Petia Koprinkova-Hristova	Bulgarian Academy of Sciences, Bulgaria
Praveen Srivastava	Bits Pilani, India
Raees Khan	B.B.A. University, Lucknow, India
Rajan Vaish	University of California, USA
Rubita Sudirman	Universiti Teknologi Malaysia, Malaysia
S.D. Katebi	Shiraz University, Shiraz, Iran
Sami Habib	Kuwait University, Kuwait
Sasanko Gantayat	GMR Institute of Technology, India
Satish Chandra	Jaypee University of Information Technology, India
Sattar Sadkhan	University of Babylon, Iraq
Satya Kumara	Udayana University, Bali, Indonesia
Satya Ghrera	Jaypee University of Information Technology, India
Sayed A. Hadei	Tarbiat Modares University, Iran
Shubhalaxmi Kher	Arkansas State University, USA
Shuliang Sun	Fuqing Branch of Fujian Normal University, P.R. China
Siti Mariyam Shamsuddin	Universiti Teknologi Malaysia, Malaysia
Smriti Srivastava	Netaji Subhas Institute of Technology, India
Sotiris Kotsiantis	University of Patras, Greece
Sriparna Saha	IIT Patna, India

Suchitra Sueeprasan	Chulalongkorn University, Thailand
Sung-Bae Cho	Yonsei University, Korea
Teruaki Ito	University of Tokushima, Japan
Theodoros Kostis	University of the Aegean, Greece
Usha Banerjee	College of Engineering Roorkee, India
Vaclav Satek	Brno University of Technology, Czech Republic
Vatsavayi Valli Kumari	Andhra University, India
Veronica Moertini	Parahyangan Catholic University, Bandung, Indonesia
Viranjay Srivastava	Jaypee University of Information Technology, Shimla, India
Visvasuresh Victor Govindaswamy	Texas A&M University-Texarkana, USA
Vivek Sehgal	Jaypee University of Information Technology, India
Wan Hussain Wan Ishak	Universiti Utara Malaysia, Malaysia
Xu Han	Univ. of Rochester, USA
Yahya Elhadj	Al-Imam Muhammad Ibn Sau Islamic University, Saudi Arabia
Yu-N Cheah	Universiti Sains Malaysia, Malaysia
Zaliman Sauli	Universiti Malaysia Perlis, Malaysia

Organising Committee (RMK Engineering College)

Chief Patron

R.S. Munirathinam

Patrons

Manjula Munirathinam
R.M. Kishore
R. Jothi Naidu
Yalamanchi Pradeep
DurgaDevi Pradeep
Sowmya Kishore

Advisory Committee

T. Pitchandi
M.S. Palanisamy
Elwin Chandra Monie
N.V. Balasubramanian
Sheshadri
K.A. Mohamed Junaid
K.K. Sivagnana Prabu

Convener

K. Chandrasekaran

Secretary

K.L. Shunmuganathan

ISI'12 Logo

Host Institution

Technical Sponsors

Contents

Data Mining, Clustering and Intelligent Information Systems

Computer Networks and Distributed Systems

Mining Top-K Frequent Correlated Subgraph Pairs in Graph Databases

Li Shang and Yujiao Jian

Abstract. In this paper, a novel algorithm called KFCP(top K Frequent Correlated subgraph Pairs mining) was proposed to discover top-k frequent correlated subgraph pairs from graph databases, the algorithm was composed of two steps: co-occurrence frequency matrix construction and top-k frequent correlated subgraph pairs extraction. We use matrix to represent the frequency of all subgraph pairs and compute their Pearson's correlation coefficient, then create a sorted list of subgraph pairs based on the absolute value of correlation coefficient. KFCP can find both positive and negative correlations without generating any candidate sets; the effectiveness of KFCP is assessed through our experiments with real-world datasets.

1 Introduction

Graph mining has been a significant research topic in recent years because of numerous applications in data analysis, drug discovery, social networking and web link analysis. In view of this, many traditional mining techniques such as frequent pattern mining and correlated pattern mining have been extended to the case of graph data. Previous studies mainly focus on mining frequent subgraph and correlated subgraph, while little attention has been paid to find other interesting patterns about frequent correlated subgraph.

There is one straightforward solution named FCP-Miner [8]to the problem mentioned above, FCP-Miner algorithm employs an effective "filter and verification" framework to find all frequent correlated graphs whose correlation with a query graph is no less than a given minimum correlation threshold.However, FCP-Miner

Li Shang · Yujiao Jian
Lanzhou University, P.R. China
e-mail: lishang@lzu.edu.cn, 18993177580@189.cn

A. Abraham and S.M. Thampi (Eds.): Intelligent Informatics, AISC 182, pp. 1–8.
springerlink.com

has several drawbacks: First, the number of candidates of FCP-Miner algorithm are large, since it processes a new subgraph f by CGSearch [7] to obtain its candidate set. Second, it is difficult for users to set an appropriate correlation threshold for each specific query graph, since different graph databases have different characteristics. Finally, FCP-Miner is not complete, due to the use of the skipping mechanism, this method cannot avoid missing some subgraph pairs.

To address these problems, in this paper, we propose an alternative mining algorithm KFCP for discovering the top-k frequent correlated subgraph pairs. The main contributions of this paper are briefly summarized as follows.

1. We propose an alternative mining task of finding the top-k frequent negatively and positively correlated subgraph pairs from graph databases, which allows users to derive the k most interesting patterns. It is not only significant, but also mutually independent and containing little redundancy.
2. We propose an efficient algorithm KFCP by constructing a co-occurrence frequency matrix. The method avoids the costly generation of a large number of candidate sets.
3. We show that KFCP is complete and correct; extensive experiments demonstrate that the approach is effective and feasible.

The remainder of this paper is organized as follows. Section 2 reports the related work. In section 3, basic concepts are described. Section 4 introduces our algorithm KFCP in detail and Section 5 shows the experimental results on two real datasets. Finally, we draw conclusions in section 6.

2 Related Work

Correlation mining attracts much attention; it plays an essential role in various types of databases, such as market-basket data [1, 2, 3, 4], multimedia data [5], stream data [6], and graph data [7, 8]. For market-basket data [1, 2, 3, 4], a number of correlated measures were proposed to discover all correlated items, including the chi-square χ^2 test [1], h-confidence [2], Pearson's correlation coefficient [3], etc. All the above works set a threshold for the correlation measure, except [4], which studied the top-k mining. For multimedia data, correlated pattern mining based on multimedia data [5] has been proposed to discover such cross-modal correlations. In the content of stream data, lagged correlation [6] has been presented to investigate the lead-lag relationship between two time series. On the correlation mining in graph mining, there are many previous researches on correlation discovery; CGS [7] has been proposed for the task of correlation mining between a subgraph and a given query graph. The work of [8] aimed to find all frequent subgraph pairs whose correlation coefficient is at least a given minimum correlation threshold.

3 Basic Concepts

Definition 1. (Pearson's correlation coefficient). Pearson's correlation coefficient for binary variables is also known as the " ϕ correlation coefficient". Given two graphs A and B, the Pearson's correlation coefficient of A and B, denoted as ϕ(A,B), is defined as follows:

$$\phi(A,B) = \frac{sup(A,B) - sup(A)sup(B)}{\sqrt{sup(A)sup(B)(1-sup(A))(1-sup(B))}} \tag{1}$$

The range of ϕ(A,B) falls within [-1,1]. If ϕ(A,B) is positive, then A and B are positively correlated, it means that their occurrence distributions are similar; otherwise, A and B are negatively correlated, in other words, A and B rarely occur together.

Definition 2. (Top-K Frequent Correlated subgraph pairs discovery). Given a graph database GD, a minimum support threshold σ and an integer k, we need to find the top k frequent subgraph pairs with the highest absolute value of correlations.

Definition 3. (Co-occurrence Frequency matrix).Given a frequent subgraph set, $F = \{g_1, g_2, \cdots, g_n\}$, co-occurrence frequency matrix, denoted as X, X= (x_{ij}),where for i=1,2,$\cdots$,n and j=1,2,$\cdots$,n.

$$x_{ij} = \begin{cases} freq(g_i, g_j),\ i \neq j \\ freq(g_i),\ i = j \end{cases} \tag{2}$$

Obviously, X is an n$\times$ n symmetric matrix, due to the symmetry, we need retain only the upper triangle part of matrix.

4 KFCP Algorithm

In this section, we describe the details of KFCP which consists of two steps: co-occurrence frequency matrix construction and top-k frequent correlated subgraph pairs extraction.

4.1 Co-occurrence Frequency Matrix Construction

In the co-occurrence frequency matrix construction step, KFCP starts by generating frequent subgraphs set F, then counts the frequency for each subgraph pair (g_i, g_j) of frequent subgraph set F. When the transaction of frequent subgraphs set F is n, the co-occurrence frequency matrix is basically an n$\times$n matrix, where each entry represents the frequency counts of 1- and 2- element of F. The co-occurrence frequency matrix is constructed based on definition 3 by scanning the database once.

Example 1. Fig.1 shows a graph database GD. $|GD|$=10. For minimum support threshold σ=0.4, we can obtain frequent subgraph set F=$\{g_1,g_2,g_3,g_4,g_5\}$, as shown in Fig.2. Then,we can construct the co-occurrence frequency matrix by scanning each one of the transaction graphs. Considering the graph transaction 4(G_4), we increment its count in the matrix depicted by Fig.3, Similarly, we also increment other transaction graphs count, thus we can construct co-occurrence frequency matrix X shown in Fig.4. All of the off-diagonal elements are filled with the joint frequency of co-occurrence of subgraph pairs. For example, the element (x_{34}) indicates the joint frequency of the subgraph pairs (g_3,g_4). On the other hand, every diagonal entry of matrix is filled with the occurrence frequency of the single element set. When σ is varied from 0.4 to 0.3, KFCP generates some new frequent subgraph pairs such as (g_1,g_6), we will increment the count of cell (g_1,g_6) to maintain uniformity of the co-occurrence frequency matrix and this can be done without spending extra cost.

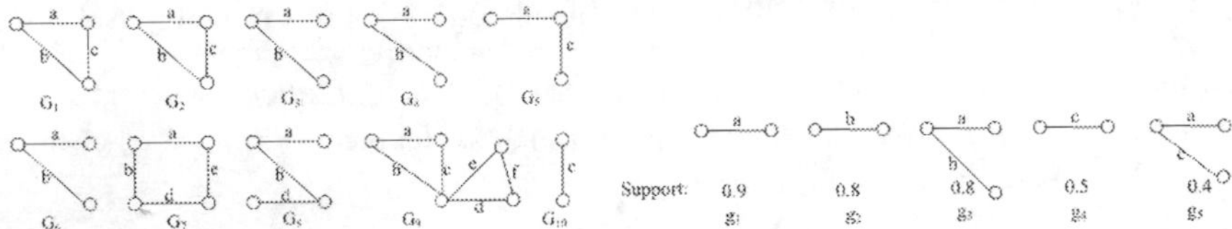

Fig. 1 A Graph Database GD

Fig. 2 Frequent subgraph set

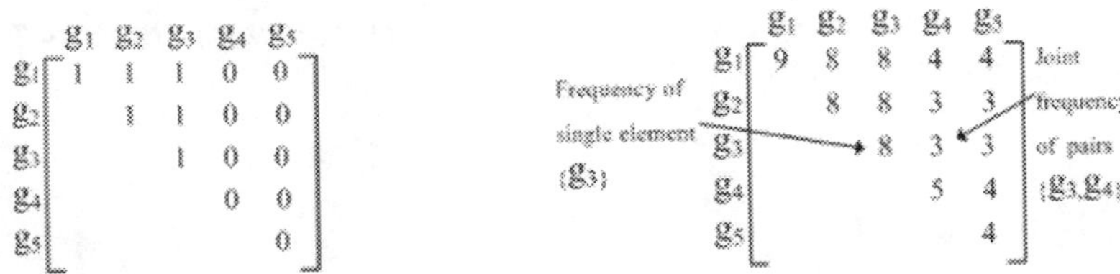

Fig. 3 Frequency matrix of transaction 4

Fig. 4 Co-occurrence frequency matrix X

Table 1 All pairs with their correlation coefficient

Pairs	(g_1,g_2)	(g_1,g_3)	(g_1,g_4)	(g_1,g_5)	(g_2,g_3)	(g_2,g_4)	(g_2,g_5)	(g_3,g_4)	(g_3,g_5)	(g_4,g_5)
Correlation	0.667	0.667	-0.333	0.272	1	-0.5	-0.102	-0.5	-0.102	0.816

4.2 *Top-k Frequent Correlation Subgraph Pairs Extraction*

Once the co-occurrence frequency matrix has been generated, frequency counts of all 1- and 2- element set can be computed fast. Using these frequency counts, KFCP computes the ϕ correlation coefficient of all the frequent subgraph pairs, then extracts the k mostly correlated pairs based on the $|\phi|$.

Example 2. According to the matrix element shown in Fig.4, to compute $\phi(g_3,g_4)$, we note that freq(g_3)=8, freq(g_4)=5, and freq(g_3,g_4)=3, we get sup(g_3)=8/10, sup(g_4)=5/10, and sup(g_3,g_4)=3/10, using equation (1) above, we get $\phi(g_3,g_4)$=-0.5, ϕ correlation coefficients for other subgraph pairs can be computed similarly. Table 1 shows ϕ correlation coefficients for all subgraph pairs. Suppose k=6, with the result in table 1, we know that the absolute value of 6-th pair's correlation coefficient ($|\phi(TL[k])|$)is 0.5, through checking the $|\phi(g_i,g_j)|$ to determine whether it can be pruned or not,four subgraph pairs (g_1,g_4),(g_1,g_5) ,(g_2,g_5),(g_3,g_5) will be deleted,we are able to obtain the top-6 list.

4.3 Algorithm Descriptions

In the subsection, we show the pseudocode of the KFCP algorithm in ALGORITHM 1. KFCP accepts the graph database GD, minimum support σ and an integer k as input, it generates list of top-k strongly frequent correlated subgraph pairs, TL, as output. First, Line 1 initializes an empty list TL of size k. Line 2 enumerates all frequent subgraph pairs by scanning the entire database once. Line 3 constructs co-occurrence frequency matrix.Line 4-9 calculates the correlation coefficient for each surviving pairs from the frequent subgraph set and pushes the pair into the top-k list if the correlation coefficient of the new pair is greater than the k-th pair in the current list.

ALGORITHM 1. KFCP Algorithm

Input: GD: a graph database
σ: a given minimum support threshold
k: the number of most highly correlated pairs requested
Output: TL: the sorted list of k frequent correlation subgraph pairs
1. initialize an empty list TL of size k;
2. scan the graph database to generete frequent subgraph set F (with input GD and σ);
3. construct co-occurrence frequency matrix;
4. for each subgraph pair $(g_i,g_j) \in$F do
5. compute $\phi(g_i,g_j)$;
6. if $|\phi(g_i,g_j)| > |\phi(TL[k])|$ then
7. add subgraph pair $\{(g_i,g_j),|\phi|\}$ into the last position of TL;
8. sort the TL in non-increasing order based on absolute value of their correlation coefficient;
9. Return TL;

Here, we analyze KFCP algorithm in the area of completeness and correctness.

Theorem 1. *The KFCP algorithm is complete and correct.*

Proof. KFCP computes the correlation coefficient of all the frequent subgraph pairs based on exhaustive search,this fact guarantees that KFCP is complete in all aspects. KFCP creates a sorted list of subgraph pairs based on the absolute value of correlation coefficient and prunes all those subgraph pairs whose absolute value of correlation coefficient lower than the k-th pair; this fact ensures KFCP is correct.

5 Experimental Results

Our experiments are performed on a PC with a 2.1GHz CPU and 3GB RAM running Windows XP. KFCP and FCP-Miner are implemented in C++. There are two real datasets we tested: PTE[1]and NCI [2].PTE contains 340 graphs,the average graph size is 27.4. NCI contains about 249000 graphs, we randomly select 10000 graphs for our experiments, the average graph size is 19.95.

Since FCP-Miner is dependent on a minimum correlation threshold θ, in order to generate same result by FCP-Miner we set the θ with the correlation coefficient of the k-th pair from the top-k list generated by KFCP. Fig.5 shows the comparison between KFCP and FCP-Miner on PTE dataset with different values of k, we can obtain the correlation coefficient of the k-th pair shown in Table 2. As the increasing of k, KFCP keeps stable running time,but the performance of FCP-Miner decreases greatly, since when k is large,FCP-Miner cannot avoid generating large number of candidates. Fig.6 displays that performance comparison between KFCP and FCP-Miner on NCI dataset with different support threshold, we vary σ from 0.3 to 0.03, the running time for enumerating all frequent subgraph pairs increases greatly,so the performances of KFCP and FCP-Miner decrease greatly. We also analyze the completeness of KFCP by recording the following experimental findings, as reported in Table 3:(1)% of excess: the percentage of excess pairs by KFCP,calculted as (total number of pairs obtained by KFCP/ total number of pairs obtained by FCP-Miner-1);(2)avg ϕ of excess: the average ϕ value of the excess pairs.We create six NCI datasets, with sizes ranging from 1000 to 10000 graphs, the values of σ and k are fixed at 0.05 and 40, respectively, when k=40, we set the θ with the ϕ of the 40-th pair from the top-k list generated by KFCP. Thus, we obtain θ=0.8. The results verify that FCP-Miner may miss some frequent correlated subgraph pairs, but KFCP is complete.The experimental results confirm the superiority of KFCP in all cases.

Table 2 The correlation coefficient of the k-th pair at varing k

K	10	20	30	40	50	60	70	80	90	100
ϕ of the k-th pair(θ)	0.95	0.92	0.88	0.85	0.82	0.76	0.74	0.7	0.68	0.65

[1] http://web.comlab.ox.ac.uk/oucl/research/areas/machlearn/PTE/.

[2] http://cactus.nci.nih. gov/ncidb2/download.html.

Table 3 The completeness of KFCP compared to FCP-Miner

the size of NCI	1000	2000	4000	6000	8000	10000
% of excess	2.5%	2.1%	1.7%	2.1%	3.3%	4.9%
avg ϕ of excess	0.82	0.82	0.82	0.82	0.82	0.82

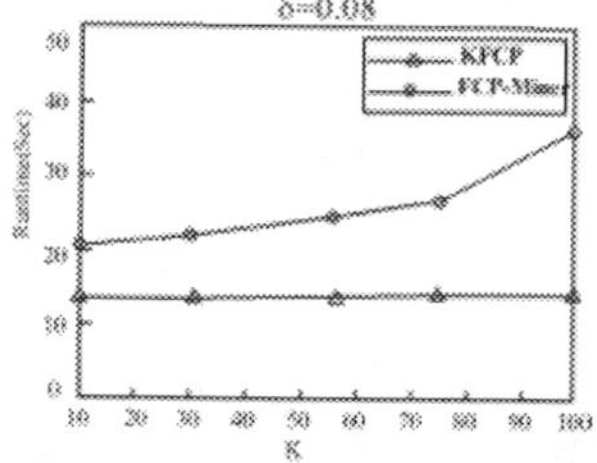

Fig. 5 Runtime comparision on PTE dataset

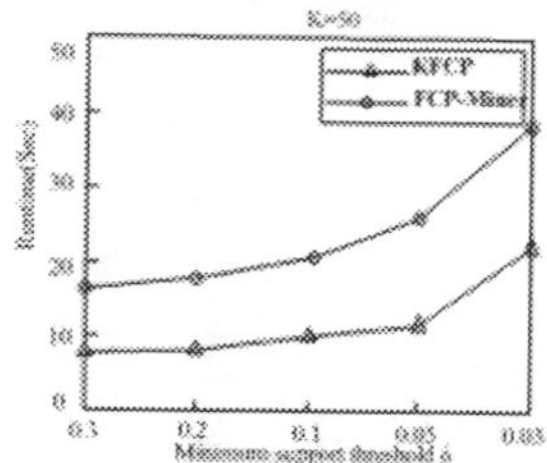

Fig. 6 Runtime comparision on NCI dataset with different support threshold

6 Conclusions

In this paper, we present an algorithm KFCP for the frequent correlation subgraph mining problem. Comparing to existing algorithm FCP-Miner, KFCP can avoid generating any candidate sets. Once the co-occurrence frequency matrix is constructed, the correlation coefficients of all the subgraph pairs are computed and k numbers of top strongly correlated subgraph pairs are extracted very easily. Extensive experiments on real datasets confirm the efficiency of our algorithm.

Acknowledgements. This work was supported by the NSF of Gansu Province grant (1010RJZA117).

References

1. Morishita, S., Sese, J.: Traversing itemset lattice with statistical metric pruning. In: Proc. of PODS, pp. 226–236 (2000)
2. Xiong, H., Tan, P., Kumar, V.: Hyperclique pattern discovery. DMKD 13(2), 219–242 (2006)
3. Xiong, H., Shekhar, S., Tan, P., Kumar, V.: Exploiting a support-based upper bound of Pearson's correlation coefficient for efficiently identifying strongly correlated pairs. In: Proc. ACM SIGKDD Internat. Conf. Knowledge Discovery and Data Mining, pp. 334–343. ACM Press (2004)
4. Xiong, H., Brodie, M., Ma, S.: Top-cop: Mining top-k strongly correlated pairs in large databases. In: ICDM, pp. 1162–1166 (2006)
5. Pan, J.Y., Yang, H.J., Faloutsos, C., Duygulu, P.: Automatic multimedia cross-modal correlation discovery. In: Proc. of KDD, pp. 653–658 (2004)

6. Sakurai, Y., Papadimitriou, S., Faloutsos, C.: Braid: Stream mining through group lag correlations. In: SIGMOD Conference, pp. 599–610 (2005)
7. Ke, Y., Cheng, J., Ng, W.: Correlation search in graph databases. In: Proc. of KDD, pp. 390–399 (2007)
8. Ke, Y., Cheng, J., Yu, J.X.: Efficient Discovery of Frequent Correlated Subgraph Pairs. In: Proc. of ICDM, pp. 239–248 (2009)

Evolutionary Approach for Classifier Ensemble: An Application to Bio-molecular Event Extraction

Asif Ekbal, Sriparna Saha, and Sachin Girdhar

Abstract. The main goal of Biomedical Natural Language Processing (BioNLP) is to capture biomedical phenomena from textual data by extracting relevant entities, information and relations between biomedical entities (i.e. proteins and genes). Most of the previous works focused on extracting binary relations among proteins. In recent years, the focus is shifted towards extracting more complex relations in the form of bio-molecular events that may include several entities or other relations. In this paper we propose a classifier ensemble based on an evolutionary approach, namely differential evolution that enables extraction, i.e. identification and classification of relatively complex bio-molecular events. The ensemble is built on the base classifiers, namely Support Vector Machine, nave-Bayes and IBk. Based on these individual classifiers, we generate 15 models by considering various subsets of features. We identify and implement a rich set of statistical and linguistic features that represent various morphological, syntactic and contextual information of the candidate bio-molecular trigger words. Evaluation on the BioNLP 2009 shared task datasets show the overall recall, precision and F-measure values of 42.76%, 49.21% and 45.76%, respectively for the three-fold cross validation. This is better than the best performing SVM based individual classifier by 4.10 F-measure points.

1 Introduction

The past history of text mining (*TM*) shows the great success of different evaluation challenges based on carefully curated resources. Relations among biomedical entities (i.e. proteins and genes) are important in understanding biomedical

Asif Ekbal · Sriparna Saha · Sachin Girdhar
Department of Computer Science and Engineering,
Indian Institute of Technology Patna, India
e-mail: {asif,sriparna,sachin}@iitp.ac.in

A. Abraham and S.M. Thampi (Eds.): Intelligent Informatics, AISC 182, pp. 9–15.
springerlink.com

phenomena and must be extracted automatically from a large number of published papers. Similarly to previous bio-text mining challenges (e.g., LLL [1] and BioCreative [2]), the BioNLP'09 Shared Task also addressed bio-IE, but it tried to look one step further toward finer-grained IE. The difference in focus is motivated in part by different applications envisioned as being supported by the IE methods. For example, BioCreative aims to support curation of PPI databases such as MINT [3], for a long time one of the primary tasks of bioinformatics. The BioNLP'09 shared task contains simple events and complex events. Whereas the simple events consist of binary relations between proteins and their textual triggers, the complex events consist of multiple relations among proteins, events, and their textual triggers. The primary goal of BioNLP-09 shared task [4] was aimed to support the development of more detailed and structured databases, e.g. pathway or Gene Ontology Annotation (GOA) databases, which are gaining increasing interest in bioinformatics research in response to recent advances in molecular biology.

Classifier ensemble is a popular machine learning paradigm.

We assume that, in case of weighted voting, weights of voting should vary among the various output classes in each classifier. The weight should be high for that particular output class for which the classifier is more reliable. Otherwise, weight should be low for that output class for which the classifier is not very reliable. So, it is a very crucial issue to select the appropriate weights of votes for all the classes in each classifier. Here, we make an attempt to quantify the weights of voting for each output class in each classifier. A Genetic Algorithm (GA) based classifier ensemble technique has been proposed in [5] for determining the proper weights of votes in each classifier. This was developed aiming named entity recognition in Indian languages as well as in English. In this paper we propose a single objective optimization based classifier ensemble technique based on the principle of differential evolution [6], an evolutionary algorithm that proved to be superior over GA in many applications. We optimize F-measure value, which is a harmonic mean of recall and precision both. The proposed approach is evaluated for event extraction from biomedical texts and classification of them into nine predefined categories, namely *gene expression*, *transcription*, *protein catabolism*, *phosphorylation*, *localization*, *binding*, *regulation*, *positive regulation* and *negative regulation*. We identify and implement a very rich feature set that incorporates morphological, orthographic, syntactic, local contexts and global contexts as the features of the system. As a base classifiers, we use Support Vector Machine, naïve-Bayes and instance-based leaner IBk. Different versions of these diverse classifiers are built based on the various subsets of features. Differential evolution is then used as the optimization technique to build an ensemble model by combining all these classifiers. Evaluation with the BioNLP 2009 shared task datasets yield the recall, precision and F-measure values of 42.76%, 49.21% and 45.76%, respectively for the three-fold cross validation. This is better than the best performing SVM based individual classifier by 4.10 F-measure points.

2 Proposed Approach

The proposed differential evolution based classifier ensemble method is described below.

String Representation and Population Initialization: Suppose there N number of available classifiers and M number of output classes. Thus, the length of the chromosome (or vector) is $N \times M$. This implies $D = N \times M$, where D represents the number of real parameters on which optimization or fitness function depends. D is also dimension of vector $x_{i,G}$. Each chromosome encodes the weights of votes for possible M output classes for each classifier. Please note that chromosome represents the available classifiers along with their weights for each class. As an example, the encoding of a particular chromosome is represented below, where M = 3 and O = 3 (i.e., total 9 votes can be possible): 0.59 0.12 0.56 0.09 0.91 0.02 0.76 0.5 0.21 The chromosome represents the following ensemble: The weights of votes for 3 different output classes in classifier 1 are 0.59, 0.12 and 0.56, respectively. Similarly, weights of votes for 3 different output classes are 0.09, 0.91 and 0.02, respectively in classifier 2 and 0.76, 0.5 and 0.21, respectively in classifier 3. We use real encoding that randomly initializes the entries of each chromosome by a real value (r) between 0 and 1. Each entry of chromosome whose size is D, is thus, initialized randomly. If the population size is P then all the P number of chromosomes of this population are initialized in the above way.

Fitness Computation: Initially, all the classifiers are trained using the available training data and evaluated with the development data. The performance of each classifier is measured in terms of the evaluation metrics, namely recall, precision and F-measure. Then, we execute the following steps to compute the objective values.

1) Suppose, there are total M number of classifiers. Let, the overall F-measure values of these M classifiers for the development set be F_i, $i = 1 \ldots M$, respectively.
2) The ensemble is constructed by combining all the classifiers. Now, for the ensemble classifier the output label for each word in the development data is determined using the weighted voting of these M classifiers' outputs. The weight of the class provided by the ith classifier is equal to $I(m,i)$. Here, $I(m,i)$ is the entry of the chromosome corresponding to m^{th} classifier and i^{th} class. The combined score of a particular class for a particular word w is:
$f(c_i) = \sum I(m,i) \times F_m, \quad \forall m = 1 : M \ \& \ op(w,m) = c_i$ Here, $op(w,m)$ denotes the output class provided by the mth classifier for the word w. The class receiving the maximum combined score is selected as the joint decision.
3) The overall recall, precision and F-measure values of the ensemble classifier are computed on the development set. For single objective approach, we use F-measure value as the objective function, i.e. f_0 = F-measure. The main goal is to maximize this objective function using the search capability of DE.

Mutation: For each target vector $x_{i,G}$; $i = 1,2,3,\ldots,NP$, a mutant vector is generated according to $v_{i,G+1} = x_{r1,G} + F(x_{r2,G} - x_{r3,G})$, where $r1$, $r2$, $r3$ are the random indexes and belong to $\{1,2,\ldots,NP\}$. These are some integer values, mutually different and $F > 0$. The randomly chosen integers $r1$, $r2$ and $r3$ are also chosen to be different from the running index i, so that NP must be greater or equal to four to allow for this condition. F is a real and constant factor 2 $[0,2]$ which controls the amplification of the differential variation $(x_{r2,G} - x_{r3,G})$. The $v_{i,G+1}$ is termed as the donor vector.

Crossover: In order to increase the diversity of the perturbed parameter vectors, crossover is introduced. This is well-known as the recombination. To this end, the trial vector:
$u_{i,G+1} = (u_{1i,G+1}, u_{2i,G+1}, \ldots, u_{Di,G+1})$ is formed, where

$$u_{j,i,G+1} = v_{j,i,G+1} \text{ if } (randb(j) \leq CR) \text{ or } j = rnbr(i) \quad (1)$$

$$= x_{j,i,G} \text{ if } (randb(j) > CR) \text{ and } j \neq rnbr(i) \quad (2)$$

for $j = 1,2,\ldots,D$. In Equation 1, $randb(j)$ is the jth evaluation of an uniform random number generator with outcome belongs to $[0,1]$. CR is the crossover constant belongs to $[0,1]$ which has to be determined by the user. $rnbr(i)$ is a randomly chosen index x belongs to $\{1,2,\ldots,D\}$ which ensures that $u_{i,G+1}$ gets at least one parameter from $v_{i,G+1}$.

Selection: To decide whether or not it should become a member of generation G+1, the trial vector $u_{i,G+1}$ is compared to the target vector $x_{i,G}$ using the greedy criterion. If vector $u_{i,G+1}$ yields a smaller cost function value than $x_{i,G}$, then $x_{i,G+1}$ is set to $u_{i,G+1}$, otherwise, the old value $x_{i,G}$ is retained.

Termination Condition: In this approach, the processes of mutation, crossover (or, recombination), fitness computation and selection are executed for a maximum number of generations. The best string seen up to the last generation provides the solution to the above classifier ensemble problem. Elitism is implemented at each generation by preserving the best string seen up to that generation in a location outside the population. Thus on termination, this location contains the best classifier ensemble.

3 Features for Event Extraction

We identify and use the following set of features for event extraction. All these features are automatically extracted from the training datasets without using any additional domain dependent resources and/or tools.

• **Context words:** We use preceding and succeeding few words as the features. This feature is used with the observation that contextual information plays an important role in identification of event triggers.

•**Root words:** Stems of the current and/or the surrounding token(s) are used as the features of the event extraction module. Stems of the words were provided with the evaluation datasets of training, development and test.
• **Part-of-Speech (PoS) information:** PoS information of the current and/or the surrounding tokens(s) are effective for event trigger identification. PoS labels of the tokens were provided with the datasets.
• **Named Entity (NE) information:** NE information of the current and/or surrounding token(s) are used as the features. NE information was provided with the datasets.
• **Semantic feature:** This feature is semantically motivated and exploits global context information. This is based on the content words in the surrounding context. We consider all unigrams in contexts $w_{i-3}^{i+3} = w_{i-3} \dots w_{i+3}$ of w_i (crossing sentence boundaries) for the entire training data. We convert tokens to lower case, remove stopwords, numbers, punctuation and special symbols. We define a feature vector of length 10 using the 10 most frequent content words. Given a classification instance, the feature corresponding to token t is set to 1 if and only if the context w_{i-3}^{i+3} of w_i contains t.
• **Dependency features:** A dependency parse tree captures the semantic predicate-argument dependencies among the words of a sentence. Dependency paths between protein pairs have successfully been used to identify protein interactions. In this work, we use the dependency paths to extract events. We use the McClosky- Charniak parses which are converted to the Stanford Typed Dependencies format and provided with the datasets. We define a number of features based on the dependency labels of the tokens.
• **Dependency path from the nearest protein:** Dependency relations of the path from the nearest protein are used as the features.
• **Boolean valued features:** Two boolean-valued features are defined using the dependency path information. The first feature checks whether the current token's child is a proposition and the chunk of the child includes a protein. The second feature fires if and only if the current token's child is a protein and its dependency label is OBJ.
•**Shortest path:** Distance of the nearest protein from the current token is used as the feature. This is an integer-valued feature that takes the value equal to the number of tokens between the current token and the nearest protein.
• **Word prefix and suffix:** Fixed length (say, n) word suffixes and prefixes may be helpful to detect event triggers from the text. Actually, these are the fixed length character strings stripped either from the rightmost (for suffix) or from the leftmost (for prefix) positions of the words. If the length of the corresponding word is less than or equal to n-1 then the feature values are not defined and denoted by ND. The feature value is also not defined (ND) if the token itself is a punctuation symbol or contains any special symbol or digit. This feature is included with the observation that event triggers share some common suffixes and/or prefixes. In this work, we consider the prefixes and suffixes of length up to four characters.

4 Datasets and Experimental Results

We use the BioNLP-09 shared task datasets. The events were selected from the GENIA ontology based on their significance and the amount of annotated instances in the GENIA corpus. The selected event types all concern protein biology, implying that they take proteins as their theme. The first three event types concern protein metabolism that actually represents protein production and breakdown. *Phosphorylation* represents protein modification event whereas *localization* and *binding* denote fundamental molecular events. *Regulation* and its sub-types, *positive* and *negative* regulations are representative of regulatory events and causal relations. The last five event types are universal but frequently occur on proteins. Detailed biological interpretations of the event types can be found in Gene Ontology (GO) and the GENIA ontology. From a computational point of view, the event types represent different levels of complexity.

Training and development datasets were derived from the publicly available event corpus [7]. The test set was obtained from an unpublished portion of the corpus. The shared task organizers made some changes to the original GENIA event corpus. Irrelevant annotations were removed, and some new types of annotation were added to make the event annotation more appropriate. The training, development and test datasets have 176,146, 33,937 and 57,367 tokens, respectively.

4.1 Experimental Results

We generate 15 different classifiers by varying the feature combinations of 3 different classifiers, namely Support Vector Machine (SVM), K-Nearest Neighbour (IBk) and Naïve Bayesian classifier. We determine the best configuration using development set. Due to non-availability of gold annotated test datasets we report the final results on 3-fold cross validation. The system is evaluated in terms of the standard recall, precision and F-measure. Evaluation shows the highest performance with a SVM-based classifier that yields the overall recall, precision and F-measure values of 33.17%, 56.00% and 41.66%, respectively.

The dimension of vector for our experiment is 15*19 = 285 where 15 represents number of classifiers and 19 represents number of output classes. We construct an ensemble from these 15 classifiers. Differential Evolution (DE) based ensemble technique is developed that determines the appropriate weights of votes of each class in each classifier. When we set population size, P=100, cross-over constant, CR=1.0 and number of generations, G = 50, with increase in F over range $[0,2]$, we get the highest recall, precision and F-value of 42.90%, 47.40%, 45.04%, respectively.

We observe that for $F < 0.5$ the solution converges faster. With 30 generations we can reach to optimal solution (for case when $F < 0.5$). For $F = 0.0$, the solution converges at the very beginning. We observe the highest performance with the settings P=100, F=2.0, CR=0.5 and G=150. This yields the overall recall, precision and F-measure values of 42.76%, 49.21% and 45.76%. This is actually an improvement of 4.10 percentage of F-measure points over the best individual base classifier, i.e. SVM.

5 Conclusion and Future Works

In this paper we have proposed differential evolution based ensemble technique for biological event extraction that involves identification and classification of complex bio-molecular events. The proposed approach is evaluated on the benchmark dataset of BioNLP 2009 shared task. It shows the F-measure of 45.76%, an improvement of 4.10%.

Overall evaluation results suggest that there is still the room for further improvement. In this work, we have considered identification and classification as one step problem. In our future work we would like to consider identification and classification as a separate problem. We would also like to investigate distinct and more effective set of features for event identification and classification each. We would like to come up with an appropriate feature selection algorithm. In our future work, we would like to identify arguments to these events.

References

1. Nédellec, C.: Learning Language in Logic -Genic Interaction Extraction Challenge. In: Cussens, J., Nédellec, C. (eds.) Proceedings of the 4th Learning Language in Logic Workshop, LLL 2005, pp. 31–37 (2005)
2. Hirschman, L., Krallinger, M., Valencia, A. (eds.): Proceedings of the Second BioCreative Challenge Evaluation Workshop. CNIO Centro Nacional de Investigaciones Oncológicas (2007)
3. Chatr-aryamontri, A., Ceol, A., Palazzi, L.M., Nardelli, G., Schneider, M.V., Castagnoli, L., Cesareni, G.: MINT: the Molecular INTeraction database. Nucleic Acids Research 35(suppl. 1), 572–574 (2007)
4. Kim, J.-D., Ohta, T., Pyysalo, S., Kano, Y., Tsujii, J.: Overview of BioNLP 2009 shared task on event extraction. In: BioNLP 2009: Proceedings of the Workshop on BioNLP, pp. 1–9 (2009)
5. Ekbal, A., Saha, S.: Weighted Vote-Based Classifier Ensemble for Named Entity Recognition: A Genetic Algorithm-Based Approach. ACM Trans. Asian Lang. Inf. Process. 10(2), 9 (2011)
6. Storn, R., Price, K.: Differential Evolution A Simple and Efficient Heuristic for Global Optimization over Continuous Spaces. J. of Global Optimization 11(4), 341–359 (1997), http://dx.doi.org/10.1023/A:1008202821328, doi:10.1023/A:1008202821328
7. Kim, J.-D., Ohta, T., Tsujii, J.: Corpus annotation for mining biomedical events from literature. BMC Bioinformatics 9, 10 (2008)

A Novel Clustering Approach Using Shape Based Similarity

Smriti Srivastava, Saurabh Bhardwaj, and J.R.P. Gupta

Abstract. The present research proposes a paradigm for the clustering of data in which no prior knowledge about the number of clusters is required. Here shape based similarity is used as an index of similarity for clustering. The paper exploits the pattern identification prowess of Hidden Markov Model (HMM) and overcomes few of the problems associated with distance based clustering approaches. In the present research partitioning of data into clusters is done in two steps. In the first step HMM is used for finding the number of clusters then in the second step data is classified into the clusters according to their shape similarity. Experimental results on synthetic datasets and on the Iris dataset show that the proposed algorithm outperforms few commonly used clustering algorithm.

Keywords: Clustering, Hidden Markov Model, Shape Based similarity.

1 Introduction

Cluster analysis is a method of creating groups of objects or clusters in such a way that the objects in one cluster are very similar to each other while the objects in different clusters are quite different. Data clustering algorithms could be generally classified into the following categories [1]: Hierarchical clustering, Fuzzy clustering, Center based clustering, Search based clustering, Graph based clustering, Grid based clustering, Density based clustering, Subspace clustering, and Model based clustering algorithms. Every clustering algorithm is based on the index of similarity or dissimilarity between data points. Many authors have used the distances as the index of similarity. Commonly used distances are Euclidean distance, Manhattan distance, Minkowski distance, and Mahalanobis distance [2] . As shown in [3] the distance functions are not always adequate for capturing correlations among the objects. It is also shown that strong correlations may still exist among a set of objects even if they are far apart from each other. HMMs are the dominant models for the sequential data. Although HMMs have extensively used in speech recognition,

Smriti Srivastava · Saurabh Bhardwaj · J.R.P. Gupta
Netaji Subhas Institute of Technology, New Delhi – 110078, India
e-mail: {bsaurabh2078,jairamprasadgupta}@gmail.com,
ssmriti@yahoo.com

A. Abraham and S.M. Thampi (Eds.): Intelligent Informatics, AISC 182, pp. 17–27.
springerlink.com

pattern recognition and time series prediction problems but they have not been widely used for the clustering problems and only few papers can be found in the literature. Many researchers have used single sequences to train the HMMs and proposed different distance measures based on a likelihood matrix obtained from these trained HMMs. Clustering of sequences using HMM, were introduced in [4] . In this a (Log-Likelihood) LL based scheme for automatically determining the number of clusters in the data is proposed. A similarity based clustering of sequences using HMMs is presented in [5]. In this, a new representation space is built in which each object is described by the vector of its similarities with respect to a predeterminate set of other objects. These similarities are determined using LL values of HMMs. A single HMM based clustering method was proposed in [6] , which utilized LL values as the similarity measures between data points. The method was useful for finding the number of clusters in the data set with the help of LL values but it is tough to actually obtain the data elements for the clusters as the threshold for the clusters was estimated by simply inspecting the graph of LL.

The present research proposes an HMM based unsupervised clustering algorithm which uses the shape similarity as a measure to capture the correlation among the objects. It also automatically determines the number of clusters in the data. Here the hidden state information of HMM is utilized as a tool to obtain the similar patterns among the objects.

The rest of the paper is organized as follows. Sect. 2 briefly describes the HMM. Sect. 3 details the proposed shape based clustering paradigm. In Sect. 4 Experimental results are provided to illustrate the effectiveness of proposed model. Finally, conclusions are drawn in Sect. 5.

2 Hidden Markov Model

Hidden Markov Model (HMM) [7][8] springs forth from Markov Processes or Markov Chains. It is a canonical probabilistic model for the sequential or temporal data It depends upon the fundamental fact of real world, "Future is independent of the past and given by the present". HMM is a doubly embedded stochastic process, where final output of the system at a particular instant of time depends upon the state of the system and the output generated by that state. There are two types of HMMs: Discrete HMMs and Continuous Density HMMs. These are distinguished by the type of data that they operate upon. Discrete HMMs (DHMMs) operate on quantized data or symbols, on the other hand, the Continuous Density HMMs (CDHMMs) operate on continuous data and their emission matrices are the distribution functions. HMM Consists of the following parameters

O {O1,O2,… ,OT }	:	Observation Sequence
Z {Z1, Z2,…,ZT }	:	State Sequence
T	:	Transition Matrix
B	:	Emission Matrix/Function
π	:	Initialization Matrix
λ(T, B, π)	:	Model of the System
$\mathbb{Q}$	:	Space of all state sequence of length T

$m\{m_{q1}, m_{q2}, \ldots m_{qT}\}$	:	Mixture component for each state at each time
$c_{il}, \mu_{il}, \sum_{il}$	:	Mixture component (i state and l component)

There are three major design problems associated with HMM:

Given the Observation Sequence {O1, O2, O3,.., OT} and the Model λ(T, B, π), the first problem is the computation of the probability of the observation sequence P (O|λ).The second is to find the most probable state sequence Z {Z1, Z2,.., ZT},

The third problem is the choice of the model parameters λ (T, B, π), such that the probability of the Observation sequence, P (O|λ) is the maximum.

The solution to the above problems emerges from three algorithms: Forward, Viterbi and Baum-Welch [7].

2.1 Continuous Density HMM

Let O = {O1,O2,..,OT } be the observation sequence and Z {Z1, Z2,…,ZT}be the hidden state sequence. Now, we briefly define the Expectation Maximization (EM) algorithm for finding the maximum-likelihood estimate of the parameters of a HMM given a set of observed feature vectors. EM algorithm is a method for approximately obtaining the maximum a posteriori when some of the data is missing, as in HMM in which the observation sequence is visible but the states are hidden or missing. The Q function is generally defined as

$$Q(\lambda,\lambda') = \sum_{q\varepsilon p} \log P(0, z \mid \lambda) P(0, z \mid \lambda') \tag{1}$$

To define the Q function for the Gaussian mixtures, we need the hidden variable for the mixture component along with the hidden state sequence. These are provided by both the E–step and the M-step of EM algorithm given

E Step:

$$\mathrm{Q}(\lambda,\lambda') = \sum_{z\varepsilon p} \sum_{m\varepsilon M} \log P(O, z, m \mid \lambda) P(O, z, m \mid \lambda') \tag{2}$$

M Step:

$$\lambda' = \arg m_{\lambda} ax[Q(\lambda,\lambda')] + constra\,\mathrm{int} \tag{3}$$

The optimized equations for the parameters of the mixture density are

$$\mu_{il} = \frac{\sum_{t=1}^{T} O_t P(z_{t=1,} m_{z_t t} = 1 \mid O_t \lambda')}{\sum_{t=1}^{T} P(z_{t=1,} m_{z_t t} = 1 \mid O_t \lambda')} \tag{4}$$

$$\sum_{il} = \frac{\sum_{t=1}^{T} (O_t - \mu_{il})(O_t - \mu_{il})^T P(z_t = i, m_{z_t t} = 1 \mid O_t \lambda')}{\sum_{t=1}^{T} P(z_t = i, m_{z_t t} = 1 \mid O_t \lambda')} \tag{5}$$

$$c_{il} = \frac{\sum_{t=1}^{T} P(z_t = i, m_{z_t t} = 1 | O_t \lambda')}{\sum_{t=1}^{T} \sum_{l=1}^{M} P(z_t = i, m_{z_t t} = 1 | O_t \lambda')} \tag{6}$$

3 Shape Based Clustering

Generally different distance functions such as Euclidean distance, Manhattan distance, and Cosine distance are employed for clustering the data but these distance functions are not always effective to capture the correlations among the objects. In fact, strong correlations may still exist among a set of objects even if their distances are far apart from each other as measured by the distance functions. Fig.1 shows '4' objects with '5' attributes among a set of 300 objects which were allotted in different clusters when the segmental k-means applied to partition them in six clusters. As it is clear from the Fig.1 that these objects physically have the same pattern of shape and also have the strong correlation among each other which is shown with the help of correlation matrix between the '4' data elements in Table 1. So by taking the motivation from here in the present research we have extended the basic concept of Shape Based Batching (SBB) procedure as introduced in [9],[10]. Earlier it was shown that by carefully observing the datasets and their corresponding log-likelihoods (LL), it is possible to find the shape of the input variation for certain value of log-likelihood but further it is found that to detect the shape by simply observing is not always easy. Moreover, in some datasets it is very difficult to determine the threshold for the batch allocation. Although the states are hidden, for many practical applications there is often some physical significance attached to the states of the model. In the present research it is found that the patterns of objects corresponding to any particular state of HMM is highly correlated and have different pattern or uncorrelated with the objects corresponding to any other state, so here the concept of SBB is modified and in this modified SBB the shape is a function of the state and not of the log likelihoods.

Table 1 Correlation among different row vectors

	Data-1	Data-2	Data-3	Data-4
Data-1	1.000	0.988	0.955	0.905
Data-2	0.988	1.000	0.989	0.959
Data-3	0.955	0.989	1.000	0.990
Data-4	0.905	0.959	0.990	1.000

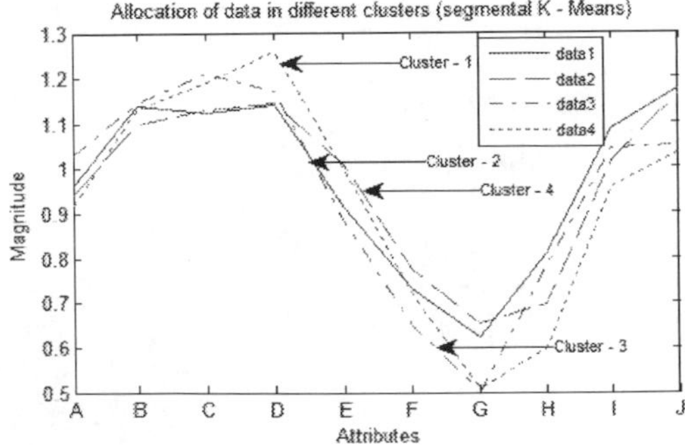

Fig. 1 Clustering results with the segmental K- means

Here unsupervised clustering algorithm is proposed in which important thing to note is that the numbers of clusters are not fixed, and the algorithm automatically decides the number of clusters. The whole procedure is as shown in Fig. 2. First of all the number of clusters in the data set are obtained. The steps for obtaining the number of clusters are as follows. Estimate the HMM model parameters λ(T,B,π) for the entire input dataset using Baum–Welch/Expectation maximization algorithm, for the appropriate values of the state 'Z' and mixture components 'm'. Once the HMM has been trained, the forward algorithm is used to compute the value of P(O|λ) which can then be used to calculate the LL of each row of the dataset . Now by sorting the LL values in the ascending (descending) order we can get the clear indication regarding the number of clusters in the dataset.

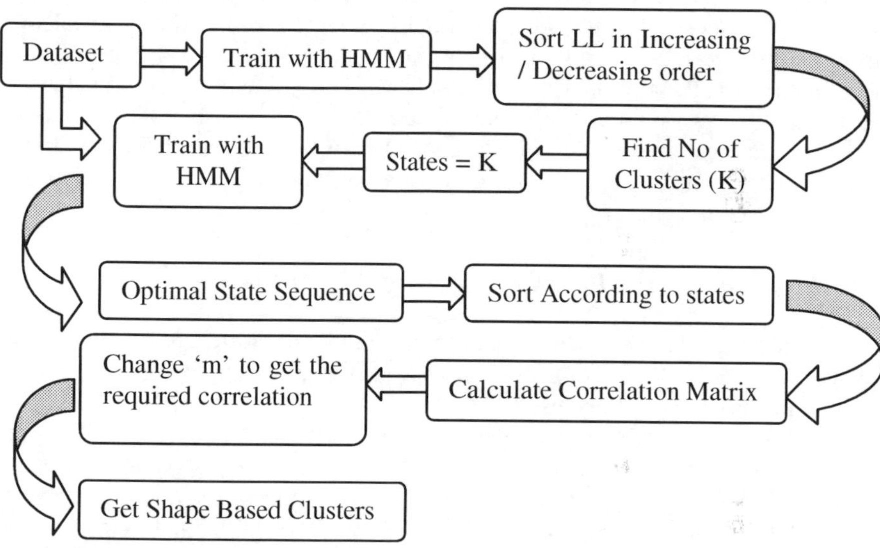

Fig. 2 Procedure for Shape Based Clustering

Now after getting the information about the number of clusters initialize the value of the parameters of the HMM. This includes initialization of transition matrix 'T', initialization matrix 'π' and the mixture component 'm' for each state. Take the number of states as equal to the number of clusters .The Continuous Density Hidden Markov Model (CDHMM) is trained using Baum Welch/ Expectation maximization algorithm for the entire input dataset After freezing the HMM parameters the next step is to find the optimal state sequence, with the help of 'Viterbi algorithm' by taking the entire input dataset as the 'D' dimensional observation vector sequence. Now the observation sequence and the corresponding optimal state sequence is obtained. After doing this one important thing is observed that the data vectors which are associated with the same state have identical shape while the data vectors with different states have no similarity in their shapes. So once the optimal value of hidden state sequence is deduced the next step is to put the data into clusters according to their state. Now each cluster have almost identical shape but by simply observing the clusters it is difficult to find the required shape based similarity, so an attempt is made to get the appropriate values of ‘Z’ and ‘m’ for the required shape based clusters by calculating the value of correlation coefficient among the data vectors into the clusters. Here the Pearson R model [11] comes handy for finding the coherence (correlation) among a set of objects. The correlation between the two objects ‘x1’ and ‘x2’ is defined as:

$$\frac{\sum(x_1-\overline{x_1})(x_2-\overline{x_2})}{\sqrt{\sum(x_1-\overline{x_1})^2 x \sum(x_2-\overline{x_2})^2}} \tag{7}$$

Where $\overline{x_1}$ and $\overline{x_2}$ are the mean of all attribute values in ‘x1’ and ‘x2’, respectively. It may be noted that Pearson R correlation measures the correlation between two objects with respect to all the attribute values. A large positive value indicates a strong positive correlation while a large negative value indicates a strong negative correlation. Now the correlation coefficients can be used as a threshold value of the similarity between the data vectors in the clusters and by using this value as a threshold the appropriate value of ‘Z’ and ‘m’ can be determined for the shape based clusters. Using these basic criteria, an algorithm was developed which arranged the data into clusters.

3.1 Steps for Shape Based Clustering Algorithm

Step 1: Take the number of states equal to the number of clusters and estimate the HMM parameters $\lambda(T, B, \pi)$ for the entire input dataset by taking the appropriate value of the mixture components ‘m’.

Step 2: Calculate the optimal value of hidden state sequence with the help of “Viterbi Algorithm” by taking the input as a ‘D’ dimensional observation vector.

Step 3: Rearrange the complete dataset according to their state values.

Step 4: Calculate correlation matrix by using the Pearson R model as in (7).

Step 5: Change the value of ‘m’ and repeat the steps 1-4 until the required tolerance of correlation is achieved.

The effectiveness of the proposed model can be demonstrated by taking the Iris plants database. The data set contains ‘3’ classes of ‘50’ instances each, where each class refers to a type of Iris plant. Fig.3 shows the patterns of Iris data before clustering. Now as a first step entire Iris data is trained with the help of Baum–Welch/Expectation maximization. Once the HMM has been trained, the forward algorithm is used to calculate the LL of each row of the dataset. Fig.3 shows the graph of LL vales sorted in ascending order. As it is clear from the Fig.3 that we can get the information regarding the number of clusters, but to choose the threshold value for allocating the data to the clusters by simply watching the LL graph (Fig.4) is not possible. This is the main drawback in previous approaches which is now removed in the present research. After getting the information regarding the number of clusters the shape based clustering approach is applied as described earlier. After applying step 1-step 5 of proposed algorithm Table 2 is obtained. The

Table 2 States and LL values of Iris data

Attributes	5.1	4.9	4.7	4.6	5.0	6.3	5.8	7.1	6.3	6.5	7.0	6.4	6.9	5.5	6.5
	3.5	3.0	3.2	3.1	3.6	3.3	2.7	3.0	2.9	3.0	3.2	3.2	3.1	2.3	2.8
	1.4	1.4	1.3	1.5	1.4	6.0	5.1	5.9	5.6	5.8	4.7	4.5	4.9	4.0	4.6
	0.2	0.2	0.2	0.2	0.2	2.5	1.9	2.1	1.8	2.2	1.4	1.5	1.5	1.3	1.5
No.	1	2	3	4	5	51	52	53	54	55	101	102	103	104	105
LL	0.6	0.1	0.2	0.1	0.5	335.5	214.2	309.7	**158.1**	298.1	162.4	142.5	181.9	109.3	**158.8**
States	1	1	1	1	1	3	3	3	3	3	2	2	2	2	2

Table 3 Actual parameters of the model

Sigma(:,:,1)				Mean		
0.2706	0.0833	0.1788	0.0545	5.936	5.006	6.589
0.0833	0.1064	0.0812	0.0405	2.770	3.428	2.974
0.1788	0.0812	0.2273	0.0723	4.262	1.462	5.553
0.0545	0.0405	0.0723	0.0487	1.327	0.246	2.026
Sigma(:,:,2)				Initial Matrix		
0.1318	0.0972	0.0160	0.0101	0.000	1.000	0.000
0.0972	0.1508	0.0115	0.0091			
0.0160	0.0115	0.0396	0.0059	States =3		
0.0101	0.0091	0.0059	0.0209			
Sigma(:,:,3)				Transition Matrix		
0.4061	0.0921	0.2972	0.0479	1.000	0.000	0.000
0.0921	0.1121	0.0701	0.0468	0.000	0.980	0.020
0.2972	0.0701	0.3087	0.0477	0.020	0.000	0.980
0.0479	0.0468	0.0477	0.0840			

description of this table is as follows: Row 1 to Row 4 shows the 4 attribute values of the IRIS data. Row 5 shows the number of data vector, Row 6 shows the LL values corresponding to data vectors and Row 7 shows the optimized state values associated with that particular data vector. Due to the limitation of page width it is not possible to show the complete table. So only '5' values of each class is shown. The values of LL in the table are only displayed to show the effectiveness of our method over LL based clustering method. As it is clear from the table that the LL values between the data element '54' is almost equal to the LL value of data element '105' but these two elements belong to two different clusters. Hence it can be said that the LL based clustering method is not adequate, while it is clear from the Table 2 that the 'states' clearly partition the data accurately and in this dataset (Iris dataset) the misclassification is zero meaning we are getting 100 % accuracy. The plots of three clusters obtained after the application of proposed algorithm are as shown in Fig. 5 and the actual parameters of the model are shown in the Table 3.

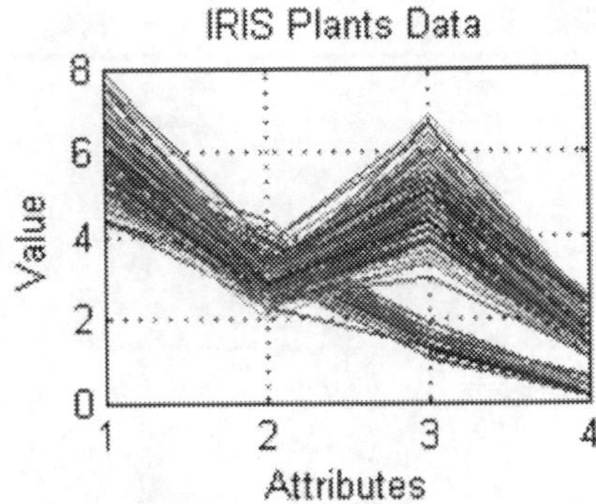

Fig. 3 Iris Plant Data Patterns

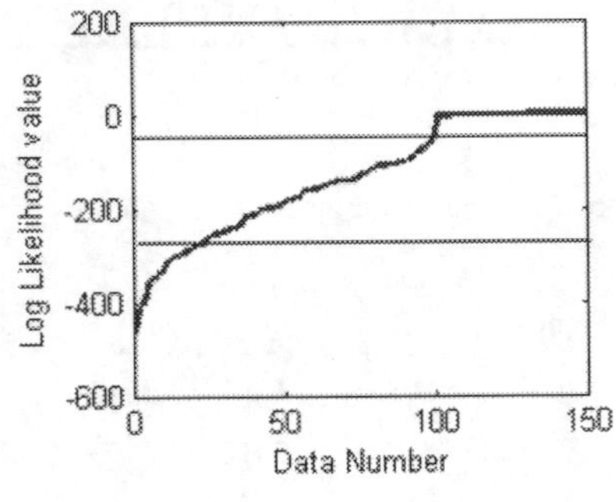

Fig. 4 Iris Data LL Values

4 Experimental Results

To show the effectiveness of the proposed method it is applied on both the synthetic data and the real world data.

Table 4 Parameters for synthetic data generation

Class -1								
	1/3	1/3	1/3		1/3		μ_1=1	σ_1^2=0.6
T =	1/3	1/3	1/3	π =	1/3	B =	μ_2=3	σ_2^2=0.6
	1/3	1/3	1/3		1/3		μ_3=5	σ_3^2=0.6
Class -2								
	1/3	1/3	1/3		1/3		μ_1=1	σ_1^2=0.5
T =	1/3	1/3	1/3	π =	1/3	B =	μ_2=3	σ_2^2=0.5
	1/3	1/3	1/3		1/3		μ_3=5	σ_3^2=0.5
Class -3								
	1/3	1/3	1/3		1/3		μ_1=1	σ_1^2=0.4
T =	1/3	1/3	1/3	π =	1/3	B =	μ_2=3	σ_2^2=0.4
	1/3	1/3	1/3		1/3		μ_3=5	σ_3^2=0.4

4.1 Synthetic Data

The description of the synthetic data is given in [5]. The data contains 3 classes. The training set is composed of 30 sequences (of length 400) from each of the three classes generated by 3 HMMs. The parameters of the synthetic are as shown in the Table 4. The comparison of results with the previous approaches is as shown in Table 5. The results of the starting three rows are taken from [5].

4.2 Real Data

We have tested the proposed approaches on classical Iris Plants Database. The data set contains 3 classes of 50 instances each, where each class refers to a type of iris plant (Irisvirginica, Irisversicolor, Irissetosa). The dataset consists of the following four attributes: sepal length, sepal width, petal length, and petal width. The comparison of results with the previous approaches is as shown in Table 6. The data in the column of Errors lists the numbers of data points which are classified wrongly. The starting four rows of the table are taken from [12].

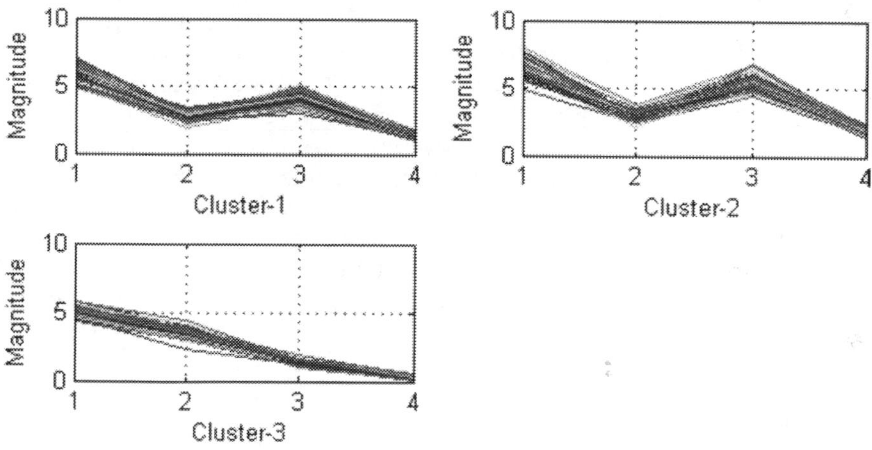

Fig. 5 Iris Data Cluster

Table 5 Comparison of previous methods

Learning Algorithm	Accuracy (%)
ML_{OPS} [5]	95.7
1 – NN on S_T [5]	98.9
1 – NN on S_T [5]	98.9
Shape Based Clustering	98.888

Table 6 Comparison of previous methods

Learning Algorithm	Error	Accuracy (%)
FCM(Fuzzy c -means)	16	89.33
SWFCM(Sample Weighted Robust Fuzzy c -means)	12	92
PCM(Possibilistic c -means)	50	66.6
PFCM(Possibilistic Fuzzy c -means)	14	90.6
Shape Based Clustering	15	100

5 Conclusion

This present research proposes a novel clustering approach based on the shape similarity. The paper shows that the distance functions are not always adequate for clustering of data and strong correlations may still exist among the data points even if they are far apart from each other. The method is applied in a two phase sequential manner. In the first phase, HMM is applied on the dataset to yield HMM parameters assuming a certain number of states and gaussian mixtures. Then the log likelihood values are obtained from the forward algorithm. The sorted log likelihood values give the clear indication regarding the number of clusters into the dataset. Next the shape based clustering algorithm is applied to cluster the dataset. The method overcomes the problem of finding the threshold value in LL based clustering algorithms. The proposed method is tested on real (Iris data) as well as on the synthetic dataset. The results of simulation are very encouraging; the method gives 100% accuracy on Iris dataset while about 99% accuracy on the synthetic test data. Further the shortcoming of previous HMM based clustering approaches in which the number of HMMs required were equal to that of number of sequences/classes [4] [5] is removed by utilizing only single HMM for clustering and hence reducing the computational time and the complexity considerably.

References

[1] Gan, G., Ma, C., Wu, J.: Data Clustering: Theory, Algorithms, and Applications. Society for Industrial and Applied Mathematics, Philadelphia (2007)

[2] Xu, R., Wunsch, D.I.: Survey of clustering algorithms. IEEE Transactions on Neural Networks 16(3), 645–678 (2005)

[3] Wang, H., Pei, J.: Clustering by Pattern Similarity. Journal of Computer Science and Technology 23(4), 481–496 (2008)

[4] Smyth, P.: Clustering sequences with hidden Markov models. Advances in Neural Information Processing Systems 9, 648–654 (1997)

[5] Bicego, M., Murino, V., Figueiredo, M.A.: Similarity-based classification of sequences using hidden Markov models. Pattern Recognition 37(12), 2281–2291 (2004)

[6] Hassan, R., Nath, B.: Stock market forecasting using hidden markov model. In: Proceedings of the Fifth International Conference on Intelligent Systems Design and Application, pp. 192–196 (2005)

[7] Rabiner, L.R.: A tutorial on hidden Markov models and selected applications in speech recognition. IEEE (77), 257–286 (1989)

[8] Blimes, J.A.: A gentle tutorial of the EM algorithm and its application to parameter estimation for gaussian mixture and hidden markov models. Berkeley, California: International Computer Science Institute Technical Report ICSI-TR-97-021 (1998)

[9] Srivastava, S., Bhardwaj, S., Madhvan, A., Gupta, J.R.P.: A Novel Shape Based Batching and Prediction approach for Time series using HMMs and FISs. In: 10th International Conference on Intelligent Systems Design and Applications, Cairo, Egypt, pp. 929–934 (2010)

[10] Bhardwaj, S., Srivastava, S., Madhvan, A., Gupta, J.R.P.: A Novel Shape Based Batching and Prediction approach for Sunspot Data using HMMs and ANNs. In: India International Conference on Power Electronics, New Delhi, India, pp. 1–5 (2011)

[11] Maes, U.S.: Social Information Filtering: Algorithms for automating word of mouth. In: ACM CHI, pp. 210–217 (1995)

[12] Xia, S.-X., Han, X.-D., Liu, B., Zhou, Y.: A Sample-Weighted Robust Fuzzy C-Means Clustering Algorithm. Energy Procedia (13), 3924–3931 (2011)

Knowledge Discovery Using Associative Classification for Heart Disease Prediction

M.A. Jabbar, B.L. Deekshatulu, and Priti Chandra

Abstract. Associate classification is a scientific study that is being used by knowledge discovery and decision support system which integrates association rule discovery methods and classification to a model for prediction. An important advantage of these classification systems is that, using association rule mining they are able to examine several features at a time. Associative classifiers are especially fit to applications where the model may assist the domain experts in their decisions. Cardiovascular deceases are the number one cause of death globally. An estimated 17.3 million people died from CVD in 2008, representing 30% of all global deaths. India is at risk of more deaths due to CHD. Cardiovascular disease is becoming an increasingly important cause of death in Andhra Pradesh. Hence a decision support system is proposed for predicting heart disease of a patient. In this paper we propose a new Associate classification algorithm for predicting heart disease for Andhra Pradesh population. Experiments show that the accuracy of the resulting rule set is better when compared to existing systems. This approach is expected to help physicians to make accurate decisions.

Keywords: Andhra Pradesh, Associative classification, Data mining, Heart disease.

1 Introduction

The major reason that the data mining has attracted great deal of attention in the information industry in the recent years is due to the wide availability of huge amounts of data and imminent need for turning such data into useful information

M.A. Jabbar
JNTU Hyderabad
e-mail: jabbar.meerja@gmail.com

B.L. Deekshatulu
Distinguished fellow IDRBT, RBI Govt of India

Priti Chandra
Senior Scientist Advanced System Laboratory, Hyderabad

A. Abraham and S.M. Thampi (Eds.): Intelligent Informatics, AISC 182, pp. 29–39.
springerlink.com

and knowledge. The information gained can be used for applications ranging from business management, production control, and market analysis to emerging design and science exploration and health data analysis. Data mining, also known as knowledge discovery in data bases (KDD), is the process of automatically discovering useful information in large data repositories [1].Association rule mining and classification are analogous tasks in data mining, with the exception that classification main aim is to build a classifier using some training instances for predicting classes for new instance, while association rule mining discovers association between attribute values in a data set. Association rule uses unsupervised learning where classification uses supervised learning .The majority of traditional classification techniques use heuristic-based strategies for building the classifier [2].In constructing a classification system they look for rules with high accuracy. Once a rule is created, they delete all positive training objects associated with it. Thus these methods often produce a small subset of rules, and may miss detailed rules that might play an important role in some cases. The heuristic methods that are employed by traditional classification technique often use domain independent biases to derive a small set of rules, and therefore rules generated by them are different in nature and more complex than those that users might expect or be able to interpret [3]. Both classification rule mining and association rule mining are indispensable to practical applications. Thus, great savings and convenience to the user could result if the two mining techniques can somehow be integrated.

Associative classifications (AC) is a recent and rewarding technique that applies the methodology of association into classification and achieves high classification accuracy, than traditional classification techniques and many of the rules found by AC methods can not be discovered by traditional classification algorithms. This generally involves two stages.

1) Generate class association rules from a training data set.
2) Classify the test data set into predefined class labels.

The various phases in Associative classification are Rule generation, Rule pruning, Rule ranking, and Rule sorting, Model construction and Prediction. The rule generation phase in Associative classification is a hard step that requires a large amount of computation. A rich rule set is constructed after applying suitable rule pruning and rule ranking strategies. This rule set which is generated from the training data set is used to build a model which is used to predict test cases present in the test data set.

Coronary heart disease (CHD) is epidemic in India and one of the major causes of disease burden and deaths. Mortality data from the Registrar General of India shows that cardiovascular diseases are a major cause of death in India now. Studies to determine the precise causes of death in Andhra Pradesh have revealed that cardiovascular diseases cause about 30% in rural areas [4].Medical diagnosis is regarded as an important yet complicated task that needs to be executed accurately and efficiently. The automation of this system should be extremely advantageous. Medical history of data comprises of a number of tasks essential to diagnosis particular disease. It is possible to acquire knowledge and information concerning a disease from the patient -specific stored measurement as far as medical data is concerned. Therefore data mining has developed into a vital domain in health care [5]. A classification system can assist the physician to examine a patient. The system can predict if the patient is likely to have a certain disease or present incompatibility with some treatments. Associative classification is better alternative for predictive analysis [6].This paper proposed a new associative classification method. Considering the classification model, the physician can make a better decision.

Basic concepts in Associative classification and heart disease are discussed in section 2, 3 and common algorithms surveyed in Section 4.Section 5 describes our proposed method. Experimental results and comparisons are demonstrated in section 6. We will conclude our final remarks in Section 7.

2 Associative Classification

According to [7] the AC problem was defined as Let a training data set T has M distinct Attributes A1, A2 ...Am and C is a list of class labels. The number of rows in D is denoted | D |. Attributes could be categorical or continuous. In case of categorical attributes, all possible values are mapped to a set of positive integers. For continuous attributes, a discreteisation method is first used to transform these attributes into categorical cases.

Definition-1:- An item can be described as an attribute name A_i and its value a_i, denoted (A_i, a_i)

Definition-2:- A row in D can be described as a combination of attribute names A_i and values a_{ij} , plus a class denoted by C_j.

Definition -3:- An item set can be described as a set of items contained in a training data.

Definition -4:- A rule item r is of the Form < item set->c) where c C is the class.

Definition -5:- The actual occurrence (actoccr) of a rule r in D is the no. of rows in D that match the item set defined in r.

Definition -6:- The support count (supp. Count) of rule item r < item set, c> is the No. of rows in D that matches r's item set, and belongs to a class c.

Definition -7:- The occurrence of an item set I in T is the no. of rows in D that match I.

Definition -8:- A rule r passes the min supp threshold if (supp count (r) >= min-supp)

Definition -9:-A rule r passes min.confidence threshold if (sup. Count (r) / actoccr (r)) >= min.confidence

Definition -10:- An item set I that passes the min .supp threshold is said to be a frequent item set.

Definition -11:- Associate classification rule is represented in the Form (item set →c) where antecedent is an item set and the consequent is a class.

The main task of AC s to discover a subset of rules with significant supports and higher confidence subset is then used to build an automated classifier that could be used to predict the classes of previously unseen data.

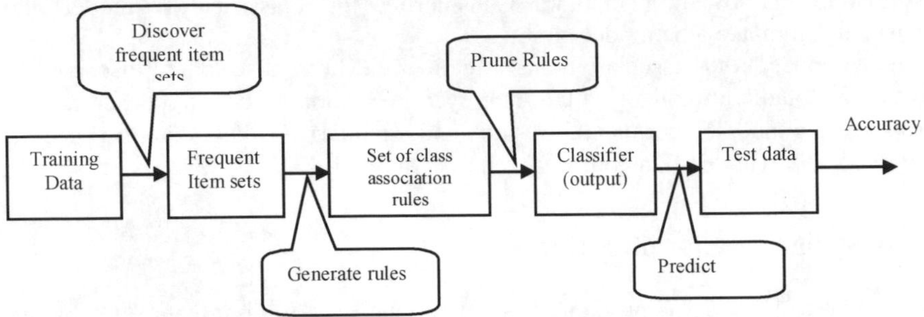

Fig. 1 Steps in Associative Classification

Table 1 A Training Data set

Row id	A	B	C	Class Label
1	a1	b2	c1	c1
2	a2	b1	c2	c0
3	a3	b3	c3	c1
4	a2	b2	c0	C0

3 Heart Disease

Coronary heart disease is the single largest cause of death in developed countries and is one of the main contributors to disease burden in developing countries. According to WHO an estimated 17.3 million people died from CVD in 2008, representing 30% of all global deaths .Of these deaths, an estimated 7.3 million were due to coronary heart disease and 6.2 million were due to stroke. By 2030 almost 23.6 million people will die from CVD's mainly from heart disease and stroke [8].Coronary heart disease (CHD) is epidemic in India and one of the Major causes of disease burden and deaths. Mortality data from the Registrar general of India shows that CVD are a Major cause of death in India and in Andhra Pradesh 30% deaths in rural areas. The term heart disease encompasses the diverse diseases that affect the heart. Cardiovascular disease or heart diseases are a class of disease that involves the heart or blood vessels. Cardiovascular disease results in severe illness, disability, and death. Narrowing of the coronary arteries results in the reduction of blood and oxygen supply to the heart and leads to the coronary heart disease. Myocardial infractions, generally known as heart attacks, and angina pectoris or chest pain are encompassed in the CHD. A sudden blockage of a coronary artery, generally due to a blood clot results in a heart attack, chest pains arise when the blood received by the heart muscles is inadequate [9]. Over 300 risk factors have been associated with coronary heart disease and stroke. The major established risk factors are 1) Modifiable risk factors 2) Non-modifiable risk factors 3) Novel risk factors [8].

The following features are collected for heart disease prediction in Andhra Pradesh based on the data collected from various corporate hospitals and opinion from expert doctors.

1) Age 2) Sex 3) Hypertension 4) Diabetic 5) Systolic Blood pressure 6) Diastolic Blood pressure 7) Rural / Urban.Comprehensive and integrated action is the means to prevent and control cardio vascular diseases.

4 Related Work

One of the first algorithms to use an association rule mining approach for classification was proposed in [10] and named CBA. CBA implement the famous Apriori algorithm [11] in order to discover frequent item sets.

Classification based on multiple association rules (CMAR) adopts the FP-growth ARM Algorithm [12] for discovering the rules and constructs an FP-Tree to mine large databases efficiently [13]. It consists of two phases, rule generation and classification. It Adopts FP-growth algorithm to scan the training data to find complete set of rules that meet certain support and confidence thresholds. Classification based on predictive association rules (CPAR) is a greedy method proposed by [14]. The Algorithm inherits the basic idea of FOIL [15] in rule generation and integrates it with the features of AC.Accurate and effective, multi class, multi label associative classification was proposed in [7]. A new approach based on information gain is proposed in [16] where the attribute values that are more informative are chosen for rule generation. Numerous works in literature related with heart disease have motivated our work. Some of the works are discussed below.

Cluster based association rule mining for heart attack prediction was proposed in [17].Their method is based on digit sequence and clustering. The entire data base is divided into partitions of equal size. Each partition will be called as cluster. Their approach reduces main memory requirement since it considers only a small cluster at a time and it is scalable and efficient.

Intelligent and effective heart attack prediction system using data mining and artificial neural net work was proposed in [18].They employed the multilayer perception neural network with back – propagation as the training algorithm. The problem of identifying constrained association rules for heart disease prediction was studied in [19]. These constraints are introduced to decrease the number of patterns.

Enhanced prediction of heart disease with feature subset selection using genetic algorithm was proposed in [20]. The objective of their work is to predict accurately the presence of heart disease with reduced number of attributes.

We propose a better strategy for associative classification to generate a compact rule set using only positively correlated rules, thereby the less significant rules are eliminated from the classifier. Informative attribute centric rule generation produces a compact rule set and we go for an attribute selection approach. We used Gini Index measure as a filter to reduce number of item sets ultimately generated. This classifier will be used for predicting heart disease.

5 Proposed Method

Most of the associate classification Algorithms adopt Apriori candidate generation step for the discovery of the frequent rule items. The main drawback in terms of mining efficiency of almost all the AC algorithms is that they generate large number of candidate sets, and they make more than one pass over the training data set to discover frequent rule items, which causes high I/O overheads. The search space for enumeration of all frequent item sets is 2^m which is exponential in m, where m, number of items.

Two measures support and confidence are used to prune the rules. Even after pruning the infrequent items based on support and confidence, the Apriori [11] association rule generation procedure, produces a huge number of association rules. If all the rules are used in the classifier then the accuracy of the classifier would be high but the building of classification will be slow.

An Informative attribute centered rule generation produces a compact rule set. Gini Index is used as a filter to reduce the number of candidate item sets. In the proposed method instead of considering all the combinations of items for rule generation, Gini index is used to select the best attribute. Those attributes with minimum Gini index are selected for rule generation.

$$Gini(t) = 1 - \sum_{i=0}^{c-1} [p(i/t)]^2 \tag{1}$$

We applied our proposed method on heart disease data to predict the chances of getting heart disease. Let us consider a sample training data set given in table 2

Table 2 Example Training data

Gender	Car type	Shirt size	class
M	Family	Small	C0
M	Sports	medium	C0
M	Sports	medium	C0
M	Sports	large	C0
M	Sports	Extra large	C0
M	Sports	Extra large	C0
F	Sports	Small	C0
F	Sports	Small	C0
F	Sports	medium	C0
F	luxury	large	C0
M	family	large	C1
M	family	Extra large	C1
M	family	medium	C1
M	Luxury	Extra large	C1
F	Luxury	Small	C1
F	Luxury	Small	C1
F	Luxury	medium	C1
F	Luxury	medium	C1
F	Luxury	medium	C1
F	luxury	large	C1

After calculating Gini index of each attribute car type has the lowest Gini index. So car type would be the better attribute. The rules generated like the following are considered for classifier.

1) Car type=sports, shirt size=small, gender = male->class C0
2) Car type=sports, shirt size=medium, gender = female->class C0
3) Car type=luxury, shirt size=small, gender = female->class C1
4) Car type=luxury, shirt size=small, gender=male->class C1

Proposed Algorithm:
Input: Training data set T, min-support, min- confidence
Output: Classification Rules.

1) n ← no. of Attributes
C- Number of classes,
Classification rule $X \rightarrow c_i$
2) For each attribute A_i calculate Gini weighted average where

$$Gini(t) = 1 - \sum_{i=0}^{c-1} [p(i/t)]^2$$

3) Select best attribute
Best Attribute = Minimum (Weighted Average of Gini (attribute))
4) For each t in training data set
i) $(X \rightarrow c_i)$ = candidates Gen. (Best attribute, T)
ii) If (Support $(X \rightarrow c_i)$ > min-support and min.confidence $(x \rightarrow (i))$
iii) Rule set ← $(X \rightarrow c_i)$
5) From the generated association classification rules, test the rules on the test data and find the Accuracy.

In our proposed method, we have selected the following attributes for heart disease prediction in Andhra Pradesh.

1) Age 2) Sex 3) Hypertension 4) Diabetic 5) Systolic BP 6) Dialectic BP 7) Rural / Urban .We collected the medical data from various corporate hospitals and applied our proposed approach to analyze the classification of heart disease patients

6 Results and Discussion

We have evaluated the accuracy of our proposed method on 9 data sets from SGI Repository [21]. A Brief description about the data sets is presented in Table 3. The accuracy is obtained by hold out approach [22], where 50% of the data was randomly chosen from the data set and used as training data set and remaining 50% data was used as the testing data set. The training data set is used to construct a model for classification. After constructing the classifier; the test data set is used to estimate the classifier performance. Class wise distribution for each data set is presented from Table 4-10

Accuracy Computation: Accuracy measures the ability of the classifier to correctly classify unlabeled data. It is the ratio of the number of correctly classified data over the total number of given transaction in the test data-set.

$$\textbf{Accuracy} = \frac{\text{Number of objects correctly Classified}}{\text{Total No. of objects in the test set.}}$$

Table 11 and Fig. 2 Represents the classification rate of the rule sets generated by our algorithm. Table 12 represents the accuracy of various algorithms on different data sets. Table 13 shows the size of the rules set generated by our algorithm, CBA, C4.5.The table indicates that classification based association rule methods often produce larger rules than traditional classification techniques. Table 14 shows the classification rules generated by our method, when we applied on heart disease data sets.

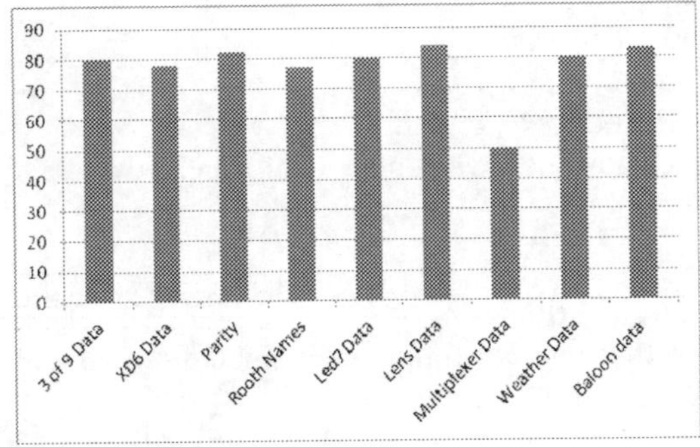

Fig. 2 Accuracy of Various Data sets

Table 3 Data set Description

Data Sets	Transactions	Items	Classes
3 of 9 Data	150	9	2
XD6 Data	150	9	2
Parity	100	10	2
Rooth Names	100	4	3
Led7 Data	100	7	10
Lens Data	24	9	3
Multiplexer Data	100	12	2
Weather Data	14	5	2
Baloon data	36	4	2

Table 4 Class distribution for weather Data

Class	Frequency	Probability
Yes	9	9/14=0.64
No	5	5/14=0.36

Table 5 Class distribution for lens Data

Class	Frequency	Probability
Hard Contact lenses	4	0.16
soft Contact lenses	5	0.28
No Contact lenses	15	0.625

Table 6 Class distribution for Balloon Data

Class	Frequency	Probability
True	4	0.44
False	5	0.55

Table 7 Class distribution for Multiplexer Data

Class	Frequency	Probability
true	55	0.55
false	45	0.45

Table 8 Class distribution for Parity Data

Class	Frequency	Probability
true	58	0.58
false	42	0.42

Table 9 Class distribution for XD6 data

Class	Frequency	Probability
true	41	0.27
false	109	0.72

Table 10 Class distribution for 3 of 9 data

Class	Frequency	Probability
true	73	0.48
false	77	0.51

Table 11 Accuracy of various data sets by Our algorithm

Data set	Accuracy
3 of 9 Data	80
XD6 Data	78
Parity	82
Rooth Names	77
Led7 Data	80
Lens Data	84
Multiplexer Data	50
Weather Data	80
Baloon data	83

Table 12 Accuracy of various algorithms on different data sets

Data Set	C4.5	Naïve Bayes	CBA	Our Approach
Contact lenses	83.33	70.83	66.67	84
Led7	73.34	73.15	71.10	80
Vote Data	88.27	87.12	87.39	88.5
Weather Data	50.0	57.14	85	80

Table 13 Rule Set generated by various algorithms on different data sets

Data Set	C4.5	CBA	Our Proposed Method
Contact lenses	4	6	13
Led7	37	50	29
Vote Data	4	40	30
Weather Data	-	-	5
Balloon Data	5	3	13

Table 14 Rules Generated for Heart Disease Prediction are

Sl.no	Rules
1	Age>45,gender=male,systolicBP>120,Urban people→Heart Disease
2	Age>55,gender=female,Hypertension→Heart Disease
3	Age>45,gender=male,Hypertension,Diabetic people→Heart Disease
4	Age>55,gender=female,Hypertension,Diabetic people→ Heart Disease
5	Age>55,gender=female,Hypertension,systolic BP>120→ Heart Disease

Rule 1 says that persons with age group above 45 and who lives in urban areas of Andhra Pradesh, they have high chance of getting heart disease.

Rule 2 states that females in Andhra Pradesh with age group above 55 and hypertension have a high probability of having heart disease.

Rule 3 to 5 confirms heart disease for risk factors like hypertension, diabetic. These rules characterize the patients with heart disease.

7 Conclusions and Future Work

In this paper, we have applied associative classification algorithm to medical health data to explore risk factors associated with heart disease. Associate classifiers are especially fit to applications where the model may assist the domain experts in their decisions. There are many domains such as medical, where the maximum accuracy of the model is desired. The model is applied for Andhra Pradesh population. Andhra Pradesh is at risk of more deaths due to CHD. Coronary heart disease can be handled successfully if more research is encouraged in this area. In the near feature we will investigate extraction of patterns associated with Heart Disease using associative classification and evolutionary algorithms.

References

[1] Tan, P.N., Steinbach, Kumar, V.: Introduction to Data Mining. Pearson Education (2006)

[2] Written, I., Frank, K.: Data Mining: Practical machine Learning tools and techniques with java implementations. Morgan Kaufmann, San Francisco (2010)

[3] Pazzani, M., Mani, S., et al.: Beyond concise and colorful learning intelligible rules. In: Proceedings of the KDD, pp. 235–238. AAAI Press, Menlo Park (1923)
[4] Gupta, R.: Recent trends in coronary heart disease epidemiology in India. Indian Heart Journal, B4–B18 (2008)
[5] Shilou, S., Bamidic, P.D., Maglareras, N., Papas, C.: Mining association rules from clinical data bases an intelligent diagnosis process in health care. Study of Health Technology, pp. 1399–1403 (2001)
[6] Vyas, R.J., et al.: Associative classifiers for predictive analysis: Comparative performance Study. In: 2nd UCSIM European Symposium on Computer Modeling and Simulations (2008)
[7] Thabtah, F.C., Peng, Y.: MMAC: A new multi class, multi-label associative classification approach. In: ICDM, pp. 217–214 (2004)
[8] WHO Report on non communicable diseases (September, 2011)
[9] Patil, S.B., et al.: Extraction of significant patterns from heart disease warehouses for heart attack prediction. IJCSNS 9(2) (February 2009)
[10] Liu, B., Hsu, W., Ma, Y.: Integrating classification and association rule mining. In: KDD 1998, NY, pp. 80–86 (1998)
[11] Agarwal, Srikant, R.: Fast algorithm from mining association rule. In: Proceedings of the 20th International Conference on Very Large Data Bases, pp. 487–499 (2003)
[12] Li, W., Han, J., Pei, J.: CMAR: Accurate classification based on multiple class association rules. In: Proceedings of the ICDM, pp. 363–366 (2001)
[13] Han, J., Pei, J., Yin, Y.: Mining frequent item sets without candidate generation. In: Proceedings of ACM SIGMOD, pp. 1–12 (2000)
[14] Yin, Y., Han, J.: CPAR: Classification Based on predictive association rules. In: SDM (2003)
[15] Quinlan, J.R., Cameran, R.M.: FOIL: A Midterm Report. In: Brazdil, P.B. (ed.) ECML 1993, vol. 667, pp. 3–20. Springer, Heidelberg (1993)
[16] Chan, G., Lanyu, H.L.: A New approach to classification based on association rule mining. Decision Support System, 674–689 (2006)
[17] Jabbar, M.A., Chandra, P., Deekshatulu, B.L.: Cluster Based association rule mining for heart attack Prediction. JATIT 32(2) (October 2011)
[18] Patil, S.B., et al.: Intelligent and effective heart attack prediction system using data mining and artificial neural network. Europian Journal of Scientific Research 31(4) (2009)
[19] Ordonez, C.: Improving Heart Disease Prediction using constrained association Rule. Seminar Presentation, Tokyo (2004)
[20] Ambarasi, M., et al.: Enhanced Prediction of Heart Disease with Feature subset selection using Genetic Algorithm. IJESI 2(10) (2010)
[21] http://www.sgi.com/tech/mlc/db
[22] Han, J., Kamber, M.: Data Mining: Concepts and Techniques. Morgan Kaufmann Publishers, Newyork (2001)

An Application of K-Means Clustering for Improving Video Text Detection

V.N. Manjunath Aradhya and M.S. Pavithra

Abstract. In the present work, we explore an extensive applications of Gabor filter and K-means clustering algorithm in detection of text in an unconstrained complex background and regular images. The system is a comprehensive of four stages: In the first stage, combination of wavelet transforms and Gabor filter is applied to extract sharpened edges and textural features of a given input image. In the second stage, the resultant Gabor output image is grouped into three clusters to classify the background, foreground and the true text pixels using K-means clustering algorithm. In the third stage of the system, morphological operations are performed to obtain connected components, then after a concept of linked list approach is in turn used to build a true text line sequence. In the final stage, wavelet entropy is imposed on an each connected component sequence, in order to determine the true text region of an input image. Experiments are conducted on 101 video images and on standard ICDAR 2003 database. The proposed method is evaluated by testing the 101 video images as well with the ICDAR 2003 database. Experimental results show that the proposed method is able to detect a text of different size, complex background and contrast. Withal, the system performance outreaches the existing method in terms of detection accuracy.

Keywords: Wavelet Transform, Gabor filter, K-means clustering, linked list approach, Wavelet Entropy.

V.N. Manjunath Aradhya
Department of Information Science and Engineering,
Dayananda Sagar College of Engineering, Bangalore
e-mail: aradhya.mysore@gmail.com

M.S. Pavithra
Department of Master of Computer Applications,
Dayananda Sagar College of Engineering, Bangalore
e-mail: pavithraspurthi@yahoo.com

A. Abraham and S.M. Thampi (Eds.): Intelligent Informatics, AISC 182, pp. 41–47.
springerlink.com

1 Introduction

Texture, Color and Shape based multimedia database registering and retrieving requires a task of text detection in images, video etc. Text detection is the process of determining the location of text in an image. Text region detection is mainly based on texture and dominant orientation. Text regions are detected either by analyzing the edges of the candidate regions or by using textural properties of an image. It provides primal information for text extraction and verification. Though many efforts have been devoted to, it remains a challenge due to variations of background, font of a text image. Text Information Extraction techniques can be broadly divided into two classes: i) region based and ii) texture based methods. In Region based methods connected components(CC) or edges are found on the basis of their perceptive difference with the background. This is followed by merging of the CCs or edges to get the text bounding boxes. In Texture based methods text in an image has distinct textural properties that gets distinguish from the background [1].

Most of the proposed text detection methods use text features, color, edge and texture information. So to extract a text from an image and discriminate it from the background, many researchers have applied heuristic rules based on empirical constraints and other few researchers have used machine-learning methods trained on real data. Aradhya et.al.[2] describe the text detection method using wavelet transform and Gabor filter. Kaushik et.al.[3] propose an approach for text detection, using morphological operators and Gabor wavelet. Phan et. al.[4] describe an efficient text detection based on the Laplacian operator. Shivakumara et.al.[5] proposed a Wavelet Transform Based technique for video text detection. Recently, few research works have carried out on K-means and connected component analysis in the domain of text detection in video images. Shivakumara et.al.[6] describe a method based on the Laplacian in the frequency domain for video text detection.

From the literature study, it is clear that, though the concept of K-means algorithm and a connected component analysis have used in many of the text detection approaches, the detection accuracy of the text region can still be improved without missing any data. By sustaining the development of the system [2], we propose a system with the combination of Gabor filter and K-means clustering is extensively used to detect the true text region accurately in attaining better detection rate with a very few missing data in numbers.

The remaining of our paper is structured as follows: In Section 2, as per the stages the proposed method is described. Section 3 presents the experimental results and performance evaluation on considered datasets, and finally in Section 4 conclusions are drawn.

2 Proposed Methodology

Proposed method is an improvised work of a robust multilingual text detection approach based on transforms and wavelet entropy [2]. An efficient texture feature information is extracted by applying wavelet transforms and Gabor filter as described in [2]. The resulted Gabor output image is grouped into three clusters to classify the background, foreground and the sharpened texture edges obtained by applying the K-means clustering. In the next stage, morphological operations are performed to obtain connected components, then after a concept of linked list approach is in turn used to build a true text line sequence. In the final stage, wavelet entropy is imposed on an each connected component sequence in order to determine the true text region of an input image. The complete text detection procedure of our work is explained in the following subsections.

2.1 An Integrated Approach of Gabor Filter and K-Means Clustering for an Efficient Text Region Classification

The work performed using Wavelet transform and Gabor based method [2] is employed in our proposed method. In this we selected an average image of details of three orientation such as horizontal, vertical and diagonal images. The obtained detail information represents the sharpen edges of an image in all three orientations and this is subsequently used by Gabor filter to extract the textural information.

Gabor filter is optimally localized as per the uncertainty principle in both the spatial and frequency domain. That is the Gabor filter is highly selective in both position and frequency. This results in sharper texture boundary detection as in [2]. The main purpose of applying K-means clustering to a resultant Gabor image is to classify a highest energy class as a text candidates and the remaining classes as a uncertainty and non-text pixels. In the present work we considered three clusters to classify objects based on the feature set.

Choosing K is often an ad hoc decision based on prior knowledge, assumptions and practical experience [11]. Likewise we practically worked on choosing K value. Initially we set the value K=2 and observed true text pixels including false deteceted blocks. The obtained text classification result is shown in Figure 1(b) for the input image Figure 1(a). When we set the value K=3, we observed the true text regions are well classified compared to two clusters set in the first demonstration and results obtained to the same input image is shown in Figure 1(c) accordingly. The main idea behind applying K-means clustering by choosing a value K=3 for resultant Gabor image is, it considerably reduces the non-text pixels and efficiently classifies the text region from the background.

2.2 Morphological Operations and an Implementation of a Linked List Approach

To the obtained K-means clustering resultant image, we applied morphological operations to get a connected components of true text pixels. A concept of linked list approach [7] is then used to build a true text line sequence in order to get a sequence

Fig. 1 (a)Sample Input Image of 101 video images (b) K-means clustering images obtained for an input image when K=2 (c) K-means clustering image obtained for an input image when K=3

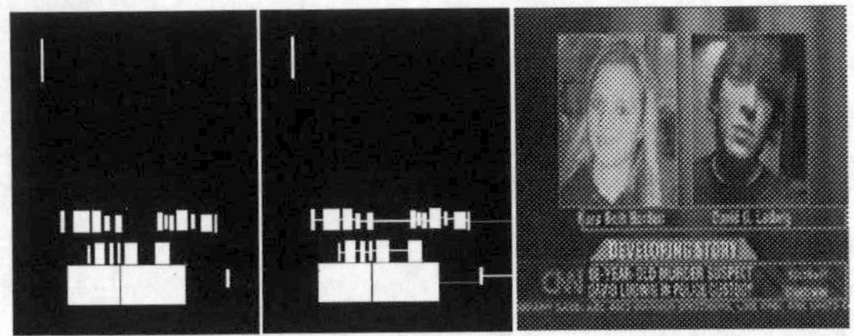

Fig. 2 (a) Resultant images obtained after applying morphological operations and as a sequence of true text line sequence of components after applying linked list approach (b) Resulted image of truly detected text region

of connected components to detect a sequence of true text regions of an input image. The results obtained for these stages are shown in Figure 2(a).

2.3 Wavelet Entropy

From the obtained sequence of connected components, we imposed the wavelet entropy to the corresponding region of a sequence of connected components in an input image, inorder to extract true text region as well to eliminate falsy blocks of an image. Then we extracted an energy information from an input image of the regions specified. Average energy of all the regions specified in the input image is fixed as threshold α. If the specified sequence of a text region $\geq \alpha$, where α is the threshold, it is considered as a text region or else considered as a non-text region. Figure 2(b) shows the text region obtained from the above mentioned procedure.

3 Experimental Results and Performance Evaluation

The proposed system is tested on two datasets. Firstly, a dataset of 101 video images provided by [4] comprising news programmers, sport video and movie clips. The dataset also includes both graphic text and seen text of different languages, e.g.

English, Chinese and Korean in the dataset. Second,the most cited ICDAR 2003 dataset [12], which contains images with text of varying sizes and positions. In order to evaluate the performance of proposed method, we used the following criteria:

- Truly Detected Block(TDB): A detected block that contains a text line, partially or fully.
- False Detected Block(FDB): A detected block that does not contain text.
- Text Block with Missing Data(MDB): A detected block that misses some characters of a text line (MDB is a subset of TDB).

Table 1 shows the results obtained for existing and proposed method on the database provided by [4]. For each image in the dataset we manually count the Actual Text Blocks(ATB). The performance measures defined as follows:

- Detection Rate (DR) = TDB / ADB
- False Positive Rate (FPR) = FDB /(TDB + FDB)
- Miss Detection Rate (MDR) = MDB / TDB

Table 1 shows the comparative study of proposed and existing methods. From this table is clear that the obtained TDBs are more i.e. the system detects more number of true text blocks, FDBs are sustained as in Transforms and Gabor based method, which indicates that there exists few alarms. MDBs are considerably reduced, which shows that the miss detection of text blocks are very few in number. We compared the proposed method with the existing text detection methods such as Edge-based [8], Gradient-based [9], Uniform-colored [10], Laplacian [4] and Transforms and Gabor based [2] methods. In order to evaluate the performance of the proposed method we considered 101 test images provided by [4]. From Table2, it is clear that the proposed method has higher DR and lesser MDR compare to existing methods and FPR is sustained as of Transforms & Gabor based method. The main goal of the proposed system is to achieve highest DR by detecting true text blocks of an image, we reached DR=98.9%, MDR=3.0% though FPR=13.7% which is sustained as of Transforms & Gabor based method. By the conduct of experiment, it is proved that the propose method exhibits higher detection rate and considerably lesser miss detection rate than the existing methods.

We also evaluated proposed method on a standard ICDAR 2003 dataset. Table3, shows obtained results and performance evaluation with the existing Transforms and Gabor based method on standard ICDAR 2003 dataset. The resultant text detection

Table 1 Results obtained for the dataset of 101 video imaages of[4]

Method	ATB	TDB	FDB	MDB
Edge-based[8]	491	393	86	79
Gradient-based[9]	491	349	48	35
Uniform-colored[10]	491	252	95	94
Laplacian[4]	491	458	39	55
Transform & Gabor based[2]	491	481	78	53
Proposed method	491	486	78	15

Table 2 Performance results obtained on dataset[8]

Method	DR	FPR	MDR
Edge-based[8]	80.0	18.0	20.1
Gradient-based[9]	71.1	12.1	10.0
Uniform-colored[10]	51.3	27.4	27.4
Laplacian[4]	93.3	7.9	7.9
Transform& Gabor based[2]	97.9	13.9	11.0
Proposed method	98.9	13.7	3.0

image of ICDAR 2003 dataset image is shown in Figure 3(b) for an input image shown in Figure 3(a). The vital part of our proposed method is that classifying the resultant Gabor image into three clusters by applying K-means clustering algorithm. With this we could able to detect true text regions effectively.

Table 3 Measures and Performance results obtained on ICDAR2003

Method	ATB	TDB	FDB	MDB	DR	FPR	MDR
Transform & Gabor based[2]	124	119	74	23	95.96	38.34	19.3
Proposed method	124	120	34	3	96.7	22	2.5

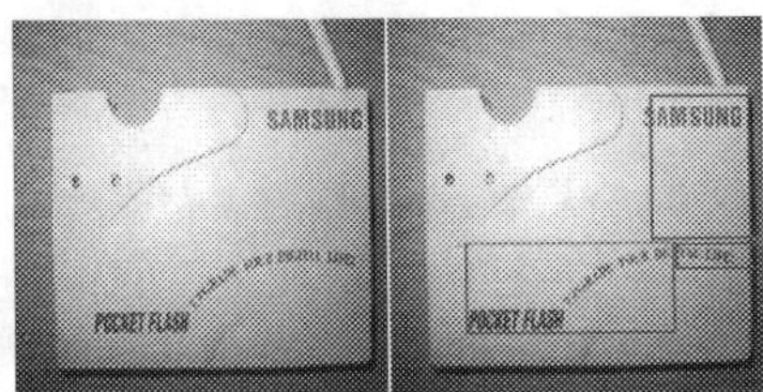

Fig. 3 (a) An input image of ICDAR 2003 dataset (b) The resulted text detection image

4 Conclusion

The proposed system is a development of an efficient text detection approach able to detect text of multilingual languages of different fonts, contrast and in unconstrained background. The key concept of our system is to detect true text region without missing any data, which is performed extensively by using the combination of Gabor filter and K-means algorithm. A concept of wavelet entropy which is used in our previous work [2] is applied to a result of the above mentioned combination of concepts to detect a true text region of an image. Experiments are conducted on two different datasets comprising of challenging images and varying background images: (1) standard ICDAR 2003 dataset, (2) dataset of 101 video images. The present improvised proposed system performance analysis has done on dataset of 101 video images and standard ICDAR 2003 dataset. The proposed system exhibits better text detection with drastically decreasing the missing of data in a exact text region detection.

References

1. Chowdhury, S.P., Dhar, S., Das, A.K., Chanda, B., McMenemy, K.: Robust Extraction of Text from Camera Images. In: The Proceedings of 10th International Conference on Document Analysis and Recognition, pp. 1280–1284 (2009)
2. Manjunath Aradhya, V.N., Pavithra, M.S., Naveena, C.: A Robust Multilingual Text Detection Approach Based on Transforms and Wavelet Entropy. In: The Proceedings of Procedia Technology (Elsevier) 2nd International Conference on Computer, Communication, Control and Information Technology (C3IT-2012), Hooghly, West Bengal, India (in press, 2012)
3. Kaushik, N., Sarathi, D., Mittal, A.: Robust Text Detection in images using morphological operations and Gabor wavelet. In: The Proceedings of EAIT, pp. 153–156 (2006)
4. Phan, T., Shivakumara, P., Tan, C.: A Laplacian method for video text detection. In: The Proceedings of 10th International Conference on Document Analysis and Recognition, pp. 66–70 (2009)
5. Shivakumara, P., Phan, T., Tan, C.: A Robust Wavelet Transform Based Technique for Video Text Detection. In: The Proceedings of 10th International Conference on Document Analysis and Recognition, pp. 1285–1289 (2009)
6. Shivakumara, P., Phan, T., Tan, C.: A Laplacian Approach to Multi-Oriented Text Detection in Video. IEEE Transactions on Pattern Analysis and Machine Intelligence 33, 412–419 (2011)
7. Naveena, C., Manjunath Aradhya, V.N.: A linked List Approach for Handwritten Textline Segmentation. Journal of Intelligent Systems (in press, 2012)
8. Liu, C., Wang, C., Dai, R.: Text Detection in Images Based on Unsupervised Classification of Edge-based Features. In: ICDAR, pp. 610–614 (2005)
9. Wong, E.K., Chen, M.: A new robust algorithm for video text extraction. In: The Proceedings of First Asian Conference on Pattern Recognition, ACPR, vol. 36, pp. 1397–1406 (2003)
10. Mariano, V.Y., Kasturi, R.: Locating Uniform Colored Text in Video Frames. In: 15th ICPR, vol. 4, pp. 539–542 (2000)
11. Teknomo, K.: Kardi Teknomo's Tutorials (2006); Available via Kardi Teknomo, `http://people.revoledu.com/kardi/tutorial/kMean/index.html`
12. Lucas, S.M., Panaretos, A., Sosa, L., Tang, A., Wong, S., Young, R.: ICDAR 2003 Robust Reading competitions. In: ICDAR 2003, pp. 682–687 (2003)

Integrating Global and Local Application of Discriminative Multinomial Bayesian Classifier for Text Classification

Emmanuel Pappas and Sotiris Kotsiantis

Abstract. The Discriminative Multinomial Naive Bayes classifier has been a center of attention in the field of text classification. In this study, we attempted to increase the prediction accuracy of the Discriminative Multinomial Naive Bayes by integrating global and local application of Discriminative Multinomial Naive Bayes classifier. We performed a large-scale comparison on benchmark datasets with other state-of-the-art algorithms and the proposed methodology gave better accuracy in most cases.

1 Introduction

Text classification has been an important application since the beginning of digital documents. Text Classification is the assignment of classifying a document under a predefined category. Sebastiani gave a nice review of text classification domain [17].

In this study, we attempted to increase the prediction accuracy of the Discriminative Multinomial Naive Bayes [19] by integrating global and local application of Discriminative Multinomial Naive Bayes classifier. Finally, we performed a large-scale comparison with other state-of-the-art algorithms on benchmark datasets and the proposed methodology had enhanced accuracy in most cases.

A brief description of data pre-processing of text data before machine learning algorithms can be applied is given in Section 2. Section 3 describes the most well known machine learning techniques that have been applied in text classification. Section 4 discusses the proposed method. Experiment results of the proposed

Emmanuel Pappas
Hellenic Open University, Greece
e-mail: mpappas@net314.eu

Sotiris Kotsiantis
Department of Mathematics, University of Patras, Greece
e-mail: sotos@math.upatras.gr

A. Abraham and S.M. Thampi (Eds.): Intelligent Informatics, AISC 182, pp. 49–55.
springerlink.com

method with other well known classifiers in a number of data sets are presented in section 5, while brief summary with further research topics are given in Section 6.

2 Data Preprocessing

A document is a sequence of words [2]. So each document is typically represented by an array of words. The set of all the words of a data set is called vocabulary, or feature set. Not all of the words presented in a document are useful in order to train the classifier [13]. There are worthless words such as auxiliary verbs, conjunctions and articles. These words are called stop-words. There exist many lists of such words which can be removed as a preprocess task. Stemming is another ordinary preprocessing step. A stemmer (which is an algorithm which performs stemming), removes words with the same stem and keeps the stem or the most general of them as feature [17].

An auxiliary feature engineering choice is the representation of the feature value [26]. Frequently, a Boolean indicator of whether the word took place in the document is satisfactory. Other possibilities include the count of the number of times the word is presented in the document, the frequency of its occurrence normalized by the length of the document, the count normalized by the inverse document frequency of the word.

The aim of feature-selection methods is the reduction of the dimensionality of the data by removing features that are measured irrelevant [3]. This transformation procedure has a number of advantages, such as smaller dataset size, smaller computational requirements for the text classification algorithms and considerable shrinking of the search space. Scoring of individual words can be carried out using some measures, such as document frequency, term frequency, mutual information, information gain, odds ratio, χ^2 statistic and term strength [5], [15], [18]. What is universal to all of these feature-scoring methods is that they bring to a close by ranking the features by their independently determined scores, and then select the top scoring features. Forman presented benchmark comparison of twelve metrics on well known training sets [3]. Since there is no metric that performs constantly better than all others, researchers often combine two metrics [6].

Feature Transformation varies considerably from Feature Selection approaches, but like them its purpose is to reduce the feature set size [26]. This approach compacts the vocabulary based on feature concurrencies. Principal Component Analysis is a well known method for feature transformation [23]. In the text mining community this method has been also named Latent Semantic Indexing (LSI) [1].

3 Machine Learning Algorithms

After feature selection and transformation the documents can be without difficulty represented in a form that can be used by a ML algorithm. Many text classifiers have been proposed in the literature using different machine learning techniques such as Naive Bayes, Nearest Neighbors, and lately, Support Vector Machines. Although many approaches have been proposed, automated text classification is still a major area of research mainly because the effectiveness of current automated text classifiers is not perfect and still needs improvement.

Naive Bayes is often used in text classification applications and experiments because of its simplicity and effectiveness [8]. However, its performance is often degraded because it does not model text well. Schneider addressed the problems and show that they can be resolved by some simple corrections [16]. In [25], an auxiliary feature method is proposed as an improvement to simple Bayes. It determines features by a feature selection method, and selects an auxiliary feature which can reclassify the text space aimed at the chosen features. Then the corresponding conditional probability is adjusted in order to improve classification accuracy.

Mccallum and Nigam [14] proposed the NB-Multinomial classifier with good results. Klopotek and Woch presented results of empirical evaluation of a Bayesian multinet learner based on a new method of learning very large tree-like Bayesian networks [9]. The study suggests that tree-like Bayesian networks can handle a text classification task in one hundred thousand variables with sufficient speed and accuracy.

In learning Bayesian network classifiers, parameter learning often uses Frequency Estimate (FE), which determines parameters by computing the appropriate frequencies from dataset. The major advantage of FE is its competence: it only needs to count each data point once. It is well-known that FE maximizes likelihood and therefore is a characteristic generative learning method. In [19], the authors proposed an efficient and effective discriminative parameter learning method, called Discriminative Frequency Estimate (DFE). The authors' motivation was to turn the generative parameter learning method FE into a discriminative one by injecting a discriminative element into it. DFE discriminatively computes frequencies from dataset, and then estimates parameters based on the appropriate frequencies. They named their algorithm as Discriminative Multinomial Bayesian Classifier.

Several authors have shown that support vector machines (SVM) provide a fast and effective means for learning text classifiers [7], [10], [21], [24]. The reason for that is SVM can handle exponentially many features, because it does not have to represent examples in that transformed space, the only thing that needs to be computed efficiently is the similarity of two examples.

kNN is a lazy learning method as no model needs to be built and nearly all computation takes place at the classification stage. This prohibits it from being applied to large datasets. However, k-NN has been used to text categorization since the early days of its research [4] and is one of the most effective methods on the Reuters corpus of newswire stories – a benchmark corpus in text categorization.

A problem of supervised algorithms for text classification is that they normally require high-quality training data to build an accurate classifier. Unfortunately, in many real-world applications the training sets present imbalanced class distributions. In order to deal with this problem, a number of different approaches such as sampling have been proposed [12], [20].

4 Proposed Methodology

The proposed model simple trains a Discriminative Multinomial Bayesian Classifier (DMNB) classifier during the train process. For this cause, the training time of the model is that of simple DMNB. During the classification of a test document the model calculate the probabilities each class and if the probability of the most possible class is at least two times the probability of the next possible class then the decision is that of global DMNB model. However, if the global DMNB is not so sure e.g. the probability of the most possible class is less than two times the probability of the next possible class; the model finds the k nearest neighbors using the selected distance metric and train the local simple DMNB classifier using these k instances. Finally, in this case the model averages the probabilities of global DMNB with local DMNB classifier for the classification of the testing instance. It must be mentioned that local DMNB classifier is only used for a small number of test documents and for this reason classification time is not a big problem. Generally, the proposed ensemble is described by pseudo-code in Fig 1.

Training:
Build Global DMNB in all the training set
Classification:
1. Obtain the test document
2. Calculate the probabilities of belonging the document in each class of the dataset.
3. If the probability of the most possible class is at least two times the probability of the next possible class then the decision is that of global DMNB model else
    a. Find the k(=50) nearest neighbors using the selected distance metric (Manhattan in our implementation)
    b. Using as training instances the k instances train the local DMNB classifier
    c. Aggregate the decisions of global DMNB with local DMNB classifier by averaging of the probabilities for the classification of the testing instance.

Fig. 1 Integrating Global and Local Application of Naive Bayes Classifier (IGLDMNB)

Combining instance-based learning with DMNB is inspired by improving DMNB through relaxing the conditional independence assumption using lazy learning. It is expected that there are no strong dependences within the k nearest neighbors of the test instance, although the attribute dependences might be strong in the whole dataset. Fundamentally, we are looking for a sub-space of the instance space in which the conditional independence assumption is true or almost true.

5 Comparisons and Results

For the purpose of our study, we used well-known datasets from many domains text datasets donated by George Forman/Hewlett-Packard Labs (http://www.hpl.hp.com/personal/George_Forman/). These data sets were hand selected so as to come from real-world problems and to vary in characteristics.

For our experiments we used Naive Bayes Multinomial algorithm and Discriminative Multinomial Naive Bayes classifier. The Sequential Minimal Optimization (or SMO) algorithm was the representative of the Support Vector Machines in out study. It must be mentioned that we used for the algorithms the free available source code by the book [23]. In order to calculate the classifiers' accuracy, the whole training set was divided into 10 mutually exclusive and equal-sized subsets and for each subset the learner was trained on the union of all of the other subsets. Then, the average value of the 10-cross validation was calculated.

In Table 1, we present the average accuracy of each classifier. In the same tables, we also represent with "*v*" that the proposed IGLDMNB algorithm *looses* from the specific algorithm. That is, the specific algorithm performed statistically better than IGLDMNB according to t-test with $p<0.05$. Furthermore, in Table 1, "*" indicates that IGLDMNB performed statistically better than the specific classifier according to t-test with $p<0.05$. In all the other cases, there is no significant statistical difference between the results (*Draws*).

Table 1 Comparing the proposed algorithm with other well known algorithms

Data-set	IGLDMNB	DMNB	SMO	NB-Multinomial
oh0	92.14	91.23	81.96*	89.03*
oh10	84.58	83.81	74.86*	81.24*
oh15	85.22	84.77	72.72*	83.78
re0	83.99	83.78	75.47*	80.38
re1	83.10	82.86	74.29*	83.35
tr11	89.27	86.23*	74.17*	84.79*
tr12	91.06	86.91*	74.46*	83.05*
tr21	92.52	91.93	79.46*	63.37*
tr23	93.29	91.17*	74.12*	71.55*
tr41	96.97	96.36	87.02*	94.42*

The proposed method is significantly more accurate than single NB-Multinomial in 7 out of the 10 data sets, while it has not significantly higher error rates than NB-Multinomial in any data set. Moreover, the proposed algorithm is significantly more accurate than SMO algorithm in all data sets. Finally, the proposed method is significantly more accurate than simple DMNB [21] in 3 out of the 10 data sets, while it has not significantly higher error rates than DMNB in any data set.

In brief, we managed to improve the performance of the Discriminative Multinomial Bayesian Classifier obtaining better accuracy than other well known classifiers. We have implemented the proposed algorithm in a software tool (see Fig. 2). The tool expects the training set as an Attribute-Relation File Format. The class attribute must be in the last column. After the training of the model (from few seconds to few minutes to complete), one is able to predict the class of the new text.

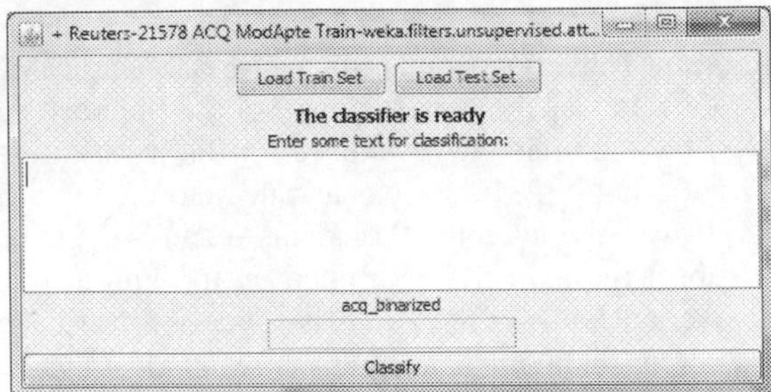

Fig. 2 A screenshot of the implemented tool

6 Conclusion

The text classification problem is a machine learning research topic, specially given the vast number of documents available in the form of web pages and other electronic texts like discussion forum postings, emails, and other electronic documents [22]. In this work, we managed to improve the performance of the Discriminative Multinomial Bayesian Classifier. We performed a large-scale comparison with other a state-of-the-art algorithms on 10 standard benchmark datasets and we took better accuracy in most cases. Reuters Corpus Volume I (RCV1) is an archive of over 800,000 manually categorized newswire stories recently made available by Reuters, Ltd. for research purposes [11]. Using this collection, we can compare more extensively the proposed algorithm.

References

1. Cardoso-Cachopo, A., Oliveira, A.L.: An Empirical Comparison of Text Categorization Methods. In: Nascimento, M.A., de Moura, E.S., Oliveira, A.L. (eds.) SPIRE 2003. LNCS, vol. 2857, pp. 183–196. Springer, Heidelberg (2003)
2. Figueiredo, F., Rocha, L., Couto, T., Salles, T., Gonçalves, M.A., Meira Jr., W.: Word co-occurrence features for text classification. Information Systems 36(5), 843–858 (2011)
3. Forman, G.: Feature selection for text classification. In: Computational Methods of Feature Selection, pp. 257–276. Chapman and Hall/CRC (2007)
4. Guo, G.D., Wang, H., Bell, D., Bi, Y.X., Greer, K.: Using kNN model for automatic text categorization. Soft Computing 10(5), 423–430 (2006)
5. Feng, G., Guo, J., Jing, B.-Y., Hao, L.: A Bayesian feature selection paradigm for text classification. Information Processing Management (2011) ISSN 0306-4573, 10.1016/j.ipm.2011.08.002
6. Chen, J., Huang, H., Tian, S., Qu, Y.: Feature selection for text classification with Naïve Bayes. Expert Systems with Applications 36(3), Part I, 5432–5435 (2009)
7. Joachims, T.: Learning to classify text using support vector machines. Kluwer Academic, Hingharn (2002)

8. Kim, S.-B., Rim, H.-C., Yook, D., Lim, H.-S.: Effective Methods for Improving Naive Bayes Text Classifiers. In: Ishizuka, M., Sattar, A. (eds.) PRICAI 2002. LNCS (LNAI), vol. 2417, pp. 414–423. Springer, Heidelberg (2002)
9. Kłopotek, M.A., Woch, M.: Very Large Bayesian Networks in Text Classification. In: Sloot, P.M.A., Abramson, D., Bogdanov, A.V., Gorbachev, Y.E., Dongarra, J., Zomaya, A.Y. (eds.) ICCS 2003. LNCS, vol. 2657, pp. 397–406. Springer, Heidelberg (2003)
10. Leopold, E., Kindermann, J.: Text Categorization with Support Vector Machines. How to Represent Texts in Input Space? Machine Learning 46, 423–444 (2002)
11. Lewis, D., Yang, Y., Rose, T., Li, F.: RCV1: A New Benchmark Collection for Text Categorization Research. Journal of Machine Learning Research 5, 361–397 (2004)
12. Liu, Y., Loh, H.T., Sun, A.: Imbalanced text classification: A term weighting approach. Expert Systems with Applications 36, 690–701 (2009)
13. Madsen, R.E., Sigurdsson, S., Hansen, L.K., Lansen, J.: Pruning the Vocabulary for Better Context Recognition. In: 7th International Conference on Pattern Recognition (2004)
14. Mccallum, A., Nigam, K.: A Comparison of Event Models for Naive Bayes Text Classification. In: AAAI 1998 Workshop on Learning for Text Categorization (1998)
15. Ogura, H., Amano, H., Kondo, M.: Feature selection with a measure of deviations from Poisson in text categorization. Expert Systems with Applications 36, 6826–6832 (2009)
16. Schneider, K.-M.: Techniques for Improving the Performance of Naive Bayes for Text Classification. In: Gelbukh, A. (ed.) CICLing 2005. LNCS, vol. 3406, pp. 682–693. Springer, Heidelberg (2005)
17. Sebastiani, F.: Machine Learning in Automated Text Categorization. ACM Computing Surveys 34(1), 1–47 (2002)
18. Shang, W., Huang, H., Zhu, H., Lin, Y.: A novel feature selection algorithm for text categorization. Expert Systems with Applications 33, 1–5 (2007)
19. Su, J., Zhang, H., Ling, C., Matwin, S.: Discriminative Parameter Learning for Bayesian Networks. In: ICML 2008 (2008)
20. Sun, A., Lim, E., Liu, Y.: On strategies for imbalanced text classification using SVM: A comparative study. Decision Support Systems 48(1), 191–201 (2009)
21. Vikramjit, M., Wang, C.-J., Banerjee, S.: Text classification: A least square support vector machine approach. Applied Soft Computing 7(3), 908–914 (2007)
22. Yu, B.: An evaluation of text classification methods for literary study. Literary and Linguistic Computing 23(3), 327–343 (2008)
23. Witten, I., Frank, E., Hall, M.: Data Mining: Practical Machine Learning Tools and Techniques, 3rd edn. Morgan Kaufmann (2011) ISBN 978-0-12-374856-0
24. Zhang, W., Yoshida, T., Tang, X.: Text classification based on multi-word with support vector machine. Knowledge-Based Systems 21(8), 879–886 (2008)
25. Zhang, W., Gao, F.: An Improvement to Naive Bayes for Text Classification. Procedia Engineering 15, 2160–2164
26. Zhang, W., Yoshida, T., Tang, X.: A comparative study of TF*IDF, LSI and multi-words for text classification. Expert Systems with Applications 38(3), 2758–2765 (2011)

Protein Secondary Structure Prediction Using Machine Learning

Sriparna Saha, Asif Ekbal, Sidharth Sharma,
Sanghamitra Bandyopadhyay, and Ujjwal Maulik

Abstract. Protein structure prediction is an important component in understanding protein structures and functions. Accurate prediction of protein secondary structure helps in understanding protein folding. In many applications such as drug discovery it is required to predict the secondary structure of unknown proteins. In this paper we report our first attempt to secondary structure predication, and approach it as a sequence classification problem, where the task is equivalent to assigning a sequence of labels (i.e. helix, sheet, and coil) to the given protein sequence. We propose an ensemble technique that is based on two stochastic supervised machine learning algorithms, namely Maximum Entropy Markov Model (MEMM) and Conditional Random Field (CRF). We identify and implement a set of features that mostly deal with the contextual information. The proposed approach is evaluated with a benchmark dataset, and it yields encouraging performance to explore it further. We obtain the highest predictive accuracy of 61.26% and segment overlap score (SOV) of 52.30%.

1 Introduction

Predicting protein structure and function from amino acid sequences has a key role in molecular biology. Proteins are made of long chains of amino acid residues.

Sriparna Saha · Asif Ekbal · Sidharth Sharma
Department of Computer Science and Engineering,
Indian Institute of Technology Patna, India
e-mail: {sriparna, asif, sidharth}@iitp.ac.in

Sanghamitra Bandyopadhyay
Machine Intelligence Unit, Indian Statistical Institute, Kolkata, India
e-mail: sanghami@isical.ac.in

Ujjwal Maulik
Department of Computer Science and Engineering, Jadavpur University, Kolkata, India
e-mail: ujjwal_maulik@yahoo.com

A. Abraham and S.M. Thampi (Eds.): Intelligent Informatics, AISC 182, pp. 57–63.
springerlink.com

A gene encodes a specific amino acid sequence. After translation it folds into a unique three-dimensional conformation. The problem of protein structure prediction is to predict this conformation (the protein's tertiary structure) from the amino acid sequence (the protein's primary structure). The structure of a protein determines the biological function of a protein. Thus structure prediction is an important step for predicting function of that protein. Potential applications of structure prediction range from elucidation of cellular processes to vaccine design. As the number of known protein sequence is increasing due to various genome and other sequencing projects, the problem of protein secondary structure prediction is becoming more important [Thorton(2001)].

In this paper we report our preliminary work on protein secondary structure prediction, where the overall problem is cast as a sequence classification problem. Given a sequence of protein sequence, the task is to assign correct labels, i.e. helix, sheet, and coil. As base classifiers we use two stochastic learning algorithms, namely Maximum Entropy Markov Model (MEMM) and Conditional Random Field (CRF). We identify a set of features that mostly deal with the contextual information. Based on the different feature subsets, several classification models are built using these two algorithms. Thereafter these models are then combined together into a final system using a weighted voting approach. The system is evaluated in terms of two prediction metrics, namely *accuracy* and *segment overlap score* (SOV). The proposed approach is evaluated on the benchmark dataset of *RS126*, and we observed the highest prediction accuracy (Q_3) of 61.26% and segment overlap score (i.e. *SOV*) of 52.30%.

2 Proposed Approach

The problem of protein secondary structure prediction belongs to the larger domain of classification problems. The task in such problems is to observe some context (read residue) $b \in B$ and predict its class (read segment) $a \in A$ accurately. This involves constructing a classifier $B \rightarrow A$. Mathematically, it is the same as implementing a conditional probability distribution p, so that the probability $p\left(\frac{a}{b}\right)$ is the probability of residue a given some context (which may be a collection of residues and other "features") b. This classifier is built on a probabilistic model rather than a deterministic one because the "context"is never large enough to reliably specify a deterministic $f : B \rightarrow A$.

Here we use two such probabilistic models, the Maximum Entropy Markov Model (or, MEMM) and the Conditional Random Fields (CRFs). Both of these models treat the protein secondary structure prediction problem as the so-called input-output problem where the task is to take an input sequence and produce an output sequence containing the labels of the corresponding input sequence. For the problem at hand, the cardinality of the set of residues is 20. The secondary structure definition program DSSP that we use assigns these residues to eight different classes. For the sake of simplicity of implementation, we reduce these to 3 classes, namely Helix (H), Sheet (E) and Coil(C).

The maximum entropy Markov model (MEMM) estimates probabilities based on the principle of making as few assumptions as possible, other than the constraints imposed. It agrees with the maximum likelihood distribution, and has the exponential form

$$P(t|h) = \frac{1}{Z(h)} exp(\sum_{j=1}^{n} \lambda_j f_j(h,t)) \quad (1)$$

where, t is the structure type, h is the context (or history), $f_j(h,t)$ are the features with associated weight λ_j and $Z(h)$ is a normalization function.

We use the OpenNLP Java based MaxEnt package[1] for the computation of the values of the parameters λ_j. We use the Generalized Iterative Scaling [Darroch and Ratcliff(1972)] algorithm to estimate the MaxEnt parameters.

Maximum Entropy Markov Models serve well as probabilistic frameworks in wide variety of application but they suffer from the so-called label bias problem where the classifier tends to be more biased towards states with fewer outgoing transitions. This is due to the fact that states with low-entropy next state distributions are less likely to notice an observation sequence. A more robust model, Conditional Random Field [Lafferty et al(2001)Lafferty, McCallum, and Pereira] is defined as a random field globally conditioned on the sequences to be labeled. This global model can efficiently avoid the demerits of MEMM.

Conditional Random Field (CRF) [Lafferty et al(2001)Lafferty, McCallum, and Pereira] is an undirected graphical model, a special case of which corresponds to conditionally trained probabilistic finite state automata. Being conditionally trained, CRF can easily incorporate a large number of arbitrary, non-independent features while still having efficient procedures for non-greedy finite-state inference and training. We use C^{++} based CRF^{++} package [2], a simple, customizable, and open source implementation of CRF for segmenting or labeling sequential data.

Initially, a number of models are built based on MEMM and CRF. Each of the MEMM and CRF is fed with a number of features that are extracted automatically from the given training data. We build a number of models of these classifiers by considering various feature subsets. Finally these models are combined together into a final system by defining appropriate voting mechanisms, i.e. majority voting or weighted voting. In case of majority voting the label proposed by the majority of the models is assigned to the corresponding sequence. In case of weighted voting, the weights are calculated based on the accuracy of the classifiers.

From the training data we extract a feature vector consisting of the features and class label. Now, we have a training data in the form (W_i, T_i), where, W_i is the ith protein sequence and its feature vector and T_i is its corresponding output class. For MEMMs we use the traditional neighborhood window features surrounding to the current protein sequence. In our analysis, we also incorporate dynamic features wherein we use the predicted output of the previous two context observations as a

---

[1] http://maxent.sourceforge.net/

[2] http://crfpp.sourceforge.net

part of the context of current observation. This is done in order to incorporate the fact that the residues belonging to same segments tend to lie in proximity. For CRF, we consider various combinations from the set of feature templates as given by,
$F_1 = \{w_{i-m}, \ldots, w_{i-1}, w_i, w_{i+1}, \ldots, w_{i+n}$; Combination of w_{i-1} and w_i; Combination of w_i and w_{i+1}; B (bigram feature template)$\}$

The bi-gram feature template computes the feature combinations of the current and previous tokens.

3 Datsets, Metrics and Evaluation Results

We experiment with the benchmark dataset of RS126. It consists of 126 proteins with sequence similarity less than 25%. In order to perform three-fold cross validation we generate three different subsets. One is withheld for testing while the remaining two are used as the training sets. This process is repeated three times to perform three-fold cross validation.

Data Encoding: DSSP is the most widely used secondary structure definition program that assigns residues to eight different segment classes based on hydrogen bonding patterns. These eight classes are reduced to three for the sake of simplicity of implementation. The H, G and I are mapped to H; E is mapped to E and the rest are mapped to C, where $H, E, C \in SegmentClasses$. This is motivated by the fact that for the data set used in the study, this reduction minimizes number of discrete states.

Evaluation Metrics: For our analysis, we use two standard metrics of accuracy measurement, the per residue accuracy where we compare every predicted segment for each residue with the solved segment determined by secondary structure definition program. The result is just a linear one to one comparison of predicted and solved structure. However, this accuracy metric has some intrinsic flaws as pointed out by [Zemla et al(1999)Zemla, Venclovas, Fidelis, and Rost]. So we have used the modified Segment overlap method illustrated in Zemla et al(1999)Zemla, Venclovas, Fidelis, and Rost. In our observation, the accuracy determined by this method is less than the per-residue accuracy but this metric is quite useful over the former.

Experiments and Discussions: We report the results of 3-fold cross validation. Table 1 presents the results of MEMM by considering only the contextual information of the current protein sequence. In table, $MEMM_{static5}$ represents MEMM with a static feature set of previous two and next two sequences, i.e. $w_{i-2}^{i+2} = w_{i-2} \ldots w_{i+2}$ and $MEMM_{static7}$ represents the MEMM with a static feature set of window size 7. We then introduce the dynamic information into models that incorporate the output labels of the previous two elements. Results are reported in Table 2 that shows significant drop in the overall performance. Thereafter, we evaluate the CRF based model. Table 3 reports the results of CRF classifier, $CRF_{static7}$ represents the CRF model with features set that considers a context of previous three and next three elements. The second model also incorporates the *bigram* feature that computes the

Table 1 Results of MEME-I

	PR	SOV
$MEMM_{static5}$	57.349	44.534
$MEMM_{static7}$	55.358	44.295

Table 2 Results of MEME-II

	PR	SOV
$MEMM_{dynamic5}$	36.990	27.549
$MEMM_{dynamic7}$	38.975	7.3205

combinations of the current and previous elements in sequence. The second model does not perform well, may be due to the lack of enough evidences in training. Comparisons between MEMM and CRF show the superiority of the later over the former. In Table 4 we report the evaluation figures of CRF by considering the bi-gram (i.e. two residues in sequence) or tri-gram (i.e. three residues in sequence) combination of all adjacent residues. The first row does not include the bi-gram feature combination whereas the second one includes the bi-gram feature combination.

Table 3 Results of CRF-I

	PR	SOV
$CRF_{static7}$	54.627	33.323
$CRF_{dynamic7}$	52.425	27.05

Table 4 Results of CRF-II

	PR	SOV
$CRF_{cumstatic7}$	56.853	41.907
$CRF_{cumdynamic7}$	56.824	48.074

Ensemble of the Classifiers

We constructed 8 different models of MEMM and CRF by varying the different subsets of features. These models are shown below:

1. Model-1: MEMM with previous two and next two residues, i.e. $w_{i-2}^{i+2} = w_{i-2} \dots w_{i+2}$ of w_i.
2. Model-2:MEMM with previous three and next three residues, i.e. $w_{i-3}^{i+3} = w_{i-3} \dots w_{i+3}$ of w_i.
3. Model-3:MEMM with previous two and next two residues, i.e. $w_{i-2}^{i+2} = w_{i-2} \dots w_{i+2}$ of w_i, and output labels of previous two residues, i.e. $t_{i-2}t_{i-1}$.

4. Model-4: MEMM with previous three and next three residues, i.e. $w_{i-3}^{i+3} = w_{i-3} \ldots w_{i+3}$ of w_i, and output labels of previous two residues, i.e. $t_{i-2}t_{i-1}$.
5. Model-5: CRF with previous three and next three residues, i.e. $w_{i-3}^{i+3} = w_{i-3} \ldots w_{i+3}$ of w_i.
6. Model-6:CRF with previous three and next three residues, i.e. $w_{i-3}^{i+3} = w_{i-3} \ldots w_{i+3}$ of w_i; and the bigram feature template that computes the feature combination of the current and previous residues.
7. Model-7:CRF with previous three and next three residues, i.e. $w_{i-3}^{i+3} = w_{i-3} \ldots w_{i+3}$ of w_i; combination of w_{i-1} and w_i.
8. Model-8: CRF with previous three and next three residues, i.e. $w_{i-3}^{i+3} = w_{i-3} \ldots w_{i+3}$ of w_i; combination of w_{i-1} and w_i, output label of the previous residue, i.e. t_{i-1} .

These models are combined together into a final system by majority and weighted voting methods. In majority voting, the final label is determined from the majority of the models' outputs. Random selection is used for resolving ties.

Where Acc_{Voting} is the accuracy of model ensembled by taking majority voting among different models. Acc_{Voting} achieved the PR and SOV values of 50.22%, and 34.99%, respectively.

Performance in this model shows that there is a considerable difference among the outputs as predicted by different machine learning frameworks or even same machine learning frameworks under different feature sets. It is to be noted that these systems could even perform better when they act stand-alone. In the weighted voting construction, we consider the conditional probabilities of each class as the corresponding weight. The conflicts are again resolved by random selection among ties. We conduct the following two set of weighted voting experiments:

1. Set-1: Each classifier assigns weight to each of the three possible classes (i.e. E, H and C) for each residue. Thus, in total we have 24 outputs for each residue. For each of the three classes we have eight confidence scores. These are summed and normalized by dividing with 8 (i.e. maximum possible confidence weights). Finally, for each residue, we have three normalized scores, each one for three different classes. Based on the weight, the final class label is assigned to the corresponding residue.
2. Set-II: In each classifier, we assign a class label to a residue depending upon the highest confidence value. We form an ensemble of eight classifiers by considering only these winning classes. The final class label is determined from these eight confidence scores.

Where $Acc_{WeightedVoting}$ is the accuracy measured by conducting a weighted voting among participating frameworks. The weight is the corresponding confidence value. $Acc_{WeightedVoting}$ achieved the PR and SOV values of 60.93%, and 51.66%, respectively.

Where Acc_{WWC} is the accuracy measured by conducting a weighted voting among the winners of participating frameworks. Acc_{WWC} achieved the PR and SOV values of 61.26%, and 52.30%, respectively.

Results with these weighted voting mechanisms suggest that the ensemble that was constructed by considering only the wining classes performs better. It is to be noted that individual model sometimes performs better in comparison to the ensemble formed by majority voting approach.

4 Conclusion

Protein structure prediction is an important component in understanding protein structures and functions. In this paper we have reported our preliminary work on protein secondary structure predication using the supervised machine learning approaches. We have proposed an ensemble technique that is based on two stochastic supervised machine learning algorithms, namely Maximum Entropy Markov Model (MEMM) and Conditional Random Field (CRF). We have identified and implemented a set of features that mostly deal with the contextual information and previous knowledge. The proposed approach is evaluated with a benchmark dataset, namely RS126. We obtain the highest predictive accuracy of 61.26% and segment overlap score (SOV) of 52.30%. In future we would like to investigate some biologically motivated set of features.

References

[Darroch and Ratcliff(1972)] Darroch, J., Ratcliff, D.: Generalized Iterative Scaling for Log-linear Models. Ann. Math. Statistics 43, 1470–1480 (1972)

[Lafferty et al(2001)Lafferty, McCallum, and Pereira] Lafferty, J., McCallum, A., Pereira, F.: Conditional random fields: Probabilistic framework for segmenting and labelling sequence data. In: 18th International Conference on Maching Learning, pp. 282–289. Morgan Kaufmann, San Franciso (2001)

[Thorton(2001)] Thorton, J.M.: From genome to function. Science 292, 2095–2097 (2001)

[Zemla et al(1999)Zemla, Venclovas, Fidelis, and Rost] Zemla, A., Venclovas, C., Fidelis, K., Rost, B.: A modified definition of sov, a segment-based measure for protein secondary structure prediction assessment. PROTEINS: Structure, Function, and Genetics 34, 220–223 (1999)

Refining Research Citations through Context Analysis

G.S. Mahalakshmi, S. Sendhilkumar, and S. Dilip Sam

Abstract. With the impact of both the authors of scientific articles and also scientific publications depending upon citation count, citations play a decisive role in the ranking of both researchers and journals. In this paper, we propose a model to refine the citations in a research article verifying the authenticity of the citation. This model will help to eliminate author-centric and outlier citations thereby refining the citation count of research articles.

1 Introduction

This model is intended to serve as an environment for calculating the context relevant citations and thereby its count and ultimately finding an article's citation impact. The journal impact factor [17] was designed mainly to compare the citation impact of one journal with other journals. A journals impact factor is based on two factors:

- *numerator* denoting the number of citations in the current year to any items published in the journal in the previous two years
- *denominator* denoting the number of substantive articles in the last two years.

The numerator is not always accurate due to guest citations or false citations. This system aids in overcoming this problem by finding the appropriate numerator count by evaluating the citations made for context relevance. In this paper, we propose to find the context relevancy between the cite and the base paper it cites by extracting a citation context around the cite placeholder. We extend the notion of semantic similarity between two entities to consider inherent relationships between concepts/words appearing in the citation contexts extracted, using fuzzy

G.S. Mahalakshmi · S. Dilip Sam
Department of Computer Science & Engineering, Anna University, Chennai

S. Sendhilkumar
Department of Information Science & Technology, Anna University, Chennai

A. Abraham and S.M. Thampi (Eds.): Intelligent Informatics, AISC 182, pp. 65–71.
springerlink.com

cognitive maps. Extracting entity description by referring to an ontology such as Wordnet [http://wordnet.princeton.edu/] and Wikipedia [http://en.wikipedia.org/wiki/Main_Page], has been used in earlier frameworks. We build on such earlier techniques and use our evolved domain specific ontology for computing similarity. Our similarity computation using fuzzy cognitive maps provides a closer resemblance to the workings of the human mind.

2 Related Work

2.1 Research Impact Analysis

The impact factor of a journal is calculated as the frequency with which its published papers are cited up to two years after publication [19]. It is a hidden process [2–4] and is highly sensitive to the categorisation of papers as 'citeable' (e.g., re-search-based) or 'frontmatter' (e.g., editorials and commentary) [6]. Attempts to replicate or to predict the reported values have generally failed [5–8]. With the emergence of more sophisticated metrics for journals [3, 4, 9, 10], the job of assessing the importance of specific papers is becoming an important area of research. However, impact factor—or any other journal-based metric for that matter—cannot escape an even more fundamental problem: it is simply not designed to capture qualities of individual papers [16].

An article with higher citations need not actually be concluded as the best, because more authors will lead to more self-citations [15].For finding a journal's impact factor, the time frame for citations is a major criterion to be accounted for. Initial-ly, the time frame for impact factor calculation was assumed as 2. In later years another factor called 'Eigen factor' evolved to solve this issue [Impact Factor Cochrane Database of Systematic Reviews (CDSR),2009]. However even in this methodology, the self-citation counts are eliminated. In addition, related indices like immediacy index, cited half-life and aggregate impact factor are considered. These indices suit well but only for the journal as a whole rather than for an individual article of the respective journal. However a newer index, namely 'h-index,' is coined by Hirsch to solve this issue [16] [18].Though h-index was very useful, it does not account for the number of authors of an article. It possesses no significance to the authors, their positions or contributions in the article, etc [18].The h-index is merely a natural number and it does not include the self-citations that the particular article has gained. This is similar to the issue addressed in the 'Eigen factor'. Other issues found with the h-index are that it does not account for article content. It just measures the article productivity i.e. page count. The scientific impact of a publication can be determined not only based on the number of times it is cited but also based on the citation speed with which its content is noted by the scientific community. This is tackled through speed index [18], which is the number of months that have elapsed since the first citation. Further empirical studies in different disciplines based on larger data sets will be needed to test the validity of the citation speed index. Many of the disadvantages that have been discussed in connection with the h-index and its variants surely apply to the citation speed index [18]. Seglen has shown that citation rate used in Thomson ISI is proportional to the article length [19]. It is not true in many cases as citation rate does not depend on article length. In addition, a journal's impact factor should be decided by the impact created by individual articles.

2.2 Fuzzy Cognitive Maps

Fuzzy Cognitive Map is a directed and weighted graph of concepts and relationships between the concepts. FCM is derived from Cognitive Map. Based on the CM structure, FCM was proposed by Kosko [2]. As a great improvement compared to CM, FCM introduces fuzzy quantitative relationship between concepts to describe the weight of the causal relationship. FCM has iterative characteristic. In FCM, the arcs are not only directed to show the direction of causal relations, but also accompanied by a quantitative weight within the interval [0, 1]. In general, FCMs have been found useful in many applications: administrative sciences, game theory, information analysis, popular political developments, electrical circuits analysis, cooperative man machines, distributed group-decision support and adaptation and learning, etc. [2]. An FCM represents the whole system in a symbolic manner, just as humans have stored the operation of the system in their brains, thus it is possible to help man's intention for more intelligent and autonomous systems. FCMs model the possible worlds as collection of classes and causal relations between classes. In this system FCM based document similarity measure has been used.

3 Citation Analysis

3.1 Wordnet Based Methods to Identify Domain Specific Words

This process takes as input the bigrammed and trigrammed documents. There is a possibility that the bi-grams and tri-grams collected may contain incomplete words or hyphenated words which must to be removed Thus we take each and every word and check if it is a valid synset. If it is found to be valid synset possessing some meaning, we consider those words. Those words that does not prove to be a valid sysnet are considered junk and hence to be removed. This is achieved with the help of Wordnet. There is a possibility that the bi-grams and tri-grams collected may contain author names, country names and other words that are irrelevant to the domain considered. Thus the set of author names comprising all the authors' names is collected from the metadata of the corpus. Those words that found to have a match with this list are considered to be junk and they are removed. Thus we have a set of words which are nouns having valid meanings. These words can be used for creating concepts in Ontology.

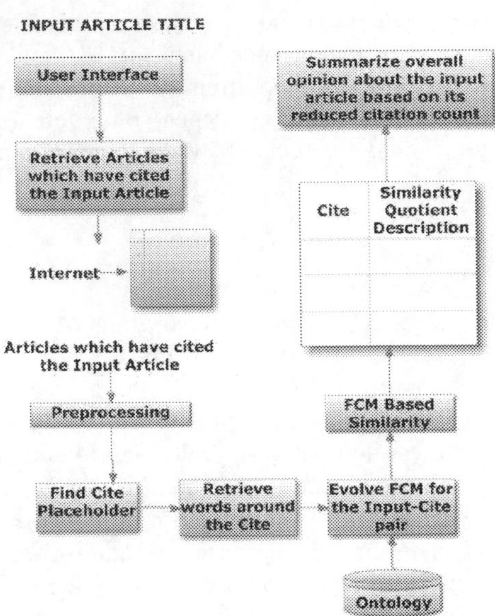

Fig. 1 Refining Citations through Context Analysis

3.2 Evolving FCM

The context that is retrieved is stemmed and stop words are removed. Keywords are identified and they are mapped to the knowledge tree (Ontology) that we are using. The keywords that are identified in a document undergo mapping. These concepts are represented in the form of cognitive maps which represent the documents structure where the concepts act as nodes and the link between each other concept as edge. The fuzziness represents how two documents (base and the citation) considered say Ci are closely related to each other. FCM Based Similarity Matrices are constructed for each FCM where the rows and columns of each matrix are the nodes found in both the FCMs. Now these matrices hold the FCMs evolved for both the documents. Now cosine similarity measure is implemented to find the similarity for these 2 matrices and a similarity quotient is identified.

3.3 Polarization and Sentiment Analysis

For each of the citation context Concci of each of the citation Ci to the base paper, sentiment analysis is done to find the sentiment of the citation context towards the base paper. This is done by maintaining a list of training words that includes positive, negative and neutral sentiment words [1] together with certain rules for analysing the sentiment. Citation context retrieved is compared against this training list of words and sentiment analysis is done.

3.4 Detection of Citation Outlier

Citation outliers are those citations that are relevant to the base paper but not greatly relevant to it. Base paper might be a outlier to its citation. eg, If the base article talks about data mining, one of its citation talks about opinion mining, then the base paper is considered to be an outlier to its citation.

Latent Dirichlet Allocation LDA is used for topic modelling [17] .in LDA, each document may be viewed as a mixture of various topics. A probability distribution is found based on Gibbs sampling and distribution of a content over various topics is identified. In our proposed system, contents of the base paper and all its citations are topic modelled and distribution of the base paper and each of its cites across various topics is identified. Now the topic across which citations are distributed widely is compared against the topics across which the base paper is distributed. If they are found to be the same or found 50% similar minimally, then the citations are found to be apt, else the base paper is considered to be an outlier for that citation. Rating the article, our proposed system aims to analyse the article based on the citation impact it has got. Citation impact is analysed mainly based on the context relevancy, sentiment analysis and outlier detection.

4 Results and Observation

4.1 Data Set

A set of 1000 documents is considered from the corpus and tested on the system. Results were tabulated and they are validated by manually identified values and the system is found to return decent results.

Table 1 Refinement of Citations in the Machine Learning Domain

Article ID	Total References	Total Cites	Reduced References	Reduced Cites	Quality References
1	10	20	3	7	30%
2	17	31	6	10	35.29%
3	12	23	2	2	16.67%
4	18	30	9	24	50%
5	12	4	6	12	50%
6	19	28	7	14	36.84%
7	15	20	8	18	53.33%
8	16	20	5	16	31.25%
9	15	6	4	8	26.67%
10	16	25	9	15	56.25%

4.2 Refining Citations

Table 1 shows the reduction in the number of citations in the machine learning domain. In a sample of ten papers, more than half the citations and references were found to be outliers in the papers. From the table, it is evident that on an average only 40 percent of the references, citations in a scientific publication are relevant. From these observations it can be inferred that citations alone can't be used a parameter for measuring research impact.

Figure 2 summarizes the percentage of quality citations (relevant) across five domains.

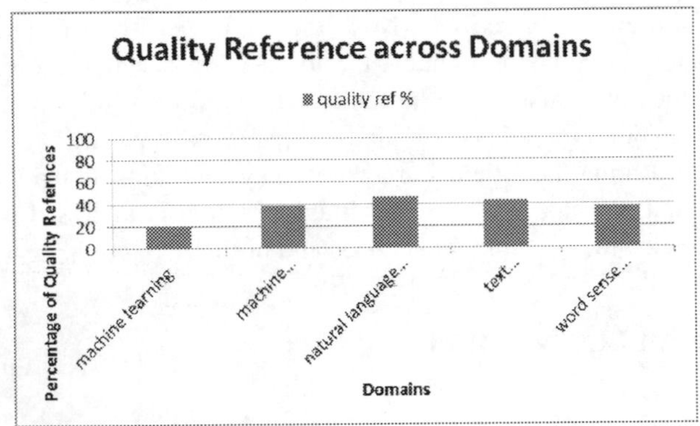

Fig. 2 Percentage of Quality references across domains

5 Conclusion and Future Work

From the initial experiments, it is evident that the citation count is not the most reliable measure to quantify impact of an article or an author. It is required to analyse the content of research articles and identify further parameters to arrive at relevant references and citations in the article. In future, this model will be refined to use the citation sentiment analysis to study the integrity in the cites and thereby using the same for refining the citations.

Acknowledgments. This work was funded by the Center for Technology Development and Transfer, Anna University Chennai under Approval no. CTDT-1/2360/RSS/2011 for Innovative project by Young faculty members under research support scheme.

References

[1] Aguilar, J.: Adaptive Random Fuzzy Cognitive Maps. In: Garijo, F.J., Riquelme, J.-C., Toro, M. (eds.) IBERAMIA 2002. LNCS (LNAI), vol. 2527, pp. 402–410. Springer, Heidelberg (2002)

[2] Burgess, C., Livesay, K., Lund, K.: Explorations in context space: words, sentences, discourse. Discourse Processes 25, 211–257 (1998)

[3] Foltz, P.W., Kintsch, W., Landauer, T.K.: The measurement of textual coherence with latent sematic analysis. Discourse Processes 25(2&3), 285–307 (1998)

[4] Gouws, S.: Evaluation and Development of Conceptual Document Similarity Metrics with Content-based Recommender Applications. Master‘s thesis, Stellenbosch University, South Africa (2003)

[5] He, Q., Pei, J., Kifer, D., Mitra, P., Giles, C.L.: Context-aware citation recommendation. In: Rappa, M., Jones, P., Freire, J., Chakrabarti, S. (eds.) WWW, pp. 421–430 (2010)

[6] Islam, A., Inkpen, D.Z.: Semantic text similarity using corpus-based word similarity and string similarity. TKDD 2(2) (2008)

[7] Janez Brank, D.M., Grobelnik, M.: A survey of ontology evaluation techniques (June 2005)

[8] Jiang, J.J., Conrath, D.W.: Semantic similarity based on corpus statistics and lexical taxonomy. CoRR, vol. cmp-lg/9709008 (1997)

[9] Landauer, T.K., Dumais, S.T.: Latent semantic analysis. Scholarpedia 3(11), 43–56 (2008)

[10] Leacock, C., Chodorow, M.: WordNet: An Electronic Lexical Database - Combining local context and WordNet similarity for word sense identification. In: Wordnet: An Electronic Lexical Database, ch. 11, pp. 265–283. MIT Press (1998)

[11] Li, Y., McLean, D., Bandar, Z., O‘Shea, J., Crockett, K.A.: Sentence similarity based on semantic nets and corpus statistics. IEEE Trans. Knowl. Data Eng. 18(8), 1138–1150 (2006)

[12] Mihalcea, R., Corley, C., Strapparava, C.: Corpus-based and knowledge based measures of text semantic similarity. In: AAAI, pp. 775–780. AAAI Press (2006)

[13] Resnik, P.: Using information content to evaluate semantic similarity in a taxonomy. In: IJCAI, pp. 448–453 (1995)

[14] Rodríguez, M.A., Egenhofer, M.J.: Determining semantic similarity among entity classes from different ontologies. IEEE Trans. Knowl. Data Eng. 15(2), 442–456 (2003)

[15] Varelas, G., Voutsakis, E., Raftopoulou, P., Petrakis, E.G.M., Milios, E.E.: Semantic similarity methods in wordnet and their application to information retrieval on the web. In: Bonifati, A., Lee, D. (eds.) WIDM, pp. 10–16. ACM (2005)

[16] Hirsch, J.E.: An index to quantify an individual's scientific research output. Proc. Natl. Acad. Sci. USA 102(46), 16569–16573 (2009)

[17] Garfield, E.: The Thomson Reuters impact factor (1994), http://thomsonreuters.com/products_services/science/free/essays/impact_factor/ (last accessed August 30, 2011)

[18] Bornmanna, L., Daniel, H.-D.: The citation speed index: A useful biblio-metric indicator to add to the h index. Journal of Informetrics 4, 444–446 (2010)

[19] Seglen, P.O.: Why the impact factor of journals should not be used for evaluating research. BMJ 314, 497–502 (1997)

Assessing Novelty of Research Articles Using Fuzzy Cognitive Maps

S. Sendhilkumar, G.S. Mahalakshmi, S. Harish, R. Karthik, M. Jagadish, and S. Dilip Sam

Abstract. In this paper, we compare and analyze the novelty of a scientific paper (text document) of a specific domain. Our experiments utilize the standard Latent Dirichlet Allocation (LDA) topic modeling algorithm to filter the redundant documents and the Ontology of a specific domain which serves as the knowledge base for that domain, to generate cognitive maps for the documents. We report results based on the distance measure such as the Euclidean distance measure that analyses the divergence of the concepts between the documents.

1 Introduction

We are facing an ever increasing volume of research publications. These have brought challenges for the analysis of novelty in these texts. Our main objective is to rate and find the novel information present in a journal of a specific domain. A large set of documents (corpus) of a specific domain is maintained so that the input journal of that domain is compared with those journals/documents to get the novelty score. The documents that are similar to the input document are found. After the similar documents are found, all those documents along with the input document are subjected to mapping by referring over Knowledge Base (Ontology) of that domain. The generated maps are Fuzzy Cognitive Maps (FCM) and they contain the required information to perform novelty computation. The corresponding maps of two documents are compared to find the divergence between two documents. To find the novel regions/parts of the input document, we have proposed a new measure to calculate the novelty score. Finding novelty in the scientific documents requires a knowledge base of a specific domain to analyse the concepts present in the documents. We used Ontology as the knowledge base for the domain to know the hierarchy of concepts and their relationships.

S. Sendhilkumar
Department of Information Science & Technology, Anna University, Chennai

G.S. Mahalakshmi · S. Harish · R. Karthik · M. Jagadish · S. Dilip Sam
Department of Computer Science & Engineering, Anna University, Chennai

A. Abraham and S.M. Thampi (Eds.): Intelligent Informatics, AISC 182, pp. 73–79.
springerlink.com

2 Related Work

2.1 Sentence-Level Approach

Sentence Level Novelty Detection aims at finding relevant and novel sentences given a query/topic and a set of documents. The cosine similarity is the widely used metric in the sentence level approach [9]. In Xiaoyan Li and Bruce Crofts work on Answer updating approach to Novelty Detection [2], they treated new information as new answers to questions that represented user's information requests (query). In Flora S. Tsais work on Novelty Detection for text documents [5], the named entities are assigned weights by using two different metrics, If the number of unique entities exceeded a particular threshold, the sentence was declared as novel. The other model such as vector space model [6] [1], Graph based text representation [4] ,Language model [3] ,Overlap relations [13] etc. are implemented for sentence level Novelty detection.

2.2 Document-Level Approach

In Zhang et als work on novelty and redundancy detection in adaptive filtering[12], the cosine metric and a mixture of probabilistic language models which is shown to be effective are used. Flora S. Tsais proposed D2S: Document-to-sentence framework for novelty detection [8] in which the novelty score of each sentence is determined to compute the novelty score of the document based on a fixed threshold.

3 System Design and Implementation

3.1 Distributional Similarity

By using LDA, the topic distributions over a number of documents are obtained. In this model each document d is represented by a topic distribution Θ_d. Kullback-Leibler divergence is a non-symmetric or distributional similarity measure of the difference between two probability distributions P and Q. Here it is used to the difference between topic distributions of two documents [12] which is one way to measure the redundancy of one document given another.

$$R(dt|di) = KLDiv(\Theta_{dt}, \Theta_{di}) = \sum(t_i|\theta_{dt})\log(P(t_i|\theta_{di})/P(t_i|\theta_{dt}) \quad (1)$$

Where R is the redundancy of document dt over document di and Θd is the topic model for document d and is a multinomial distribution. The documents that have least divergence value with the input document are selected so that redundant documents are filtered out and the documents that are more similar to the input document are selected for further processing.

3.2 Ontology Mapping

We view the ontology as the concept tree of that domain where each node represents a concept and their parent and child nodes represent their generalized and specialized form respectively. The concepts information include the level information of the concept in the tree i.e. height and depth of the concept in the tree, the occurring concepts (term) frequency in the document, the path to the root node of the concept etc. based on the needs of the novelty score computation. The map-ping process is per formed for the input document and for each document that is found during the similarity process and corresponding mappings are generated in xml format.

3.3 Divergence Analysis

The hopping distance between two nodes (concepts) represents the closeness of concepts to each other. For example, the hopping distance between the "RSSI measures" and "Route Request process" is 2 which have a common parent node "Route discovery". For further computation, Concept matrices are constructed for each document where the rows and columns of each matrix are the concepts found in both the documents and their corresponding matrix values are the hopping distance values between the nodes.

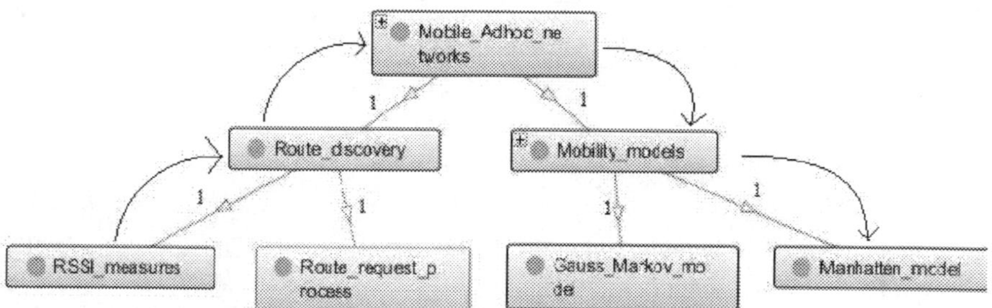

Fig. 1 Mapping snapshot to demonstrate hopping distance

3.4 Novelty Score Estimation

Each concept found in a document is assigned weight based on the number of other concepts which are present under the same category. For example, in a document, if there are n concepts associated with a concept X or those that come under the same category as concept X, the weight assigned to the concept X will be n+1. The summation of all the concepts weights gives the total value of the document. Consider there are n concepts that have been spoken in a document. The total weight W(d) of the document is given by

$$\mathrm{W(d)} = \sum_{C_i}^{n} W(C_i) \tag{2}$$

Where $W(C_i)$ is the weight of the concept C_i. After the concepts in each document are weighted, then those documents are compared with the input document to compute the novelty score and those concepts that contributed to the novelty score. When a concept of the input document is found in the comparing document and if its weight is the minimum weight or equal to one, then that concept is ignored and its weight is deducted. But if its weight is more than the minimum weight, then its related or associated concepts are considered and are checked for their occurrence in the comparing document. If found, the number of those concepts found in the comparing document are used to deduct the weight of the comparing concept. This is repeated for all the concepts in the input document. The ratio of the new weight of the document to the initial weight of the document gives the novelty score. Consider $W(d|d_t)$ is the total weight of the computed concepts in the document d with respect to document d_t.

$$\text{Novelty score}(d|d_t) = W(d|d_t)/W(d) \tag{3}$$

Where W(d) is the initial weight of the document. The weights of concepts that are below or equal to the minimum weight are ignored and those that are above the minimum weight are considered to be the combination of concepts that contributed to the novelty score. These concepts are considered to be the novelty regions of the input document with respect to the comparing document. Once the concepts that contributed to the novelty of the document have been found, their term occurrence is searched in the document. The sections or paragraphs that contain those terms are retrieved to summarize the novelty regions of the document.

4 Results and Observation

Initial experiments were done to find the change in the novelty scores by introducing a survey paper in the corpus. We considered a survey paper of wireless sensor networks domain whose concepts matched the concepts defined in the Ontology at 7.2%. It is introduced in the dataset to observe if the novelty scores of each document changes. Similarly we observed the changes with respect to another survey paper which matched at 11.4% in the Ontology.

From the table 1, we observe that when a survey paper is introduced, the number of documents for which the novelty has been reduced depends on the number of concepts in the survey paper that are matched to the concepts defined in ontology. The number of novelty reduced documents when compared with survey pa-per(wireless sensor networks) matching 7.2% of concepts mapped in Ontology is less than that of when compared with survey paper(wireless sensor networks) matching 11.4% of concepts mapped in Ontology.

The table 2 shows the variation of novelty score of 10 research papers in the wire-less sensor networks domain. From the table 2, we can identify 2 cases which ex-plain the impact of survey papers on the research papers of that domain.

Table 1 Novelty reduced Documents in each domain

Sub-Domain Name	Number of Documents	Number of novelty reduced documents after comparison / Overall reduction in the Novelty (%)			
		With Survey Paper 1		With Survey Paper 2	
Wireless networks	66	24	4.8%	40	9.2%
Wireless Sensor	54	28	7.18%	34	12.79%
Adhoc	57	25	5.8%	39	14.3%
Multimedia	28	12	3.8%	16	13.1%
Peer to Peer	45	28	6.2%	26	7.8%
Cloud Computing	8	4	0.73%	3	3.9%
Network Traffic	82	30	4.72%	38	6.59%
Security	94	25	5.26%	41	5.42%
QOS	64	29	6.2%	27	10.7%
Mobile	31	13	4.81%	15	12.77%
Web service	33	14	3.03%	14	3.12%
Optical Network	57	21	5.52%	31	7.85%
Biometrics	19	4	6.34%	6	12.11%
Others	247	86	3.9%	96	5.68%

Table 2 Novelty scores for sample documents in Wireless Sensor Networks domain

Document ID	Novelty Score		
	Initial	Addition of Survey Paper 1	Addition of Survey Paper 2
1	0.86017	0.779661	0.533898
2	0.527851	0.527851	0.527851
3	0.652174	0.304348	0.304348
4	0.608374	0.490148	0.608374
5	0.426667	0.426667	0.266667
6	0.875	0.791667	0.441667
7	0.974217	0.961326	0.810313
8	0.86017	0.779661	0.533898
9	0.757895	0.610526	0.6
10	0.86017	0.779661	0.533898

1. The novelty score of the research paper reduces after comparing with survey papers when more concepts or concepts with more depth (more associated concepts) are matched with survey papers than any others.
2. The novelty score of the research papers does not reduce after comparing with survey papers when less number of concepts or concepts with less depth(less associated concepts) are matched with survey papers.

5 Conclusion and Future Work

On the basis of the studies, it is concluded that the system has shown promising results in identifying the relevant documents and filtering redundant documents with respect to a given document, using LDA and Kullback Leibler Divergence. The use of Ontology(Knowledge base) of a specific domain has enabled the system to measure the novelty of a document belonging to that particular domain and has highly contributed in the proper analysis of each document which in turn effects the overall performance of the system. Since measuring the novelty of a document is domain specific, the system requires a well-defined ontology for that particular domain. Thus the novelty present in a research paper of a specific domain is measured with respect to top relevant documents and given as a novelty score for the research paper. The concepts in the document that contributed to the novelty score are summarized to the user.

Our definition of novelty in research publication is based on the contribution of new combination of concepts and its depth. It can be further enhanced by analysing the contribution of each concept at the sentence level. The sentence level approach analyse the definition of concepts explored at deeper level which includes part of speech tagging, identifying entities etc. By improving the definition of concepts in the Ontology, the results can be enhanced.

Acknowledgments This work was funded by the Center for Technology Development and Transfer, Anna University Chennai under Approval no. CTDT-1/2360/RSS/2011 for Innovative project by Young faculty members under research support scheme.

References

[1] Allan, J., Wade, C., Bolivar, A.: Retrieval and novelty detection at the sentence level, pp. 314–321. ACM (2003), http://doi.acm.org/10.1145/860435.860493

[2] Croft, X.L.W.: Answer updating approach to novelty detection. ACM (2004), http://dx.doi.org/10.3115/1220575.1220665

[3] Fernandez, R.T.: The effect of smoothing in language models for novelty detection. In: Future Directions in Information Access, FDIA 2007 (2007), http://www-gsi.dec.usc.es/_dlosada/fdia07.pdf

[4] Gamon, M.: Graph-based text representation for novelty detection, pp. 17–24. ACM (2006), http://dl.acm.org/citation.cfm?id=1654758.1654762

[5] Kok Wah Ng, L.C., Tsai, F.S., Goh, K.C.: Novelty detection for text documents using named entity recognition. IEEE (2007)

[6] Schiffman, B., McKeown, K.R.: Context and learning in novelty detection, pp. 716–723. ACM (2005), http://dx.doi.org/10.3115/1220575.1220665

[7] Stokes, N., Carthy, J.: First story detection using a composite document representation, pp. 1–8. ACM (2001), http://dx.doi.org/10.3115/1072133.1072182

[8] Tsai, F.S., Zhang, Y.: D2S: Document-to-sentence framework for novelty detection, vol. 29(2), pp. 419–433. ACM (November 2011), http://dx.doi.org/10.1007/s10115-010-0372-2

[9] Tsai, M.-F., Feng Tsai, M., Hsi Chen, H.: Some similarity computation methods in novelty detection. NIST (2002)

[10] Yang, Y., Zhang, J., Carbonell, J., Jin, C.: Topic-conditioned novelty detection, pp. 688–693. ACM (2002), http://doi.acm.org/10.1145/775047.775150

[11] Zhang, J., Ghahramani, Z.: A probabilistic model for online document clustering with application to novelty detection. In: Neural Information Processing Systems (NIPS), vol. 17 (2004)

[12] Zhang, Y., Callan, J., Minka, T.: Novelty and redundancy detection in adaptive filtering, pp. 81–88. ACM (2002), http://doi.acm.org/10.1145/564376.564393

[13] Zhao, L., Zhang, M., Ma, S.: The nature of novelty detection, vol. 9(5), pp. 521–541. ACM (November 2006), http://dx.doi.org/10.1007/s10791-006-9000-x

Towards an Intelligent Decision Making Support

Nesrine Ben Yahia, Narjès Bellamine, and Henda Ben Ghezala

Abstract. This paper presents an intelligent framework that combines case-based reasoning (CBR), fuzzy logic and particle swarm optimization (PSO) to build an intelligent decision support model. CBR is a useful technique to support decision making (DM) by learning from past experiences. It solves a new problem by retrieving, reusing, and adapting past solutions to old problems that are closely similar to the current problem. In this paper, we combine fuzzy logic with case-based reasoning to identify useful cases that can support the DM. At the beginning, a fuzzy CBR based on both problems and actors' similarities is advanced to measure usefulness of past cases. Then, we rely on a meta-heuristic optimization technique i.e. Particle Swarm Optimization to adjust optimally the parameters of the inputs and outputs fuzzy membership functions.

1 Introduction

Decision making (DM) whenever and wherever it is happening is crucial to organizations' success. DM represents process of problem resolution based on four phases as it is proposed by [1] and revisited by [2]: intelligence (problem identification), design (solutions generation), choice (solution selection) and review (revision phase).

In order to make correct decisions, we use the Case based reasoning (CBR) technique to support DM. CBR is a means of solving a new problem by reusing or adapting solutions of old similar problems [3]. It solves a new problem by retrieving, reusing, and adapting past solutions to old problems that are closely similar to the current problem [4]. The most important part of a case-based reasoning system

Nesrine Ben Yahia · Narjès Bellamine · Henda Ben Ghezala
RIADI Laboratory, National School of Computer Sciences, University Campus Manouba, 2010, Tunisia
e-mail: {Nesrine.benyahia,Narjes.bellamine}@ensi.rnu.tn,
Henda.benghezala@ensi.rnu.tn

A. Abraham and S.M. Thampi (Eds.): Intelligent Informatics, AISC 182, pp. 81–87.
springerlink.com

is the selection of the similarity measurement, because, if the one selected is inappropriate, the system will generate erroneous outputs. To address this problem, Main et al. in [5] explain how fuzzy logic applies to CBR.

Fuzzy logic is similar to the process of human reasoning and it lets people compute with words [6]. It is useful when the information is too imprecise to justify and when this imprecision can be exploited to realize better rapport with reality [6].

Thus, in this paper we combine the use of fuzzy logic and case-based reasoning. Then, in order to make more efficiency, an optimal design of membership functions of fuzzy sets is desired [7]. Therefore, we rely on a meta-heuristic optimization technique i.e. Particle Swarm Optimization to adjust the parameters of the inputs and outputs fuzzy membership functions. We select this technique thanks to its ability to provide solutions efficiently with only minimal implementation effort and its fast convergence [8].

The outline of this paper is as follows. In section two, we begin with a brief background to illustrate the crucial concepts involved in case-based reasoning and fuzzy logic followed by a discussion of the usefulness of combining these two techniques to support decision making. In section three, we present our fuzzy case-based reasoning process. In section four, PSO is used to optimize the proposed process by optimizing the membership functions of fuzzy sets.

2 Fuzzy Case-Based Reasoning Supported Decision Making

2.1 Overview of Case Based Reasoning and Fuzzy Logic

Thanks to Case-based reasoning (CBR), we can learn from past experience and avoid repeating mistakes made in the past. CBR process is based on four steps: (a) identifying key features, (b) retrieving similar cases, (c) measuring case similarity and selecting best ones (d) modifying and adapting the existing solution to resolve the current problem [9]. According to classic or crisp logic each proposition must either be true or false, however fuzzy logic [6] is a solution to represent and treat the uncertainty and imprecision using linguistic terms represented by membership functions. Fuzzy logic process is based on three phases: fuzzification, fuzzy inference, and defuzzification. Fuzzification transforms the crisp values into fuzzy values and maps actual input values into fuzzy membership functions. The membership function of a fuzzy set A is defined by μ_A(x) and it represents the degree of membership of x to the fuzzy set A. The second phase in the fuzzy logic process involves the inference of the input values. The fuzzy inference system generates conclusions from the knowledge-based fuzzy rule set of IF-THEN linguistic statements. The final phase is the defuzzification where fuzzy results are converted into crisp values.

There are several advantages of using fuzzy logic techniques to support the case-based reasoning in the retrieval stage [10]. First, it simplifies comparison by converting numerical features into fuzzy terms. Second, fuzzy sets allow simultaneous indexing of a case on a single feature in different sets with different degrees of membership which increases the flexibility of case matching.

2.2 Similarity Measurement in CBR

In this paper, we consider a case i as a set of (A_i, P_i, Alt_i, S_i) where A_i represents the actor involved in the case (who lived the experience), P_i represents the problem, Alt_i represents the different proposed alternatives to solve P_i and S_i represents the choice among alternatives.

Then, to measure similarity, we distinguish between two types of similarity: one between current situation problem and problems saved in past cases and another between current situation actor and actors saved in past cases. Similarity between two elements i and j (two problems or two actors) is computed using the function CalculSimilarity(element i, element j) described in Fig. 1.

```
Float function CalculSimilarity(element i, element j) {
W(i, j) = 0;
For each element attribute a do
If a is nominal, mono-valued and i.a = j.a then
W(i, j) = W(i, j) +1;
Else if a is nominal, multi-valued then
            For each value i.a.v = 1...k do
            For each value j.a.w = 1...m do
              If i.a.v = j.a.w then W(i, j) = W(i, j) +1;
              End if
            Break;
            End for
            End for
Else if a is continuous then
W(i, j) = W(i, j) +1 – α |i.a – j.a|;
End if
End for
Return W(i, j);
}
```

Fig. 1 Calculus of weights

For example, we will apply this function to measure similarities between actors' attributes (an element consists in an actor) then we consider that each actor A, is characterized by the set of attributes (languages, nationality, topics of interest, age). For the nominal attributes (Languages, Topics of interest and Nationality) the weight is incremented by 1 for each similar attributes but for the continuous attribute (Age), we set α to 0.035 (α =1/28 where 28 is the difference between the minor age 25 and the greater one 53). In addition, we suppose given an actor $A_{current}$ characterized by a set of attributes as follows ((Arabic, English, French), (Tunisian), (CSCW, KM, DM), 25), Table 1 illustrates the compute of the similarities between attributes of past actors cases and attributes of $A_{current}$. Problems' similarities are calculated with the same manner. The global similarity between current and past cases is then the

Table 1 Calculus of similarities between $A_{current}$ and actors saved in past cases

Languages	Nationality	Topics of interest	Age	Similarity with $A_{current}$
Arabic, French	Tunisian	IR, DB, KM	30	0.53
Arabic, French	French	AI, KM	28	0.4
English, French	French	CSCW, BPM	42	0.32
Arabic, English, Spanish	Spanish	SOA, SMA	35	0.21
Arabic, English, French	Tunisian	KM, DM	25	1
English, French	French	DB, SGBD	53	0.12
French, Spanish	Spanish	SE, OS	29	0
French, Spanish	French	WWW	45	0.1
Arabic, English, French	English	PL, CSCW	35	0.51

sum of the similarity between current and past problems with the similarity between current and past actors. As these values are crisp, each case is classified as useful or not useful for the current situation. In fact, if we propose a threshold t then each case with the degree of similarity exceeds t is classified as useful else it is not useful. In next section, we rely on fuzzy logic to fuzzy these values and represent uncertain cases.

2.3 Using Fuzzy Logic to Support CBR

In the proposed fuzzy logic process, two inputs are considered which are likeness and closeness and one output which consists of usefulness. Likeness and closeness variables represent respectively the values of problems' and actors' similarities. Each variable has three linguistic values: likeness has the values (low, medium, high), closeness has the values (minor, average, major) and usefulness has the values (poor, good, excellent). In this paper, we use triangular membership functions feature, which is specified by three parameters, as it is characterized by a mathematical simplicity. The inputs membership functions parameters, in our case, are a_{ij}, b_{ij}, c_{ij} where i $\in$ [1..2] represents i^{th} linguistic variable and j $\in$ [1..3] represents j^{th} linguistic value of ith linguistic variable and the outputs membership functions parameters a_j, b_j, c_j where j $\in$ [1..3].

The fuzzy inference system makes conclusions using the knowledge-based fuzzy rules which is based on a set of IF-THEN linguistic statements. In this paper, the MIN-MAX inference method is used. It consists in using the operators min and max for respectively AND and OR. Table 2 illustrates the matrix inference of our fuzzy control system.

Once the functions are inferred and combined, they are defuzzified into a crisp output which drives the system. In our case, defuzzification of the output variable usefulness is based on Center Of Gravity defuzzification method.

Table 2 The proposed fuzzy inference system

Likeness/Closeness	minor	average	major
low	poor	poor	good
medium	poor	good	excellent
high	excellent	excellent	excellent

3 Using Particle Swarm Optimization in FCBR

3.1 *Overview of Particle Swarm Optimization*

Particle Swarm Optimization (PSO) is introduced in [11] as a new evolutionary and metaheuristic computation technique inspired by social behavior simulation of fish schooling. We select this technique thanks to its ability to provide solutions efficiently with only minimal implementation effort and its fast convergence [8]. The population in PSO is called a swarm where the individuals, the particles, are candidate solutions to the optimization problem in the multidimensional search space. For each iteration t, every particle i is characterized by its position $x_i(t)$ and its velocity $v_i(t)$ which are usually updated synchronously in each iteration of the algorithm. Each particle converges towards positions that had optimized fitness function values in previous iterations by adjusting its velocity according to its own flight experience and the flight experience of other particles in the swarm. There are two kinds of position, $Pbest_i$ that represents personal best position of particle i and G_{best} that represents the global best position obtained by all particles [12]. In this paper, we choose the PSO version with constriction factor for its speed of convergence [13]. In this case, the position $x_i(t+1)$ and velocity $v_i(t+1)$ at the iteration t+1 for particle i are calculated using following formulas:

$$x_i(t+1) = x_i(t) + v_i(t) \tag{1}$$

$$V_i(t+1) = K*(v_i(t) + c_1*r_1*(Pbest_i - x_i(t+1)) + c_2*r_1*(G_{best} - x_i(t+1)) \tag{2}$$

Where K: the constriction factor given by K = 2/ | 2 - c - $\sqrt{c^2}$ - 4c |
With c=c_1 + c_2 and c>4.
r_1 and r_2: random numbers between 0 and 1.
c_1: self confidence factor and c_2: swarm confidence factor.

3.2 *Using PSO to Optimize the Proposed FCBR*

In this approach, we rely on PSO to optimally adjust the inputs' parameters: a_{ij}, b_{ij}, c_{ij} where i ∈ [1..2], j ∈ [1..3] and the output's membership functions parameters a_j, b_j, c_j where j ∈ [1..3]. The fitness function, in our case, consists of the error function of cases retrieval which must be minimized. The evolution of PSO Algorithm can be summed up in the following pseudo code:

1. Initial swarm population is composed of 30 particles, Number of Iterations is 1000, c_1= 2.8 and c_2=1.3. The choice of the population size, the values of c_1 and c_2 are based on the study done in [14] and initial positions and velocities of particles are randomly generated.
2. The fitness of each particle is calculated.
3. The local best of each particle Pbest_i and the global best G_{best} values are updated.
4. If the maximum iteration number is reached, the best set of parameters is given.
5. Otherwise, position and velocity of each particle are updated using (1) and (2) and we loop to 2.

4 Tests

In this section, we aim to test the proposed fuzzy logic process with setting different values of input variables (likeness and closeness) and getting values of output variable (usefulness) as it is shown in Table 3.

Table 3 Data tests

Likeness	Closeness	Usefulness
2.0	5.0	2.5000000000000053
5.5	5.5	5.603448275862117
5.2	2.1	3.180850532598447
2.5	4.6	2.500000000000006
1.5	8.3	5.000000000000018
3.9	1.8	2.5000000000000213

5 Conclusion

In this paper, we present a mixed framework that combines case-based reasoning (CBR), fuzzy logic and particle swarm optimization to build an intelligent decision support model. CBR is used to search and retrieve past useful cases that can help in the current problem resolution. The similarity measurement is based on both problems and actors similarities. Then in order to support CBR, fuzzy logic is used to avoid instable cases retrieval in case of uncertainty and imprecision. In order to make more efficiency, we rely on Particle Swarm Optimization to adjust the parameters of the inputs and outputs fuzzy membership functions.

References

1. Simon, H.: The New science of management decision. Prentice Hall, Englewood Cliffs (1977)
2. Zaraté, P.: Des Systèmes Interactifs d'Aide la Décision Aux Systèmes Coopératifs d'Aide la Décision: Contributions conceptuelles et fonctionnelles. HDR dissertation, INP Toulouse (2005)
3. Riesbeck, C.K., Schank, R.C.: Inside Case-Based Reasoning. Lawrence Erlbaum Associates, New Jersey (1989)
4. Aamodt, A., Plaza, E.: Case-based reasoning: foundational issues, methodological variations and system approaches. AI Communications 7, 39–59 (1994)
5. Main, J., Dillon, T.S., Khosla, R.: Use of fuzzy feature vectors and neural vectors for case retrieval in case based systems. In: Biennial Conference of the North American Fuzzy Information Processing Society, NAFIPS 1996, pp. 438–443. IEEE, New York (1996)
6. Zadeh, L.A.: Fuzzy Logic = Computing with Words. IEEE Transactions on Fuzzy Systems 4(2), 103–111 (1996)
7. ShengZhou, Y., Lai, L.Y.: Optimal design for fuzzy controllers by genetic algorithms. IEEE Transactions on Industry Applications 36(1), 93–97 (2000)
8. Parsopoulos, K.E., Vrahatis, M.N.: Particle Swarm Optimization and Intelligence: Advances and Applications. Information Science Reference (an imprint of IGI Global), United States of America (2010)
9. Kolodner, J.: Case-Based Reasoning. Morgan Kaufmann, California (1993)
10. Jeng, B.C., Liang, T.P.: Fuzzy indexing and retrieval in case-based systems. Expert Systems with Applications 8(1), 135–142 (1995)
11. Eberhart, R.C., Kennedy, J.: New optimizer using particle swarm theory. In: Proceedings of the 6th International Symposium on Micro Machine and Human Science, Nagoya, Japan, pp. 39–43 (1995)
12. Yisu, J., Knowles, J., Hongmei, L., Yizeng, L., Kell, D.B.: The Landscape Adaptive Particle Swarm Optimizer. Applied Soft Computing 8, 295–304 (2008)
13. Clerc, M.: The Swarm and the Queen: Towards A Deterministic and Adaptive Particle Swarm Optimization. In: Proceedings of the Congress of Evolutionary Computation, Washington, DC, pp. 1951–1957 (1999)
14. Carlisle, A., Dozier, G.: An Off-The-Shelf PSO. In: Proceedings of the Particle Swarm Optimization Workshop, Indianapolis, Ind., USA, pp. 1–6 (2001)

An Approach to Identify n-wMVD for Eliminating Data Redundancy

Sangeeta Viswanadham and Vatsavayi Valli Kumari

Abstract. Data Cleaning is a process for determining whether two or more records defined differently in database, represent the same real world object. Data Cleaning is a vital function in data warehouse preprocessing. It is found that the problem of duplication /redundancy is encountered frequently when large amounts of data collected from different sources is put in the warehouse. Eliminating redundancy in the data warehouse resolves conflicts in making wrong decisions. Data cleaning is also used to solve problem of "wastage of storage space". One way of eliminating redundancy is by retrieving similar records using tokens formed on prominent attributes. Another approach is to use Conditional Functional Dependencies (CFD's) to capture the consistency of data by combining semantically related data. Existing work on data cleaning do not deal with the case of multi-valued attributes. This paper deals with nesting based weak multi-valued dependencies (n-wMVD) which can handle multi-valued attributes and redundancy removal. Our contributions are of two fold (i) An approach to convert the given database to wMVD (ii) Implementation of n-wMVD to eliminate redundancy. The applicability of our approach was tested. The results are encouraging and are presented in the paper.

Keywords: Conditional Functional Dependencies (CFD), weak Multi-valued Dependencies (wMVD), nesting based weak Multi-valued Dependencies (n-wMVD).

1 Introduction

Data warehouses collect large amounts of data from a variety of sources. They load and refresh continuously, so that the probability of some sources containing

Sangeeta Viswanadham
Pydah College of Engg & Tech, Visakhapatnam
e-mail: `sangeetaviswanadham@yahoo.com`

Vatsavayi Valli Kumari
Andhra University, Andhra Pradesh, India
e-mail: `vallikumari@gmail.com`

A. Abraham and S.M. Thampi (Eds.): Intelligent Informatics, AISC 182, pp. 89–97.
springerlink.com

redundant information can be eliminated. Further, even data warehouses are used for decision making, so the correctness of the data is vital to avoid inferring wrong conclusions. [2]

Data cleaning is a complex process in data warehousing that deals in detecting and removing inconsistencies (redundancies) to improve data quality. Data cleaning should be performed together with schema related transformations. [3] Data cleaning determines redundancy by considering any two tuples to see whether both refer to the same real world object. Then it either combines the tuples to represent the consolidated information if they are similar or retains only one of them if they are exactly same.

The existing techniques for data cleaning are multi-fold. One way is achieved by Tokenization. This allows finding tokens on prominent attributes. The work that depends on the data may not be efficient in identifying exact duplicates. Conditional Functional Dependencies (CFDs) can identify inconsistencies by forming appropriate application specific integrity constraints. Violations of CFDs result in redundancy and can be detected using SQL. This is a constraint based method of improving data quality. It can deal with the situation with single valued dependencies but not with multi-valued dependencies.

This paper introduces an extension of CFDs referred to as weak Multi-Valued Dependencies (wMVD) to deal with multi-valued attributes and n-wMVDs in capturing inconsistencies.

The remaining paper is organized into sections, where section 2 discusses related work, section 3 introduces how to eliminate redundancy, section 4 explains methodology to reduce redundancy, section 5 deals with experimental analysis and section 6 concludes the paper.

2 Related Work

Existing technique for removing redundancy is token management [3]. In this method tokens are formed using attributes of the database to get all similar records together. Token management includes

(i) Selection and Ranking of fields. (ii) Formation of Tokens based on Ranking.
(iii)Sort the database using the tokens (iv) Detection of duplicates and elimination.

It uses pre-determined threshold named Similarity Match Count (SMC) to decide on a match between to input records.

(a) If (SMC = 1.0) is a perfect match
(b) Else if (SMC<=0.99 and SMC >=0.67) then is a near match
(c) Else if (SMC<=0.66 and SMC >=0.33) then it may match
(d) Else it do not match (no match)

This cannot be done in unstructured databases.

According to [4] an unstructured database can be handled by forming application specific integrity constraints called Conditional Functional Dependencies (CFD's).

Definition 1 (CFD's): Consider a relation schema R defined over a fixed set of attributes attr(R).

A CFD Φ on R is a pair (R : X → Y, Tp), where

(i) X, Y are sets of attributes from (R),

(ii) R : X → Y is a standard FD, referred to as the FD embedded in Φ; and

(iii) Tp is a tableau with all attributes in X and Y, referred to as the pattern tableau of Φ, where for each A in X or Y and each tuple t ∈Tp, t[A] is either a constant 'a' in the domain dom(A) of A, or an unnamed variable '_ '. If A appears in both X and Y, we use $t[A_L]$ and $t[A_R]$ to indicate the A field of t corresponding to A in X and Y , respectively. We write Φ as (X → Y, Tp) when R is clear from the context.

A first step for data cleaning is the efficient detection of constraint violations in the data. Given an instance I of a relation schema *R* and a set Σ of CFDs on *R*, it is to find all the inconsistent tuples in I, i.e., the tuples that (perhaps together with other tuples in *I*) violate some CFD in Σ.

3 Eliminating Redundancy

This paper introduces an extension of CFDs referred to as weak Multi-Valued Dependencies (wMVD) to deal with multi-valued attributes. wMVDs are capable of capturing inconsistencies using Nesting.

Our contributions in this paper are in two phases. In the first phase we prove that the given database is in wMVD by considering conditions. In second phase we use the concept of nesting [5] to eliminate redundancy in the database.

The overall architecture of our system is shown in Figure 1. The system is divided into two phases.

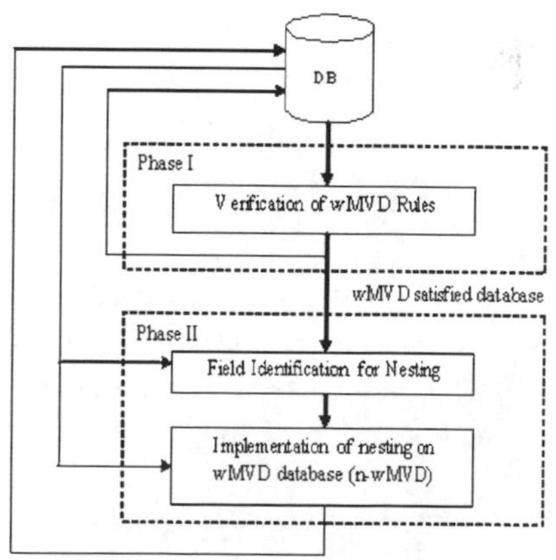

Fig. 1 A model for achieving n-wMVD

***Phase I*:** In this phase, first we consider a database and if satisfies conditions to prove it is wMVD.

Definition 2 (wMVD): Let R be a relation schema and X, Y are sets of attributes, where X, Y ⊆ R. A weak multivalued dependency (wMVD) is an expression X-w->Y and the relation R is said to satisfy the wMVD X-w->Y,

if and only if for all tuples t1, t2, t3 $\in$ R such that
t1[XY] = t2[XY] and t1[X(R-Y)] = t3[X(R-Y)]
there is some tuple t4 $\in$ R such that
t4[XY] = t3[XY] and t4[X(R-Y)] = t2[X(R-Y)]

Example: Suppose we have the relation schema Flat Relation (DANCE) with attributes Course (C), Lady (L) and Gentleman (G). The intention is to record information about who partners up with whom in which course. Naturally, some pairs do not change their partners at all, but sometimes pairs switch. The following is a hypothetically small relation over DANCE shown in Table 1.We can prove that the Table 1. Flat relation satisfies wMVD.

Let us assume first 5 Tuples in Flat Relation are mapped as follows
2→t1, 3→t4, 4→t3 and 5→t2
and attributes X → C, Y → L and (R-XY) →G
then the following conditions hold good
t1(CL) = t2(CL) t4(CL) = t3(CL)
t1(G) = t3(G) t4(G) = t2(G)
C-w->L is proved.

In this relation, the pairs (Lorena,Flavio) and (Flor,Ruy) have switched at least once while the pair (Racquel,Jose) always dance together. It is evident that this relation does not satisfy the MVD Course->>Lady, for instance because Racquel never dances with Flavio. However, the relation does satisfy the weak MVD Course-w-> Lady.

Phase II: In this phase, we prove that nesting in wMVD (n-wMVD) helps in reducing redundancy and wastage of space.

The first normal form condition of relational model is restrictive for some applications.

Definition 3 (First Normal Form): A relation is said to be in First Normal Form (1NF) if and only if each attribute of the relation is atomic. More simply, to be in 1NF, each column must contain only a single value and each row must contain the same number of columns.

Complex-valued data models can help us to overcome several limitations of relational data model in designing many practical data base applications. One of the complex-value data model is the nested relational model in which an attribute contains an atomic value or a nested relation.

Definition 4 (Nesting): Nesting is a fundamental operation for the nested relational data model, in which data tuples that have matching values on some fixed attribute set can be represented as a single nested tuple by collecting set of the different tuple values on the remaining attributes. This concept can be explained by using Table 1 as input and result is shown in Table 3.

4 Methodology

This section includes two algorithms to convert the given flat database to nested database to reduce redundancy.

Algorithm 1. It deals with the conversion of flat database to wMVD satisfied database.

Algorithm 2. It deals with the conversion of wMVD database to n-wMVD database.

Algorithm 1 returns a database which satisfies wMVD. We can clearly observe from the given input database (Table 1) contains 17 tuples. After verifying (Table 1) with the algorithm 1 a resultant table produced 14 tuples as an output which is wMVD database shown in (Table 2). The tuples which do not satisfy the given four conditions are removed and remaining tuples are retained in the database. It is clearly observed that a tuple (Latin, Racquel, Jose), (Raquel, Jose) do not pair with others. Such a tuple is retained in the database. Whereas tuple (Latin, Flor, Martin), "Flor" is paired with others but "Martin" never paired with others. Such tuples are removed as they are not satisfying wMVD conditions. So, all such tuples are deleted from the database.

Algorithm 1: Flat relation to wMVD conversion

Input : A Flat Relation or Database (DB)
Output : Database table which satisfy weak multi-valued dependency (WDB)
Assumptions : Cursor (CR) Variables = { C1, T_1, T_2, T_3, T_4, X } ,
DB Attributes C (Course) , L (Lady) , G (Gent)
' N' represents total no of records in the DB
Method : Entire Database is scanned for identifying the pairs of tuples satisfying wmvd

Select the first record from the DB into Cursor X
Begin
  Open Cursor X
  Fetch a record from the Cursor X into T_1
  For I in 1 to N do
    For each tuple (C1.T_K) in DB do
      If (T_1.C= C1.T_K) and (T_1.L = C1.T_K.L) then
        If (T_1.G= C1.T_K.G) then
          Continue to select next record from the C1.T_K;
        Else
        If (T_1.C = C1.T_K.C and T_1.L = C1.T_K.L and T_1.G<> C1.T_K.G) then
          Store the attribute values of C1.T_K into corresponding ones of T_2
        End if
        End if
      End if
          For each tuple (C1.T_K) in DB do
            If (T_1.G<> C1.T_K.G) then
              Continue to select next record from C1.T_K
            Else

```
    If (T1.C = C1.TK.C and T1.G = C1.TK.G and T2.L <> C1.TK.L) then
      Store the attribute values of C1.TK into corresponding ones of T3
          End if
        End if
      For each tuple (C1.TK) in DB do
       If (T3.C = C1.TK.C and T3.L = C1.TK.L and T2.G = T1.G ) then
         Store the attribute values of C1.TK into corresponding ones of T4
        End if
          If (T1.C = T2.C and T1.L = T2.L and T1.G = T2.G and
                                   T1.G = T3.G and T2.G = T4.G) then
           Insert all the four tuples (T1, T2, T3, T4) values into the Database
          End if
         End for
        End for
      End for
         Fetch the next record from the Cursor X into T1
      End for
       Insert the distinct tuples ( L should have exactly
       one relation with G and vice versa ) into the WDB
    End Begin
```

Table 1 Flat Relation

T-Id	Course(C)	Lady(L)	Gentleman(G)
1	Latin	Racquel	Jose
2	Latin	Lorena	Flavio
3	Latin	Flor	Ruy
4	Latin	Flor	Flavio
5	Latin	Lorena	Ruy
6	Swing	Dulcinea	Quixote
7	Swing	Dulcinea	Sancho
8	Swing	Theresa	Sancho
9	Swing	Theresa	Quixote
10	Street	Beatrice	Dante
11	Street	Eve	Tupac
12	Street	Queen L.	DMX
13	Street	Eve	DMX
14	Street	Queen L.	Tupac
15	Latin	Flor	Martin
16	Swing	Dulcinea	DavidSon
17	Street	Eve	Quixote

Table 2 wMVD Relation

T-Id	Course(C)	Lady(L)	Gentleman(G)
1	Latin	Racquel	Jose
2	Latin	Lorena	Flavio
3	Latin	Flor	Ruy
4	Latin	Flor	Flavio
5	Latin	Lorena	Ruy
6	Swing	Dulcinea	Quixote
7	Swing	Dulcinea	Sancho
8	Swing	Theresa	Sancho
9	Swing	Theresa	Quixote
10	Street	Beatrice	Dante
11	Street	Eve	Tupac
12	Street	Queen L.	DMX
13	Street	Eve	DMX
14	Street	Queen L.	Tupac

The concept of nesting is a useful notion in the study of non-first-normal-form relations [6].

Let U be a set of attributes and R a relation over U. In this Relation an attribute will either be an ordinary attribute or a nested set of attributes. Let X, Y attributes of R and subsets of U.

Let $\Phi \neq X \subseteq U$ and Y=U-X

$$\text{NESTx}(U, R) = (Ux, Rx)$$

where Ux is a set of attributes, equal to (U-X).

Rx= $\{t \mid \exists$ a tuple $u \in \pi_y(R)$ such that $t[Y] = u$ and $t[X]=\{v[X] \mid v \in R$ and $v[Y]=u\}\}$.

Example

Consider the relation

(U={C, L, G}, R) in Table 1.

$\text{Nest}_G(U, R)=(\{C, L, G'\}, R_x)$

where G' is a multi-valued data of G

R_x is a nested relation which is shown in Table 3

Algorithm 2 [1] returns a database which satisfies n-wMVD. The output of Algorithm 1 (Table 3) is given as an input to Algorithm 2. It converts the wMVD database (Table 3) into n-wMVD database (Table 2).

Table 3 n-wMVD Relation

Tuple-Id	Course(C)	Lady(L)	Gentleman(G)
1	Latin	Racquel	{Jose}
2	Latin	Lorena	{Flavio, Ruy}
3	Latin	Flor	{Flavio, Ruy}
4	Swing	Dulcinea	{Quixote, Sancho}
5	Swing	Theresa	{Quixote, Sancho}
6	Swing	Beatrice	{Dante}
7	Street	Eve	{Tupac, DMX}
8	Street	Queen L.	{Tupac, DMX}

5 Experimental Analysis

In this section, we present our findings about the performance of our schemes for reducing redundancies over wMVD databases.

Setup: For the experiments, we used postgres SQL on Windows XP with 1.86 GHz Power PC dual CPU with 3GB of RAM.

Data: Our experiments used Adult Database (UCI) with a few synthetic attributes addition.

There are two alternative evaluation strategies for the comparison of flat and nested relations. 1. Time for Query Execution. 2. Size of the Relation.

Time for Query Execution: We studied the impact of reduced redundancy in the nested relation by varying time. In this we considered number of tuples whose value varied from 100 to 700 tuples in 100 increments can be observed in Table 4 and Fig.2.

Table 4 Times for Query Execution

No.of Tuples	Flat Relation Time(msec)	NestedRelation Time(msec)
100	0.605	0.157
200	1.249	0.19
300	1.659	0.225
400	2.218	0.27
500	3.061	0.33
600	3.242	0.357
700	4.3	0.37

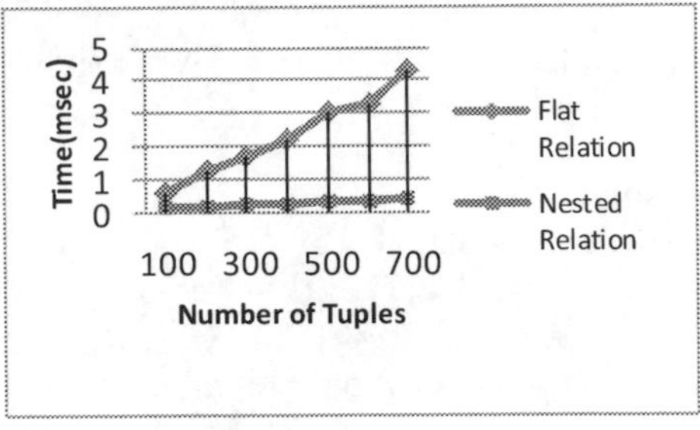

Fig. 2 Graph representing Query Execution time

Size of the Relation: In this experiment we investigated how the redundancy is reduced by considering the space occupied by the flat and nested relations on the disk by varying number of tuples can be seen in Table 5 and Fig. 3.

Table 5 Size of the Relation

No. of Tuples	Flat Relation Size(KB)	Nested Relation Size(KB)
100	16	8
200	24	9
300	32	9.5
400	40	10
500	48	16
600	56	20
700	64	15

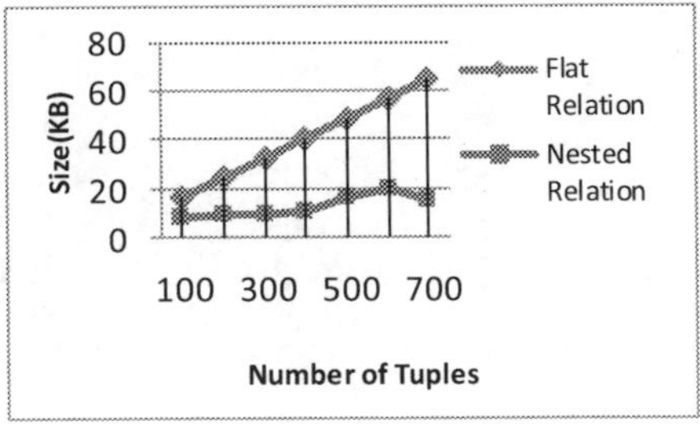

Fig. 3 Graph representing size of the Relation

6 Conclusions

We presented, the nesting based wMVD's (n-wMVD's) proved to be good in eliminating redundancy in generalized databases. We considered a single attribute for nesting and showed the experiments resulting in space and time saving of nested relation over a flat relation. There is naturally much more to be done, first to remove redundancy on fly databases (incremental databases), second application of nesting on different unstructured and semi structured data and third extending of nesting on multiple attributes and to verify with different attribute combinations.

References

1. Viswanadham, S., Kumari, V.V.: Eliminating Data Redundancy using Nesting based wMVD. In: 2012 4th International Conference on Electronics Computer Technology, ICECT. IEEE Publications (April 2012)
2. Rahm, E., Hong, H.D.: Data Cleaning: Problems and Current Approaches. In: IEEE Techn. Bulletin on Data Engineering, University of Leipzig, Germany (2000)
3. Ezeife, C.I., Ohanekwu, T.E.: Use of Smart Tokens in Cleaning Integrated Warehouse Data. International Journal of Data Warehousing & Mining (April-June 2005)
4. Fan, W., Kementsietsidis, A.: Conditional Functional Dependencies for Capturing Data Inconsistencies. ACM Transactions on Data Base Systems (TODS) 33(2) (June 2008)
5. Hartmann, S., Link, S.: On Inferences of Weak Multivalued Dependencies. Fundamenta Informaticae 92, 83–102 (2009), doi:10.3233/FI-2009-0067
6. Fischer, P., Van Gucht, D.: Weak Multivalued Dependencies. In: PoDS Conference. ACM (1984)
7. Korth, H., Roth, M.: Query Languages for Nested Relational Databases. In: Abiteboul, S., Schek, H.-J., Fischer, P.C. (eds.) NF2 1987. LNCS, vol. 361, pp. 190–204. Springer, Heidelberg (1989)
8. Fagin, R.: Multivalued Dependencies and a New Normal Form for Relational Databases. Trans. ACM Database Syst. (1977)
9. Hartmann, S., Link, S.: Characterising nested database dependencies by fragments of propositional logic. Annals of Pure and Applied Logic Journal of Science Direct 152, 84–106 (2008)

Comparison of Question Answering Systems

Tripti Dodiya and Sonal Jain

Abstract. Current Information retrieval systems like Google are based on keywords wherein the result is in the form of list of documents. The number of retrieved documents is large. The user searches these documents one by one to find the correct answer. Sometimes the correct or relevant answer to the searched keywords is difficult to find. Studies indicate that an average user seeking an answer to the question searches very few documents. Also, as the search is tedious it demotivates the user and he/she gets tired if the documents do not contain the content which they are searching for. Question-answering systems (QA Systems) stand as a new alternative for Information Retrieval Systems. This survey has been done as part of doctoral research work on "Medical QA systems". The paper aims to survey some open and restricted domain QA systems. The surveyed QA systems though found to be useful to obtain information showed some limitations in various aspects which should resolved for the user satisfaction.

Keywords: Information retrieval systems, Question Answering system, Open QA systems, Closed QA systems.

1 Introduction

Current search engines are based on keywords wherein the result is in the form of list of documents. The number of retrieved documents is large. For instance: querying about obesity results in more than 186,000,000 documents. The user searches these documents one by one to find the correct answer. Sometimes the correct or relevant answer is difficult to find. Studies indicate that an average user seeking an answer to the question searches very few documents. Also, as the search is tedious, it demotivates the user and gets tired if the documents do not contain the content which they are searching for.

Tripti Dodiya
GLS Institute of Computer Applications, Ahmedabad
e-mail: triptidodiya@glsica.org

Sonal Jain
GLS Institute of Computer Technology, Ahmedabad
e-mail: sonal@glsict.org

A. Abraham and S.M. Thampi (Eds.): Intelligent Informatics, AISC 182, pp. 99–107.
springerlink.com

QA systems, unlike information retrieval (IR) systems, can automatically analyze a large number of documents and generate precise answers to questions posed by users. These systems employ Information Extraction (IE) and Natural Language Processing (NLP) techniques to provide relevant answers. QA systems are regarded as the next step beyond search engines [8].

While the research on automated QA in the field of Artificial Intelligence (AI) dates back to 1960s, more research activities involving QA within the IR/IE community have gained momentum by the campaigns like TREC evaluation which started in 1999 [10]. Since then techniques have been developed for generating answers for three types of questions supported by TREC evaluations, namely, factoid questions, list questions, and definitional questions.

2 Literature Survey

QA systems are classified in two main parts [8]: (a) open domain QA system (b) restricted domain QA system.

Open domain QA systems deal with questions about nearly everything and can only rely on general ontology and world knowledge [8]. Alternatively, unlimited types of questions are accepted in open domain question answering system.

Restricted domain QA systems deal with questions under a specific domain (for example, biomedicine or weather forecasting) and can be seen as an easier task because NLP systems can exploit domain-specific knowledge frequently formalized in ontology. Alternatively, limited types of questions (or questions related to a particular domain) are accepted in restricted domain system.

Some of the open domain QA systems and restricted domain QA systems are surveyed as part of the research work on "Medical QA Systems".

2.1 START(SynTactic Analysis Using ReversibleTransformations)

START is the world's first Web-based open QA system developed by Boris Katz and associates in December 1993 [3]. Currently, the system answers questions about places, movies, people, dictionary definitions, and more. START parses incoming questions, matches the queries created from the parse trees against its knowledge base (KB) and presents the appropriate information segments to the user. It uses a technique "natural language annotation" which employs natural language sentences and phrases "annotations" as descriptions of content that are associated with information segments at various granularities. An information segment is retrieved when its annotation matches an input question. This technique allows START to handle variety of questions from different domains [3].

START system was tested using some sample questions related to various domain, and most of them were answered. Resluts were precise and few of them were supported with images. Figure 1 shows the question and the result displayed along with the source. However, dissatisfaction was observed in some cases. For instance, output for question "*procedure for kidney stone removal*" was unsuccessful as shown in Figure 2.

START's reply

===> what is obesity

Obesity

Obesity is a medical condition in which excess body fat has accumulated to the extent that it may have an adverse effect on health, leading to reduced life ex and/or increased health problems.[1][2] Body mass index (BMI), a measurement which compares weight and height, defines people as overweight (pre-obese BMI is between 25 and 30 kg/m^2, and obese when it is greater than 30 kg/m^2.[3]

Source: Wikipedia

Fig. 1 START QA result along with source

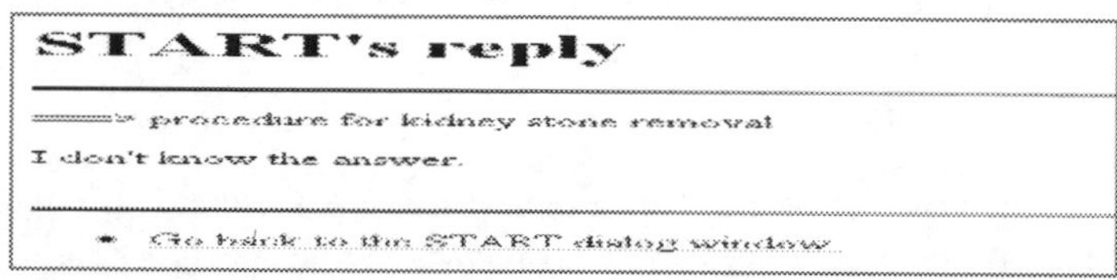

Fig. 2 START QA result for restricted domain question

Limitations

The survey results concluded that START answered to almost many questions pertaining to different domains, but when questions are more related to restricted domain, many were unsuccessful.

START can be accessed at http://start.csail.mit.edu/

2.2 *Aqualog*

Venessa lopez et al. in [11,12] discusses Aqualog as a portable QA system that takes queries in natural language, an ontology as input, and returns answers drawn from knowledge bases (KBs), which instantiate the input ontology with domain-specific information. The techniques used in Aqualog are:

- It makes use of the GATE NLP platform in linguistic component
- String metrics algorithms
- WordNet (open domain ontology)
- Novel ontology-based similarity services for relations and classes, to make sense of user queries with respect to the target knowledge base.

AquaLog is coupled with a portable and contextualized learning mechanism to obtain domain-dependent knowledge by creating a lexicon [11,12]. It is portable as its architecture is completely independent from specific ontology's and knowledge representation systems. It is also portable with respect to knowledge representation, because it uses a modular architecture based on a plug-in mechanism to access information about an ontology, using an OKBC-like protocol. AquaLog uses KMi ontology and is coupled with the platform WebOnto , a web server which contains the knowledge models or ontologies in an OCML format.

Limitations

- AquaLog does not handle temporal reasoning. The results are not proper for questions formed with words like: yesterday, last year etc.
- It does not handle queries which require similarity or ranking reasoning. For instance: "*what are the top/most successful researchers*?".
- It does not handle genitives. For example: What's, Project's etc.

2.3 MedQA (Askhermes)

MedQA developed by Lee et al. [7] is a biomedical QA system. It is developed to cater to the needs of practicing physicians. The system generates short text answers from MEDLINE collection and the Web. The current system implementation deals with definitional questions, for example: "*What is X?*". Figure 3 shows sample question "kidney stone removal" with output extracted from Google, PubMed and OneLook along with time taken to search the result.

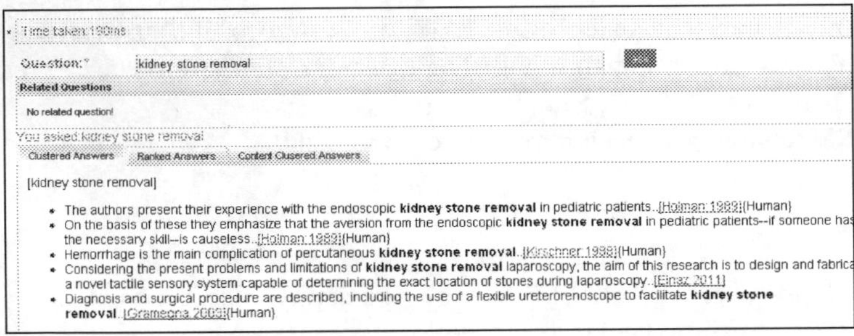

Fig. 3 MedQA results with the posted question

Limitations

Minsuk Lee et al. in [2,7] concluded that MedQA out-performed three online information systems: Google, OneLook, and PubMed in two important efficiency criteria; a) time spent and b) number of actions taken for a physician to identify a definition. However, some limitations observed are:

- Capacity is limited due to its ability to answer only definitional question.
- Another important issue is speed. As the QA systems needs to process large documents and incorporates many computational intensive components it consumes more time.
- It does not capture semantic information which plays an important role for answer extraction and summarization.

MedQA can be accessed at www. askhermes.org/MedQA/

2.4 *HonQA*

HonQA developed by HON (Health On the Net) foundation is a multilingual biomedical system useful for both patients and health professionals [13]. It is restricted domain and has multilingual support in English, French and Italian languages. Figure 4 shows the advanced search option with selection of the language and source for the search. By default the search is done in the database of certified websites.

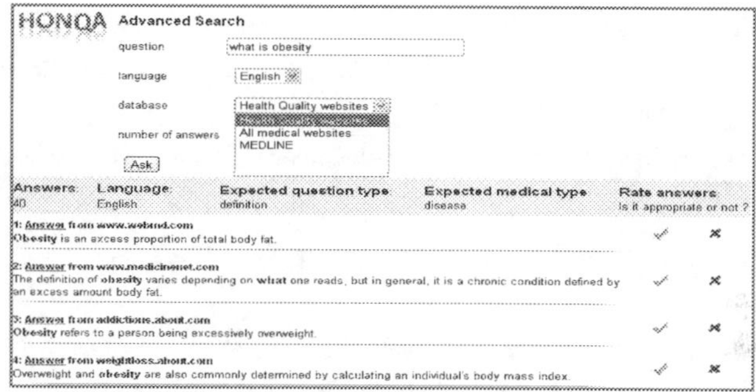

Fig. 4 Search results of HonQA

The results displays multiple definitions, source, question type and medical type. The user is also asked to rate the answer for further reference.

Limitations

The survey results shows that HonQA results are short definitional answers, whereas the users expect more details pertaining to the query. Time taken to extract the answers is high.

HonQA can be accessed at www.hon.ch/QA/

2.5 *QuALiM*

Michael kaisser et al. in [6] describes the demo of QuALiM open domain QA system. The results are supplemented with relevant passages. In [6] Wikipedia is used as search engine. QuALiM uses linguistic methods to analyse the questions and candidate sentences in order to locate the exact answers [5].

Limitations

Michael kaisser et al. in [6] concluded with some issues faced regarding the delay caused due to search engines API's, post processing delay and results fetched from Wikipedia that needs to be resolved.

3 QA Systems in Other Languages

QA Systems generally have been developed in English as majority of documents available on the internet are in English. However, there are examples of QA systems that have been developed in foreign languages like Arabic, Spanish etc.

Table 1 QA systems in other languages

Name	Language	Details
AQUASYS [9]	Arabic	Fact based questions
QARAB [1]	Arabic	Extracts answers from Arabic newspapers
GeoVAQA [4]	Spanish	Voice Activated Geographical Question Answering System

Table 2 Comparative study of open and restricted domain QA systems

System observed [paper ref.] / Features	Start [3]	Aqualog [11,12]	MedQA [2,7]	HonQA [13]	Qualim [5,6]
Open/Restricted domain	Open	Restricted	Restricted	Restricted	Open
Language used for Implementation	Common LISP	Java	Perl	Not specified	Java
Wordnet support	Yes	Yes	No	Not specified	Yes
Ontology support	Yes	Yes	No	Not specified	No
Ontology portability	Not required	Yes	No	No	Not required
Supports multilingual documents	Yes	No	No	Yes	No
Language used to display result	English	English	English	English, Italian, French	English

4 Evaluation and Results

For the evaluation, sample questions were generated for open and restricted domain QA systems. The medical QA systems were evaluated by doctors and their satisfaction level with 5 point likert scale was recorded. Restricted domains were tested with MedQA, HonQA and START while for open domain START was used. Some sample questions were not answered which we considered as dissatisfactory for the result. Table 3 lists some sample questions.

Table 3 Sample questions input to QA systems

Open Domain	Restricted Domain
Who is the president of India	What is length of large intestine
When was Gandhiji born	What is the function of pancreas
Who invented electricity	What is normal blood pressure in adult
Which is fastest flying bird	Which are the reproductive organs of human body
Which country has the largest army	How to calculate BMI
Who is world's fastest runner	How many bones are there in human body
How many planets are there in the milky way	When was penicillin invented

The evaluation results in case of restricted and open domain systems are given below.

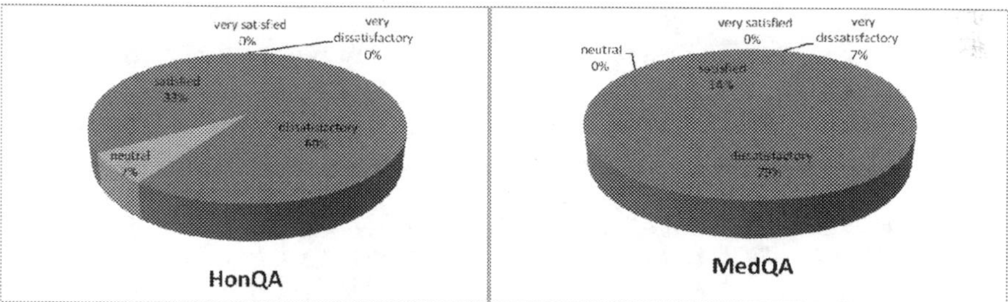

Fig. 5 Results of HonQA and MedQA for restricted domain sample questions

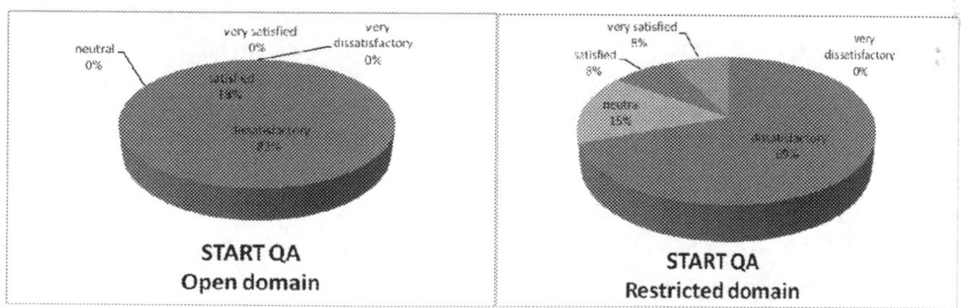

Fig. 6 Results of START QA for restricted and open domain sample questions

Looking at the above results, we can conclude that the user satisfaction level in case of restricted and open domain QA systems is less. As per the oral inputs given by the users, the dissatisfaction was also because of the format and the content provided in the results.

5 Conclusion

QA systems represent the next step in information access technology. By delivering precise answers they can more effectively fulfill users' information needs. Corresponding to the growth of information on the web, there is a growing need for QA systems that can help users better utilize the ever-accumulating information. The surveyed QA systems though found to be useful to obtain information, showed some limitations in various aspects which can be resolved for the user satisfaction. Also, continued research toward development of more sophisticated techniques for processing NL text, utilizing semantic knowledge, and incorporating logic and reasoning mechanisms, will lead to more useful QA systems.

6 Future Scope

In the survey, many of the limitations discussed needs to be overcome. The results should be formatted to be more understandable to the user. Speed is an important issue. Obviously, the higher is the speed, the greater is the user satisfaction. It makes a challenge for QA systems to deliver optimal response times.

References

[1] Hammo, B., Abu-Salem, H., Lytinen, S., Evens, M.: QARAB: A Question Answering System to Support the Arabic Language. In: ACL 2002 Workshop on Computational Approaches to Semitic Languages, Philadelphia, PA, pp. 55–65 (July 2002)

[2] Yu, H., Lee, M., Kaufman, D., Ely, J., Osheroff, J.A., Hripcsak, G., Cimino, J.: Development, implementation, and a cognitive evaluation of a definitional question answering system for physicians. Journal of Biomedical Informatics 40, 236–251 (2007)

[3] `http://start.csail.mit.edu/start-system.html`

[4] Luque, J., Ferrés, D., Hernando, J., Mariño, J.B., Rodríguez, H.: Geovaqa: A voice activated geographical question answering system

[5] Kaisser, M., Becker, T.: Question Answering by Searching Large Corpora with Linguistic Methods. In: The Proceedings of the 2004 Edition of the Text Retrieval Conference, TREC 2004 (2004)

[6] Kaisser, M.: The QuALiM question answering demo: supplementing answers with paragraphs drawn from Wikipedia. In: Proceedings of the 46th Annual Meeting of the Association for Computational Linguistics on Human Language Technologies: Demo Session, Columbus, Ohio, June 16, pp. 32–35 (2008)

[7] Lee, M., Cimino, J., Zhu, H.R., Sable, C., Shanker, V., Ely, J., Yu, H.: Beyond Information Retrieval-Medical Question Answering. In: AMIA Annu. Symp. Proc., pp. 469–473 (2006)

[8] Kangavari, M.R., Ghandchi, S., Golpour, M.: Information Retrieval: Improving Question Answering Systems by Query Reformulation and Answer Validation. World Academy of Science, Engineering and Technology 48, 303–310 (2008)

[9] Bekhti, S., Rehman, A., Al-Harbi, M., Saba, T.: AQUASYS: An Arabic Question-Answering system based on extensive question analysis and answer relevance scoring. International Journal of Academic Research 3(4), 45–54 (2011)
[10] Athenikosa, S.J., Hanb, H.: Biomedical question answering: A survey. Computer Methods and Programs in Biomedicine 99, 1–24 (2010)
[11] Lopez, V., Motta, E.: Aqualog: An ontology-portable question answering system for the semantic web. In: Proceedings of the International Conference on Natural Language for Information Systems, NLDB, pp. 89–102 (2004)
[12] Lopez, V., Uren, V., Motta, E., Pasin, M.: AquaLog: An ontology-driven question answering system for organizational semantic intranets. Web Semantics: Science, Services and Agents on the World Wide Web 5, 72–105 (2007)
[13] http://www.hon.ch

Transform for Simplified Weight Computations in the Fuzzy Analytic Hierarchy Process

Manju Pandey, Nilay Khare, and S. Shrivastava

Abstract. A simplified procedure for weight computations from the pair-wise comparison matrices of triangular fuzzy numbers in the fuzzy analytic hierarchy process is proposed. A transform T:R3→R1 has been defined for mapping the triangular fuzzy numbers to equivalent crisp values. The crisp values have been used for eigenvector computations in a manner analogous to the computations of the original AHP method. The objective is to retain both the ability to capture and deal with inherent uncertainties of subjective judgments, which is the strength of fuzzy modeling and the simplicity, intuitive appeal, and power of conventional AHP which has made it a very popular decision making tool.

Keywords: Fuzzy, AHP, Triangular Fuzzy Number, Fuzzy Synthetic Extent, Weight Vector, Eigenvector, Decision Making, Optimization and Decision Making.

1 Introduction

Conventional AHP treats decision making problems as follows [1, 2, 3, 4]. All decision making problems have at least one objective or goal, set of more than one alternative or option (from which a choice of the best alternative has to be made), and a set of criteria (and possibly sub-criteria) against which these alternatives are to be compared. In conventional AHP, first, the problem objective(s) and the criteria to be considered are defined. Second, the problem is arranged into a hierarchy with the problem objective or goal at the top level, criteria and sub-criteria at the

---

Manju Pandey
Department of Computer Applications
NIT Raipur, Chhattisgarh, India
e-mail: manjutiwa@gmail.com

Nilay Khare · S. Shrivastava
Department of Computer Science and Engineering
MANIT Bhopal, Madhya Pradesh, India
e-mail: nilay.khare@rediffmail.com, scs_manit@yahoo.com

A. Abraham and S.M. Thampi (Eds.): Intelligent Informatics, AISC 182, pp. 109–117.
springerlink.com

intermediate levels, and the alternatives or options at the final level. Fig.1 shows the hierarchy for a decision making problem with 5 criteria and 5 options or alternatives. Third, pair-wise comparisons of the criteria are made and the (normalized) eigenvector of the pair-wise comparison matrix is computed to prioritize the criteria. Fourth, for each criterion, pair-wise comparisons of the alternatives are made and the (normalized) eigenvector of the pair-wise comparison matrix is computed for ranking the alternatives with respect to a particular criterion. Fifth, weighted sum of ranks of each alternative with respect to different criteria and the corresponding criteria priorities is computed to determine overall ranks of alternatives. Last step, the alternatives are ranked in the order of rank/cost ratio.

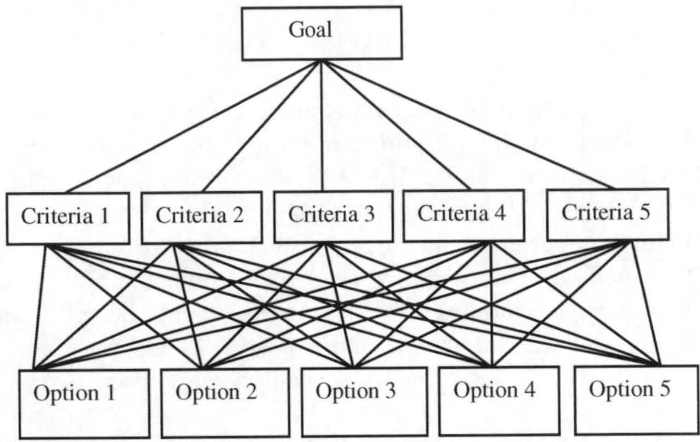

Fig. 1 Arrangement of Goal, Criteria, and Options in AHP

Conventional AHP uses a 9-point ratio scale called Saaty's scale which is used by the decision makers for assigning criteria and alternative weights during the pair-wise comparisons. The simplicity, intuitive appeal and power of AHP as a decision making tool is beautifully illustrated with a hypothetical example in [5].

Fuzzy logic was propounded by Lotfi Asker Zadeh with the objective of mathematically handling situations with inherent uncertainties and imprecision and subjective matters which are not readily amenable to mathematical modeling [6]. Fuzzy AHP is an application of the extent analysis method [7, 8, 9]. The scale for choosing preferences is based on triangular fuzzy numbers (TFNs). For prioritizing criteria and alternatives, fuzzy AHP relies on the computation of synthetic fuzzy extent values from pair-wise comparison matrices. The degree of possibility concept is used for determining the order relationship between triangular fuzzy numbers. Computations are based on fuzzy number arithmetic [10, 11, 12, 13] and fuzzy addition, subtraction, multiplication, and inverse operations are defined. The original AHP method, however, uses matrix eigenvector computations in the prioritization steps, and simple ordering and arithmetic of real numbers.

In this paper, a transform $\mathbf{T}:\mathbf{R}^3 \rightarrow \mathbf{R}^1$ has been defined for mapping TFNs to equivalent crisp numbers. The transform is an attempt to represent the value of the

TFN closely. The crisp equivalents of the triangular fuzzy numbers are then used for matrix eigenvector computations and for ordering in a manner analogous to the original AHP method. It is empirically shown through some numerical examples that the priority vectors match closely those obtained from Chang's fuzzy AHP.

2 Triangular Fuzzy Numbers

A triangular fuzzy number depicted in Fig. 2 is a triplet *(l,m,u)* where $l,m,u \in \mathbf{R}$, i.e., $(l,m,u) \in \mathbf{R}^3$. The triangular fuzzy number [10][11][12][13] is defined as follows:

$$t = \begin{cases} 0, x < l \\ \dfrac{x-l}{m-l}, l \le x \le m \\ \dfrac{u-x}{u-m}, m \le x \le u \\ 0, x > u \end{cases}$$

The "value" of the TFN is "close to" or "around" *m* [13]. Membership is linear on both sides of m and decreases to zero at *l* for values less than *m* and at *u* for values greater than *m*.

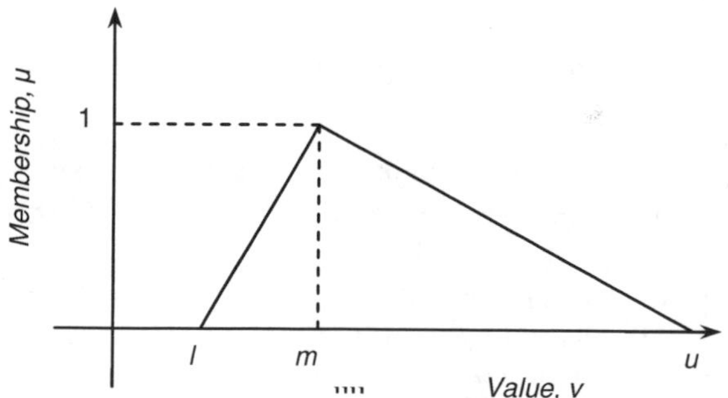

Fig. 2 Triangular Fuzzy Number

3 Transform

The origin of the axis is translated to *(m,1)* and rotated clockwise by 90° as shown in Fig.3. The transformed positive X-axis is now along the negative Y-axis in the original system and the transformed positive Y-axis is along positive X-axis of the original system.

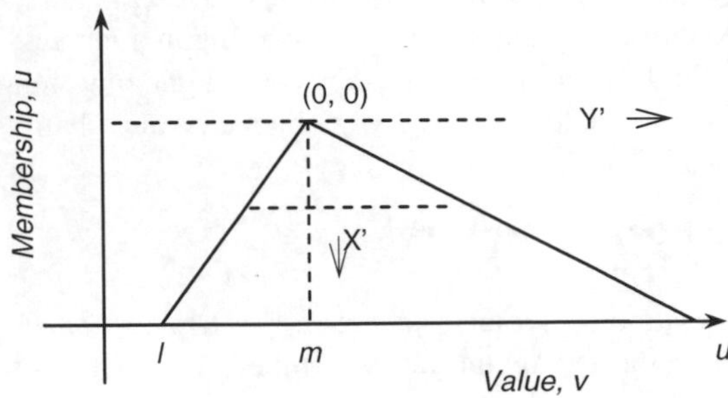

Fig. 3 Axis translated to (m,1) and rotated clockwise by 90º

Coordinates of the key points in the transformed system are in Table 1 below

Table 1 Coordinates with reference to the new coordinate system

Coordinate in Old System	Transformed Coordinate
(0,0)	*(1,-m)*
(l,0)	*(1, l-m)*
(m,0)	*(1,0)*
(u,0)	*(1,u-m)*
(m,1)	*(0,0)*

Following computations have been performed with reference to the new coordinate system.

- Equation of the membership line to the right of the core m is $y_R=(u-m)x$
- Equation of the membership line to the left of the core m is $y_L=(l-m)x$
- Difference is $y_R - y_L = (u-m)x - (l-m)x=(u-l)x$
- Corresponding membership is x
- Product of membership and difference functions is $x(u-l)x$.
- Averaging the product

$$\int_0^1 x(u-l)xdx = \frac{(u-l)}{3}$$

The equivalent crisp value is: (core value + average value), or,

$$v_c = m + \frac{(u-l)}{3}.$$

For example, the crisp value corresponding to the triangular fuzzy number (1,4,10) is $4 + \frac{(10-1)}{3} = 7$.

4 Method

The proposed method involves computing the normalized eigenvectors of the crisp matrices corresponding to a given TFN pair-wise comparison matrix. The elements of the crisp matrix are the crisp transforms of the corresponding elements of the TFN matrix.

For comparison of results we demonstrate the method on three TFN matrices taken from literature [14, 15, 16]. For the TFN matrices considered, the reader is referred to the original papers which are publicly accessible on the internet. TFN matrices are mentioned in Tables 2 of [14], 1 of [15], and 2 of [16] respectively.

4.1 Numerical Example No. 1

From the TFN matrix of [14] we derive the following matrix of equivalent crisp values using the formula developed in the paper, i.e., $v_c = m + \frac{(u-l)}{3}$

$$\begin{bmatrix} 1 & 1.333 & 1.833 & 2.333 \\ 1.443 & 1 & 1.333 & 1.833 \\ 0.837 & 1.443 & 1 & 1.33 \\ 0.59 & 0.837 & 1.443 & 1 \end{bmatrix}$$

Normalized eigenvector corresponding to the real eigenvalue is computed as $\begin{bmatrix}0.31 & 0.28 & 0.23 & 0.19\end{bmatrix}^T$. Here values have been rounded to two decimal places. Difference vector between this priority vector and the priority vector computed using Chang's synthetic extent method in [14] is $\begin{bmatrix}-0.01 & 0.01 & 0.01 & 0.00\end{bmatrix}^T$ and the root mean squared difference is 0.00866.

4.2 Numerical Example No. 2

From the TFN matrix of [15] we derive the following matrix of equivalent crisp values using the formula developed in the paper

$$\begin{bmatrix} 1 & 0.85 & 0.6 & 2.283 & 1.113 \\ 2.34 & 1 & 1.003 & 2.893 & 1.7 \\ 2.816 & 1.85 & 1 & 3.46 & 2.28 \\ 0.77 & 0.523 & 0.44 & 1 & 0.683 \\ 1.43 & 0.983 & 0.71 & 2.196 & 1 \end{bmatrix}$$

Normalized eigenvector corresponding to the real eigenvalue is $\begin{bmatrix}0.16 & 0.25 & 0.32 & 0.09 & 0.18\end{bmatrix}^T$. Here values have been rounded to two decimal places. Difference vector between this priority vector and the priority vector computed using Chang's synthetic extent method in [15] is $\begin{bmatrix}-0.02 & -0.02 & 0.00 & 0.04 & 0.01\end{bmatrix}^T$ and the root mean squared difference is 0.0224.

4.3 Numerical Example No. 3

From the TFN matrix of [16] we derive the following matrix of equivalent crisp values using the formula developed in the paper

$$\begin{bmatrix} 1 & 2.053 & 1.48 & 2.117 & 2.167 \\ 0.847 & 1 & 3.953 & 1.223 & 3.48 \\ 1.263 & 0.433 & 1 & 1.437 & 1.307 \\ 0.963 & 1.193 & 1.19 & 1 & 1.093 \\ 1.047 & 0.633 & 1.227 & 1.253 & 1 \end{bmatrix}$$

Normalized eigenvector corresponding to the real eigenvalue is $\begin{bmatrix}0.259 & 0.274 & 0.154 & 0.164 & 0.149\end{bmatrix}^T$. Here values have been rounded to three decimal places. Difference vector between this priority vector and the priority vector computed using Chang's synthetic extent method in [16] is $\begin{bmatrix}0.013 & 0.001 & -.013 & 0.000 & -0.001\end{bmatrix}^T$ and the root mean squared difference is 0.0082.

5 Advantages over Chang's Method

In Chang's method, priority vectors are computed from the pair-wise comparison matrices of triangular fuzzy numbers using the following steps. First, the fuzzy synthetic extents are computed by summing all rows which is then divided by the total sum of rows for normalization.

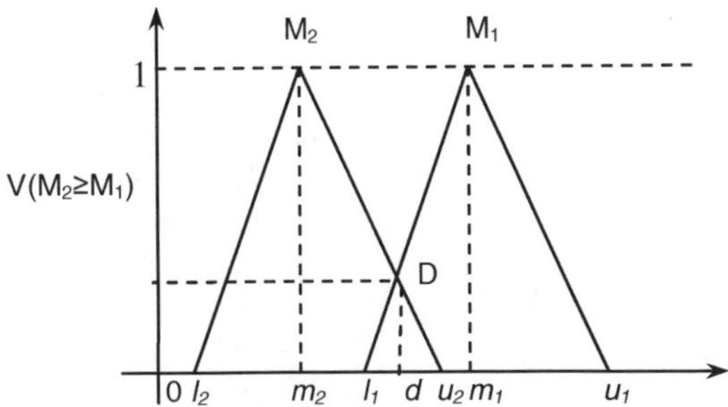

Fig. 4 Ordering of two triangular fuzzy numbers in Chang's method

In the second step, these fuzzy synthetic extent values are ordered by computing the degrees of possibility, V, of each fuzzy number being greater than the other in the fuzzy synthetic extent vector.

The fuzzy synthetic extents S_i of the first step are computed for a matrix $\widetilde{P}_{mn}$ of triangular fuzzy numbers using the equation

$$S_i = \frac{\sum_{j=1}^{n} \tilde{P}_{ij}}{\sum_{i=1}^{m} \sum_{j=1}^{n} \tilde{P}_{ij}}$$

In the second step, if M_1 and M_2 are two triangular fuzzy numbers as shown in Fig. 4 above, then the degree of possibility, V, of $M_2 \geq M_1$ is given by

$$V(M_2 \geq M_1) = \begin{cases} 1, & \text{if } m_2 = m_1 \\ 0, & \text{if } l_1 = u_2 \\ \dfrac{(l_1 - u_1)}{(m_2 - u_2) - (m_1 - l_1)}, & \text{otherwise} \end{cases}$$

For a *m* x *n* matrix of triangular fuzzy numbers, Chang's method involves *m* x *(n-1)* fuzzy addition operations and *m* fuzzy division operations for computing the fuzzy synthetic extents. Also in the ordering step, it involves *m(m-1)/2* degree of possibility, i.e., V calculations.

Chang's method has the following limitations:

- The method requires knowledge of fuzzy sets and fuzzy arithmetic involving triangular fuzzy numbers.
- The method involves ordering fuzzy numbers, that is, the ranking of fuzzy numbers based on the computation of degree of possibility. This is not intuitive.
- It does not extend classical AHP method of Saaty for priority vector computations.

Considering the above, the advantages of the method of this paper over Chang's method can be summarized as follows:

- The method does not require any knowledge of fuzzy arithmetic involving triangular fuzzy numbers.
- The method involves ordering numbers on the real line, which is intuitive, has a definite geometric interpretation, and is well understood.

- After the transform is applied to individual elements of the triangular fuzzy number matrix, computation proceeds exactly as in Saaty's classical analytic hierarchy process method.

Therefore, the method is easy to follow and adopt for a decision maker making the transition from classical AHP to fuzzy AHP. The results of the method closely match those of Chang's method as has been demonstrated in three numerical examples.

6 Conclusions

The transform and method discussed in this paper concern a simplified technique for computing the priority vectors from the pair-wise comparison matrices of triangular fuzzy numbers in the context of the fuzzy analytic hierarchy method. The resulting priority vectors closely match those obtained from Chang's fuzzy synthetic extent analysis which is the backbone of the fuzzy AHP method. It can be seen that with the sole exception of ranks 3 and 4 of numerical example 3 which get interchanged between the method developed in this paper and Chang's method, the order is preserved, and no other discrepancy in the order relation of the priorities is observed. In the exception mentioned also, it needs to be noted that the priorities assigned to 3 and 4 are the same if truncated to two decimal places by Chang's method, and therefore these ranks should be expectedly closer. The similar results are obtained by a rather simple transformation of the triangular fuzzy numbers in the pair-wise comparison matrices to equivalent crisp numbers and then proceeding in a manner similar to conventional AHP. Unlike Chang's method, the method does not require knowledge of fuzzy arithmetic, involves ordering of crisp real numbers in place of ordering fuzzy numbers, and involves priority vector computations using the familiar method of conventional AHP. Present method therefore has the simplicity, intuitive appeal and power of conventional AHP while retaining the ability to capture and deal with subjective information which is characteristic of fuzzy modeling.

References

[1] Saaty, T.L.: Decision making with the analytic hierarchy process. Int. J. Serv. Sci. 1(1), 83–98 (2008)

[2] Saaty, T.L.: The Analytic Hierarchy Process. McGraw Hill, New York (1980)

[3] Saaty, T.L.: Decision Making for Leaders: The Analytic Hierarchy Process for Decisions in a Complex World. Wadsworth, Belmont (1982)

[4] Saaty, T.L.: How to Make a Decision: the Analytic Hierarchy process. Interfaces 24(6), 19–43 (1994)

[5] Haas, R., Meixner, O.: An Illustrated Guide to the Analytic Hierarchy Process (2006), `http://www.fakr.noaa.gov/sustainablefisheries/sslmc/july-06/ahptutorial.pdf` (accessed April 27, 2012)

[6] Zadeh, L.A.: Fuzzy Sets. Inf. Control 8, 338–353 (1965)

[7] Chang, D.Y.: Applications of the extent analysis method on fuzzy AHP. Eur. J. Oper. Res. 95(3), 649–655 (1996)

[8] Zhu, K.J., Jing, Y., Chang, D.Y.: A discussion on Extent Analysis Method and applications of fuzzy AHP. Eur. J. Oper. Res. 116(2), 450–456 (1999)

[9] Wang, Y.M., Luo, Y., Hua, Z.: On the extent analysis method for Fuzzy AHP and its applications. Eur. J. Oper. Res. 186(2), 735–747 (2008)

[10] Kauffman, A., Gupta, M.M.: Introduction to Fuzzy Arithmetic – Theory and Applications. Van Nostrand Reinhold Company, New York (1985)

[11] Haans, M.: Applied Fuzzy Arithmetic – An Introduction with Engineering Applications. Springer, Heidelberg (2005)

[12] Verma, A.K., Srividya, A., Prabhu Gaonkar, R.S.: Fuzzy-Reliability Engineering Concepts and Applications. Narosa Publishing House Pvt. Ltd., New Delhi (2007)

[13] Nguyen, H.T., Walker, E.A.: A First Course in Fuzzy Logic, 3rd edn. CRC Press (2006)

[14] Kilic, H.S.: A Fuzzy AHP Based Performance Assessment System for the Strategic Plan of Turkish Municipalities. Int. J. Bus. Manag. 3(2), 77–86 (2011), `http://www.sobiad.org/eJOURNALS/journal_IJBM/2011.html` (accessed on April 27, 2012)

[15] Meixner, O.: Fuzzy AHP Group Decision Analysis and its Application for the Evaluation of Energy Sources. In: Proc. of the 10th Int. Symp. on the Analytic Hierarchy/Netw. Process Multi-criteria Decis. Mak., U of Pitt, PA, USA, pp. 1–14 (2009), `http://www.isahp.org/2009Proceedings/index.html` (accessed on April 27, 2012)

[16] Kabir, G., Hasin, M.A.A.: Multiple criteria inventory classification using fuzzy analytic hierarchy process. Int. J. Ind. Engg. Comput. 3(2), 123–132 (2011), `http://growingscience.com/ijiec/Vol3/Vol3No2.html` (accessed on April 27, 2012)

Parameterizable Decision Tree Classifier on NetFPGA

Alireza Monemi, Roozbeh Zarei,
Muhammad Nadzir Marsono, and Mohamed Khalil-Hani

Abstract. Machine learning approaches based on decision trees (DTs) have been proposed for classifying networking traffic. Although this technique has been proven to have the ability to classify encrypted and unknown traffic, the software implementation of DT cannot cope with the current speed of packet traffic. In this paper, hardware architecture of decision tree is proposed on NetFPGA platform. The proposed architecture is fully parameterizable to cover wide range of applications. Several optimizations have been done on the DT structure to improve the tree search performance and to lower the hardware cost. The optimizations proposed are: a) node merging to reduce the computation latency, b) limit the number of nodes in the same level to control the memory usage, and c) support variable throughput to reduce the hardware cost of the tree.

Keywords: Data Mining, Machine Learning, Search Tree.

1 Introduction

One of the most critical factors of network management and surveillance tasks is to identify network traffic accurately and rapidly. Classical methods of identifying network applications based on detecting well known port numbers are not reliable anymore [8]. On the other hand, some other approaches such as deep packet inspection (DPI) as well as statistical approaches [4] have been proposed. DPI solutions have high accuracy but need constant updates of signatures. Besides, these solutions cannot classify the encrypted packets, such as P2P applications. Statistical

Alireza Monemi · Roozbeh Zarei · Muhammad Nadzir Marsono · Mohamed Khalil-Hani
Faculty of Electrical Engineering, Universiti Teknologi Malaysia,
81310 Johor Bahru, Malaysia
e-mail: {monemi, roozbeh.zarei}@fkegraduate.utm.my,
{nadzir, khalil}@fke.utm.my

A. Abraham and S.M. Thampi (Eds.): Intelligent Informatics, AISC 182, pp. 119–128.
springerlink.com

approaches which are based on machine learning are more suitable in the term of encryption, privacy, and protocol obfuscation.

Traffic classification based on machine learning can be divided into two categories: supervised [10] and unsupervised [15] classifications. Y. Wang [14] compared the both methods for traffic classification and showed that supervised machine learning based on decision trees (DTs) provides higher accuracy with faster classifying speed. However, there are some limitations as almost all the proposed methods were implemented in software [2, 3] and capable to classify traffic offline. Reference [15] shows that the above implementations are not able to keep up with line speed for online traffic classification due to their requirement on computation and storage. In this scenario, in order to make DT classifier suitable for online traffic classification, the advantage of hardware implementation is used to improve the performance of the DT.

In this paper, the hardware DT classifier prototyped for NetFPGA[6] is proposed. The NetFPGA is a versatile platform which allows the development of rapid prototype of Gigabit rate network applications on line-rate. The proposed DT architecture is fully parameterizable in the term of the throughput, number of features, features size, tree depth and maximum required node in the same level.

The remainder of this paper is organized as follows: Section 2 introduces some related works which used DTs. Section 3 discusses the hardware architecture of decision tree as well as the required optimization. In Section 4, we present our DT hardware architecture targeting NetFPGA board. The synthesis results of the implementation is given in Section 5. Finally, we conclude our work in Section 6.

2 Related Works

In the past few years, several traffic classifications based on machine learning have been proposed [11]. These approaches offer high accuracy and are able to classify both unknown and encrypted traffic. Almost all of the proposed methods discussed only the classification accuracy. The performance and timing issue were not discussed due to the software implementation. Hence, these approaches are only suitable for offline traffic classification due to the performance reasons. Reference [15] studied the performance of five machine learning (Naïve Bayes (NBD, NBK), C4.5 Decision Tree (C4.5), Bayesian Network (BayesNet), Naïve Bayes Tree (NBTree)) traffic classifiers. It shows that machine learning based traffic classification are slow and are incapable to keep up with line rates for online traffic classification.

There are only a few articles discuss the hardware implementation of DTs [1, 7, 12, 13]. In [7], the basic approach to implement DT in hardware is to implement all DT nodes in a single module. The main drawback of this approach is the low throughput since new instances cannot be inserted until the previous instance has been classified. Other hardware DTs based on the equivalence between DTs and threshold networks are presented in [1]. This work provides a high-throughput classification since the signals have to propagate through only two levels, irrelevant of the depth of the original DT. However, both mentioned approaches/architectures for hardware realization of DTs require a considerable amount of hardware resources. To increase the throughput, [13] proposed a new pipeline architectures, single module per level, which is specified for implementing oblique decision tree. Reference [12] used the pipeline architecture for implementing axis-parallel DT specific for packet classification.

In contrast to these related works, we use the advantages of the above approaches to implement the axis-parallel decision tree for NetFPGA projects. Our goal is to have a prototype DT which is fully parameterizable, minimized in term of the hardware cost and can work with maximum available line speed of the NetFPGA.

3 Decision Tree

Decision trees are one of the most efficient predictive models in machine learning which introduced first by Breiman et al in 1984 [5]. Decision tree is a classifier in the form of a tree structure which includes either *terminal nodes* (leaf node) indicating classifications or *non-terminal* nodes (root node and decision nodes) contain feature test conditions. Instances are classified by being directed from the parent of the tree to a terminal node which specifies the classification of the instance.

Based on the method used on making the decision at a node, the decision trees are categorized in two main groups: axis-parallel and oblique. In the first group, only one feature is used at each non-terminal node for making the decision, whereas in the second group, more than one feature are needed. In this work, we focus on the implementation of axis-parallel decision tree (DT) in hardware. A DT with depth of 6 is illustrated in Fig. 1(a).

The method which is used to implement a DT in hardware is to map the tree nodes inside block RAMs and locating each individual block RAM in a single module. The decision tree is created by connecting these modules in a pipeline chain.

3.1 Basic DT Design

In the basic design, each level of the DT is mapped into one single module and the tree structure is created by connecting these modules in a pipeline chain. In this structure, each single module is able to process one instance in each clock cycle. This results in the throughput of one instance per one clock cycle. The process latency is equal to the tree depth which contains the last leaf node. In order to map one level of DT on hardware, all nodes located in that level are mapped in one block RAM. Inside this block RAM, the two child nodes connected to the same parent node are mapped in 2 consecutive memory addresses. The size of each module block RAM S_{BRAM} and the total required memory M_{Total} for implementing a DT are obtained from equations (1b) and (1c).

$$W = M_{FW} + \lceil log_2(N_F) \rceil + \lceil log_2(M_{NSL}) \rceil \quad (1a)$$
$$S_{BRAM} = 2N_{node}(W) \quad \text{bit} \quad (1b)$$
$$M_{Total} = T_{node}(W) \quad \text{bit} \quad (1c)$$

where W is the block RAM data width, T_{node} is the total number of nodes inside the DT, N_{node} is the number of existed nodes in a level of DT, M_{FW} is the maximum feature width in bit, N_F is number of input features, and M_{NSL} is the maximum number of node in the same level of DT.

3.2 DT Design Optimization

3.2.1 Node Merging

To decrease the number of parallel block RAMs and reduce the computation latency, the node merging algorithm is used, where the two consecutive stages of a DT is merged into one stage. As illustrated in Fig. 1(b) each parent merged node can have up to four child merged nodes. This optimization results in the reduction of the number of pipeline modules as well as the latency by half while the total memory required remains unchanged.

In a merged node, the input features are compared by three compare values in the same time. These three compare values are the compare values of the three nodes inside a merged node. The four child merged nodes are stored in four consecutive memory addresses inside the module block RAM. The size of the block RAM in each module is defined in equation (2). Note that the total required memory is defined in equation (2) is approximately equal to the basic design obtained from equation (1c) for the same DT.

$$W = 3M_{FW} + \lceil 3log_2(N_F) \rceil + \lceil log_2(M_{NSL}) \rceil \quad (2a)$$
$$S_{BRAM} = 4N_{m\_node}(W) \quad \text{bit} \quad (2b)$$
$$M_{Total} = (T\_node/3)(W) \quad \text{bit} \quad (2c)$$

where N_{m_node} is the number of merged nodes in the same stage of DT.

The differences of our proposed node merging algorithm with the one in [12] is that we merged even the parent nodes which one or both of their child nodes are leafs. Thus, a merged node can reach to a leaf in any of its tree nodes, in contrast to [12] in which a node has to be classified just by the parent node. This optimization reduces the maximum memory size for implementing a DT.

To understand it better, imagine a complete tree which all of its nodes are classified in the 8th level is implemented in hardware. This tree has 127 leafs in its lowest level. In our proposed design, this tree needs a total of 3 modules in pipeline chain while the lowest-level module block RAM just needs 64 rows. The architecture in [12] cannot merge the level 7 and 8 into one stage because all the nodes in the 7th level are connected to the leafs. Hence, it needs to add another module to the

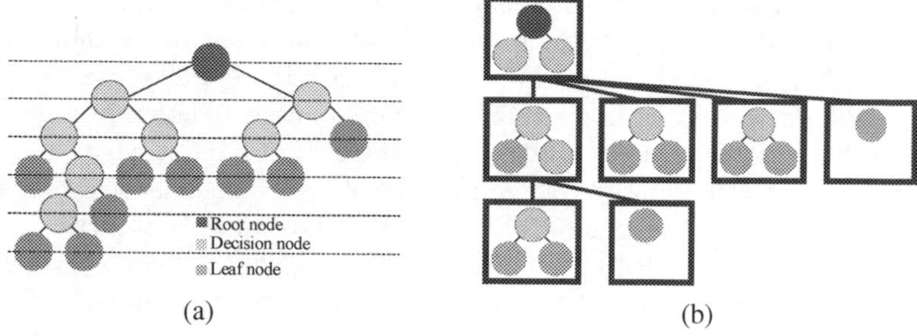

Fig. 1 (a) A Simple DT with Depth of 6. (b) The Merged DT

pipeline chain and place one individual leaf in one row of the block RAM (which now needs 127 rows to store all the leafs). The pseudo code of the node merging algorithm is explained in Algorithm 1.

Algorithm 1. Pseudo code of the merged node

```
if features_in[f_sel1] > cmp_v1 then
    if features_in[f_sel2] > cmp_v1 then
        RAM_addr = addr_in
    else
        RAM_addr = addr_in + 1
    end if
else
    if fsel2 == 0 then {//the first child node is a leaf}
        if features_in[f_sel3] > cmp_v3 then
            RAM_addr = addr_in
        else
            RAM_addr = addr_in + 1
        end if
    else
        if features_in[f_sel3] > cmp_v3 then
            RAM_addr = addr_in + 2
        else
            RAM_addr = addr_in + 3
        end if
    end if
end if
```

3.2.2 Limiting Number of Nodes in the Same Level

In both basic and Merged designs, the block RAM address size of each module is equal to the number of existed nodes/merged nodes inside that level/merged level. Hence, the total required memory is equal to the number of tree nodes multiply by the total memory needed for implementing a single node. The problem of this structure is that the DT is not able to be updated with a new tree without FPGA reconfiguration since the number of nodes inside each level of a DT may vary due to the training data set used to build that DT. One solution to above problem is to impediment a complete DT in hardware. A complete DT is a tree which all of the instances are classified in last level. Hence, no matter how many nodes are in each level, the tree is able to be placed in hardware. The problem with this new design is that the tree depth cannot exceed a limited number due to lack of available memory inside the FPGA device. To address this problem, the maximum number of nodes inside one stage of the DT is considered not to exceed than a fixed number (M_{NSL}). It is possible because in reality not all the instances are classified in the last DT stage. Most of nodes are ended to the leaf in the middle of the DT mostly depends on the training data set used to build the tree. In our proposed design, we allow the DT users to define this value as a parameter in the Verilog program. This value can be estimated by testing several training data sets in building the DT. The block RAM size of each module and the total memory needed is obtained from equations (3b) and (3c).

$$W = 3M_{FW} + \lceil 3log_2(N_F) \rceil + \lceil log_2(M_{NSL}) \rceil \tag{3a}$$

$$S_{BRAM} = \begin{cases} 4^n(W) & 4^n < M_{NSL} \\ M_{NSL}(W) & \text{elsewhere} \end{cases} \quad \text{bit} \tag{3b}$$

$$M_{Total} = \sum_{n=1}^{n=D/2} S_{BRAM} \quad \text{bit} \tag{3c}$$

where D is the tree depth and n is the merged level index.

3.2.3 Support Variable Throughput

The original DT structure has the ability to classify one instance per one clock cycle. Therefore, a new instance can enter in every clock cycle. In this structure, the total DT hardware cost is directly related to the tree depth. In our design, given the n throughput, then we merge the n modules of the pipeline chain in one merged module. The merged module share the computational part (e.g. multiplexer, adder and registers) of one single module. Its block RAM size equals to the total for all n modules. For this purpose, the *addr_in* part of each node except for the $n-th$ module is updated with the address of the next node inside the merged module. A new instance and the nodes values are entered to a merged module in every n clock cycles. The instance features are latched for n clock cycles while for the next $n-1$ clock cycles, the input node values are fed back to the merged module from its own block RAM. This optimization reduces the hardware cost by the factor of approximately $\frac{1}{n}$ while not affecting memory usage and computation latency.

4 NetFPGA Implementation

NetFPGA [6] is a low cost configurable hardware platform for processing network in line-rate. The platform consists of the Xilinx Virtex II-Pro 50 FPGA running at 125 MHz, four 1Gbps Ethernet ports, SRAM, DRAM, and PCI interface to communicate with the host computer. NetFPGA open source hardware and software programs are available in [9]. The NetFPGA comes with four main open-source reference designs (NIC, switch, router and hardware accelerated Linux router). Most of the projects have been done by modification or development on one of these main reference designs.

The NetFPGA reference designs have the limitation of processing 64-bits of the packet data in each clock cycle which causes the minimum number of clocks needed to process a single packet inside one NetFPGA pipeline module is 16 clock cycles. Therefore, if the feature instances are selected from an individual packet, the maximum needed throughput would be 1 instance per 16 clock cycles. Another possible way to send feature instances to the NetFPGA is through packetizing of instances and sending them via UDP packets. Since, only 64-bit of data is processed in each clock cycle. The minimum clock cycles between receiving two consecutive instances varies based on the total length of features.

4.1 Basic Design on NetFPGA

Fig. 2 illustrates the functional block diagram of the basic decision tree. As it shown the mapping of a single node in memory, each node is converted to three values: the compare value *cmp_v*, the feature selector *f_sel*, and the memory address *addr_in*. The *f_sel* value is fed to a multiplexer to select the appropriate feature from feature instances. The selected feature is compared with the *cmp_v*. If the feature value is bigger than the *cmp_val*, the next node memory address is the *addr_in*. Otherwise it is *addr_in* + 1. The read value from the memory determine the *node_in* data of the next module. If a node reaches to a leaf node, the *f_sel* carries the zero value (*done* pin is asserted) while the *cmp_v* contains the class number.

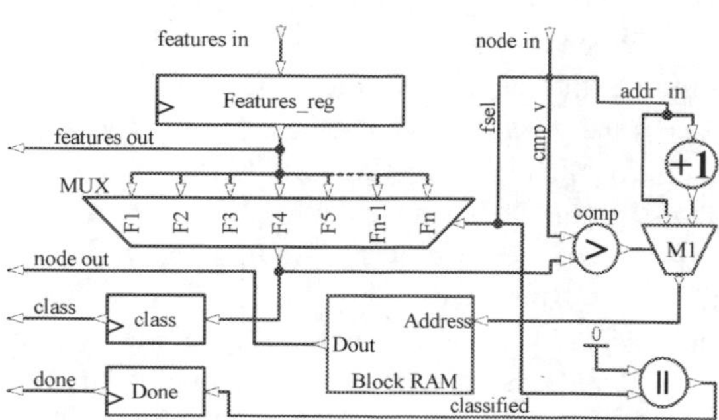

Fig. 2 The Function Block Diagram of one Tree Level in Basic Design

4.2 Our Proposed Architecture

A binary search tree which is mapped into a pipeline structure is illustrated in Fig. 3. Every module has four inputs, *F_in*, *node_in*, *valid_in*, and *next_instance_rd*, and five outputs, *F_out*, *node_out*, *valid_out*, *class*, and *done*. The pipeline structure is made by connecting *F_out* and *node_out* port of a single module to the neighbor module *F_in* and *node_in* ports. The *next_instance_rd* signal is connected to all pipeline modules and controls the number of clock cycles which a single instance remains inside a module. A single module of the pipeline structure is illustrated in Fig. 4. The instance features which are given to a module by *features_in* port are captured in a register when the *next_instance_rd* is asserted. The output of this register is connected to the *features_out* port which provides the instance features for the next neighbor module. The *features_reg* is connected to three parallel multiplexers. These multiplexers are controlled by *f_sel* taken from the node data. The node data includes six parts, three node compare values (*cmp_v*), three node feature selectors (*f_sel*), and the address (*addr_in*). The selected features are compared with the *cmp_v* values and create three *cmp* signal. The *cmp* is set if the value of input features is bigger than *cmp_v* value otherwise is reset.

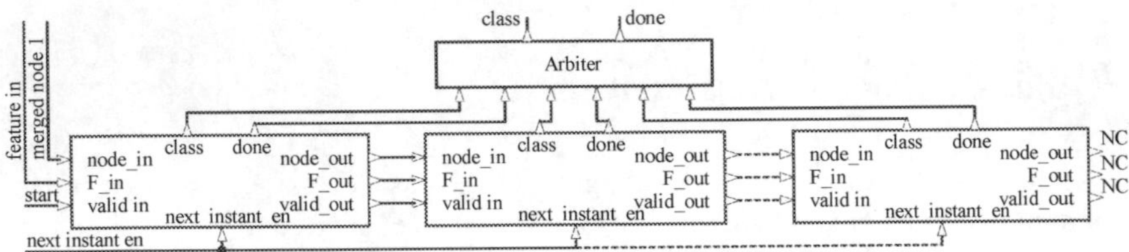

Fig. 3 Functional Block Diagram of the Pipeline DT

The four child nodes are stored in four consecutive memory addresses, the address of next child node is obtained by using these three signals ($cmp1, cmp2, cmp3$) and the feature selector of the second child node. The *node_in* port is used to receive the node data from the previous module in pipeline chain one clock cycle after the *next_instance_rd* is asserted. Otherwise the node data is provided by internal block RAM. When an instance is classified and *valid_reg* is assigned the *done* pin is asserted. At the same time, the *valid_out* pin is reset to inform next module not to process the *feature_in* values anymore. A leaf is reached when one of the three *f_sel*s contains the zero value. In this case, the (*cmp_v*) of the node with zero *f_sel* value contains the instance class.

In order to achieve on-the-fly update, we use a dual-port RAM. The port A of memory is connected to DT and another port is connected to the host computer. The *Mem_sel* pin and its inverted signal are used as the most significant bit of port A and B addresses, respectively. Using this structure, the DT is able to read nodes data from one half of memory while the host computer is able to update new DT on another half of the memory.

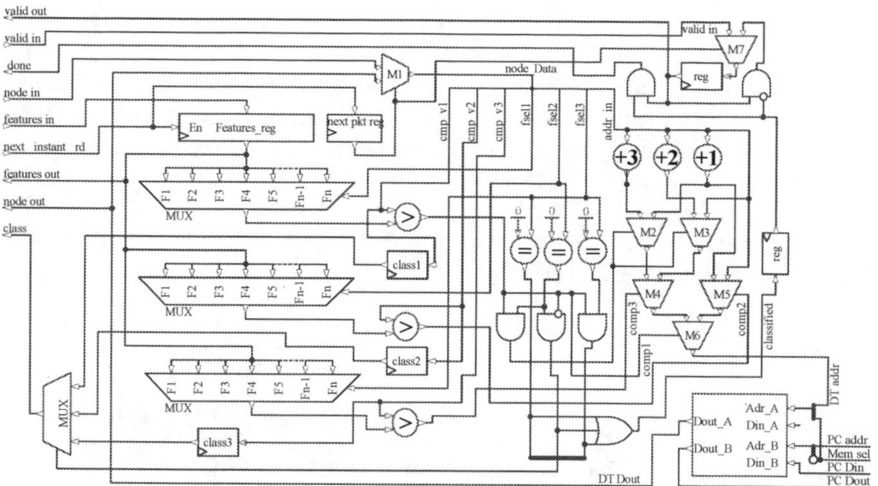

Fig. 4 Functional Block Diagram of the Merged Node

5 Resources and Timing on NetFPGA

Table 1 illustrates the hardware cost for the implementation of a DT with the depth of 32, input features number of 15, maximum feature width of 32 bit, and the M_{NSL} equals to 127 on Vertex 2vp50ff1152-7. The hardware cost is compared for different clocks per instance. As expected, the hardware cost is reduced approximately by half when the clock per instance is doubled. The difference in memory usage for different clocks per packet is because of fixes size internal block RAMs. The required operating frequency is provided by dividing the NetFPGA core clock frequency (125 Mhz) by two using internal Digital Clock Manager (DCM).

Table 1 Device utilization comparison for different clks per instance on the Vertex 2vp50ff1152-7

clks per instance	Number of Slices	Number of BRAMs	Maximum Frequency
2	3956 / 23616 (17%)	32 /232(14%)	68 MHz
4	1997 / 23616 (8.4%)	25 /232(11%)	67.6 MHz
8	1297 / 23616 (5.5%)	28 /232(12%)	64 MHz

6 Conclusion

In this paper, a fully parameterizable DT based on NetFPGA was proposed. Three architecture optimization strategies are proposed: a) node merging to decrease the number of parallel block RAMs and computation latency, b) limiting number of nodes in the same level to control the memory usage and c) support variable throughput to decrease hardware cost. We showed that our design is fully parameterizable in terms of throughput, number of feature size, tree depth and maximum required node in the same level. Our proposed architecture can be applied for any network traffic classifications which utilize DT to classify network traffic, online. As a future work we plan to implement our DT hardware structure for detecting and mitigating network traffic on NetFPGA.

Acknowledgements. This work was supported the Ministry of Higher Education of Malaysia Fundamental Research Grant, UTM Vote No 78577.

References

[1] Bermak, A., Martinez, D.: A compact 3D VLSI classifier using bagging threshold network ensembles. IEEE Transactions on Neural Networks 14, 1097–1109 (2003)

[2] Breiman, L.: Random Forests. Machine Learning 45, 5–32 (2001)

[3] Eklund, P., Kirkby, S.: Machine learning classifier performance as an indicator for data acquisition regimes in geographical field surveys. In: Proceedings of the Third Australian and New Zealand Conference on Intelligent Information Systems, pp. 264–269 (1995)

[4] Erman, J., Mahanti, A., Arlitt, M., Cohen, I., Williamson, C.: Offline/realtime traffic classification using semi-supervised learning. Performance Evaluation 64(9-12), 1194–1213 (2007)

[5] Breiman, L., Friedman, J.H., Olshen, R.A., Stone, C.J.: Classification and regression trees. Wadsworth, Monterey (1984)

[6] Lockwood, J.W., McKeown, N., Watson, G., Gibb, G., Hartke, P., Naous, J., Raghuraman, R., Luo, J.: NetFPGA–An Open Platform for Gigabit-Rate Network Switching and Routing. In: Proceedings of 2007 IEEE International Conference on Microelectronic Systems Education (2007)

[7] Lopez-Estrada, S., Cumplido, R.: Decision Tree Based FPGA-Architecture for Texture Sea State Classification. In: IEEE International Conference on Reconfigurable Computing and FPGA's (2006)

[8] Moore, A.W., Papagiannaki, K.: Toward the Accurate Identification of Network Applications. In: Dovrolis, C. (ed.) PAM 2005. LNCS, vol. 3431, pp. 41–54. Springer, Heidelberg (2005)

[9] NetFPGA (2012), `http://www.netfpga.org/`

[10] Nguyen, T., Armitage, G.: Training on multiple sub-flows to optimise the use of machine learning classifiers in real-world IP networks. In: Proceedings of IEEE 31st Conference on Local Computer Networks, pp. 369–376 (2006)

[11] Nguyen, T., Armitage, G.: A survey of techniques for internet traffic classification using machine learning. IEEE Communications Surveys Tutorials 10(4), 56–76 (2008)

[12] Qi, Y., Fong, J., Jiang, W., Xu, B., Li, J., Prasanna, V.: Multi-dimensional packet classification on FPGA: 100 Gbps and beyond. In: International Conference on Field-Programmable Technology, FPT, pp. 241–248 (2010)

[13] Struharik, R., Novak, L.: Intellectual property core implementation of decision trees. IET, Computers Digital Techniques 3(3), 259–269 (2009)

[14] Wang, Y., Yu, S.Z.: Machine Learned Real-time Traffic Classifiers. In: Second International Symposium on Intelligent Information Technology Application, vol. 3, pp. 449–454 (2008)

[15] Williams, N., Zander, S., Armitage, G.: A preliminary performance comparison of five machine learning algorithms for practical IP traffic flow classification. Special Interest Group on Data Communication (SIGCOMM) 36(5), 5–16 (2006)

Diagnosing Multiple Faults in Dynamic Hybrid Systems

Imtiez Fliss and Moncef Tagina

Abstract. Due to their quite complex nature, Dynamic Hybrid Systems represent a constant challenge for their diagnosing. In this context, this paper proposes a general multiple faults model-based diagnosis methodology for hybrid dynamic systems characterized by slow discernable discrete modes. Each discrete mode has a continuous behavior. The considered systems are modeled using hybrid bond graph which allows the generating of residuals (Analytical Redundancy Relations) for each discrete mode. The evaluation of such residuals (detection faults step) extends previous works and is based on the combination of adaptive thresholdings and fuzzy logic reasoning. The performance of fuzzy logic detection is generally linked to its membership functions parameters.Thus, we rely on Particle Swarm Optimization (PSO) to get optimal fuzzy partition parameters. The results of the diagnosis module are finally displayed as a colored causal graph indicating the status of each system variable in each discrete mode. To make evidence of the effectiveness of the proposed solution, we rely on a diagnosis benchmark: The three-tank system.

1 Introduction

With the spread and the omnipresence of the dynamic hybrid systems, there is a great need for more efficiency, safety and reliability of these systems. The need for diagnostic tools is then crucial in this case as it is a key technology guaranteeing these criteria. In fact, correct and timely diagnosis helps the operator to take the adequate corrective actions in time. Diagnosis is the process of detecting an abnormality in the system behavior and isolating the cause or the source of this abnormality. This problem is more complex in case of multiple occurrences of faults. However, this case should be taken into account, as the performance of physical processes is

Imtiez Fliss · Moncef Tagina
SOIE Laboratory, National School of Computer Sciences-Tunisia
e-mail: {Imtiez.Fliss,Moncef.Tagina}@ensi.rnu.tn

A. Abraham and S.M. Thampi (Eds.): Intelligent Informatics, AISC 182, pp. 129–139.
springerlink.com

affected by the presence of a single fault and is severely affected in case of multiple faults. In this context, this paper deals with the diagnosis of multiple faults in dynamic hybrid systems. These systems are characterized by the exhibition of both continuous and discrete dynamic behaviors. In this work, we consider hybrid systems whose dynamic evolution is described through the succession of a number of slow discernable discrete modes[4]. Each mode is characterized by a continuous evolution of its states. The transition from one mode to another occurs when a number of constraints are checked. Faults can affect either the sequence of discrete states or the continuous behavior in each discrete mode. In this paper, we focus in the diagnosis of multiple faults affecting the continuous behavior in each discrete mode of the dynamic hybrid systems. In this context, we propose to extend previous work [8]dealing with diagnosing multiple faults in continuous systems to dynamic hybrid systems. The proposed approach relies on the extension of Analytical Redundancy Relations (ARRs), a well known residual generating approach used in continuous systems, into hybrid plants inspired from [4]. In this case, we calculate to each discrete mode the corresponding ARRs. ARRs are symbolic equations representing constraints between different known process variables (parameters, measurements and sources). The evaluation of such residuals is done using the combination of adaptive thresholdings and fuzzy logic detection. The performance of fuzzy detection is closely linked to the fuzzy membership functions. For efficiency, an optimal design of membership functions is desired[9]. Thus, we choose to use an optimization technique to adjust the parameters of the fuzzy partitions: the Particle Swarm Optimization (PSO) [3]. The result of detection step is then presented as a colored causal graph. This result is then used in the isolation step which relies on the causal reasoning and gives final findings to the operator helping him to make proper corrective actions. To test the performance of the proposed approach, we rely on a simulation of a benchmark in the diagnosis domain: the three- tank hydraulic system. The remainder of this paper is organized as follows: section two details the proposed approach. While, the third section presents and discusses the simulation results we get, the fourth section points out our contribution to the literature. Finally, some concluding remarks are made.

2 The Proposed Approach for Diagnosing Multiple Faults in Dynamic Hybrid Systems

The diagnosis result indicates whether the system is normally functioning or there are some single or multiple faults that occur. In this work, the considered systems are dynamic hybrid systems characterized by the evolution of m slow discernable discrete modes. Each i mode {i in {1,.., m}} has a continuous evolution of possible configurations. The faults that can affect such systems are either caused by inadequate evolution of discrete modes or by the continuous behavior of each discrete mode. We concentrate, in this paper, on faults affecting the continuous behavior of system variables in each discrete mode. The aim of our work, consists, first of all, in detecting the presence of faults. Such a step results in deciding if each continuous behavior of each mode is faulty or not regardless of disturbances. To get this decision, we rely on the comparison of the system behavior to a reference model. This model should respect the particularities of the dynamic hybrid system to be diagnosed. A best manner to model systems is to use a hybrid bond graph [18] which has the pros of bond graph modeling [5] and introduces the mode switching thanks to the use of Switching elements (Sw) which evolution is described as a finite state sequential automata. The result of this comparison is called residuals. To generate these residuals, several approaches can be used:parity relations [12], state estimation [15], and methods based on parameter identification [11]. In this work, we drew inspiration from the work of Cocquempot et al. [4], which defines an extension of Analytical redundancy relations for each system mode as shown in fig.1.

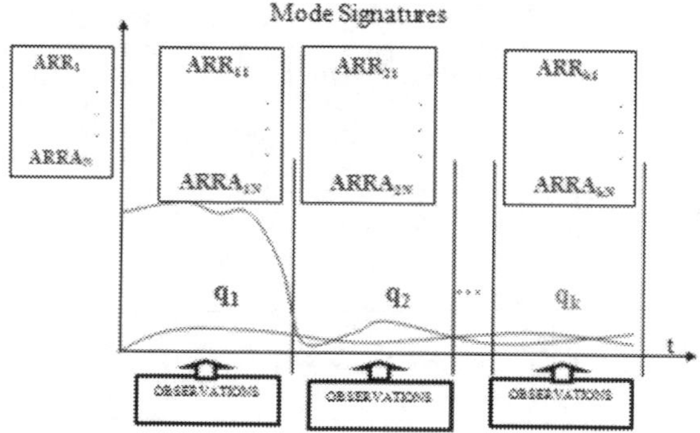

Fig. 1 Extension of Analytical Redundancy Relations for Hybrid systems [10]

For each discernable discrete mode, we generate corresponding residuals thanks to the use of bond graph modeling through the use of the procedure described in [21]. Then for each discrete mode, an evaluation of the system behavior is provided. The architecture of the whole proposed approach is described in fig.2.

In this step, we rely on previous works[8] which consists first of all in evaluating residuals using the Hôfling's adaptive thresholdings [14]. The result of such method is not too robust to disturbances, thus, we integrate the use of fuzzy logic reasoning considered as the best framework dealing with disturbances and uncertainties. Fuzzy logic fault detection consists in interpreting the residuals by generating a value of belonging to the class AL (Alarm) between 0 and 1 that allows one to decide whether the measurement is normal or not. The gradual evolution of this variable from 0 to 1 represents the evolution of the variable to an abnormal state [7]. In practical cases, fuzzy logic effectiveness is closely linked to its partitions parameters. So, to get the best results, optimal values of these parameters should be used. In this context, we rely on Particle Swarm optimization technique, which is characterized by an easy implementation and no gradient information requirement. There have been several versions of Particle Swarm Optimization. We choose in our research one

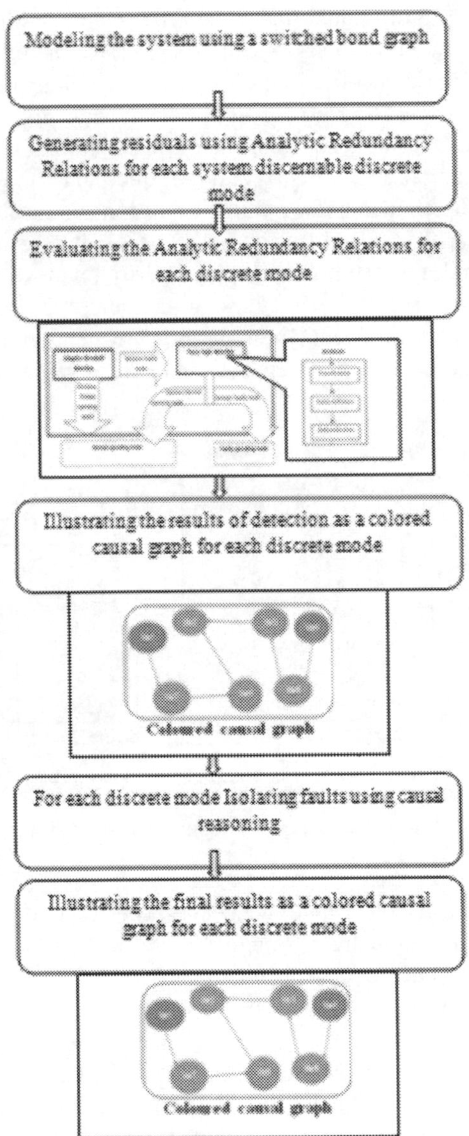

Fig. 2 The proposed approach

of the basic versions as they focus on cooperation rather than competition and are characterized by no selection:the PSO version with constriction factor. This version is also characterized by its speed of convergence [3].

For each discrete mode, the results of the detector combining the adaptive thresholdings and the fuzzy reasoning are displayed as a causal graph whose nodes are either red (suspected to be faulty) or green (normally functioning). Once at least one fault is detected, the isolation procedure is activated. This procedure is based on the causal graph reasoning and consists in looking for the source or the cause of the single or multiple detected faults. The details of such a reasoning are given in [6]. Finally and in order to assist the operator to make the proper corrective actions, the results are summed up as a colored causal graph.

3 Application of the Proposed Approach to an Industrial Process

The proposed solution is tested on a simulation of the three- tank hydraulic system.

3.1 Process Description

The considered hybrid process, shown in fig. 3, consists of three cylindrical tanks (Tank1, Tank2 and Tank3) that can be filled with two identical, independent pumps acting on the outer tanks 1 and 2.Tanks communicate through feeding valves that can assume either the completely open or the completely closed position. Pumps are controlled through on/off valves. The total number of valves is six. The liquid levels h1, h2, h3 in each tank represent continuous valued variables. The flow liquid rate from tank i to tank j is given by the following formula:

$$Q_{ij} = a_z \times S \times sgn(h_i - h_j) \times \sqrt{2 \times g \times |h_i - h_j|} \tag{1}$$

Where:

- hi (measured in meters) is the liquid level of tank i for i=1, 2, 3, respectively.
- az the outflow coefficient.
- S is the sectional area of the connecting valve.
- g the gravitational constant.

Fig. 3 The three-tank system [1]

The global purpose of the three- tank system is to keep a steady fluid level in the Tank 3, the one in the middle. A first step of our work concerns the modeling of the considered system. Thus, and as we previously mentioned, we rely on hybrid bond graph.

3.2 The System Hybrid Bond Graph Model

The hybrid bond graph model of the three- tank system is giving in fig.4. The tanks are modeled as capacitances and the valves are modeled as resistances. Msf1 and Msf2 correspond to flows volume applied to the system. 0- and 1- junctions represent respectively the common effort and common flow. The switching 1i-junctions represent idealized discrete switching element that can turn the corresponding energy connection on and off. These switching junctions are specified as a finite state sequential automata.

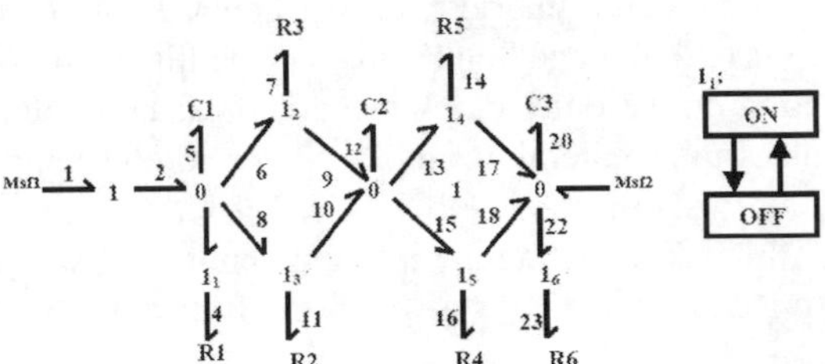

Fig. 4 The three-tank system Hybrid Bond Graph model

The levels of fluid in the tanks are all governed by continuous differential equations. These equations change as the valve configurations change. Then, a specific valve configuration determines the mode of the system. Since there are six valves, there are 64 total modes of the system. Each mode is governed by a different set of Analytic Redundancy Relations. For each mode of the three-tank system, the generation of Analytic Redundancy Relations is done directly from the bond graph model based on the procedure described in [21] and [9].

3.3 Experimental Results

The proposed approach was implemented on the three-tank hydraulic system in order to evaluate its performance. In this context, we considered as a first step forty discernable discrete modes whose evolution is given in fig.5. We performed a series of more than one hundred- forty tests (by injecting single and multiple faults) for several system functioning modes. At each test scenario, we checked if proper decision is finally given. The Simulation results we get are summed up in the following figures (fig. 6, fig.7 and fig. 8), knowing that the decision has the value of 1 in case of correct decision and 0 otherwise. According to simulation results (fig. 8) and

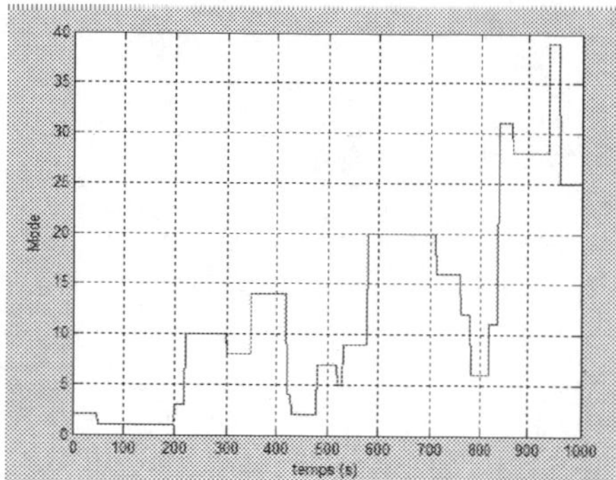

Fig. 5 Discrete modes evolution

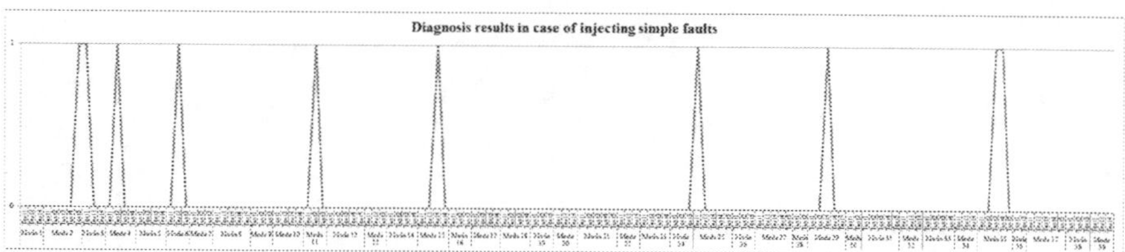

Fig. 6 Results in case of injecting simple faults

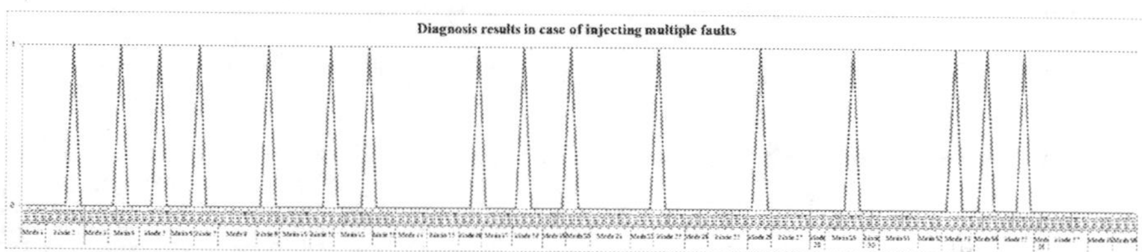

Fig. 7 Results in case of injecting multiple faults

the preliminary result got in [10], the proposed approach gives promising diagnosis results. In fact, it gives the correct decision in almost 88 % of tested cases. This result is obtained thanks to the combination of the use of adaptive thresholdings and fuzzy logic optimized by Particle Swarm Optimization technique. In fact, according to the results we get, the integration of these two detection techniques overcomes the limitations of individual strategies of each method and ameliorate significantly the diagnosis result. This can also explain by the use of causal reasoning in the localization step which analyzes the propagation paths in the graph to determine whether fault hypotheses are sufficient to account for other secondary faults, resulting from its propagation in the process over time. Then, only variables that are really faulty are announced defective [6]. The proposed approach provides us also with a representative results facilitating the operator's decision making thanks to use of colored causal graph displaying the systems variables state. For instance, in case of injecting

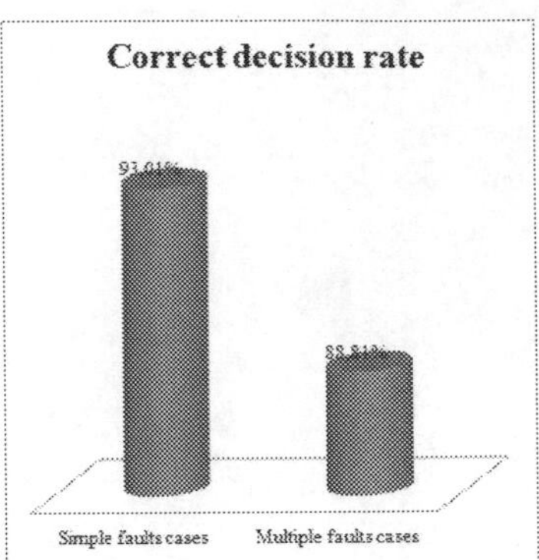

Fig. 8 Correct diagnosis rates

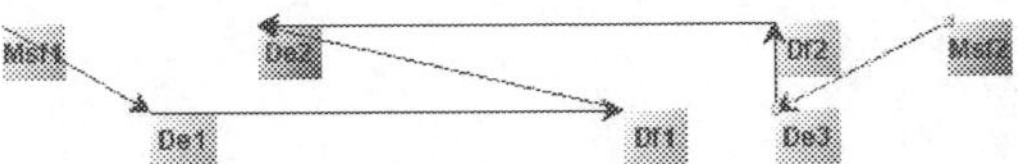

Fig. 9 Result of injecting {De2and Msf2} using the proposed diagnosing approach

multiple faults in De2 and Msf2 in the discrete mode in which all valves are in on functioning mode, the final findings are shown in fig.9.

4 Related Works and Discussion

A literature review shows that there are several approaches addressing the diagnosis of dynamic hybrid systems problem[2,13,16,17,19,20,22,23] based on extension of continuous systems approaches. We generally find two diagnosing options: works based on residual generation techniques and others based on causal reasoning. For instance, in [19], authors present a diagnosis methodology based on the use of a hybrid observer to track system behavior. The observer uses the state equations models for tracking continuous behavior in a mode, and hybrid automata for detecting and making mode transitions as system behavior evolves. Detection of mode changes requires access to controller signals for controlled jumps, and predictions of state variable values for autonomous jumps. If a mode change occurs in the system, the observer switches the tracking model (different set of state space equations), initializes the state variables in the new mode, and continues tracking system behavior with the new model. The fault detector compares the observations from the system and the predictions from the observer to look for significant deviations in the observed signals. In the same option, authors in[17] present a novel approach to monitoring and diagnosing real-time embedded systems that integrates model-based techniques using hybrid system models with distributed signature analysis. They present a framework for fault parameterization based on hybrid automata models.

The developed model is used to generate the fault symptom table for different fault hypotheses. The fault symptom table is generated off-line by simulation and is compiled into a decision tree that is used as the on-line diagnoser. The model compares observed sensor events with their expected values. When a fault occurs, the deviation from the simulated behavior triggers the decision-tree diagnoser. The diagnoser either waits for the next sensor event or queries the mode estimator to search for a particular event, depending on the next test. This approach has the advantage of detecting faults due to the continuous variables and the occurrence of disruptive events. In [4] the well known parity space approach is extended to hybrid systems in order to identify on-line the current mode and to estimate the switching instants. The fault detection consists in generating residuals between the input variables and measured outputs and analytical redundancy relations determined from the inputs and the outputs as well as their derivations, independently of the system discrete mode. The structured residuals are used to determine the current mode and to detect continuous and discrete faults. On the other hand, qualitative modeling by causal graph is proposed in [13] through a representation based on hybrid continuous causal Petri nets (HC2PN). Causal links (transitions) between continuous variables (the places) are represented through quality transfer functions quality (QFT) based on information on Gain (K), late (r) and time constant (r) (transitions). The evolution of the input variables and the qualitative response (QR) are at a QFT approximated by a piecewise affine function via a segmentation procedure. Each segment is called an episode. Crossing speed of a transition (change of marking) is a function of constant time piecewise, depending on the detected episodes, on the evolution of the marking of the upstream place and parameters of the QFT. The model HC2PN is then integrated into supervisor modeled by a Petri Nets through an interface event, forming a structure similar to Petri net models of Hybrids and monitoring approaches. The system fault detection, influencing continuous variables, is carried out asynchronously; fault location is performed by chaining backward / forward causal links between variables by using their temporal characteristics. Another approach based on causal reasoning has been proposed in [16]. This approach is based primarily on modeling the system by a hybrid bond graph model and then generates a graph of faults propagation, which can describe the temporal and causal relationships between different faults modes on one side, and observations related to another. This approach integrates the use of failure-propagation graph-based techniques for discrete-event diagnosis and combined qualitative reasoning and quantitative parameter estimation methods for parameterized fault isolation of degraded components (sensors, actuators, and plant components).

Unlike these works, the major contribution of our work consists in detecting and localizing multiple faults in dynamic hybrid systems using both generating residual reasoning and causal reasoning. In fact, we rely in the detection step on the generation of analytical redundancy relations (ARRs) in each discrete discernable mode. These ARRs are generated through the using of hybrid bond graph and are evaluated using a combination of adaptive thresholdings and fuzzy logic optimized using PSO. Exploiting fuzzy reasoning allows us to get the most accurate decision even if residuals are affected by the noise contamination and uncertainty effects. On the other hand, the results of detection step are summed up as a colored causal graph which is used in the isolation step. Indeed, isolation step consists in generating a causal propagation reasoning. As it uses a backward/forward procedure starting from an inconsistent variable localized as a faulty node: red node in the generated colored causal graph. The use of colored causal graph to display the diagnosis results facilitates the operator's understandings and helps him to make adequate corrective actions in time.

5 Conclusion

This paper addresses the problem of multiple faults in dynamic hybrid systems. These systems are assumed to have slow discernable discrete modes characterized by continuous behaviors.The continuous behaviors could be affected by single or multiple faults. Thus, we propose, in this paper, a general approach to diagnose single and multiple faults in dynamic hybrid systems relying on extension of previous works dealing with continuous systems. The proposed approach exploits the performance of combining adaptive thresholdings and fuzzy logic reasoning optimized using Particle Swarm Optimization. Experiments are based on the case of the hybrid dynamic system: three-tank hydraulic system, considered as a benchmark in the diagnosis field. They have proven the efficacy of the proposed approach. We intend in future work to consider faults affecting the discrete mode evolution of dynamic hybrid systems

References

1. Amira: Gesellschaft für angewandte Mikroelektronik, Regelungstechnik und Automation mbH. Laborversuche für Forschung und regelungstechnische Praktika, www.amira.de
2. Balluchi, A., Benvenuti, L., Di Benedetto, M.D., Sangiovanni-Vincentelli, A.L.: Design of Observers for Hybrid Systems. In: Tomlin, C.J., Greenstreet, M.R. (eds.) HSCC 2002. LNCS, vol. 2289, pp. 76–89. Springer, Heidelberg (2002)
3. Clerc, M.: The Swarm and the Queen: Towards A Deterministic and Adaptive Particle Swarm Optimization. In: Proceedings of the Congress of Evolutionary Computation, Washington, DC, pp. 1951–1957 (1999)
4. Cocquempot, V., Mezyani, T.E., Staroswieckiy, M.: Fault detection and isolation for hybrid systems using structured parity residuals. In: Asian Control Conference, ASCC 2004, New Mexico, vol. 2, pp. 1204–1212 (2004)
5. Dauphin-Tanguy, G.: Les bond graph. Hermès Sciences Publications (2000)
6. Fliss, I., Tagina, M.: Multiple faults diagnosis using causal graph. In: The Proceeding of The 6th IEEE International Multi Conference on Systems, Signals Devices, SSD 2009, Djerba, Tunisia, March 23-26 (2009)
7. Fliss, I., Tagina, M.: Multiple faults fuzzy detection approach improved by Particle Swarm Optimization. In: The 8th International Conference of Modelling and Simulation, MOSIM 2010, Hammamet, Tunisia, May 10-12, vol. 1, pp. 592–601 (2010)

8. Fliss, I., Tagina, M.: Multiple faults model-based detection and localisation in complex systems. Journal of System Decision (JDS) 20(1), 7–31 (2011) ISSN 1246-0125 (Print), 2116-7052 (Online)
9. Fliss, I., Tagina, M.: Exploiting Particle Swarm Optimization in Multiple Faults Fuzzy Detection. Journal of Computing 4(2), 80–91 (2012)
10. Fliss, I., Tagina, M.: Exploiting fuzzy reasoning Optimized by Particle Swarm Optimization and Adaptive thresholding to diagnose multiple faults in Dynamic Hybrid Systems. Accepted in 2012 International Conference on Communications and Information Technology, ICCIT 2012, Hammamet, Tunisia, June 26-28 (2012)
11. Frank, P.M., Ding, X.: Survey of robust residual generation and evaluation methods in observer-based fault detection systems. J. Proc. Control 7(6), 403–424 (1997)
12. Gertler, J.: Fault detection and isolation using parity relations. Control Engineering Practice 5(5), 653–661 (1997)
13. Gomaa, M., Gentil, S.: Hybrid industrial dynamical system supervision via hybrid continuous causal petri nets. In: IEEESMC IMACS Symposium on Discrete Events and Manufacturing Systems, CESA 1996, Lille, France, pp. 380–384. Springer (1996)
14. Hôfling, T., Isermann, R.: Fault detection based on adaptive parity equations and singleparameter tracking. Control Engineering Practice 4(10), 1361–1369 (1984)
15. Isermann, R.: Process fault detection based on modeling and estimation methods-A survey. Automatica 20(4), 387–404 (1984)
16. Karsai, G., Abdelwahed, S., Biswas, G.: Integrated diagnosis and control for hybrid dynamic systems (2003)
17. Koutsoukos, X., Zhao, F., Haussecker, H., Reich, J., Cheung, P.: Fault modeling for monitoring and diagnosis of sensor-rich hybrid systems. In: Proceedings of the 40th IEEE Conference on Decision and Control, pp. 793–801 (2001)
18. Mosterman, P.J.: Hybrid dynamic systems: a hybrid bond graph modeling paradigm and its application in diagnosis, Thesis (1997)
19. Narasimhan, V.S., Biswas, G., Karsai, G., Pasternak, T., Zhao, F.: Building Observers to Handle Fault Isolation and Control Problems in Hybrid Systems. In: Proc. 2000 IEEE Intl. Conference on Systems, Man, and Cybernetics, Nashville, TN, pp. 2393–2398 (2000)
20. Olivier-Maget, N., Hétreux, G., Le Lann, J.M., Le Lann, M.V.: Model-based fault diagnosis for hybrid systems: Application on chemical processes. Computers and Chemical Engineering 33, 1617–1630 (2009)
21. Tagina, M., Cassar, J.P., Dauphin-Tanguy, G., Staroswiecki, M.: Bond Graph Models For Direct Generation of Formal Fault Detection Systems. International Journal of Systems Analysis Modelling and Simulation 23, 1–17 (1996)
22. Vento, J., Puig, V., Serrate, R.: Fault detection and isolation of hybrid system using diagnosers that combine discrete and continuous dynamics. In: Conference on Control and Fault Tolerant Systems, Nice, France (2010)
23. Xu, J., Loh, A.P., Lum, K.Y.: Observer-based fault detection for piecewise linear systems: Continuous-time cases. In: IEEE International Conference on Control Applications, CCA 2007, Singapore, pp. 379–384 (2007)

Investigation of Short Base Line Lightning Detection System by Using Time of Arrival Method

Behnam Salimi, Zulkurnain Abdul-Malek,
S.J. Mirazimi, and Kamyar MehranZamir

Abstract. Lightning locating system is very useful for the purpose of giving exact coordinates of lightning events. However, such a system is usually very large and expensive. This project attempts to provide instantaneous detection of lightning strike using the Time of Arrival (TOA) method of a single detection station (comprises of three antennas). It also models the whole detection system using suitable mathematical equations. The measurement system is based on the application of mathematical and geometrical formulas. Several parameters such as the distance from the radiation source to the station and the lightning path are significant in influencing the accuracy of the results (elevation and azimuth angles). The signals obtained by all antennas were analysed using the LabVIEW software. Improvements in the lightning discharge locating system can be made by adopting a multi-station technique instead of the currently adopted single-station technique.

Keywords: Time of arrival, Short base line, Lightning locating system.

1 Introduction

Nowadays, the higher the rate of urbanism and building construction especially in tropical areas, the bigger concern on the safety of facility and human beings due to lightning strikes. Issues related to lightning locating systems are actively being researched. The research is very useful for the purpose of human safety and for the lightning protection system. It can also benefit the insurance companies and weather forecast organizations. Step leaders propagate electromagnetic waves in the range

---

Behnam Salimi · Zulkurnain Abdul-Malek · S.J. Mirazimi · Kamyar MehranZamir
Institute of High Voltage and High Current (IVAT), Universiti Teknologi Malaysia, 81310 Johor Bahru, Malaysia
e-mail: {sbehnam3,mkamyar3}@live.utm.my, zulk@fke.utm.my, j_mirazimi@yahoo.com

A. Abraham and S.M. Thampi (Eds.): Intelligent Informatics, AISC 182, pp. 141–147.
springerlink.com

of kHz-GHz within an electrical discharge [8]. There are several available methods to analyse and locate lightning signals such as the time of arrival [7], magnetic direction finding, and interferometry methods. One of the best techniques to improve the accuracy of the detection is to combine two or more methods as one measurement system [3]. A new technique to estimate the location of lightning strike with a better accuracy, based on the measurement of induced voltages due to lightning in the vicinity of an existing overhead telephone line is proposed [1]. In this work, the TOA method utilising three broadband antennas is used for lightning locating due to its many advantages.

2 Methods

The geometry of the installed antennas is demonstrated in Figure 1. This system consists of three circular plate antennas. They are placed 14.5 meter apart to form two perpendicular base lines. The antenna output signals were fed into a four-channel digital oscilloscope (Tektronix MSO4104) operating at 8-bit, 5Gs/s using three 50 m long coaxial cables (RG 59, 75 Ω).

The TOA method detects the electromagnetic waves arrival at the antennas and computes the time difference of arrival. To accomplish this, the detected signals should be properly captured and stored. Generally, the amount of data storage involved is huge and costly. The sequential triggering method had been used to overcome the problem [4]. Various methods can be used to analyse the captured waves. In this work, LabVIEW software based cross correlation method is implemented to calculate the time delays. The LabVIEW software has the advantage of low cost and short processing time [2].

With the help of several geometric formulas, the elevation and azimuth angles of the radiation source can be settled. Together these angles specify the locus of the radiation source [6].

2.1 Direction Analysis

The fundamental concept of this TOA technique is to determine the time delay of arrival between signals impinging on a pair of antenna. The simple TOA is composed of two antennas. Consider two broadband antennas set apart in a horizontal position on the ground by a distance d, as shown in Figure 2. The signals which come from a common source detected by antenna 1 is r1 (t) and by antenna 2 is r2 (t). Assuming that the radiation source is very far compared to the distance d, the incident angle can be expressed by:

$$\theta = \cos^{-1}\left(\frac{c.\triangle t}{d}\right) \tag{1}$$

Fig. 1 The location of 1st, 2nd and 3rd antennas

Where 'c' is the light speed in space ($3*10^8$ m/s) and it is the time delay of arrival. By applying two-antenna sensors, only one dimension localization can be obtained [5].

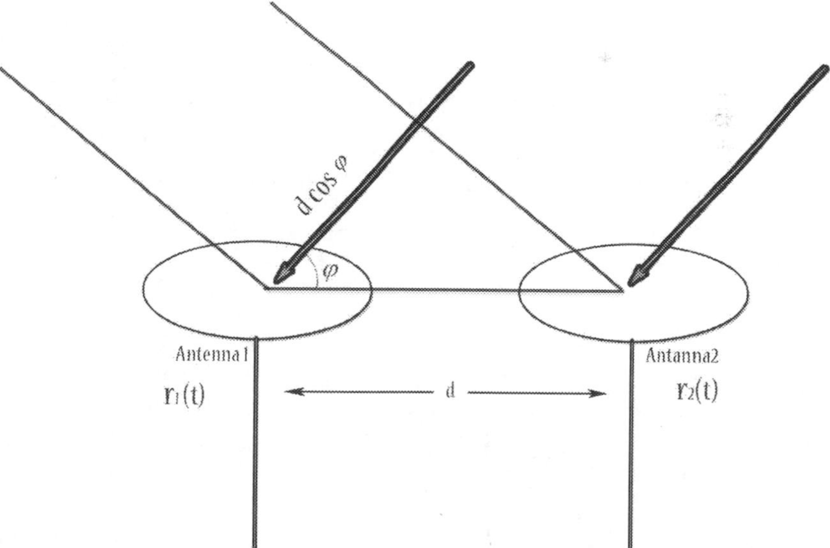

Fig. 2 Direction of radiation source estimated using two antennas sensors in TOA technique

2.2 Direction Finding by Three Antennas

By implementing two-antenna sensors, only one-dimension localization can be obtained. To provide the location in a two-dimension (2D), a third antenna should be added. This extra antenna can determine the elevation and azimuth angles. The first and second antennas form the first base line, while the second and third antennas form the second base line. These two base lines are perpendicular to each other. This is shown in Figure 3.

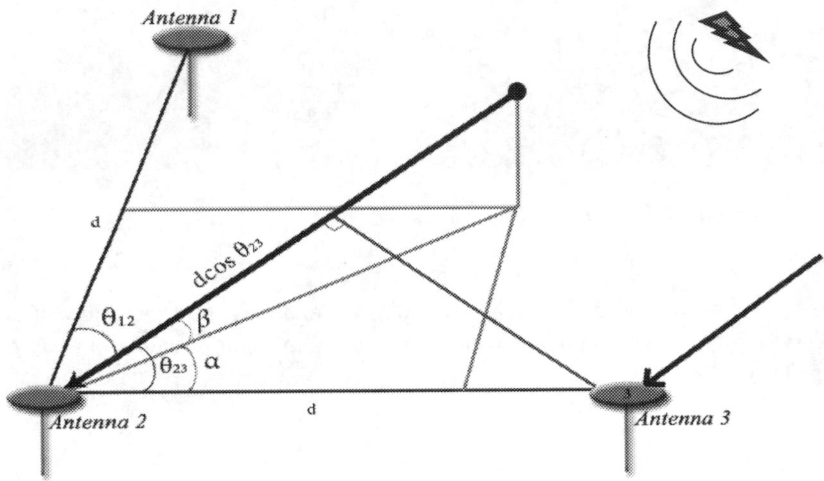

Fig. 3 The geometry of three antennas position and the radiation source direction

As can be seen in equations (2) and (3) the time differences between antennas 1, 2 and 3, are as follows:

$$t_{21} = t_2 - t_1 \tag{2}$$

$$t_{23} = t_2 - t_3 \tag{3}$$

Here, t_1, t_2, and t_3 are the arrival times of signals at antennas 1, 2, and 3, respectively. With the help of Figure 3, the incident angles of the radiation source for the base lines of antennas 2 and 1 (θ_{21}), and antennas 2 and 3 (θ_{23}), can be determined by the following relations:

$$t_{21} = \frac{d\cos\theta_{21}}{c} \tag{4}$$

$$t_{23} = \frac{d\cos\theta_{23}}{c} \tag{5}$$

It is more comprehensible to describe the lightning source in the elevation and azimuth format compared to the incident angles. Hence, the incident angles obtained by above equations were converted to elevation and azimuth. From Figure 3:

$$\cos\theta_{21} = \cos\beta \sin\alpha \tag{6}$$

$$\cos\theta_{23} = \cos\beta \cos\alpha \tag{7}$$

Using equations (4) to (7) the elevation (β) and azimuth (α) can be obtained.

$$\alpha = \tan^{-1}\left(\frac{t_{21}}{t_{23}}\right) \tag{8}$$

$$\beta = \cos^{-1}\left(\frac{c\sqrt{t_{21}^2 + t_{23}^2}}{d}\right) \tag{9}$$

3 Results

The following signals (shown in Figure 4) were captured on 16 April 2011 (during thunderstorm in a total of one hour duration of a lightning event), one lightning sample was analysed by measuring the peak voltage, the front time, and the decay time The maximum peak voltage for this event is 15.5 V. Figure 5 and 6 illustrate the calculated azimuth and elevation angles of lightning signals based on mentioned formula (α, β) in time domain. The amplitude is in degrees. Although figures 5 and 6 displays the location of radiation source, it can be observed that the quantity of degree is fluctuated between 60 to 90 and -15 to 40 and these transforms are due to

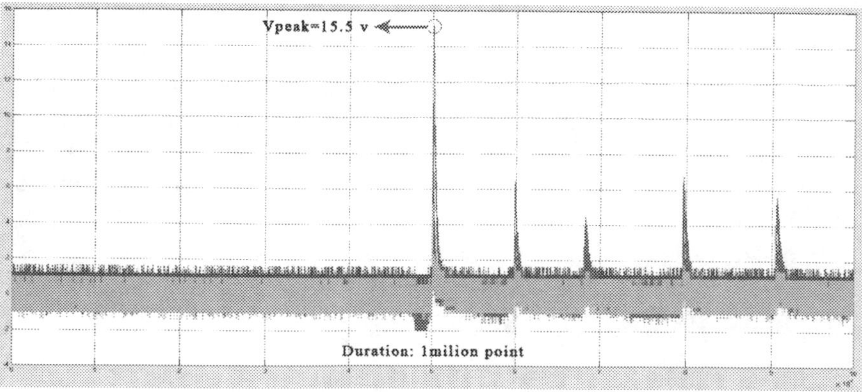

Fig. 4 Signal recorded on 16 April 2011

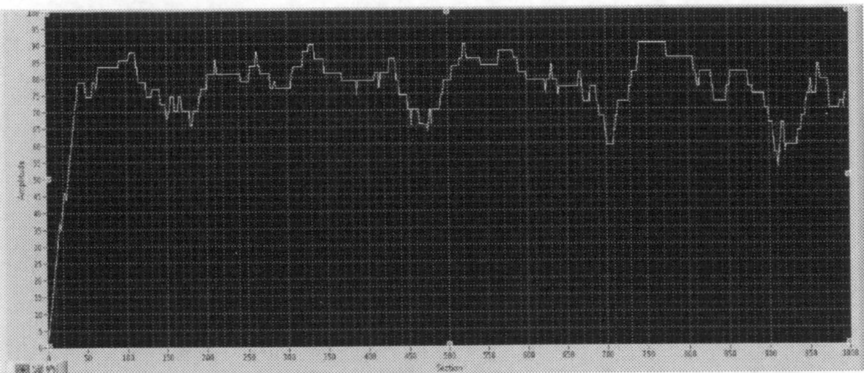

Fig. 5 The variation of the elevation of lightning discharge (using cross-correlation in time domain)

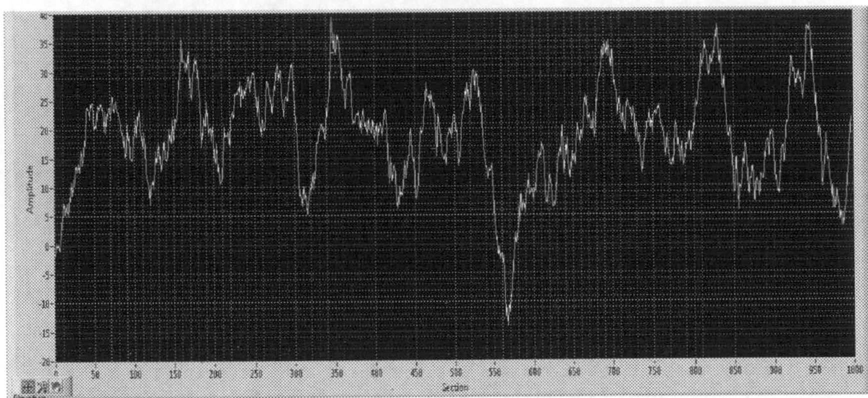

Fig. 6 The variation of the azimuth of lightning discharge (using cross-correlation in time domain)

some reasons such as rapid changing in lightning path, noise affections, and signal interferences. In this work, using mathematical simulation with LabView, it is shown that, the TOA method is roughly accurate to calculate azimuth and elevation. The distance between antennas, and also the use of long cables were cause of problems which affect the results.

4 Conclusion

A circular plate antenna system for locating the cloud-to-ground lightning strike has been utilization short base-line configuration. The time domain signal analysis was conducted to determine the time difference of the broadband VHF electromagnetic pulses detected by the sensors. The cross correlation technique was applied to maintain the high resolution analyses. From the experiment that was conducted, the TOA method is suitable for lightning locating system. However, there are many considerations before and during detection of lightning location. The lightning position due to lightning detection must be within certain area which is not affected by other signals or noises. Besides that, the use of long cables affect the result because the cable itself can become sensor which detect signals from lightning or noise. The system can be said to successfully map a cloud-to-ground lightning discharge in 2D mode.

Acknowledgements. The authors would like to thank Universiti Teknologi Malaysia (Research Vot No. 4C022) and Tenaga Nasional Berhad, Malaysia for funding and supporting this research.

References

[1] Aulia, Malek, Z.A., Adzis, Z., Novizon: A new localised lightning locating system utilising telecommunication subscriber lines. In: IEEE 2nd International Power and Energy Conference, PECon 2008, pp. 403–407 (2008)
[2] Jianhua, K., Baoqiang, W., Jie, G., Yanjie, W.: Research on lightning location method based on labview. In: 8th International Conference on Electronic Measurement and Instruments, ICEMI 2007, pp. 3-86–3-90 (2007)
[3] Kulakowski, P., Vales-Alonso, J., Egea-López, E., Ludwin, W., García-Haro, J.: Technical communication: Angle-of-arrival localization based on antenna arrays for wireless sensor networks. Computers and Electrical Engineering 36(6), 1181–1186 (2010)
[4] Mardiana, R., Kawasaki, Z.: Broadband radio interferometer utilizing a sequential triggering technique for locating fast-moving electromagnetic sources emitted from lightning. IEEE Transactions on Instrumentation and Measurement 49(2), 376–381 (2000)
[5] Mardiana, R., Meiladi, E.: A technique for lightning recontructions using short-baseline broadband time-of-arrival. In: Proceedings of the 14th Asian Conference on Electrical Discharge, November 23-25 (2008)
[6] Mashak, S.V., Afrouzi, H.N., Abdul-Malek, Z.: Simulation of lightning flash in time of arrival (toa) method by using three broadband antennas. In: 5th European Symposium on Computer Modeling and Simulation EMS, pp. 287–292. IEEE (2011)
[7] Roberts, R.: Tdoa localization techniques. IEEE P80215 Working Group for Wireless Personal Area Networks, WPANs (2004)
[8] Tantisattayakul, T., Masugata, K., Kitamura, I., Kontani, K.: Broadband vhf sources locating system using arrival-time differences for mapping of lightning discharge process. Journal of Atmospheric and Solar-Terrestrial Physics 67, 1031–1039 (2005)

Investigation on the Probability of Ferroresonance Phenomenon Occurrence in Distribution Voltage Transformers Using ATP Simulation

Zulkurnain Abdul-Malek, Kamyar MehranZamir, Behnam Salimi, and S.J. Mirazimi

Abstract. Ferroresonance is a complex non-linear electrical phenomenon that can make thermal and dielectric problems to the electric power equipment. Ferroresonance causes overcurrents and overvoltages which is dangerous for electrical equipment. In this paper, ferroresonance investigation will be carried out for the 33kV/110V VT at PMU Kota Kemuning, Malaysia using ATP-EMTP simulation. Different preconditions of ferroresonance modes were simulated to ascertain possible ferroresonance conditions in reality compare with simulated values. The effect of changing the values of series capacitor is considered. The purpose of this series of simulations is to determine the range of the series capacitance value within which the ferroresonance is likely to occur.

Keywords: Ferroresonance, EMTP, Voltage Transformers, Over-voltages, Overcurrents.

1 Introduction

The term 'ferroresonance' has appeared in publications dating as far back as the 1920s, and it refers to all oscillating phenomena occurring in an electrical circuit which contains a non-linear inductor, a capacitor and a voltage source [1, 7]. The first step in understanding the ferroresonance phenomenon is to begin with the 'resonant' condition. Resonance can be explained by using a simple RLC circuit as shown in Figure 1.

This linear circuit is resonanting when at some given source of frequency the inductive (X_L) and capacitive (X_C) reactance cancel each other out. These impedance

Zulkurnain Abdul-Malek · Kamyar MehranZamir · Behnam Salimi · S.J. Mirazimi
Institute of High Voltage and High Current (IVAT), Universiti Teknologi Malaysia, 81310 Johor Bahru, Malaysia
e-mail: zulk@fke.utm.my, {mkamyar3, sbehnam3}@live.utm.my, j_mirazimi@yahoo.com

A. Abraham and S.M. Thampi (Eds.): Intelligent Informatics, AISC 182, pp. 149–155.
springerlink.com

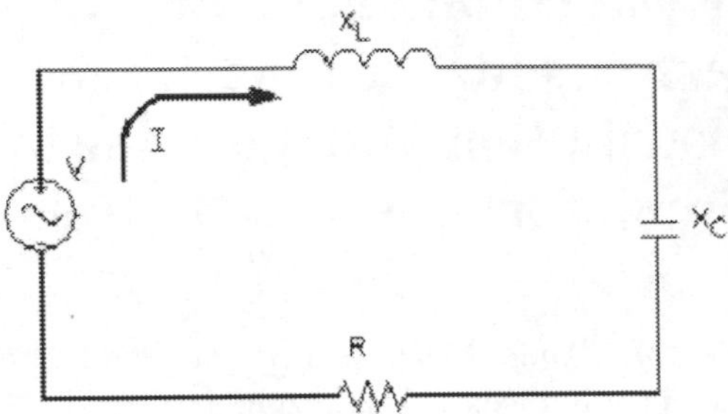

Fig. 1 RLC circuit for explaining ferroresonance

values can be predicted and change the with frequency. The current (I) in the circuit depends on the resistance (R). If this resistance is small, then the current can become very large in the RLC circuit. If the inductor in Figure 1 is replaced by an iron cored non-linear inductor, the exact values of voltage and current cannot be predicted as in a linear model. The inductance becomes nonlinear due to saturation of flux in the iron core. The understanding that ferromagnetic material saturates is very important. Ferromagnetic material has a property of causing an increase to the magnetic flux density, and therefore magnetic induction [3, 5, 6].

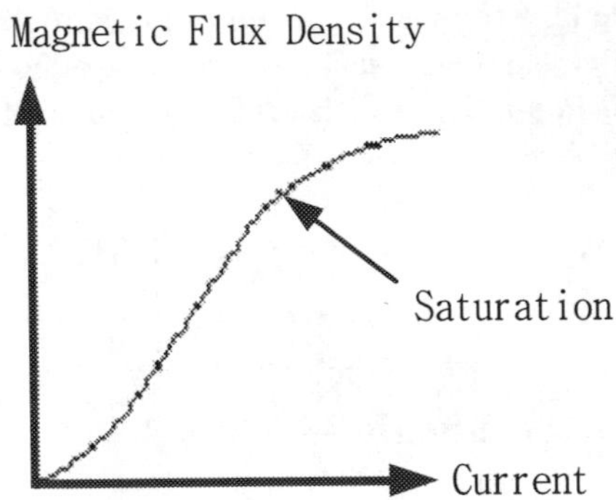

Fig. 2 Magnetization curve

As the current is increased, so does the magnetic flux density until a certain point where the slope is no longer linear, and an increase in current leads to smaller and smaller increases in magnetic flux density. This is called the saturation point. Figure 2 shows the relationship between magnetic flux density and current. As the current increases in a ferromagnetic coil, after the saturation point the inductance of the coil changes very quickly. This causes the current to take on very dangerously high values. It is these high currents that make ferroresonance very damaging. Most transformers have cores made from ferromagnetic material. This is why ferroresonance is a concern for transformer operation [2, 4].

2 Simulation Model

TNB(Tenaga Nasional Berhad, Malaysia) Transmission has had several failures of 33kV voltage transformers (VT) in the system. The simplified single line diagram for the 132/33 kV PMU Kota Kemuning is shown in Figure 3. The Engineering Services Department TNB Transmission Division reported that at 00:45 hour, PMU Kota Kemuning 3T0 (Fig. 3) tripped due to the explosion of 33 kV red phase VT. The equipment detail is shown in Table 1.

Table 1 VT detailed specifications

VT Type	UP 3311
From	V50
Accuracy Class	0.5
Primary Voltage	33/3
Secondary Voltage	110/3
Rated output	100 VA
Standard	BS 3941
Insulation Level	36/70/170 kV
Number of phase	1
Frequency	50 Hz
V.F	1.2 Cont 1.9 30 SEC

It was also reported that there were cracks on the VT and parts of it had chipped off. All three VT fuses of 33kV Incomer at the Double Busbar Switchgear had opened circuited and the screw cap contact surface with the termination bars was badly pitted.

The system arrangement shown in Figure 3 can be effectively reduced to an equivalent ferroresonant circuit as shown in Figure 4.

The sinusoidal supply voltage (e) is coupled to the VT through a series capacitor Cseries. The VT's high voltage winding shunt capacitance to ground can greatly contribute to the value of C_{shunt}. The resistor R is basically made up of the VT's equivalent magnetizing branch resistance (core loss resistance). The nonlinear inductor is represented by a nonlinear flux linkage (λ) versus current (i) curve (Fig. 4).

3 Simulation of Capacitance Precondition

The simulated circuit in ATP is shown in Figure 5. The switch can represent the circuit breaker or the disconnection of the fuse due to its operation. After the circuit breaker or the fuse opens, it is proposed that the supply voltage can still be coupled to the VT through equivalent series capacitance, C_{series}.

The voltage transformer was modeled as a nonlinear inductor in parallel with a resistance in the magnetizing branch (Rc). The circuit opening is represented by a time controlled switch. The values of all circuit components in Figure 5 were

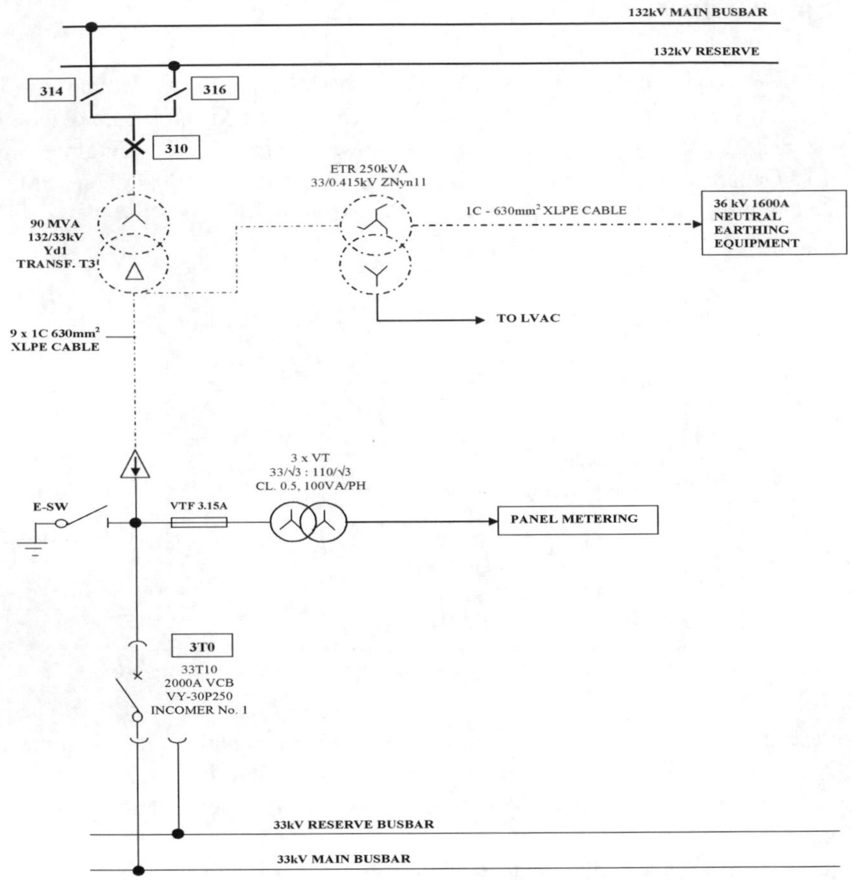

132kV MAIN BUSBAR

132kV RESERVE

314

316

310

ETR 250kVA

33/0.415kV ZNyn11

1C - 630mm² XLPE CABLE

36 kV 1600A
NEUTRAL
EARTHING
EQUIPMENT

90 MVA
132/33kV
Yd1
TRANSF. T3

TO LVAC

9 x 1C 630mm²
XLPE CABLE

3 x VT
33/√3 : 110/√3
CL. 0.5, 100VA/PH

E-SW

VTF 3.15A

PANEL METERING

3T0

33T10
2000A VCB
VY-30P250
INCOMER No. 1

33kV RESERVE BUSBAR

33kV MAIN BUSBAR

Fig. 3 Single line diagram of the substation

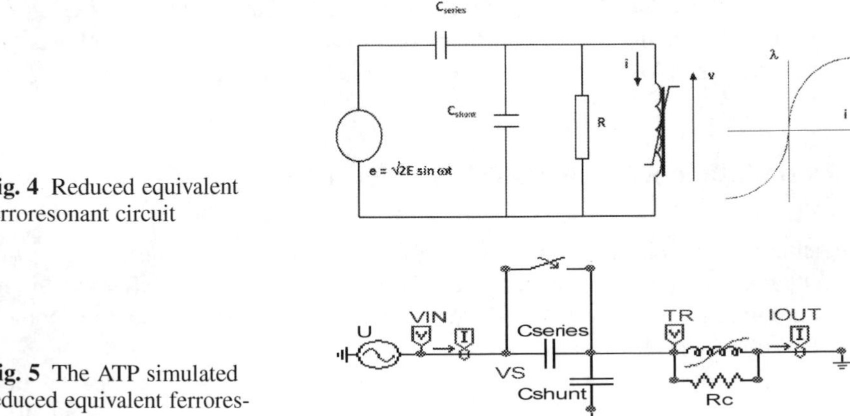

Fig. 4 Reduced equivalent ferroresonant circuit

Fig. 5 The ATP simulated reduced equivalent ferroresonant circuit

Table 2 Circuit Components Values

Parameter	Measured value
CHV-gnd	97.4 pF
DF of CHV-gnd	51.88 %
CHV-LV	640.7 pF
DF of CHV-LV	0.938 %
CLV-gnd	328.7 pF
DF of CLV-gnd	7.462 %
Rc	16.9 MΩ (calculated from open circuit test data)

Table 3 The effect of changing the value of series capacitor

Cseries (pF)	Peak Voltage at Transformer(kV) (before) (after)	Peak Current at Transformer(mA) (before) (after)	Frequency of System (Hz) (before) (after)	Ferroresonance Occur
8000	(26.944) (27.941)	(4.136) (3.947)	(50)(50)	NO
4000	(26.943) (28.895)	(4.137) (4.483)	(50)(50)	NO
2000	(26.943) (32.884)	(4.137) (5.612)	(50)(50)	NO
1500	(26.943) (35.297)	(4.135) (5.951)	(50)(50)	Yes
1000	(26.943) (52.429)	(4.136) (47.166)	(50)(50)	Yes
500	(26.944) (44.228)	(4.136) (20.813)	(50)(50)	Yes
350	(26.943) (41.609)	(4.137) (9.143)	(50)(50)	Yes
200	(26.943) (24.275)	(4.137) (3.548)	(50)(50)	NO
100	(26.943) (8.846)	(4.134) (1.245)	(50)(50)	NO
50	(26.943) (3.035)	(4.135) (5.561)	(50)(50)	NO

determined based on as far as possible the actual parameters. The following values in Table 2 were obtained from measurements made on the VT.

The simulation was carried out with a fixed value of shunt capacitor (C_{shunt}) at 97.4 pF and the value of resistance in magnetizing branch (Rc) at 16.9 MΩ. The circuit was supplied by AC source peak voltage of 26.94 kV with 50 Hz frequency. The time controlled switch was closed at 0 sec and disconnected after 0.25 sec. Table 3 shows the effect of variation in the series capacitor values. The peak voltage and the peak current at the VT were recorded before and after the switch operation. The time from 0 sec until 0.25 sec was considered as the before switch opening, and the remaining time was considered as after. The output waveforms for 1000 pF series capacitor where ferroresonance has occurred is shown in Figure 6. Figure 7 and Figure 8 show the output waveforms for 100 pF and 2000 pF series capacitor value where ferroresonance has not occurred.

It is clear from Figure 8 which shows that ferroresonance does not occur at 2000 pF of series capacitor although there is a small changes in the measured voltage and current of the system around disconnected time.

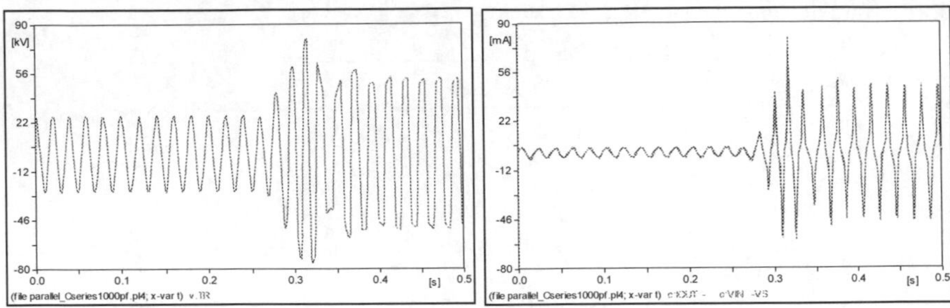

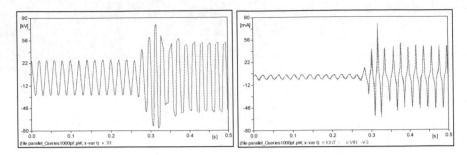

Fig. 6 The output waveform for 1000 pF series capacitor

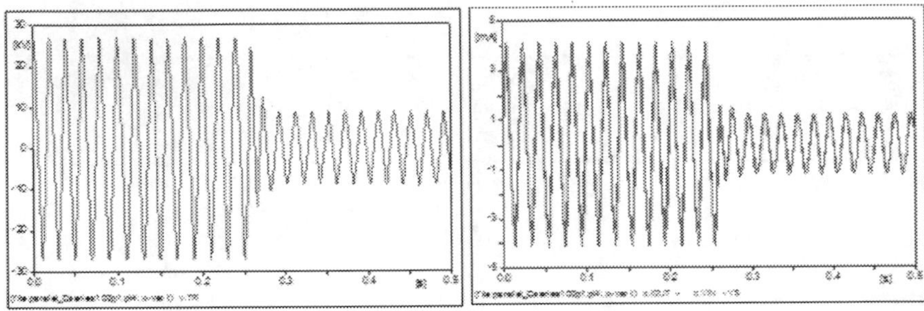

Fig. 7 The output waveform for 100 pF series capacitor

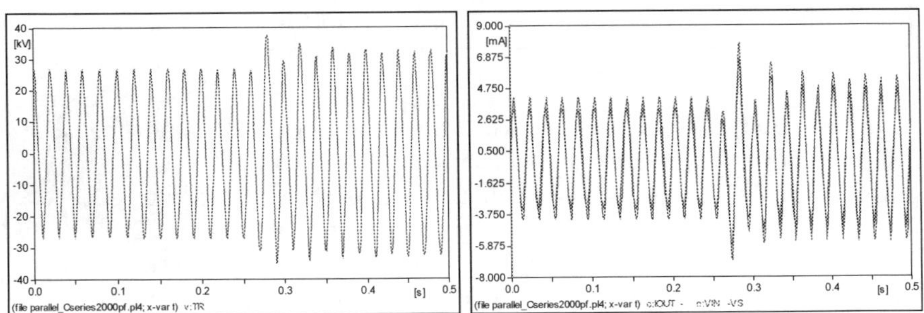

Fig. 8 The effect of changing series capacitor value for parallel RLC circuit by fixed the value of C_{shunt} at 97.4 pF and the value of Rc at 16.9 MΩ. C_{series} of 2000 pF

4 Conclusion

From the simulations it can be concluded that the value of series capacitor should be in the range of about 400 pF to 1400 pF (for 3-times magnification) in order for the ferroresonance to occur. As mentioned earlier, if the ferroresonance occurs due to a switching operation to disconnect the supply from the VT, the possible sources of this capacitor are the intercable capacitance, the busbar-VT capacitance, the opened circuit breaker capacitance, or the opened fuse capacitance. It can be said that, this range of values of series capacitance is relatively large compared to physically realised values. Therefore it can be concluded that once the VT is disconnected from the supply, there is no possibility of ferroresonance to occur since the supply voltage pre-condition cannot be physically met. In the simulation with C_{series} variation, it was found that there is no possibility of ferroresonance to occur due to disconnection of supply from the VT. This is due to the very high value of the series capacitance which is required.

Acknowledgements. Authors wish to thank Universiti Teknologi Malaysia (Research Vot No. 4C022) and Tenaga Nasional Berhad, Malaysia (TNB Engineering and TNB Distribution, Perak) for their assistance in the study.

References

[1] Ferracci, P.: Ferroresonance (2012), `http://www.schneiderelectric.com.au/sites/australia/encustomers/designers/designers.page`
[2] Charalambous, C., Wang, Z., Jarman, P., Sturgess, J.: Frequency domain analysis of a power transformer experiencing sustained ferroresonance. Generation, Transmission Distribution, IET 5(6), 640–649 (2011)
[3] Li, H., Fan, Y., Shi, R.: Chaos and ferroresonance. In: Canadian Conference on Electrical and Computer Engineering, CCECE 2006, pp. 494–497 (2006)
[4] Mokryani, G., Haghifam, M.R., Latafat, H., Aliparast, P., Abdollahy, A.: Analysis of ferroresonance in a 20kv distribution network. In: 2009 2nd International Conference on Power Electronics and Intelligent Transportation System, PEITS, vol. 1, pp. 31–35 (2009)
[5] Moses, P., Masoum, M., Toliyat, H.: Impacts of hysteresis and magnetic couplings on the stability domain of ferroresonance in asymmetric three-phase three-leg transformers. IEEE Transactions on Energy Conversion 26(2), 581–592 (2011)
[6] Rezaei-Zare, A., Mohseni, H., Sanaye-Pasand, M., Farhangi, S., Iravani, R.: Performance of various magnetic core models in comparison with the laboratory test results of a ferroresonance test on a 33 kv voltage transformer. In: Power Engineering Society General Meeting, p. 8. IEEE (2006)
[7] Tong, Z.E.: Ferroresonance experience in uk: Simulations and measurements. In: Power System Transients, IPST (2001)

Design of SCFDMA System Using MIMO

Kaushik Kapadia and Anshul Tyagi

Abstract. The aim of paper is to design SCFDMA system using SFBC and receiver diversity which provides satisfactory performance over fast fading channel environment. The performance evaluation will be checked through MATLAB R2009b simulator. There are comparisons of performance among 1x1 SCFDMA system, 2x1 SCFDMA system using SFBC, 1x2 SCFDMA system using receiver diversity and 2x2 SCFDMA system using SFBC and receiver diversity. We describe design of SCFDMA system using transmitter diversity technique and receiver diversity technique. We have compared the performance of these systems with the conventional SCFDMA system. The main focus is on design of SCFDMA system using SFBC and receiver diversity which enables desired system to combat detrimental effects of fast fading.

Keywords: 3rd Generation Partnership Project Long Term Evolution(3GPP-LTE), Binary Phase Shift Keying(BPSK), Discrete Fourier Transform(DFT), Inverse Discrete Fourier Transform(IDFT), Localized Frequency Division Multiple Access(LFDMA), Maximum Ratio Receiver Combining(MRRC), Multiple Input Multiple Output(MIMO), Orthogonal Frequency Division Multiple Access (OFDMA), Peak to average power ratio(PAPR), Single Carrier Frequency Division Multiple Access(SCFDMA), Space Frequency Block Code(SFBC).

1 Introduction

SCFDMA system is used in 3GPP-LTE and upcoming generation mobile communication. That is due to its possession of advantages of OFDMA as well as significant decrease in PAPR [1]. SCFDMA is a method of wireless communication under consideration to be deployed in future cellular systems. Advantages over OFDMA include less sensitivity to carrier frequency offsets [2,3] and lower PAPR [4]. The PAPR is not a big issue for base station as there is unlimited availability of power supply; hence in 3GPP-LTE OFDM is used for forward link

Kaushik Kapadia · Anshul Tyagi
Communication Systems, E&C Dept.
Indian Institute of Technology, Roorkee, Uttarakhand, India
e-mail: kvkapcisco@gmail.com, tyagifec@iitr.ernet.in

A. Abraham and S.M. Thampi (Eds.): Intelligent Informatics, AISC 182, pp. 157–164.
springerlink.com

communication that is from base station to mobile terminals. But PAPR creates more challenges to the instrument running on external limited power supplied batteries. For systems with high PAPR such as multicarrier system (OFDMA) there is need to design extremely linear power amplifier. But design of extremely linear power amplifier enhances cost of system to larger extent. Hence 3GPP-LTE uses SCFDMA technique for reverse link communication, that is from mobile terminals to base station. Though SCFDMA technique is able to reduce PAPR yet the need to increase the reliability of communication and data rate handling capability requires the use of MIMO techniques. SFBC is one of the MIMO techniques used for achieving diversity gain. SCFDMA may be used in downlink satellite communication, due to lack of power sources available at satellite, there is requirement to minimize value of PAPR.

SCFDMA system uses Multiuser detection algorithms for avoiding interference from neighbouring cells at cell edge [5,6]. Scheduling algorithms (channel dependant and static scheduling) are used to allocate resource to user based on channel conditions or in fixed fashion [7,8]. Power allocation to subcarriers of a specific user is constant, it follows maximum subcarrier power constraint and total power constraint.

Section 2 gives a description of 2x2 SCFDMA system using SFBC and MRRC techniques. Section 3 gives simulation results. Section 4 gives concluding remarks.

2 Design of 2x2 SCFDMA System Using SFBC and MRRC

The input bit stream $b_0, b_1, \ldots b_{M-1}$ is encoded by a channel encoder (convolutional or turbo) and modulated using (BPSK/QPSK/QAM), arranged in groups of M symbols and DFT is performed. Let $x_0, x_1, \ldots x_{M-1}$ be discrete time domain signals and $X_0, X_1, \ldots X_{M-1}$ are the discrete frequency domain signals.

$$X_k = \sum_{m=0}^{M-1} x_m e^{\frac{-j2\pi km}{M}} \quad , \; 0 \le k \le M-1 \tag{1}$$

Frequency domain M symbols are mapped to N symbols using LFDMA. Let $Y_0, Y_1, Y_2, \ldots, Y_{N-1}$ be resulting subcarrier mapped symbols and then IDFT is performed.

$$Y_l = \begin{cases} X_l\,, & 0 \le l \le M-1 \\ 0\,, & M-1 < l \le N-1 \end{cases} \tag{2}$$

$$y_n = y_{Qm+q} = \begin{cases} \frac{1}{Q} x_{(n)_{mod\,M}}, & q = 0 \\ \frac{1}{Q}\left(1 - e^{\frac{j2\pi q}{Q}}\right) \frac{1}{M} \sum_{p=0}^{M-1} \frac{x_p}{1-e^{j2\pi\left\{\frac{(m-p)}{M} + \frac{q}{Q.M}\right\}}}, & q \ne 0 \end{cases} \tag{3}$$

where,

$$N = QM$$
$$0 \le n = Qm + q \le N-1$$
$$0 \le m \le M-1$$
$$0 \le q \le Q-1$$

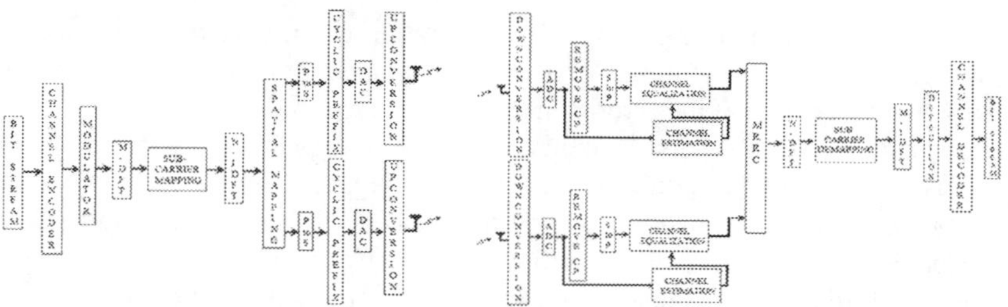

Fig. 1 (a) Transmitter of 2x2 SCFDMA system (b) Receiver of 2x2 SCFDMA system

At transmitter time domain modulated output of SCFDMA system is spatially mapped using SFBC. There are two output modulated spatially mapped streams $\mathrm{y0_n}$ and $\mathrm{y1_n}$ for antenna1 and 2 which are further processed as a stream for individual SCFDMA system. Let $y0_n = [y0_0, y0_1, y0_2, \dots, y0_{N-1}]$, $y1_n = [y1_0, y1_1, y1_2, \dots, y1_{N-1}]$ be signals in time domain transmitted from antenna1 and 2 [9,10].

Table 1 Mapping single data stream to two antenna for 2x2 (Tx-Rx)*

y_n	y_0	y_1	$\cdots$	y_{N-2}	y_{N-1}
$y0_n$(Antenna-1)	y_0	$-y_1^*$	$\cdots$	$y_{(N-2)}$	$-y_{(N-1)}^*$
$y1_n$(Antenna-2)	y_1	y_0^*	$\cdots$	$y_{(N-1)}$	$y_{(N-2)}^*$

PAPR of 2x2 (Tx-Rx)* SCFDMA system is given by:

$$PAPR = \frac{Peak\ power}{Average\ power} = \frac{\max\ \{\sum_{n=0}^{N-1}(\|y0_n\|^2+\|y1_n\|^2)\}}{\frac{1}{N}\{\sum_{n=0}^{N-1}(\|y0_n\|^2+\|y1_n\|^2)\}} \tag{4}$$

Let $PAPR_0$ be threshold PAPR. Cumulative distribution function(CDF) is given by

$$F_{PAPR}(PAPR_0) = P\{PAPR \leq PAPR_0\} \tag{5}$$

Complementary cumulative distribution function(CCDF) is given by

$$1 - F_{PAPR}(PAPR_0) = P\{PAPR > PAPR_0\} \tag{6}$$

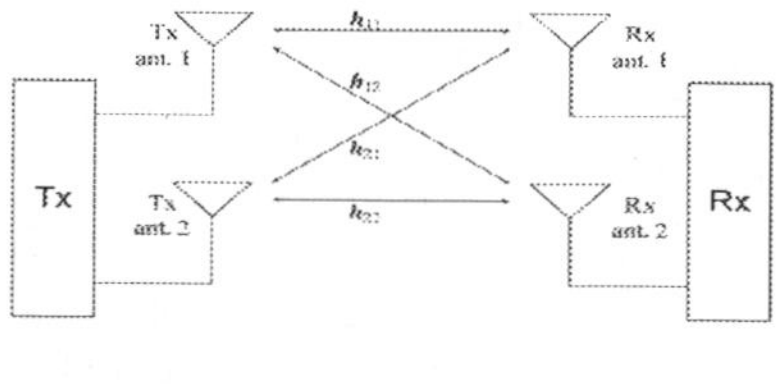

$$h_{11} = [h0_0, h0_0, h0_1, h0_1, \dots, h0_{N/2-1}, h0_{N/2-1}]$$
$$h_{12} = [h1_0, h1_0, h1_1, h1_1, \dots, h1_{N/2-1}, h1_{N/2-1}]$$
$$h_{21} = [h2_0, h2_0, h2_1, h2_1, \dots, h2_{N/2-1}, h2_{N/2-1}]$$
$$h_{22} = [h3_0, h3_0, h3_1, h3_1, \dots, h3_{N/2-1}, h3_{N/2-1}]$$

Fig. 2 2x2 (Tx-Rx)* system

Resulting time domain complex signals are parallel to serially converted and then cyclic prefix is inserted then digital to analogue conversion takes place and at last up-conversion occurs and then transmitted through antenna1 and 2.

Transmitted signal gets attenuated through channel. Let $r_0 = [r0_0, r0_1, \dots, r0_{N-1}]$ and $r_1 = [r1_0, r1_1, \dots, r1_{N-1}]$be signal received at antenna1 and 2 in time domain after down-converting and passing through analogue to digital convertor and $n_0 = [n0_0, n0_1, \dots, n0_{N-1}]$ and $n_1 = [n1_0, n1_1, \dots, n1_{N-1}]$ be samples of additive white Gaussian noise (AWGN) at antenna1 and 2.

On the two receiver side at antenna1 and 2 received symbol stream is given to two different parallel processing units one is the channel estimator and on the other side cyclic prefix is removed and serial to parallel converted for further analysis. These time domain symbols are then channel equalized using zero forcing approach using estimates of channel coefficient found by channel estimator. Channel path gains are assumed to be constant over two bit transmission period as shown in figure 2.

At receiver1 $r_{0k}$and $r_{0,k+1} = r_{0(k+1)}$be the received symbols corresponding to k^{th} and $(k+1)^{th}$ BPSK symbol period.

$$r_{0k} = h_{11k} y0_k + h_{21k} y1_k + n_{0k} \tag{7}$$

$$r_{0(k+1)} = h_{11(k+1)} y0_{(k+1)} + h_{21(k+1)} y1_{(k+1)} + n_{0(k+1)} \tag{8}$$

$$r_{0k} = h_{11k} y_k + h_{21k} y_{(k+1)} + n_{0k} \tag{9}$$

$$r_{0(k+1)} = h_{11k}(-\mathrm{y}_{\mathrm{k+1}}^*) + h_{21k}(\mathrm{y}_{\mathrm{k}}^*) + n_{0(k+1)} \tag{10}$$

Let $\tilde{y}_0$ be the estimated transmitted signal at antenna1.

$$\tilde{y}_{0k} = \frac{1}{\Delta}\{(h_{11k}^* \times r_{0k}) + (h_{21k} \times r_{0(k+1)}^*)\}, \; \Delta = \|h_{11k}\|^2 + \|h_{21k}\|^2 \tag{11}$$

Similarly estimate of y_{k+1} can be obtained by

$$\tilde{y}_{0(k+1)} = \frac{1}{\Delta}\{(h_{21k}^* \times r_{0k}) - (h_{11k} \times r_{0(k+1)}^*)\} \tag{12}$$

$$\tilde{y}_{0k} = \begin{cases} \frac{1}{\Delta}\{(h_{11k}^* \times r_{0k}) + (h_{21k} \times r_{0(k+1)}^*)\} & k = 0,2,4, \dots M-2 \\ \frac{1}{\Delta}\{(h_{21(k-1)}^* \times r_{0(k-1)}) - (h_{11(k-1)} \times r_{0k}^*)\} & k = 1,3,5, \dots M-1 \end{cases} \tag{13}$$

At receiver2 $r_{1k}$and $r_{1,k+1} = r_{1(k+1)}$be the received symbols corresponding to k^{th} and $(k+1)^{th}$ BPSK symbol period.

$$r_{1k} = h_{12k} y0_k + h_{22k} y1_k + n_{1k} \tag{14}$$

$$r_{1(k+1)} = h_{12(k+1)} y0_{(k+1)} + h_{22(k+1)} y1_{(k+1)} + n_{1(k+1)} \tag{15}$$

$$r_{1k} = h_{12k}y_k + h_{22k}y_{(k+1)} + n_{1k} \tag{16}$$

$$r_{1(k+1)} = h_{12k}(-y^*_{k+1}) + h_{22k}(y^*_k) + n_{1(k+1)} \tag{17}$$

Let $\tilde{y}_1$ be the estimated transmitted signal at antenna2.

$$\tilde{y}_{1k} = \frac{1}{\Delta_1}\{(h^*_{12k} \times r_{1k}) + \left(h_{22k} \times r^*_{1(k+1)}\right)\}, \ \Delta_1 = \|h_{12k}\|^2 + \|h_{22k}\|^2 \tag{18}$$

Similarly estimate of y_{k+1} can be obtained by

$$\tilde{y}_{1(k+1)} = \frac{1}{\Delta_1}\{(h^*_{22k} \times r_{1k}) - \left(h_{12k} \times r^*_{1(k+1)}\right)\} \tag{19}$$

$$\tilde{y}_{1k} = \begin{cases} \frac{1}{\Delta_1}\{(h^*_{12k} \times r_{1k}) + \left(h_{22k} \times r^*_{1(k+1)}\right)\} & k = 0,2,4,\ldots M-2 \\ \frac{1}{\Delta_1}\{\left(h^*_{22(k-1)} \times r_{1(k-1)}\right) - \left(h_{12(k-1)} \times r^*_{1k}\right)\} & k = 1,3,5,\ldots M-1 \end{cases} \tag{20}$$

These two receive antenna symbol stream combined through MRRC scheme. The two received signals on antenna1 and 2 are combined to improve reliability in case of fading environment. If received signal gets faded through one of the link other may not be faded and this decreases bit error rate compared to that of SCFDMA system. Symbols are arranged in groups of *N* and analyzed in frequency domain using DFT.

$$\tilde{r}_k = \frac{1}{2}\{\tilde{y}_{0k} + \tilde{y}_{1k}\} \tag{21}$$

$$\tilde{r}_k = \begin{cases} \frac{h^*_{11k}\ r_{0k} + h_{21k}\ r^*_{0,k+1}}{2\Delta} + \frac{h^*_{12k}\ r_{1k} + h_{22k}\ r^*_{1,k+1}}{2\Delta_1} & k = 0,2,\ldots M-2 \\ \frac{h^*_{21,k-1}r_{0,k-1} - h_{11,k-1}r^*_{0k}}{2\Delta} + \frac{h^*_{22,k-1}r_{1,k-1} - h_{12,k-1}r^*_{1k}}{2\Delta_1} & k = 1,3,\ldots M-1 \end{cases} \tag{22}$$

Symbols are arranged in groups of *N* and analyzed in frequency domain using DFT. $\tilde{R}_0, \tilde{R}_1, \tilde{R}_2, \ldots, \tilde{R}_{N-1}$ be *N*-point DFT of $\tilde{r}_0, \tilde{r}_1, \tilde{r}_2, \ldots, \tilde{r}_{N-1}$.

$$\tilde{R}_n = \sum_{l=0}^{N-1} \tilde{r}_l e^{\frac{-j2\pi nl}{N}}, \ 0 \le n \le N-1 \tag{23}$$

N symbols are de-mapped to *M* symbols required as per specified user's subcarrier. $Z_0, Z_1, \ldots, Z_{M-1}$ be de-mapped *M* symbols from $\tilde{R}_0, \tilde{R}_1, \tilde{R}_2, \ldots, \tilde{R}_{N-1}$. For first user symbols are de-mapped in following fashion and for further user value of *k* will enhance in a arirhmetic fashion.

$$Z_k = \tilde{R}_k, \quad 0 \le k \le M-1 \tag{24}$$

These *M* equalized symbols are then converted back to time domain using IDFT. $z_0, z_1, z_2, \ldots, z_{M-1}$ be *M*-point IDFT of $Z_0, Z_1, Z_2, \ldots, Z_{M-1}$.

$$z_m = \frac{1}{M}\sum_{k=0}^{M-1} Z_k e^{\frac{j2\pi km}{M}} \quad , \ 0 \le m \le M-1 \tag{25}$$

The processed symbols are then BPSK demodulated and then decoded.

$$w_m = \begin{cases} 0 & if \ \ Re(z_m) < 0 \\ 1 & if \ \ Re(z_m) \ge 0 \end{cases} , \ 0 \le m \le M-1 \tag{26}$$

Bit error rate can be calculated from comparison of output bit stream with the input bit stream. Let bit error rate be denoted by *BER*.

$$e_k = b_k - w_k \ , \ 0 \le k \le M-1 \tag{27}$$

Non zero entries in e results in error. To evaluate *BER*, let counting number of non zero entries in *e* be *Ne,i*. And performing over large number of turns N_0.

$$BER = \frac{1}{N_0}\frac{1}{M}\sum_{i=1}^{N_0} N_{e,i} \tag{28}$$

$$N_{e,i} = \sum_{k=0}^{M-1}\left|e_{i,k}\right| \tag{29}$$

Bit Error Rate for 2x2 (Tx-Rx)* SCFDMA system

$$BER = \frac{1}{N_0}\frac{1}{M}\sum_{i=1}^{N_0}\sum_{k=0}^{M-1}\begin{cases} |b_k| & if \ \ Re\left(\frac{1}{2M}\sum_{k=0}^{M-1}\sum_{l=0}^{N-1}\{\tilde{y}_{0l}+\tilde{y}_{1l}\}e^{\frac{-j2\pi kl}{N}}e^{\frac{j2\pi km}{M}}\right) < 0 \\ |b_k - 1| & if \ \ Re\left(\frac{1}{2M}\sum_{k=0}^{M-1}\sum_{l=0}^{N-1}\{\tilde{y}_{0l}+\tilde{y}_{1l}\}e^{\frac{-j2\pi kl}{N}}e^{\frac{j2\pi km}{M}}\right) \ge 0 \end{cases}$$

Bit Error Rate for 2x1 (Tx-Rx)* SCFDMA system

$$BER = \frac{1}{N_0}\frac{1}{M}\sum_{i=1}^{N_0}\sum_{k=0}^{M-1}\begin{cases} |b_k| & if \ \ Re\left(\frac{1}{M}\sum_{k=0}^{M-1}\sum_{l=0}^{N-1}\tilde{y}_l e^{\frac{-j2\pi kl}{N}}e^{\frac{j2\pi km}{M}}\right) < 0 \\ |b_k - 1| & if \ \ Re\left(\frac{1}{M}\sum_{k=0}^{M-1}\sum_{l=0}^{N-1}\tilde{y}_l e^{\frac{-j2\pi kl}{N}}e^{\frac{j2\pi km}{M}}\right) \ge 0 \end{cases}$$

$$\tilde{y}_k = \begin{cases} \frac{1}{\Delta}\{(h_{0k}^* \times r_k) + (h_{1k} \times r_{(k+1)}^*)\} & k = 0,2,4,\dots M-2 \\ \frac{1}{\Delta}\{(h_{1(k-1)}^* \times r_{(k-1)}) - (h_{0(k-1)} \times r_k^*)\} & k = 1,3,5,\dots M-1 \end{cases}$$

Bit Error Rate for 1x2 (Tx-Rx)* SCFDMA system

$$BER = \frac{1}{N_0 M}\sum_{i=1}^{N_0}\sum_{k=0}^{M-1}\begin{cases} |b_k| & Re\left(\frac{1}{M}\sum_{k=0}^{M-1}\sum_{l=0}^{N-1}\left(\frac{r_{0k}}{h_{0k}}+\frac{r_{1k}}{h_{1k}}\right)e^{\frac{-j2\pi kl}{N}}e^{\frac{j2\pi km}{M}}\right) < 0 \\ |b_k - 1| & Re\left(\frac{1}{M}\sum_{k=0}^{M-1}\sum_{l=0}^{N-1}\left(\frac{r_{0k}}{h_{0k}}+\frac{r_{1k}}{h_{1k}}\right)e^{\frac{-j2\pi kl}{N}}e^{\frac{j2\pi km}{M}}\right) \ge 0 \end{cases}$$

3 Simulation Result

The proposed system has been simulated in MATLAB R2009b.

3.1 Simulation Parameters

Table 2 Simulation Parameters

Transmitter	Modulation	BPSK
	Tx * DFT size	16
	Tx * IDFT size	512
	Subcarrier mapping	Localized
	Scheduling	Static (Round robin)
	Subcarrier spacing	15kHz
	Channel encoder	none
Channel	Bandwidth	5MHz
	Channel model	Rayleigh fast fading, flat fading (single path)
	Noise environment	(AWGN) Additive White Gaussian Noise
Receiver	Channel estimation	Perfectly known
	Number of runs	10^5

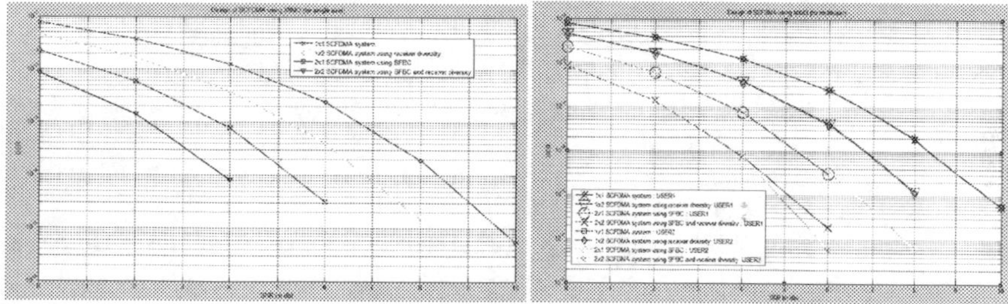

Fig. 3 BER Vs SNR plot for single and multi user SCFDMA using MIMO

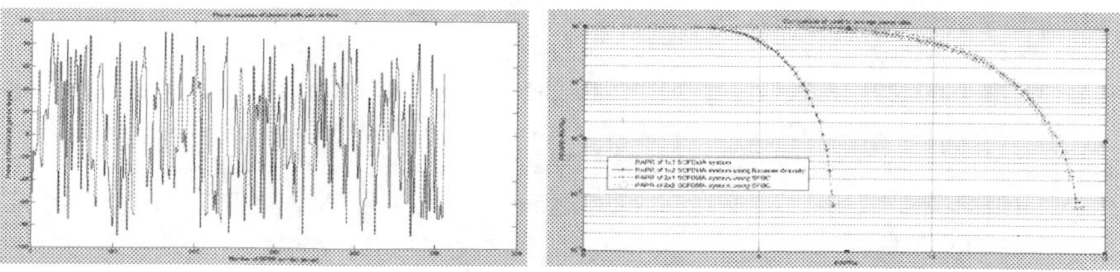

Fig. 4 Phase and amplitude of channel tap **Fig. 5** CCDF of PAPR

4 Conclusion and Future Work

BER performance: 2x2 SCFDMA system using SFBC and MRRC has the best performance. While 2x1 SCFDMA system using SFBC has better performance than 1x2 and 1x1 SCFDMA systems. While 1x2 SCFDMA system using receiver diversity has better performance than 1x1 SCFDMA system.

From simulation results it is clear that 2x2 SCFDMA system using SFBC and MRRC at an SNR 6db less can perform similar to that of 1x1 SCFDMA system. While 2x1 SCFDMA system using SFBC at an SNR 4db less can perform similar to 1x1 SCFDMA system. While 1x2 SCFDMA system using MRRC at an SNR 2db less can perform similar to 1x1 SCFDMA system.

From simulation results it is clear that at a threshold of 7db PAPR the probability of crossing the threshold PAPR is approximately 1 for 2x2 SCFDMA system using SFBC and receiver diversity and 2x1 SCFDMA system using SFBC. While the probability of crossing the threshold PAPR of 7db is approximately 10^{-3} for 1x2 SCFDMA system using receiver diversity and 1x1 SCFDMA system.

The future work may involve channel estimation for system, multiple tapped (frequency selective), channel dependent scheduling and more than two antennas.

References

Berardinelli, G., Frattasi, S., Rahman, M.I., Mogensen, P., et al.: OFDMA vs. SCFDMA: Performance Comparison in Local Area IMT-A Scenarios. IEEE Transactions on Wireless Communications 15(5), 64–72 (2008)

Myung, H.G., Goodman, D.J.: Single carrier FDMA-A New Air Interface for LTE. Wiley Series on Wireless Communications and Mobile Computing (2008)

Khan, F.: LTE for 4G Mobile Broadband-air interface technologies and performance. Cambridge University Press (2009)

Wang, Z., Ma, X., Giannakis, G.B.: OFDM or Single Carrier Transmissions? IEEE Transactions on Communications 52(3) (March 2004)

Yune, T., Choi, C., Im, G., Lim, J., et al.: SCFDMA with Iterative Multiuser Detection: Improvements on Power/Spectral Efficiency. IEEE Magazine on Communications 48(3), 164–171 (2010)

Lim, J., Choi, C., Yune, T., Im, G.: Iterative Multiuser Detection for Single-Carrier Modulation with Frequency-Domain Equalization. IEEE Letters on Communications 11(6), 471–473 (2007)

Myung, H.G., Kyungjin, O., Junsung, L., Goodman, D.J.: Channel-Dependent Scheduling of an Uplink SCFDMA System with Imperfect Channel Information. In: IEEE Conference on Wireless Communications and Networking, pp. 1860–1864 (April 2008)

Wong, I.C., Oteri, O., McCoy, W.: Optimal Resource Allocation in Uplink SCFDMA Systems. IEEE Transactions on Wireless Communications 8(5) (May 2009)

Zhang, W., Xia, X.G., Letaief, K.B.: Space Time/Frequency Coding for MIMO-OFDM in Next Generation Broadband Wireless Systems. IEEE Transactions on Wireless Communications (June 2007)

Alamouti, S.M.: A Simple Transmit Diversity Technique for Wireless Communications. IEEE Journal on Select Areas in Communications 16(8), 30–38 (1998)

Testing an Agent Based E-Novel System – Role Based Approach

N. Sivakumar and K. Vivekanandan

Abstract. Agent Oriented Software Engineering(AOSE) methodologies are meant for providing guidelines, notations, terminologies and techniques for developing agent based systems. Several AOSE methodologies were proposed and almost no methodology deals with testing issues, stating that the testing can be carried out using the existing object-oriented testing techniques. Though objects and agents have some similarities, they both differ widely. Role is an important mental attribute/state of an agent. The main objective of the paper is to propose a role based testing technique that suits specifically for an agent based system. To demonstrate the proposed testing technique, an agent based E-novel system has been developed using Multi agent System Engineering (MaSE) methodology. The developed system is tested using the proposed role based approach and found that the results are encouraging.

Keywords: Agent-Oriented Software Engineering, Multi-Agent System, Role based testing.

1 Introduction

A software development methodology refers to the framework that is used to structure, plan, and control the process of developing a software system. A wide variety of such frameworks have evolved over the years, each with its own recognized strengths and weaknesses. Now-a-days agent based systems are the solutions for complex application such as industrial, commercial, networking, medical and educational domain [1]. The key abstraction in these solutions is the agent. An "agent" is an autonomous, flexible and social system that interacts with its environment in order to satisfy its design agenda. In some cases, two or more agents should interact with each other in a Multi Agent System (MAS) to solve a

N. Sivakumar · K. Vivekanandan
Department of Computer Science and Engineering,
Pondicherry Engineering College, Puducherry, India
e-mail: {sivakumar11,k.vivekanandan}@pec.edu

A. Abraham and S.M. Thampi (Eds.): Intelligent Informatics, AISC 182, pp. 165–173.
springerlink.com

problem that they cannot handle alone. The agent oriented methodologies provide us a platform for making system abstract, generalize, dynamic and autonomous. This important factor calls for an investigation of suitable AOSE frameworks and testing techniques, to provide high-quality software development process and products.

Roles provide a well-defined interface between agents and cooperative processes [2]. This allows an agent to read and follow, normative rules established by the cooperation process even if not previously known by the agent. Their major motivation to introduce such roles is to increase the agent system's adaptability to structural changes. Several AOSE methodologies were analysed and compared and found that the strong weakness observed from almost all the methodologies were, there is no proper testing mechanism for testing the agent-oriented software. Our survey states that the agent based software are currently been tested by using Object-Oriented (OO) testing techniques, upon mapping of Agent-Oriented (AO) abstractions into OO constructs [3]. However agent properties such as Autonomy, Proactivity, and Reactivity etc., cannot be mapped into OO constructs. There arises the need for specialized testing techniques for agent based software. The main objective of the paper is to propose a testing mechanism based on agent's important mental state, the role.

The Paper is organized as follows: Section 2 describes the literature study on the existing work on agent oriented methodologies and existing testing techniques. Section 3 explains the analysis, design and implementation process of an agent based E-Novel system using MaSE methodology. Section 4 explains the proposed role based testing mechanism and its effectiveness towards agent based system.

2 Background and Related Works

Agent-oriented software engineering is a new discipline that encompasses necessary methods, techniques and tools for developing agent-based systems. Several AOSE methodologies [4] were proposed for developing software, equipped with distinct concepts and modelling tools, in which the key abstraction used in its concepts is that of an agent. Some of the popular AOSE methodologies were MASCommonKADS, MaSE, GAIA, MESSAGE, TROPOS, PROMETHEUS, ADLEFE, INGENIAS, PASSI, AOR Modeling. Very few methodologies provide validation support but fail to contribute complete testing phase. The TROPOS methodology has an agent testing framework, called eCAT. eCAT is a tool that supports deriving test cases semi-automatically. Goal oriented testing [5] contributes TROPOS methodology by providing a testing process model, which complements and strengthens the mutual relationship between goals and test cases. PROMETHEUS methodology provides only debugging support. PASSI methodology contributes only unit testing framework. INGENIAS provides basic interaction debugging support through INGENIAS Agent Framework (IAF). Table 1 clearly indicates that the existing AOSE methodologies does not support testing phase, stating that testing an agent system has been accommodated using existing traditional and object-oriented testing techniques.

Table 1 Lifecycle coverage of several AOSE methodologies [4]

Life Cycle Coverage	MAS Common KADS(1996-98)	MaSE(1999)	GAIA(2000)	MESSAGE(2001)	TROPOS(2002)	PROMETHEUS (2002)	ADLEFE(2002)	INGENIAS(2002)	PASSI(2002)	AOR Modelling (2003)
Analysis	Yes	Yes	Yes	Yes	Yes	Yes	Yes	Yes	Yes	Yes
Design	Yes	Yes	Yes	Yes	Yes	Yes	Yes	Yes	Yes	Yes
Coding	No	No	No	No	Yes	Yes	Yes	Yes	Yes	No
Testing	No	No	No	No	No	No	No	No	No	No

Role is an important mental attribute of an agent and often agent changes its roles to achieve its designated goal. Roles are intuitively used to analyze agent systems, model social activities and construct coherent and robust teams of agents. Roles are a useful concept in assisting designers and developers with the need for interaction. Generic Architecture for Information Access (GAIA) methodology [6] and Multiagent Systems Engineering (MaSE) methodology [7] were role-based methodologies for development of multi-agent systems.

3 Proposed Work

The main objective of this paper is to propose a role based testing technique that suits specifically for an agent based system. Role based testing is applied at different abstraction level such as unit, integration, system and acceptance. To illustrate the role based testing approach, an agent based E-novel system was developed using MaSE methodology. E-novel system is deployed on internet community to assist in interaction between novelists and readers. In the e-novel community, system accepts and contains a number of novels authored by various novelists. Readers simply browse to find and read novels. However, readers typically spend lot of time to browse and review a list of novels through categories and ranking. An e-novel system which is a subsystem designed from the notion of this study for this community.

E-Novel system has been developed using MASE methodology which is an iterative process. It deals the capturing the goals and refining the roles of an agent. It appears to have significant tool support. *agentTool* is a graphically based, fully interactive software engineering tool, which fully supports each step of MaSE analysis and design. Fig.1 represents goal hierarchy diagram of e-novel system designed using *agentTool*. The analysis phase involves capturing goal, Applying use cases and Refining roles (Fig 2) whereas the design phase involves Creating agent classes, Constructing conversations, Assembling agent classes (Fig 3) and System design.

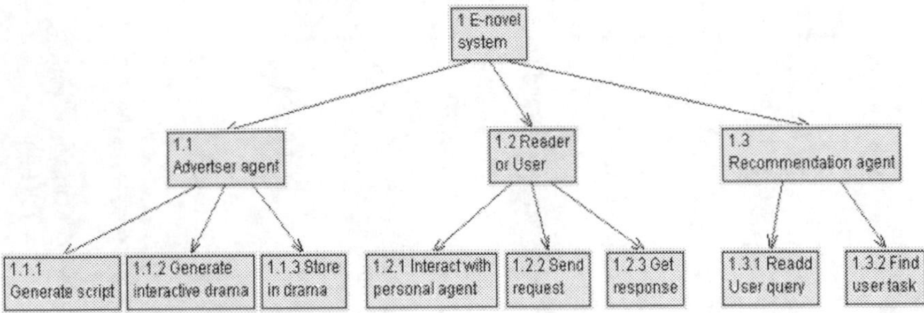

Fig. 1 Goal Hierarchy Diagram for E-Novel System

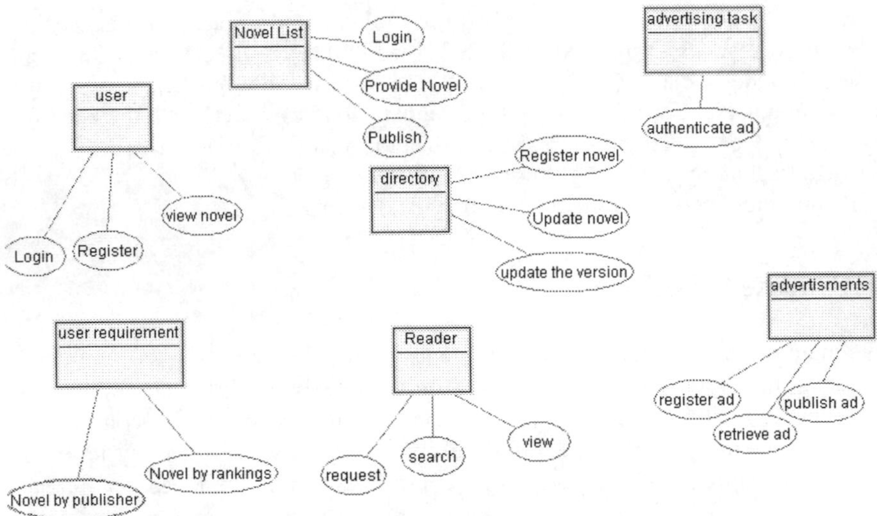

Fig. 2 Role Diagram for E-Novel System

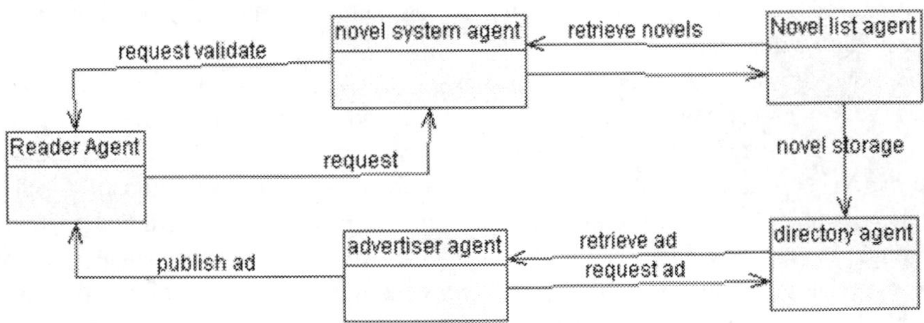

Fig. 3 Agent Template Diagram for E-Novel System

After analysis and design, the Agent-based E-Novel system is implemented using JADE (Java Agent Development Framework), a software platform that provides basic middleware layer functionalities which are independent of the specific application and which simplify the realization of distributed applications that exploit the software agent abstraction [10].

4 Testing

Testing is inseparable in software development process. Though testing is mandatory, there is a hindrance for its uptake due to the fact that the existing AOSE methodologies failed to prioritise agent-oriented testing, stating that agent systems can be tested using the existing conventional and object-oriented software testing technique. Currently testing is accomplished by mapping Agent-Oriented (AO) abstractions into OO constructs. However agent properties such as Autonomy, Proactivity, and Reactivity etc., cannot be mapped into OO constructs. This leads to the need for specialized agent-oriented software testing technique for agent-oriented software systems. In this paper, role based testing technique is proposed for effectively testing an agent based system.

4.1 Role Based Testing

Roles have been used both as an intuitive concept in order to analyse MAS and model inter-agent social activity and as a formal structure in order to implement coherent and robust agent-based software. Every individual agent has its own goal to be achieved and plans to do to fulfill the goal. In addition to goal and plan, role is one important mental state of the agent, which is defined as a set of capabilities and expected behavior. A role [8][9] can be represented as

<Goal, Responsibilities, Protocol, Permissions>

- Goal, for which the agent playing this role is responsible
- Responsibilities, Which indicates the functionalities of agents playing such roles
- Protocol, which indicates how an agent playing such role can interact with agents playing other role
- Permissions, which are a set of rights associated with the role.

Role based testing provides the full range of assurance and correctness for agents to manage the complexity of highly dynamic and unpredictable environments with a high degree of interaction and distributivity. Every Agent involved in the E-Novel system has their own roles for their accomplishment. Some roles involves only one agent and other involves more than one agent thereby interaction among agent is facilitated. Fig 5 represents the role diagram which shows the Goal-Role-Responsibility relationship.

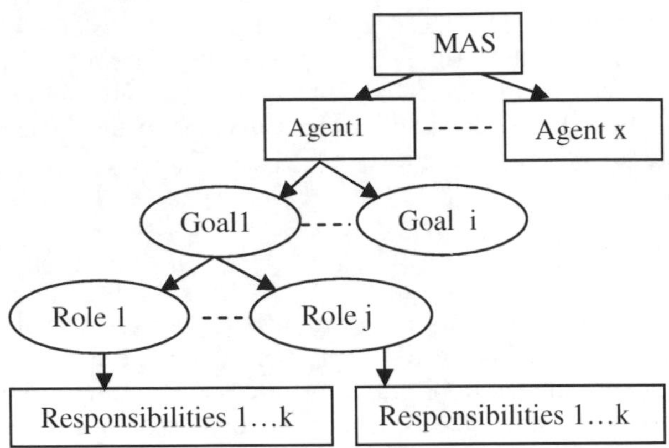

Fig. 5 Role Model

Let 'x' be the number of agents involved in Agent-based E-novel system,

$$\text{Agents} = \{A_1, A_2, A_3, \ldots, A_x\}$$

Let 'i' be the number of goals that every agent in the agent based e-novel system has to achieve.

$$\text{Goals} = \{G_1, G_2, G_3, \ldots. G_i\}$$

Let 'j' be the number of roles carried out by individual agent to accomplish every goal.

$$\text{Roles} = \{R_1, R_2, R_3, \ldots., R_j\}$$

Let 'k' be the number of functionalities to be accomplished for every role

$$\text{Responsibilities} = \{Re_1, Re_2, Re_3, \ldots. Re_k\}$$

After identifying the roles and their corresponding responsibilities of every agents involved in the MAS, test cases has to be derived to test whether the roles for the agents accomplish their task for the given set of inputs. Analyzing the Goal-Role relationship, it is found that, as long as the agent performs its role properly, the goal of the system is been achieved by default. Thus testing whether the agent performs its role properly is a challenging task. This paved way for a role-oriented testing mechanism by which the role functionalities were tested by deriving appropriate test cases. Random based test case generation technique is been used for generating test cases that suits for role based approach.

4.2 Role Schema

Roles schema provide a well-defined interface between agents and cooperative processes. This allows an agent to read and follow, normative rules established by the cooperation process even if not previously known by the agent. Their major motivation to introduce such roles is to increase the agent system's adaptability to structural changes. They formally define a role as an entity consisting of a set of required permissions, a set of granted permissions, a directed graph of service invocations, and a state visible to the runtime environment but not to other agents. Sample role schema for an agent based e-novel system is represented in Table 2.

Table 2 Sample Role Schema

Role Name : Prioritize Novel
Agent involved : Recommendation Agent
Goal: To present the best novel to the reader
Description: This role helps to prioritize the novels based on the author popularity and reader's interest.
Protocol and Activities: Analyzing No. of novels written by the author, Popularity among the readers, Novel writing skill and presentation skill,
Permissions: Read Request query, Result, Security policy, Change Result format // encrypt, Request format // decrypt
Responsibilities: *Activeness:* (count no. of novels + popularity of the novelist + writing skill + presentation skill + no. of readers read that novel *Completeness:* Suggesting best novel to the reader

4.3 Random Based Test Case Derivation

According to our approach, the role of an agent comprises the logic of the test. As every role of an agent has number of responsibilities to get satisfied, the derivation of test case focuses on the responsibilities and thereby validates whether the role hold by the agent serves the purpose. Random based test cases generation technique is applied for generating test suites. This technique generates test cases selecting the communication protocol and randomly generated messages. A sample test case is represented in Table. 3.

Table 3. Sample Test case for E-Novel system based on role

T.ID	TESTED AGENT	GOAL	ROLE	SITUATION	INPUT	EXPECTED RESULT	OBSERVED RESULT	RESULT
T1	AD Agent	To Publish Ad	Ad preference	Need for Novel	Novel name	Search for novel and publish relevant Advertisement	Novel list with appropriate Advertisement	Passed

Once every roles were identified and test cases were generated, we tested our system in JADE Test Suite and we found that correctness of the system is validated fully and thereby role based testing can be applied for testing agent-oriented system

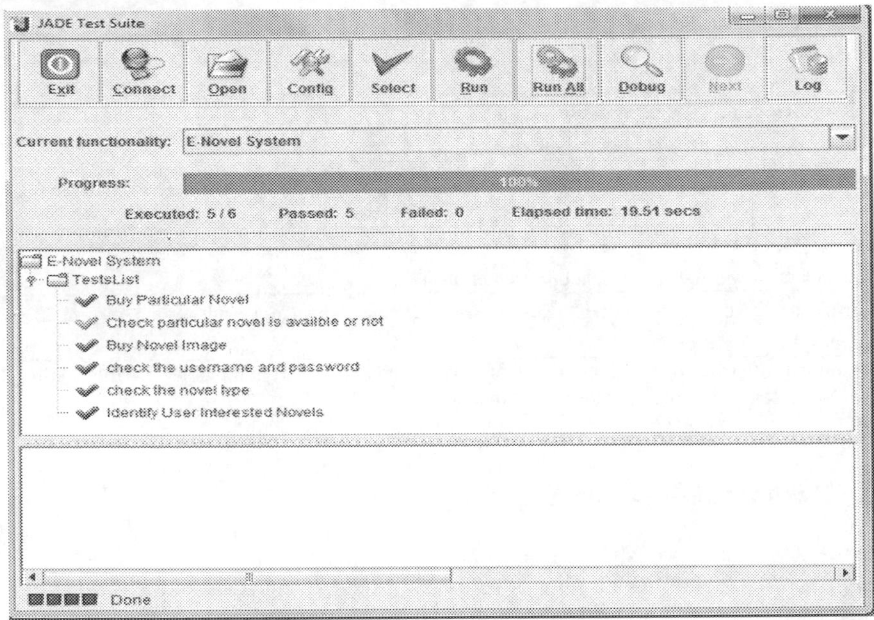

Fig. 6 Snapshot of JADE Test Suite

5 Conclusion

Testing being very important activity in Software Development Life Cycle (SDLC), there is no well defined testing technique for agent based system. None of the existing AOSE methodologies deals with testing phase stating testing can be carried out by using existing object-oriented technique. Although there is a well defined OO testing technique to test the agent based systems AO constructs cannot be mapped completely into OO constructs. Thus there arises vacancy for testing phase in the SDLC that should be filled-up. This paper deals with a new agent based testing technique i.e role based testing designed specifically for agent-oriented software so as to fit in the existing AOSE methodologies. To demonstrate our proposed testing technique, we developed an agent based E-novel system using MASE methodology and tested the system using role based approach and found that our results are encouraging. Thus the proposed testing technique performs adequately and accurately for testing the completeness of agent based system.

References

1. Shoham, Y.: Agent oriented programming (Technical Report STAN-CS-90-1335) Stanford University: Computer Science Department (1994)
2. Cabri, G., Leonardi, L., Ferrari, L., Zambonelli, F.: Role-based Software agent interaction models: a survey. The Knowledge Engineering Review 25(4), 397–419 (2010)
3. Srivastava, P.R., Anand, K.V., Rastogi, M., Yadav, V., Raghurama, G.: Extension of Object-oriented Software testing techniques to Agent Oriented software testing. Journal of Object Technology 7(8), 155–163 (2008)
4. Henderson-Sellers, B., Giorgini, P.: Agent-Oriented methodologies. Idea Group Inc. (2005)
5. Nguyen, D.C., Perini, A., Tonella, P.: A Goal-Oriented Software Testing Methodology. In: Luck, M., Padgham, L. (eds.) AOSE VIII. LNCS, vol. 4951, pp. 58–72. Springer, Heidelberg (2008)
6. Henderson, B., Giorgini, P.: The Gaia Methodology for Agent-Oriented Analysis and Design Autonomous Agent and Multi-Agent Systems, vol. 3, pp. 285–312. Kluwer Academic Publishers (2000)
7. Wood, M.F.: Multiagent system engineering: A methodology for analysis and design of muti-agent systems. Master thesis, School of Engineering, Air Force Institute of Technology, USA (2000)
8. Xu, H., Zhang, X., Patel, R.J.: Developing Role-Based Open Multi-Agent Software Systems. International Journal of Computational Theory and Practice 2(1) (June 2007)
9. Kumar, M.: Roles and Ontology for Agent Systems. Global Journal of Computer Science and Technology 11(23), Version 1.0 (December 2011)
10. Padhagam, L., WinikOff, M.: Developing Intelligent Agent Systems A practical guide. RMIT University, Melbourne

Comparative Genomics with Multi-agent Systems

Juan F. De Paz, Carolina Zato, Fernando de la Prieta, Javier Bajo, Juan M. Corchado, and Jesús M. Hernández

Abstract. The detection of the regions with mutations associated with different pathologies is an important step for selecting relevant genes. The corresponding information of the mutations and genes is distributed in different public sources and databases, so it is necessary to use systems that can contrast different sources and select conspicuous information. This work proposes a virtual organization of agents that can analyze and interpret the results from Array-based comparative genomic hybridization, thus facilitating the traditionally manual process of the analysis and interpretation of results.

Keywords: arrays CGH, knowledge extraction, visualization, multiagent system.

1 Introduction

Different techniques presently exist for the analysis and identification of pathologies at a genetic level. Along with massive sequencing, which allows the exhaustive study of mutations, the use of microarrays is highly extended. CGH arrays (aCGH) (Array-based comparative genomic hybridization) are a type of microarray that can analyze information on the gains, losses and amplifications [7] in regions of the chromosomes to detect mutations [5], [3]. Expression arrays measure the expression level of the genes. aCGH are currently used to detect relevant regions that may require deeper analysis. In these cases, it is necessary to

Juan F. De Paz · Carolina Zato · Fernando de la Prieta · Javier Bajo · Juan M. Corchado
Department of Computer Science and Automation, University of Salamanca
Plaza de la Merced, s/n, 37008, Salamanca, Spain
e-mail: {fcofds,carol_zato,fer,corchado}@usal.es

Jesús M. Hernández
IBMCC, Cancer Research Center, University of Salamanca-CSIC, Spain
e-mail: jhmr@usal.es

Jesús M. Hernández
Servicio de Hematología, Hospital Universitario de Salamanca, Spain

A. Abraham and S.M. Thampi (Eds.): Intelligent Informatics, AISC 182, pp. 175–181.
springerlink.com

work with vast amounts of information, which necessitates the creation of a system that can facilitate the automatic analysis of data that, in turn, facilitates the extraction of relevant information using different data bases. For this reason, it is necessary to automate the aCGH processing.

aCGH, also called microarray analysis, is a new cytogenetic technology that evaluates areas of the human genome for gains or losses of chromosome segments at a higher resolution than traditional karyotyping. When working with aCGH, segments of DNA (Deoxyribonucleic Acid) are selected from public genome databases based upon their location in the genome. Computer software analyzes the fluorescent signals for areas of unequal hybridization of patient versus control DNA, signifying a DNA dosage alteration (deletion or duplication). These arrays offer genome-covering resolution that can offer precise delineation of breakpoints. This is important in determining common regions of overlap and implicated genes. Due to their small target size, oligonucleotide arrays suffer from poorer signal to noise ratios that often results in a significant number of false-positive outliers. At present, tools and software already exist to analyze the data of arrays CGH, such as CGH-Explorer [2], ArrayCyGHt [12], CGHPRO [1], WebArray [8] or ArrayCGHbase [4], VAMP [6]. The problem with these tools is the lack of usability and of an interactive model. For this reason, it is necessary to create a visual tool to analyse the data in a simpler way.

The process of arrays CGH analysis is broken down into a group of structured stages, although most of the analysis process is done manually from the initial segmentation of the data. This study presents a multi-agent system [10] that defines roles to automatically perform the different stages of the analysis. In the first stage, the data are segmented [11] to reduce the number of gains or losses fragments to be analyzed. The following steps vary in terms of the type of analysis being performed and include: grouping, classification, visualization, or extraction of information from different sources. The system tries to facilitate the analysis and the automatic interpretation of the data by selecting the relevant genes, proteins and information from the previous classification of pathologies. The system provides several representations in order to facilitate the visual analysis of the data. The information for the identified genes, CNVs (Copy-number variations), pathologies etc. is obtained from public databases.

This article is divided as follows: section 2 describes our system, and section 3 presents the results and conclusions.

2 Multi-agent System

The multi-agent system designed to analyze our data is general enough that it can be adapted for other types of data analysis. The multi-agent system is divided into different layers: the analysis layer, the information management layer. The developed system receives data from the analysis of chips and is responsible for representing the data for extracting relevant segments on evidence and existing data. Working from the relevant cases, the first step consists of selecting the information about the genes and transcripts stored in the databases. This information will be associated to each of the segments, making it possible to quickly consult the data and reveal the detected alterations at a glance. The data analysis can be carried out automatically or manually.

2.1 Analysis Roles

The analysis roles contains the agents responsible for performing the actual microarray analyses. The information management layer compiles the information from the database and generates local databases to facilitate their analysis. The visualization layer facilitates the management of both the information and the algorithms; it displays the information and the results obtained after applying the existing algorithms at the analysis layer.

The agents at the analysis layer adapt to the specific class of microarray, in this case the aCGH, and within the aCGH they adapt to the different types of microarrays with which they work. To perform the data analysis, the agents are incorporated for: segmentation, Knowledge extraction, and Clustering.

The segmentation process is performed by taking into account the differential normalization for gains and losses. The segmentation process is based on the mad1dr (median absolute deviation, 1st derivative) value for each of the arrays, which determines the threshold for gains or losses that is considered relevant for each case. This metric provides a surrogate measure of experimental noise.

For this particular system, the use of chi Square was chosen because it is the technique that makes it possible to work with different qualitative nominal variables to study factor and its response. The contrast of Chi Square makes it possible to obtain as output the values that can sort the attributes by their importance, providing an easier way to select the elements. As an alternative, gain functions could be applied in decision trees, providing similar results.

2.2 Information Management Roles

Once the relevant segments have been selected, the researchers can introduce information for each of the variants. The information is stored in a local database. These data are considered in future analyses although they have to be reviewed in detail and contrasted by the scientific community. The information is shown in future analyses with the information for the gains and losses. However, because only the information from public databases is considered reliable, this information is not included in the reports.

Besides the system incorporates a role to retrieve information from UCSC (University of California Santa Cruz) and use this information to generate reports. This information is important in order to select the relevant segments.

3 Visual Analysis

A visual analysis is performed of the data provided by the system and the information recovered from the databases. New visualizations are performed in order to more easily locate the mutations, thus facilitating the identification of mutations that affect the codification of genes among the large amount of genes. Visualization facilitates the validation of the results due to the interactivity and ease of use of previous information. Existing packages such as CGHcall [9] in R do not display the results in an intuitive way because it is not possible to associate segments with regions and they do not allow interactivity.

The system provides a visualization to select the regions with more variants and relevant regions in different pathologies. The visualizations make is possible to extract information from databases using a local database.

A visual analysis is performed of the data provided by the system and the information recovered from the databases. New visualizations are performed in order to more easily locate the mutations, thus facilitating the identification of mutations that affect the codification of genes among the large amount of genes. Visualization facilitates the validation of the results due to the interactivity and ease of use of previous information. Existing packages such as CGHcall [9] in R do not display the results in an intuitive way because it is not possible to associate segments with regions and they do not allow interactivity.

The system provides a visualization to select the regions with more variants and relevant regions in different pathologies. The visualizations make is possible to extract information from databases using a local database.

4 Results and Conclusions

In order to analyze the operation of the system, different data types of array CGH were selected. The system was applied to two different kinds of CGH arrays: BAC aCGH, and Oligo aCGH. The information obtained from the BAC aCGH after segmenting and normalizing is represented in Tab. 1. As shown in the figure, there is one patient for each column. The rows contain the segments so that all patients have the same segments. Each segment is a tuple composed of three elements: chromosome, initial region and final region. The values v_{ij} represent gains and losses for segment i and patient j. If the value is positive, or greater than the threshold, it is considered a gain; if it is lower than the value, it is considered a loss.

Table 1 BAC aCGH normalized and segmented

Segment	Patient 1	Patient 2	...	Pantient n
Init-end	v_{11}	v_{12}	...	v_{1n}
Init-end	v_{21}	v_{22}	...	v_{2n}

The system includes the databases because it extracts the information from genes, proteins and diseases. These databases have different formats but basically there is a tuple of three elements for each row (chromosome, start, end, other information). Altogether, the files downloaded from UCSC included slightly more than 70,000 registries.

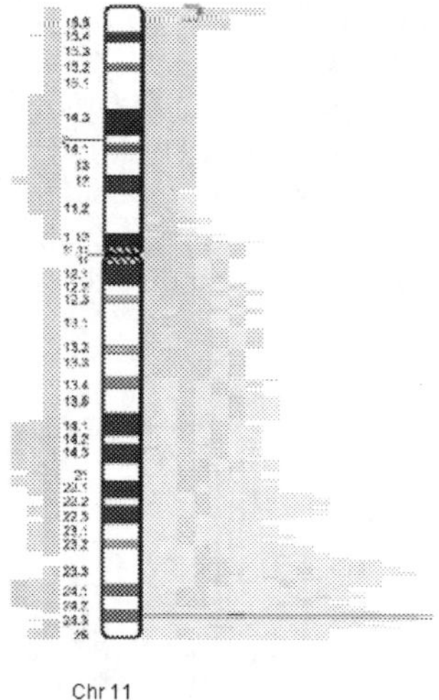

Fig. 1 Selection of segments and genes automatically

Figure 1 displays the information for 18 oligo arrays cases. Only the information corresponding to chromosome 11 is shown. The green lines represent gains for the patient in the associated region of the chromosome, while the red lines represent losses. The user can select the regions and use these highlighted regions to generate reports.

When performing the visual analysis, users can retrieve information from a local database or they can browse through UCSC. For example, figure 2 contains a

Report

File

knownGe...	knownGe...	knownGe...	knownGe...	knownGe...	knownGe...	knownGe...	keggPath...	kgXref.m...	kgXref.ge...	kgXref.de...	hgnc.hgn...	hgnc.sym...	hgnc.put
uc001qei.1	127837465	127897099	127833869	127897371	chr11	P14921	hsa04320	NM_0052...	ETS1	v-ets eryth...	HGNC:34...	ETS1	1522903
uc001qei.1	127837465	127897099	127833869	127897371	chr11	P14921	hsa05200	NM_0052...	ETS1	v-ets eryth...	HGNC:34...	ETS1	1522903
uc001qei.1	127837465	127897099	127833869	127897371	chr11	P14921	hsa05211	NM_0052...	ETS1	v-ets eryth...	HGNC:34...	ETS1	1522903
uc001qej.1	127837465	127948235	127835182	127962663	chr11	Q6N087	hsa04320	BX640634	ETS1	v-ets eryth...			
uc001qej.1	127837465	127948235	127835182	127962663	chr11	Q6N087	hsa05200	BX640634	ETS1	v-ets eryth...			
uc001qej.1	127837465	127948235	127835182	127962663	chr11	Q6N087	hsa05211	BX640634	ETS1	v-ets eryth...			
uc001qek.2	128056345	128056345	128056345	128062024	chr11			AX747861	AX747861	Homo sa...			
uc001qel.2	128066785	128066785	128066785	128071128	chr11			BC039676	BC039676	Homo sa...			
uc001qe...	128069363	128186093	128069198	128187521	chr11	Q01543		NM_0020...	FLI1	Friend leu...	HGNC:37...	FLI1	1765382
uc001qen.2	128133300	128143532	128133219	128143621	chr11			AF147318	FLI1	Homo sa...			
uc001qfc.2	128751140	128828507	128751090	128827384	chr11	Q6NT51		NM_0036...	BARX2	BarH-like ...			
uc001qfd.1	129069722	129069722	129069722	129073350	chr11			AX746800	AX746800	Homo sa...			
uc001qfe.1	129227587	129233790	129180950	129235108	chr11	Q96B21		NM_1387...	TMEM45B	transme...	HGNC:25...	TMEM45B	1247793
uc001qff.1	129227587	129233790	129225277	129235108	chr11	Q96B21		BC016153	TMEM45B	transme...	HGNC:25...	TMEM45B	1247793
uc001qfg.1	129239629	129267983	129239574	129268114	chr11	Q6P4R8-2		NM_0061...	NFRKB	nuclear fa...			
uc001qfh.1	129239629	129268449	129239574	129270662	chr11	Q6P4R8		U08191	NFRKB	nuclear fa...	HGNC:78...	NFRKB	1427843
uc001qfi.1	129239629	129267954	129239574	129270662	chr11	Q6P4R8		BC053280	NFRKB	nuclear fa...	HGNC:78...	NFRKB	1427843
uc001qfj.1	129277417	129322511	129274810	129322727	chr11	Q9NQV6		NM_1994...	PRDM10	PR domai...	HGNC:13...	PRDM10	1217587
uc001qfk.1	129277417	129322511	129274810	129322727	chr11	Q863Z2		NM_1994...	PRDM10	PR domai...			
uc001qfl.1	129277417	129322511	129274810	129322727	chr11	Q17R90		BC117415	PRDM10	PR domai...			
uc001qfm.1	129277417	129336069	129274810	129377940	chr11	NP_0646...		NM_0202...	PRDM10	PR domai...			
uc001qfn.1	129277417	129336069	129274810	129377940	chr11	NP_9554...		NM_1994...	PRDM10	PR domai...			
uc009zcg.1	127837543	127897099	127835182	127897371	chr11	Q96AC5	hsa04320	BC017314	ETS1	ETS1 prot...			
uc009zcg.1	127837643	127897099	127835182	127897371	chr11	Q96AC5	hsa05200	BC017314	ETS1	ETS1 prot...			
uc009zcg.1	127837643	127897099	127835182	127897371	chr11	Q96AC5	hsa05211	BC017314	ETS1	ETS1 prot...			
uc009zch.1	127837465	127897099	127835182	127897371	chr11	A9UL17	hsa04320	AY943926	ETS1	Ets-1 tran...			
uc009zch.1	127837465	127897099	127835182	127897371	chr11	A9UL17	hsa05200	AY943926	ETS1	Ets-1 tran...			
uc009zch.1	127837465	127897099	127835182	127897371	chr11	A9UL17	hsa05211	AY943926	ETS1	Ets-1 tran...			
uc009zci.1	128139894	128186093	128069198	128186099	chr11	Q01543-2		M93255	FLI1	Friend leu...			
uc009zcr.1	129239629	129249354	129239574	129250021	chr11	NP_0061...		AL512730	NFRKB	nuclear fa...			
uc009zcs.1	129289810	129290851	129289436	129290965	chr11			AK310354	KIAA1231	Homo sa...			

Fig. 2 Report with relevant genes

report with the information for the segment belonging to the irrelevant region shown in the previous image.

In order to facilitate the revision and learning phases for the expert, a different visualization of the data is provided. This view helps to verify the results obtained by the hypothesis contrast regarding the significance of the differences between pathologies. Figure 3 shows a dendrogram with the information of the groups. The expert can review the clusters and modify the group belong each patient selecting each patient.

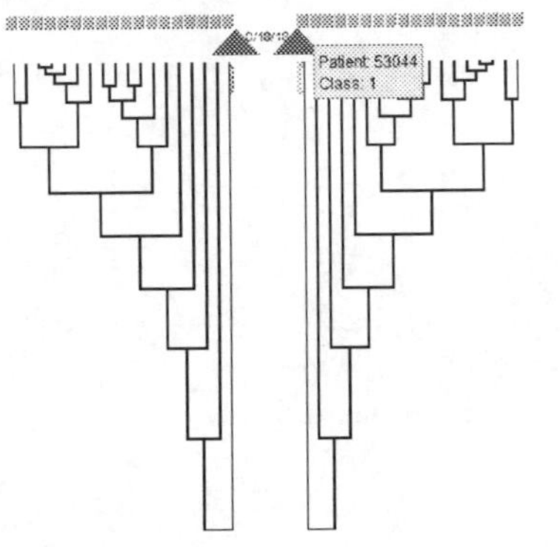

Fig. 3 Reviewing clustering process

The presented system facilitates the use of different sources of information to analyze the relevance in variations located in chromosomic regions. The system is able to select the genes, variants, genomic duplications that characterize pathologies automatically, using several databases. This system allows the management of external sources of information to generate final results. The provided visualizations make it possible to validate the results obtained by an expert more quickly and easily.

Acknowledgments. This work has been supported by the MICINN TIN 2009-13839-C03-03.

References

[1] Chen, W., Erdogan, F., Ropers, H., Lenzner, S., Ullmann, R.: CGHPRO-a comprehensive data analysis tool for array CGH. BMC Bioinformatics 6(85), 299–303 (2005)

[2] Lingjaerde, O.C., Baumbush, L.O., Liestol, K., Glad, I.K., Borresen-Dale, A.L.: CGH-explorer, a program for analysis of array-CGH data. Bioinformatics 21(6), 821–822 (2005)

[3] Mantripragada, K.K., Buckley, P.G., Diaz de Stahl, T., Dumanski, J.P.: Genomic microarrays in the spotlight. Trends Genetics 20(2), 87–94 (2004)

[4] Menten, B., Pattyn, F., De Preter, K., Robbrecht, P., Michels, E., Buysse, K., Mortier, G., De Paepe, A., van Vooren, S., Vermeesh, J., et al.: Array CGH base: an analysis platform for comparative genomic hybridization microarrays. BMC Bioinformatics 6(124), 179–187 (2006)

[5] Pinkel, D., Albertson, D.G.: Array comparative genomic hybridization and its applications in cancer. Nature Genetics 37, 11–17 (2005)

[6] Rosa, P., Viara, E., Hupé, P., Pierron, G., Liva, S., Neuvial, P., Brito, I., Lair, S., Servant, N., Robine, N., Manié, E., Brennetot, C., Janoueix-Lerosey, I., Raynal, V., Gruel, N., Rouveirol, C., Stransky, N., Stern, M., Delattre, O., Aurias, A., Radvanyi, F., Barillot, E.: VAMP: Visualization and analysis of array-CGH transcriptome and other molecular profiles. Bioinformatics 22(17), 2066–2073 (2006)

[7] Wang, P., Young, K., Pollack, J., Narasimham, B., Tibshirani, R.: A method for callong gains and losses in array CGH data. Biostat. 6(1), 45–58 (2005)

[8] Xia, X., McClelland, M., Wang, Y.: WebArray, an online platform for microarray data analysis. BMC Bionformatics 6(306), 1737–1745 (2005)

[9] Van de Wiel, M.A., Kim, K.I., Vosse, S.J., Van Wieringen, W.N., Wilting, S.M., Ylstra, B.: CGHcall: calling aberrations for array CGH tumor profiles. Bioinformatics 23(7), 892–894 (2007)

[10] Argente, E., Botti, V., Carrascosa, C., Giret, A., Julian, V., Rebollo, M.: An abstract architecture for virtual organizations: The THOMAS approach. Knowledge and Information Systems 29(2), 379–403 (2011)

[11] Smith, M.L., Marioni, J.C., Hardcastle, T.J., Thorne, N.P.: snapCGH: Segmentation, Normalization and Processing of aCGH Data Users' Guide. Bioconductor (2006)

[12] Kim, S.Y., Nam, S.W., Lee, S.H., Park, W.S., Yoo, N.J., Lee, J.Y., Chung, Y.J.: ArrayCyGHt, a web application for analysis and visualization of array-CGH data. Bioinformatics 21(10), 2554–2555 (2005)

Causal Maps for Explanation in Multi-Agent System

Aroua Hedhili, Wided Lejouad Chaari, and Khaled Ghédira

Abstract. All the scientific community cares about is understanding the complex systems, and explaining their emergent behaviors. We are interested particularly in Multi-Agent Systems (MAS). Our approach is based on three steps : observation, modeling and explanation. In this paper, we focus on the second step by offering a model to represent the cause and effect relations among the diverse entities composing a MAS. Thus, we consider causal reasoning of great importance because it models causalities among a set of individual and social concepts. Indeed, multi-agent systems, complex by their nature, their architecture, their interactions, their behaviors, and their distributed processing, needs an explanation module to understand how solutions are given, how the resolution has been going on, how and when emergent situations and interactions have been performed. In this work, we investigate the issue of using causal maps in multi-agent systems in order to explain agent reasoning.

Keywords: Multi-Agent Systems, Explanation, Reasoning, Causal Maps.

1 Introduction

Multi-agent systems are developed in various domains such as computer networks, industrial applications, process control, air traffic, simulation, etc. In spite of the rapid growth of the international interest in MAS field and the high number of developed applications, there is no global control on their execution, and no one

Aroua Hedhili · Wided Lejouad Chaari
Strategies of Optimization and Intelligent Computing Laboratory (SOIE), National School of Computer Studies (ENSI)-University of Manouba
e-mail: aroua.hedhili@gmail.com, wided.chaari@ensi.rnu.tn

Khaled Ghédira
Strategies of Optimization and Intelligent Computing Laboratory (SOIE), High School of Management (ISG) - University of Tunis ISG Tunis, 41, Rue de la Liberte, Cite Bouchoucha 2000 Le Bardo, Tunis -TUNISIE
e-mail: khaled.ghedira@isg.rnu.tn

A. Abraham and S.M. Thampi (Eds.): Intelligent Informatics, AISC 182, pp. 183–191.
springerlink.com

knows what effectively happens inside, which steps are performed, how knowledge is shared, how interactions are exchanged, how the solution path is computed, and how results are obtained. Because of this ignorance, agent behaviors are not always clearly reproducible for humans. Our objective is to give the user the possibility to become familiar with such dynamic, complex and abstract systems, to understand the way to manage the nondeterministic process, how solutions are given, how the resolution has been going on, how and when emergent situations and interactions have been performed. Explanation gives an answer to these questions. It gives information about reasoning, actions, interactions and events executed "inside" one agent or among all agents. We can consider the MAS as a set of Knowledge Based Systems (KBSs), in this case the different reasons for explanation in this area belong to MASs like, it presents a demonstration tool to show how the resolution system is well-adapted to the used knowledge or it describes the represented knowledge of the field and the used inference techniques for learning purposes or for assistance to the resolution process, etc. However, the need of explanation in MAS is related to some other reasons :

- Agent behaviors are not always clearly reproducible for humans.
- During the execution, the multi-agent system is considered as a "black box".
- Explanation fosters the control of an agent ; the expert could have a control on the agent if he/she knows "how" and "why" the agent has done an action.
- Explanation is a way to detect and understand the emergence phenomenon in MAS since Daniel Memmi [4] points out that the emergence is restricted to the problem of description and explanation and it is an example of scientific explanation.

Besides, the existing works presented in the literature are focused on explanations in kBSs and particularly expert systems [5, 7, 10]. There are a few research works related to multi-agent systems, they remain specific to some MAS applications. After a deep research, we found an explanation facility called Java Platform Annotation Processing Architecture (JPAPA) [9]. The main idea of this method is that agent software should be able to give information about itself. Using the framework JPAPA, the programmer puts information into the source code. The programmer annotates those parts of the software that are important for the end-user to understand the behavior of an agent. Annotations are similar to comments in a source code; the difference exists in the intended recipient of the annotation content, who is the end-user instead of the programmer. Also, the source code comments are skipped whereas the content of annotations is visible even at runtime. This method is specific to Java applications and it just describes how the agent has done the action. Another approach proposed in [11] considers that the agent should explain its action. The explanation is generated by recalling the situation in which the decision was made and replaying the decision under variants of the original situation. The severe limitation of this solution is the overload of the agent ; the agent should resolve a problem and give an explanation; which has an effect on the system performance.

Our work consists in establishing a generic methodology to explain reasoning in MAS based on causal maps. In this context, Chaib braa, [2], mentioned in his research that this methodology provides a foundation to explain how agents have done actions, but he did not detailed his idea. We consider that to understand and explain a complex process, we need to observe it. For this purpose, we developed first an observation module which detects the agents' events and describes each agent activity in an intelligent trace [1], the figure 1 illustrates an example of the trace.

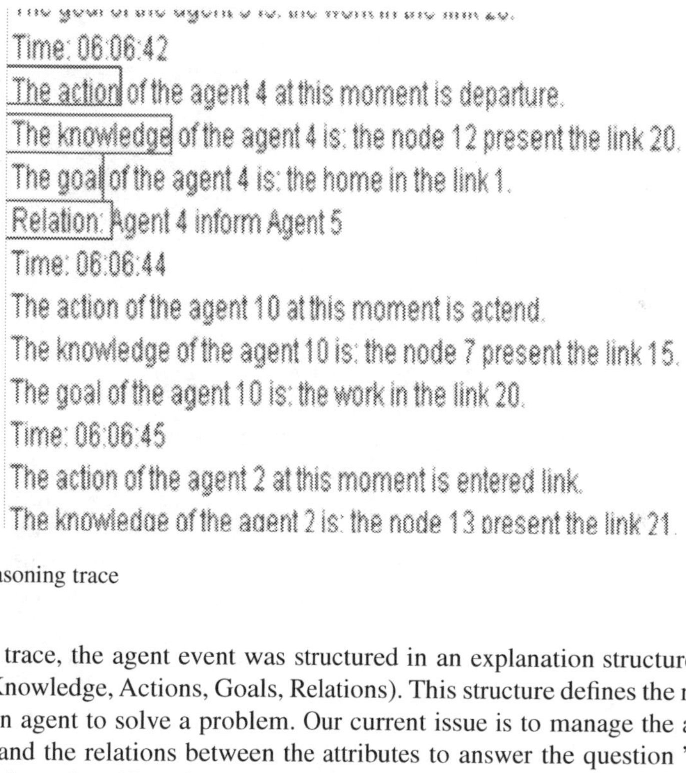

Time: 06:06:42
The action of the agent 4 at this moment is departure.
The knowledge of the agent 4 is: the node 12 present the link 20.
The goal of the agent 4 is: the home in the link 1.
Relation: Agent 4 inform Agent 5
Time: 06:06:44
The action of the agent 10 at this moment is actend.
The knowledge of the agent 10 is: the node 7 present the link 15.
The goal of the agent 10 is: the work in the link 20.
Time: 06:06:45
The action of the agent 2 at this moment is entered link.
The knowledge of the agent 2 is: the node 13 present the link 21.

Fig. 1 Reasoning trace

In this trace, the agent event was structured in an explanation structure labeled KAGR (Knowledge, Actions, Goals, Relations). This structure defines the resources used by an agent to solve a problem. Our current issue is to manage the attributes structure and the relations between the attributes to answer the question "how the action has been done?". In this paper, we highlight the use of causal maps to achieve this issue.

2 Causal Maps

Causal Maps or Cognitive Maps (CM) [2] are represented as a directed graph where the basic elements are simple. The concepts are represented as points, and the causal links between these concepts are represented as arrows between these points. The strategic alternatives; all of the various causes, effects can be considered as concept variables, and represented as points in the causal map. Causal relationships can take different values based on the most basic ones "+" (positive) such as (promotes, enhances, helps, is benefit to, etc.), "-" (negative) as (hurts, prevents, is harmful to, retards, etc.), and "0" (neutral) such as (has no effect on, does not matter for, etc.). With this graph, it is relatively easy to see how concepts are related to causal relationships and to see the overall causal relationships of one concept with another. For instance, the CM of the figure 2, explains how a PhD student makes his week planning. We consider that in a week, a PhD student (the agent) should prepare

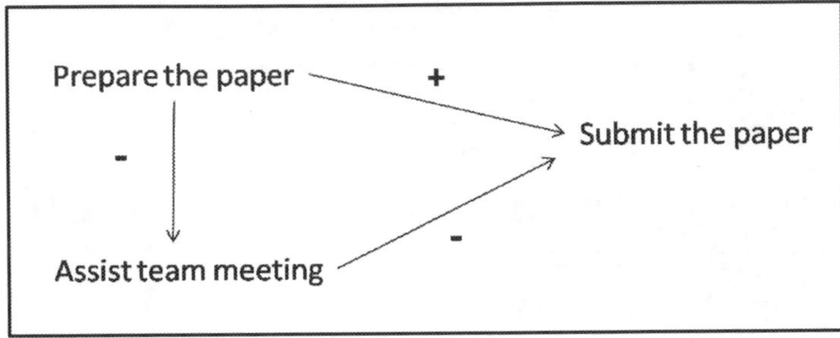

Fig. 2 A cognitive map

his paper to submit it and assist the team meeting to expose his research work. Indeed, this portion of a CM states that "Prepare the paper" is favorable to "Submit the paper" but it is harmful to "Assist team meeting". Also, "Assist team meeting" is harmful to the concept "Submit the paper". This shows that a CM is a set of concepts as "Prepare the paper", "Submit the paper", "Assist team meeting", and a set of signed edges representing causal relations like "promote(s)", "enhance(s)", "decrease(s)", etc. In fact, the real power of a causal map resides in its representation by a graph. The graph model makes then relatively easy to see how concepts are linked to causal relationships.

Usually, a CM is employed to cope with the causal reasoning. Causal reasoning is useful in multi-agent environments because it shows the presence of causal relationships and the logical effects produced when some events occur. Causal maps were widely addressed in multi-agent environment for several goals. Most of the works retrieved in the literature deal with causal maps as a mean to make a decision in distributed environment [2, 6]; to analyze or to compare the causal representation of agents for coordination or for conflict resolution [2]; to facilitate the complex phenomena for students in agent based learning systems [3]. Therefore, we find no further mention of using CM to explain a reasoning process in an intelligent, complex, and distributed systems. It is obvious for us to experiment it.

3 Reasoning Explanation Using Causal Maps

We extract the links between the KAGR attributes associated to an agent and we analyze the existent combinations of these elements in order to understand the agent reasoning. Each attribute is presented with two fields. The first field contains the number of elements in the attribute. The second field defines the elements. The first example is related to the event "LeaveHomeEvent" with the blue representation of the explanation structure. The second presents the event "TakeCarEvent" with the orange representation. Notice that, the explanation structure has two colors to indicate that it is different at the two moments.

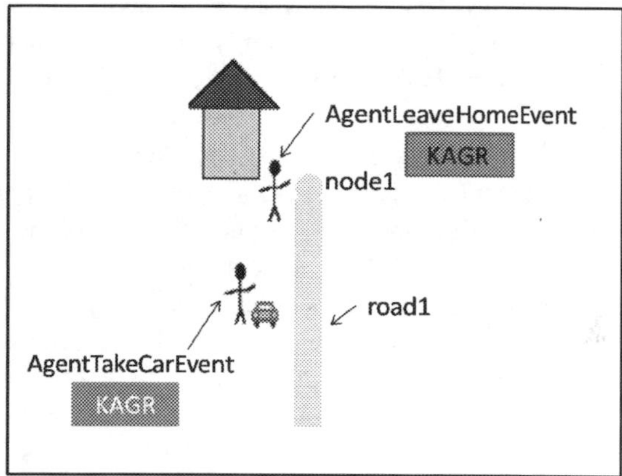

Fig. 3 Agent explanation structure

The detected event "LeaveHomeEvent" through the observation module has the following structure's attributes :

- The attribute K has the value "1" in the first field, it means that the agent has one knowledge, the second field has the definition K1="the node 1 presents the home".
- The attribute A has the value "1" in the first field, it means that the agent has done one action at the moment t. The action is described by the second field which has the definition A1="cross the road 1".
- The attribute G has the value "2" in the first field, the second field has the definition G1="go to work" as a global goal and G2=" leave the home" as an intermediate goal.
- The attribute R has the value "0".

Then, the second detected event "TakeCarEvent" has this explanation structure :

- The attribute K has the value "1" in the first field, it means that the agent has one knowledge, the second field has the definition K1="the car is in front of the house".
- The attribute A has the value "2" in the first field, it means that the agent has done two actions at a moment t. The actions are described by the second field which has the definition A1="Go straight 4 steps" and A2="send a message to agent 2".

- The attribute G has the value "2" in the first field, the second field has the definition G1="go to work" as a global goal and G2="take the car" as an intermediate goal.
- The attribute R has the value "1" in the first field, the second field has the following definition R1="the sent message to agent 2 contains: the car is not here".

Our solution consists in modeling each attribute of KAGR by a causal map. So, each agent has a causal map at four levels:

1. The first level treats the agent goals, deduced from the attribute "G" of the agent structure. This level is labeled "CM_Goals". It shows the causal relations between the agent goals. The agent goals are the concepts of the map. This level exhibits the links between sub-goals and their sequence to reach the local goal then the MAS functionalities under agent goals.
2. The second level concerns the agent actions, deduced from the attribute "A" of the agent structure, this level is labeled "CM_Actions". In this map, the concepts are the actions. CM_Actions is linked to CM_Goals to establish the actions done by the agent to achieve a goal from the first map. Moreover, through this level, the causal relations between agent actions are illustrated.
3. The third level presents the agent knowledge in a CM labeled "CM_Knowledge", deduced from the attribute "K" of the agent structure, in order to fix how the agent has done an executed action in the CM_Actions.
4. The fourth level presents the agent relations, "CM_Relations", deduced from the attribute "R" of the agent structure. This level is linked to the second one when the agent cooperates with others to execute its actions. It is also linked to the third level when the agent interacts with others to enrich its knowledge.

We notice that for each agent, four instances of a CM are respectively associated to the attributes of the explanation structure. Dependencies links are then constructed among those instances. The explanation process is performed following an incremental method starting from agents and leading to the group of agents. The global CM corresponding to the whole system is constructed thanks to the relationships between the agents CMs at a higher level. This point will be discussed in a future work since it is in progress.

4 Experimentation

The proposed approach was tested on a multi-agent application of transport network simulation MATSim [8]. This application provides a toolbox to implement large-scale agent-based transport simulations. The tested scenarios consist in achieving the goal "Go home" from a fixed position. Thus, the CM_Goals contains just this concept. The agent should follow a plan of links (streets) and nodes (crossroads) to accomplish its actions. In the first scenario, the agent leaves the node 1 to attain the home situated in the link 12, it follows a set of streets that contains the links 1, 2, 7 and 12. There are two possible actions "left link" and "entered link". The following figure illustrates the CM_Actions and the CM_Knowledge of the agent created from an XML file. The XML file contains the explanation structure KAGR of the agent.

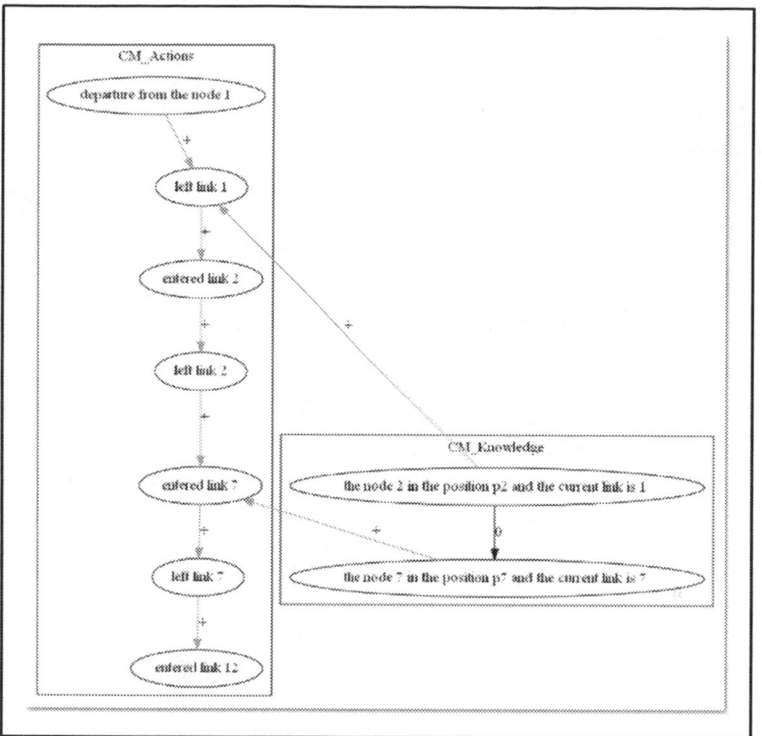

Fig. 4 Agent Causal Map 1

As shown in the figure 4 the agent accomplishes a collection of actions presented in the CM_Actions in order to achieve its goal. The link between the actions are green to mention the value '+'. In fact, we consider that the relations between these concepts have the positive value, each one promotes the other: the concerned agent will be able to realize the action "left link 1" since the action "departure from the node 1" was achieved. Moreover, the concept "the node 2 in the position p2 and the current link is 1" enhances the concept "left link 1". In this scenario, the agent uses its own knowledge, presented in the sub-graph CM_Knowledge. The relation between the knowledge concepts has the neutral value, there is no effect between them.

The second scenario presents a different agent behavior. The agent takes a car to go home. The followed plan contains the streets 1 and 2. An agent "traffic light" is also situated between the two streets. The figure 5 depicts the actions achieved in the CM_Actions. We note that the agent uses its own knowledge, in CM_Knowledge, as "the node 2 in the position p2 and the current link is 2". Besides, through CM_Relations, we deduce that the agent accomplishes the actions "stop at the traffic light" and "pass the traffic light" after receiving a message from the agent "traffic light" and it achieves the action "take the car" after receiving the message "the car is in the garage" from another agent identified "agent 1".

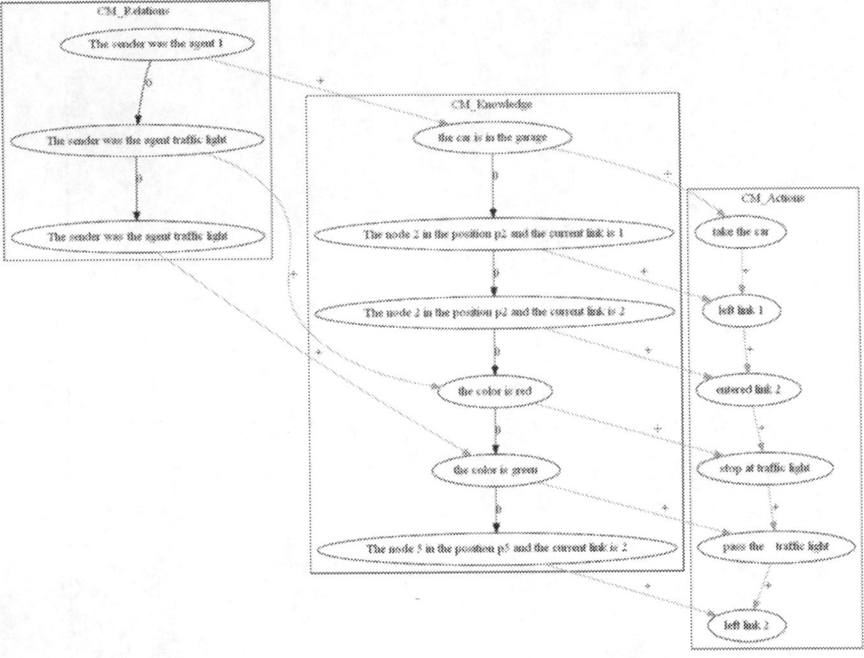

Fig. 5 Agent Causal Map 2

5 Conclusion

In this paper, we point out the issue of using causal maps to explain reasoning in MAS. The concepts of the map were retrieved directly from an intelligent trace and the value of arrows was deduced from the reasoning process done by an agent to achieve a goal. This approach constructs for each agent a sub-set of linked causal maps associated with the KAGR structure. The idea is to emphasize the individual agent role and its participation in the resolution process. In a future work, an experimentation based on the explanation of a rescue multi-agent system is going on to validate a more complex causal map structure. Then, we wish to deal with the relationships at a global level including an agent organization CM linking the sub-sets of smaller CM expressed at the agent level. Thus, we transform the approach from an explanation at fine granularity to higher granularity reflecting the switch from the agent to the group.

References

1. Hedhili, A.: Explication du raisonnement dans les systèmes multi-agents par l'observation. Report of Research Master degree. National School of Computer Studies, Tunis (2009)

2. Brahim, C.-D.: Causal Maps: Theory, Implementation and Practical Applications in Multiagent Environments. IEEE Transactions on Knowledge and Data Engineering 14, 1201–1217 (2002)
3. Basu, S., Biswas, G.: Multiple Representations to Support Learning of Complex Ecological Processes in Simulation Environments. In: Proceedings of the 19th International Conference on Computers in Education, Chiang Mai, Thailand (2011)
4. Daniel, M.: Emergence et niveaux d'explication. Journées thématiques de l'ARC (émergence et explication) (1996)
5. Dieng, R.: Explanatory Knowledge tools for expert systems. In: 2nd International Conference on Applications of A.I. to Engineering, Cambridge, M.A., USA (1987)
6. Druckenmiller, D.A., Acar, W.: Exploring agent-based simulation of causal maps: toward a strategic decision support tool. Doctoral Dissertation, Kent State University Kent, OH, USA (2005)
7. Spinelli, M., Schaaf, M.: Towards explanations for CBR-based applications. In: Hotho, A., Stumme, G. (eds.) Proceedings of the LLWA Workshop, Germany, pp. 229–233 (2003)
8. Nagel, K., Axhausen, K.W., Balmer, M., Meister, K., Rieser, M.: Agent-based simulation of travel demand, Structure and computational performance of MATSim-T. In: 2nd TRB Conference on Innovations in Travel Modeling, Portland (2008)
9. Ludwig, B., Schiemann, B., et al.: Self-describing Agents. Department of Computer Science 8. University Erlangen-Nuremberg (2008)
10. Swartout, W., Moore, J.: Explanation in second generation expert systems. In: David, J.-M., Krivine, J.-P., Simmons, R. (eds.) Second Generation Expert Systems, pp. 543–585. Springer (1993)
11. Lewis Johnson, W.: Agents that Explain Their Own Actions. In: Proceedings of the Fourth Conference on Computer Generated Forces and Behavioral Representation (1994)

Hierarchical Particle Swarm Optimization for the Design of Beta Basis Function Neural Network

Habib Dhahri*, Adel M. Alimi, and Ajith Abraham

Abstract. A novel learning algorithm is proposed for non linear modeling and identification by the use of the beta basis function neural network (BBFNN). The proposed method is a hierarchical particle swarm optimization (HPSO). The objective of this paper is to optimize the parameters of the beta basis function neural network (BBFNN) with high accuracy. The population of HPSO forms multiple beta neural networks with different structures at an upper hierarchical level and each particle of the previous population is optimized at a lower hierarchical level to improve the performance of each particle swarm. For the beta neural network consisting n particles are formed in the upper level to optimize the structure of the beta neural network. In the lower level, the population within the same length particle is to optimize the free parameters of the beta neural network. Experimental results on a number of benchmarks problems drawn from regression and time series prediction area demonstrate that the HPSO produces a better generalization performance.

1 Introduction

Architecture design of the artificial neural network can be formulated as an optimization problem. Evolutionary Algorithms (EA) like neural network (NNs)

Habib Dhahri · Adel M. Alimi
REsearch Group on Intelligent Machines (REGIM), University of Sfax, National School of Engineers (ENIS), BP 1173, Sfax 3038, Tunisia

Ajith Abraham
Faculty of Electrical Engineering and Computer Science, Technical University of Ostrava, Czech Republic

Ajith Abraham
Machine Intelligence Research Labs (MIR Labs), Scientific Network for Innovation and Research Excellence, WA, USA
e-mail: ajith.abraham@ieee.org

* Corresponding author.

A. Abraham and S.M. Thampi (Eds.): Intelligent Informatics, AISC 182, pp. 193–205.
springerlink.com

represents an evolving technology inspired by biologically motivated computational paradigms. NNs are derived from the brain theory to simulate the learning behavior of an individual, while EAs are developed from the theory of evolution to evolve whole population toward a better fitness. Although these two technologies seem quite different in the time period action, the number of involved individuals and the process scheme, their similar dynamic behaviors stimulate research on whether a synergistic combination of these two technologies may provide more problem solving power than either alone. Without the help of a proven guideline, the choice of an optimal network design for a given problem is a difficult process. In this paper, we deal with the so-called Beta Basis Function Neural Network BBFNN (Alimi, 2000) that represents an interesting alternative in which we can approximate any function. One of the most important issues in the BBF neural network applications is the network learning, i.e., to optimize the adjustable parameters, which include the centers vectors, the widths of the basis functions and the parameters forms. Another important issue is to determine the network structure or the number of BBF nodes.

Several attempts have been proposed for this, to optimize the parameters artificial neural network such us the pruning algorithm (Stepniewski, 1997) and the growing algorithm (Guang-Bin, 2005). Unfortunately, these techniques show many deficiencies and limitations (Angeline, 1994).

The applying of evolutionary algorithms to construct neural nets is also well known in the literature. The most representative algorithms include Genetic Algorithms (Sexton, 1999), Flexible Neural Trees (Chen & Yang, 2005), (Chen & Abraham, 2009) Particle Swarm Optimization (Dhahri, 2008), (Dhahri & Alimi, 2010) and the method of Differential Evolution (Subudhi, 2011).

The PSO algorithm has been shown to perform better than the Genetic Algorithm (GA) or the Differential evolution (DE) over several numerical benchmarks (Xu, 2007).

The ordinary prototype of PSO algorithm operates on a species with fixed-length particle, which is viewed as a set of potential solutions to a problem. The fixed-length PSOs that are usually thought of as optimizers operating within a fixed parameter space, variable-length PSO may be applied to design problems in which the individual can have a variable number of components such as in our application when the size of the beta basis function neural network are variable.

Building a hierarchical Beta neural system is a difficult task. This is because we need to define the architecture of the system, as well as the free parameters of each neuron.

Two approaches could be used to tackle this problem. One approach is that an expert supplies all the required knowledge for building the system. The other one is to use optimization techniques to construct the system.

The aim of the proposed paper is to extend the work (Dhahri, 2010) that focus only on optimizing the beta neural parameters with a fixed structure. In this paper we want to verify if the PSO performs in the automatic designing of the BBFNN, including not only the parameters transfer functions but also architecture. As we will see, the architecture obtained is optimal in the sense that the number of connections is minimal without losing efficiency.

In order to design optimal neural network architecture, this work proposes a new topology using hierarchical particles swarm optimization. In the upper level, the HPSO concentrates on the definition of the structure of the BBFNN. At the lower level, the PSO with a fixed dimension is applied to determine the free parameters of the beta basis function neural network. Therefore, no assumption is made about the topology of the BBFNN or the other neural information.

This work is organized as follow: Section 2 describes the basic BBF network and the concept of opposition based particle swarm optimization. The hierarchical opposition based particle swarm optimization is the subject of section 3. The set of experimental results are provided in section 4. Finally, the work is discussed concluded in section 5.

2 The Beta Basis Function Neural Network (BBFNN) and PSO Algorithm

This section first describes the basic concept of the BFNN to be designed in this study. The basic concept of PSO employed in BBFNN optimization is then described.

2.1 *The Beta Basis Function Neural Network (BBFNN)*

In this section, we want to introduce the beta basis function neural network that will be used in the remainder of this paper. The BBF Neural Network (BBFNN) is a three-layer feed-forward neural network which consists of the input layer, the hidden layer and the output layer. Each neuron of the hidden layer employs a beta basis function as nonlinear transfer function to operate the received input vector and emits the output value to the output layer. The output layer implements a linear weighted sum of the hidden neurons and yields the output value.

Besides the centre $c \in IR^n$ the beta basis function may also present a width parameter $\sigma \in IR^n$, which can be seen as a scale factor for the distance $\|x - c\|$ and the parameter forms p_i and q_i.

Fig.1 shows the schematic diagram of the BBF network with n_i inputs, n beta basis function, and one output units.

If the BBF network consists of n beta basis function and the outputs units are linear, the response of the *i*th y*i* output units is

$$y_i = w_i^T b = \sum_{i=1}^{n} w_{ij} B_i(x, c_i, \sigma_i, p_i, q_i), 1 \le i \le n_0 \qquad (1)$$

In general, five types of adjustable parameters which should be determined for BBF neural network: the centre vector $c = [c_1, ..., c_m]$, the width vector, $\sigma = [\sigma_1, ..., \sigma_m]$, the parameters former $p = [p_1, ..., p_m]$, $q = [q_1, ..., q_m]$ and W is the weight matrix $W = [w_{i0}, w_{i1}, ..., w_{im}]^T$.

The Beta basis function $B_i(x, c_i, \sigma_i, p_i, q_i)$, i =1,…,n, is defined by:

$$\beta_i(x) = \begin{cases} \left[1 + \frac{(p_i + q_i)(x - c_i)}{\sigma_i p_i}\right]^{p_i} \left[1 - \frac{(p_i + q_i)(c_i - x)}{\sigma_i q_i}\right]^{q_i} & if \quad x \in]x_0, x_1[\\ 0 \quad else & \end{cases} \tag{2}$$

Where $p_i > 0$, $q_i > 0$, x_0, x_1 are the real parameters, $x_0 < x_1$ and

$$c_i = \frac{p_i x_1 + q_i x_0}{p_i + q_i} \tag{3}$$

In the multi-dimensional case, the beta function is defined by

$$\beta(c, \sigma, p, q)(x) = \prod_{i=1}^{i=n} \beta_i(c^i, \sigma^i, p^i, q^i) \tag{4}$$

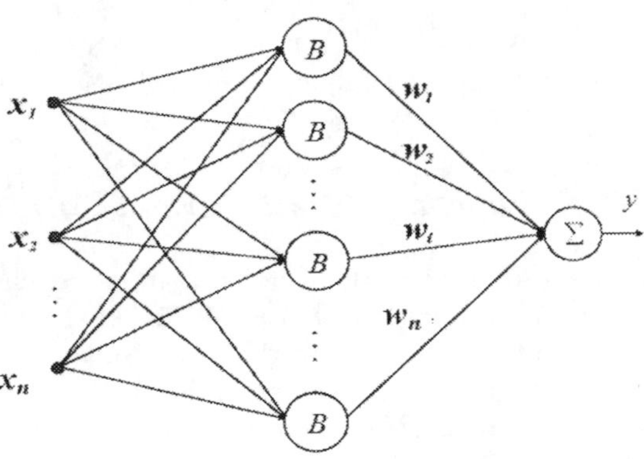

Fig. 1 The architecture of BBFNN

2.2 *The Basic Particle Swarm Optimization Algorithm*

PSO was introduced by Kennedy and Eberhart (Kennedy, 1995) is inspired by the swarming behavior of animals and human social behavior. A particle swarm is a population of particles, where each particle is a moving object that 'flies' through the search space and is attracted to previously visited locations with high fitness. In contrast to the individuals in evolutionary computation, particles neither reproduce nor get replaced by other particles. Each particle consists of a position vector x, which represents the candidate solution to the optimization problem, the fitness of solution x, a velocity vector v and a memory vector p of the best candidate solution encountered by the particle with its recorded fitness. The position of a particle is updated by

$$x_i^d = x_i^d + v_i^d \tag{5}$$

and its velocity according to

$$v_i^d = w_i v_i^d + c_1\phi_1(pbest_i^d - x_i^d) + c_2\phi_2(gbest^d - x_i^d) \tag{6}$$

where ϕ_1, ϕ_2 are uniform distributed random numbers within ϕ_{min}, ϕ_{max} (typically $\phi_{min} = 0$) and ($\phi_{max} = 1$) that determine the weight between the attraction to position $pbest_i^d$, which is the best position found by the particle and $gbest^d$ the overall best position found by all particles. A more general version of PSO considers $gbest^d$ as the best position found in a certain neighborhood of the particle, which does not generally contribute to performance improvements. Note that ϕ_1 and ϕ_2 are generated for each component of the velocity vector. Moreover, the so-called inertia weight w controls how much the particles tend to follow their current direction compared to the memorized positions $pbest_i^d$ and $gbest^d$. Instead of the inertia weight w, one can use another parameter, the so-called constriction factor χ, which is multiplied with the entire equation in order to control the overall velocity of the swarm. In the preliminary parameter tuning experiments it turned out that the tuning of the velocity update rule using the inertia weight yielded clearer and better results. Finally, the velocity of the particles is limited by a maximum velocity v_{max}, which is typically half of the domain size for each parameter in vector x_i^d. The initialization of the algorithm PSO algorithm (using randomly chosen object feature vectors from the data set), but additionally requires the initialization of the velocity vectors, which are uniformly distributed random numbers in the interval [$-v_{max}$, v_{max}]. After initialization, the memory of each particle is updated and the velocity and position update rules are applied. If a component of the velocity v_i^d of a particle exceeds v_{max} it is truncated to this value, Moreover, if a component of the new position vector is outside the domain, it is moved back into the search space by adding twice the negative distance with which it exceeds the search space and the component of the velocity vector is reversed. This process is applied to all particles and repeated for a fixed number of iterations. The optimization result is the best recorded candidate solution ($gbest^d$ in the last iteration) and fitness at the end of the run.

2.3 The Opposition Based Particle Swarm Optimization Algorithm

All the techniques based on the evolutionary algorithm generate randomly the solutions as an initial population. These candidates' solutions specified by the objective function will be changed to generate new solutions. The best individual in evolved population can be misleading because the applying of the heuristic operators (selection, mutation and crossover) or the update of the position in PSO algorithm can steer the population toward the bad solutions. Consequently, the convergence to the guess value is often computationally very expensive. To escape to these drawbacks of the evolutionary algorithms, the dichotomy search is proposed to accelerate the convergence rate and to ameliorate the generalization performance. In other word the search space is halved in two sup-spaces, in order to decrease the convergence time. Let be [a, b] the search space of the solution s and x is the solution belong to the first half. We define a concept called the opposite number to look for the guess solution in the second half. The OPSO (Dhahri & Alimi, 2010) algorithm uses the idea of opposite number to create the initial and the evolved populations to improve the performance of PSO technique. In the rest of this section, we define the concept of opposite number in single dimension and multi-dimension.

Definition: Let x be a real number in the interval [a, b], the opposing number $\overline{x}$ is defined as follows:

$$\begin{cases} \overline{x} = \alpha\left(\frac{a+b}{2}+x\right) if \quad x \prec \frac{a+b}{2} \\ \overline{x} = \alpha\left(\frac{a+b}{2}-x\right) if \quad x \succ \frac{a+b}{2} \end{cases} \qquad \alpha \in [0,1] \tag{7}$$

Definition: Let $M=(x_1, x_2, x_3, \ldots, x_n)$ be a point in the n–dimensional space, where $x_1, x_2, x_3, \ldots, x_n \in IR^n$ and $\forall i \in \{1,2\ldots n\}$, $x_i \in [a_i,b_i]$. The opposing point of M is defined by $\overline{M}\ (\overline{x}_1, \overline{x}_2 \ldots, \overline{x}n)$ where its components are defined as:

$$\begin{cases} \overline{x}_i = \alpha_i\left(\frac{a_i+b_i}{2}+x_i\right) if \quad x_i \prec \frac{a_i+b_i}{2} \\ \overline{x}_i = \alpha_i\left(\frac{a_i+b_i}{2}-x_i\right) if \quad x_i \succ \frac{a_i+b_i}{2} \end{cases} \qquad \alpha_i \in [0,1] \tag{8}$$

The initial velocities, vi(0), i = 1, . . . , NP , of all particles are randomly generated.

Step1 (*Particle evaluation*): Evaluate the performance of each particle in the population according to the beta neural system. In this paper, the evaluation function f is defined as the RMSE error .According to f, we can find individual best position pbest of each particle and the global best particle gbest .

Step 2 (*Velocity update*): At iteration t, the velocity v_i of each particle i is updated using its individual best position pbest, and the global best position, gbest. Here, the following mutation operator t is adopted

$$v_i^d = w_i v_i^d + c_1\varphi_1(pbest_i^d - x_i^d) + c_2\varphi_2(gbest^d - x_i^d) \tag{9}$$

Where c_1 and c_2 are positive constants, φ_1 and φ_2 are uniformly distributed random numbers in [0,1], and w_i controls the magnitude of v_i .

Step 3 (*Position update*): Depending on their velocities, each particle changes its position according to the crossover operator:

$$P_i^d = P_i^d + v_i^d \tag{10}$$

$$OP_i^d = OP_i^d + v_i^d \tag{11}$$

Step 4 (*End criterion*): The OPSO learning process ends when a predefined criterion is met. In this paper, the criterion is the goal or total number of iterations.

3 Hierarchical Multi-dimensional Opposition Based-PSO

In this section, an automatic design method of hierarchical beta basis function neural network is presented. The hierarchical structure is created and optimized using particle swarm optimization and the fine turning of the neuron's parameters encoded in the structure is accomplished using opposition –based particle swarm optimization algorithm.

The hierarchical Beta neural system not only provides a flexible architecture for modeling nonlinear systems, but can also reduce the neurons number. The problems in designing a hierarchical beta basis function include the following:

- selecting an appropriate hierarchical structure;
- optimizing the free parameters and the linear weights.

The Combining of the structure optimization of the BBFNN by PSO algorithm and the parameter optimization ability of the opposition based particle swarm optimization lead to the following algorithm for designing the beta basis function neural network.

---

1. Set the initial values of parameters used in the PSO. Create the initial population.
2. Do structure optimization using PSO algorithm as described in Section 2.
3. If the better BBFNN architecture found, then go to step 4), otherwise go to step 2). The criteria concerning with better structure is the fitness of each individual.
4. Parameter optimization using OPSO algorithm as described in last section. In this step, the BBFNN architecture is predetermined, and it is the best BBFNN taken from the end of run of the PSO algorithm. All of the neurons' parameters encoded in the best BBFNN will be optimized by OPSO algorithm order to decrease the fitness value.
5. If the maximum numbers of OPSO algorithm then go to step 6); otherwise go to step 4).
6. If satisfactory solution is found, then stop; otherwise go to step 2).

3.1 Higher Level Configuration

This subsection section introduces the higher level, which consists of multiple populations (species). The proposed algorithm contains multiple swarms, each of which contains *Psi* particles. A species forms a group of particles that have the same number of particles and is only responsible to optimize the architecture of the beta basis function neural network.

Furthermore, each particle is defined within the context of a topological neighborhood that is made up of itself and other particles in the multi-swarm. At each iteration t, a new velocity for a particle i is obtained by using the best position $p_{si}(t)$ and the neighborhood best position $p^{g}_{si}(t)$ of the multi-swarm. In other words, instead of having the best position and the neighborhood best position of one a swarm, we define the best position and the neighborhood best position of each swarm in the multi swarm.

Instead of operating at a fixed dimension, the HPSO algorithm is designed to seek the optima dimension. In the HPSO algorithm, the main idea is to evaluate a new velocity value for each particle according to its current velocity, the distance from the global best position. The new velocity value is then used to calculate the next position of the particle in the search space. This process is then iterated a number of times or until a minimum error is achieved. This opens the question of exactly how create the new velocity from particles with different lengths.

In order to make this hypothesis feasible, we must take into account if we pass from the dimension i to the dimension j (i < j), we add (j-i) axis of supplementary coordinates that will be orthogonal to the i axes. In other words, we project the vectors on the space where the dimension is the maximum size of the three vectors.

Fig.3 shows a sample of projection operator P which can be expressed as follows:

$$P(x_i) = [x_i \quad zeros(1, \max(x_1, x_2, x_3, x_4) - l(x_i))], \ 1 \le i \le 4 \tag{12}$$

Where [.] is vector, zeros is zeros array, max is the maximum of the four vectors and l is the length of the vector x_i. Once the projection P operator is applied, the classical variant of Particle swarm optimization PSO is used

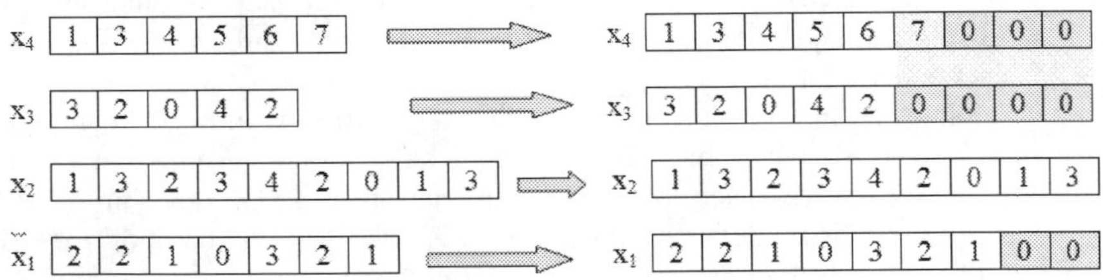

Fig. 2 Projection operator in one dimension

The selection of the best beta basis function neural network by the proposed algorithm requires the evaluation of each particle swarm. The performance – evaluation criterion used in this paper is the root mean square error (RMSE) between the desired and actual outputs.

The lower level configuration of HPSO is responsible for particle optimization. For each of sub swarm, the particles are sorted according to their performance. The best performing one is selected in this level.

4 Computational Experiments

The proposed algorithm HPSO is also compared with DE-NN and ODE-NN for the time series prediction problem. For DE and ODE parameter setting, the population size is 50, the upper and lower bounds of weights is [0, 1], the mutation constant M is 0.6, the crossover constant C is 0.5 and the jumping probability Jr is 0.3. For HPSO algorithm the population size is 50. The constant c1 and c2 are 2 and φ_1 and φ_2 are uniformly distributed random numbers in [0,1].

4.1 Nonlinear Plant Control

In this example, the plant to be controlled is expressed by

$$y_p(t+1) = \frac{y_p(t)\left[y_p(t-1)+2\right]\left[y_p(t)+2.5\right]}{8.5+\left[y_p(t)\right]^2+\left[y_p(t-1)\right]^2} + u(t) \tag{13}$$

The same plant is used in (Subudhi, 2011). The current output of the plant depends on two previous outputs values and one previous input values. The input *u(k)* was assumed to be random signal uniformly in the interval [-2, 2]. The identification model be in the form of

$$y_{pi}(t+1) = f(y_p(t), y_p(t-1)) + u(t) \tag{14}$$

Where $f(y_p(t), y_p(t-1))$ is the nonlinear function of $y_p(t)$ and $y_p(t-1)$ which will be the inputs for HPSO-BBFNN neural system identifier. The output from neural network will be y_{pi}. In this experiment, 500 training patterns are generated to train the BBFNN network and 500 for the testing data. After training, the following same test signal *u(k)* of the other compared models is used for testing the performance of BBFNN models:

$$u(t) = \begin{cases} 2\cos(2\pi t/100) & if \quad t \le 200 \\ 1.2\sin(2\pi t/20) & if \quad 200 < t \le 500 \end{cases} \tag{15}$$

The RMSE value, which is taken as the performance criterion. The parameters of HPSO were chosen as the previous example. The Fig.3 shows the actual and predicted output of the plant for the test signal with the beta BBFNN model. From the figures it is clear that desired output and the identified by HPSO is nearly the same.

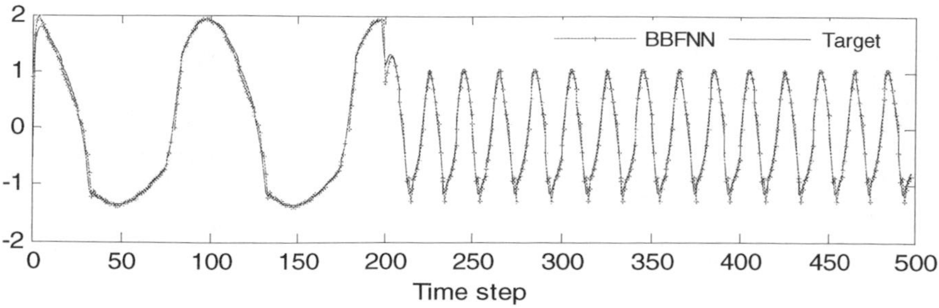

Fig. 3 HPSO identification performance

Table 1 Comparison of training and testing errors

Input	Training error (RMSE)				Testing error (RMSE)			
	DE	ODE	HMDDE	HPSO	DE	ODE	HMDDE	HPSO
y(k),y(k – 1) u(k)	0.0207	0.0190	0.0190	0.0150	0.1186	0.11370	0.110	0.010

Table 1 gives the comparison of performance HPSO for the design of Beta Basis Function Neural Network to DE based methods for Artificial Neural Network. The comparison of these methods is applied in terms of root mean squared error (RMSE). From the results it is clear that the proposed HPSO algorithm has a root mean squared error test (RMSE) of 0.10 with four beta basis function neural net. Finally it is concluded that the proposed HPSO is having better identification performance than that of the other approaches.

4.2 Box and Jenkins' Gas Furnace Problem

In this section, the proposed HPSO –BBFNN are applied to the Box-Jenkins time series data (gas furnace data) which have been intensively studied as a benchmark problem in previous literature (Chen, 2006), (Pox, 1970). The data set originally

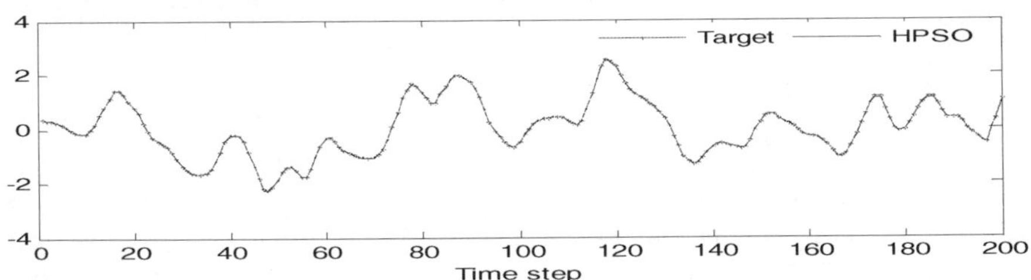

Fig. 4 Training of Box-Jenkins time series ((y (t-1), u(t-2))

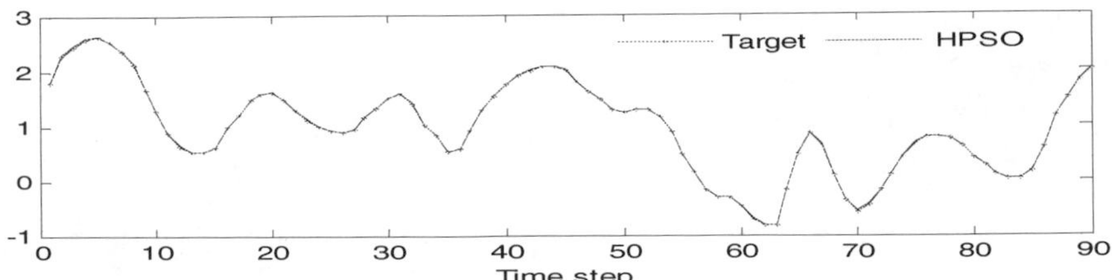

Fig. 5 Identification performance (y(t-1), u(t-2))

Table 2 Comparison of training and testing errors of Box and Jenkins

Input	Training error (RMSE)				Testing error (RMSE)			
	DE	ODE	HMDDE	HPSO	DE	ODE	HMDDE	HPSO
y(t – 1), u(t – 3)	0.1501	0.1411	**0.1328**	**0.01477**	0.4400	0.4194	**0.2276**	**0.2168**
y(t – 3), u(t – 4)	0.3402	0.2850	**0.0210**	**0.0200**	0.7838	0.7773	**0.4224**	**0.0012**
y(t – 2), u(t – 4)	0.3256	0.2898	**0.1365**	**0.1469**	0.6733	0.6602	**0.3200**	**0.2136**
y(t – 1), u(t – 2)	0.2909	0.2924	**0.1735**	**0.020**	0.4906	0.6801	**0.2334**	**0.0052**
y(t – 1), u(t – 4)	0.2991	0.2926	**0.2411**	**0.2258**	0.5430	0.5132	**0.3745**	**0.3876**
y(t – 4), u(t – 4)	0.3274	0.3428	**0.1594**	**0.2782**	12.259	0.8894	**0.4549**	**0.3101**
y(t – 2), u(t – 3)	0.2968	0.3051	**0.1702**	**0.0200**	1.1340	0.7199	**0.2700**	**0.0027**
y(t – 1), u(t – 1)	0.4638	0.4151	**0.1598**	**0.1696**	0.6183	0.6056	**0.2577**	**0.2291**
y(t – 4), u(t – 3)	0.7266	0.4301	**0.1921**	**0.2018**	1.2405	1.2771	**0.6148**	**0.2476**
y(t – 1), u(t – 6)	0.6012	0.5661	**0.6619**	**0.6253**	0.8469	0.8410	**0.6638**	**0.5762**
y(t – 3), u(t – 3)	0.5172	0.5176	**0.1600**	**0.1696**	1.0067	1.0347	**0.2521**	**0.2273**
y(t – 2), u(t – 2)	0.6314	0.6261	**0.1615**	**0.1686**	0.9889	0.9753	**0.2773**	**0.2283**
y(t – 1), u(t – 5)	0.6220	0.6303	**0.3333**	**0.3332**	0.6873	0.6518	**0.5595**	**0.5226**
y(t – 4), u(t – 5)	0.7038	0.6373	**0.0178**	**0.020**	1.0149	0.9698	**0.0203**	**0.0028**
y(t – 2), u(t – 1)	0.8934	0.6844	**0.1960**	**0.2022**	1.8368	1.2726	**0.2759**	**0.2478**
y(t – 2), u(t – 5)	0.7222	0.6804	**0.2165**	**0.2213**	0.9176	1.1808	**0.4021**	**0.3843**
y(t – 3), u(t – 5)	0.7138	0.7338	**0.1346**	**0.1567**	0.9536	1.0470	**0.2307**	**0.2233**
y(t – 3), u(t – 2)	0.8766	0.8600	**0.2128**	**0.2022**	1.8184	1.4138	**0.2760**	**0.2461**
y(t – 4), u(t – 6)	1.3988	1.1126	**0.1379**	**0.146**	1.7628	1.4677	**0.2635**	**0.2138**
y(t – 2), u(t – 6)	1.6264	1.1945	**0.3389**	**0.3334**	1.3352	1.2639	**0.5590**	**0.5264**
y(t – 4), u(t – 2)	1.1799	1.1963	**0.2152**	**0.2205**	1.6725	1.6377	**0.2737**	**0.2336**
y(t – 3), u(t – 6)	1.2063	1.2424	**0.2175**	**0.2229**	27.468	1.4641	**0.4027**	**0.3899**
y(t – 3), u(t – 1)	1.5725	1.2702	**0.2135**	**0.2203**	1.7123	1.6475	**0.2803**	**0.2535**
y(t – 4), u(t – 1)	1.4250	1.4352	**0.2270**	**0.2267**	2.0821	2.0217	**0.2695**	**0.2386**

consist of 296 data points [y(t),u(t)]. For the design of the experiment, the delayed term of the observed gas furnace process data, y(t), is used as system input variables made up of by ten terms given as follows: y(t – 1), y(t – 2), y(t – 3), y(t – 4), y(t – 5) u(t – 1), u(t – 2), u(t – 3), u(t – 4), u(t – 6). Consequently, the effective number of data points is reduced to 296 providing 100 for training and 190 samples for testing. Here we have taken two inputs for simplicity one is from furnace output and other is from furnace input so we have build 24 models of different input and output. The criterion used was the Root Mean Square Error (RMSE). Each case is trained for 2000 epochs and the number of neurons belongs to the interval [2, 10]. Table 2 gives the training and testing performances of these 24 models. As can be seen from this table, HPSO-BBFNN model is powerful for Box-Jenkins process in training. Compared with the recent results presented in (Subudhi, 2011), we can see that the proposed algorithm can achieve accuracy with a smaller number of nodes.

The training and testing gas furnace process data are shown in Fig. 4. Fig. 5 shows the predicted time series and the desired time series. From the Table 2, it is clear that HPSO is having less training and testing errors in comparison to DE, ODE, and HMDDE counterpart. The RMSE for testing turned out to be the least for 24 cases in HPSO-BBFNN approach.

5 Conclusion

This paper proposes a new hierarchical evolutionary BBFNN based on PSO algorithm. The integration of PSO and OPSO enables the BBFNN to dynamically evolve its architecture and adapts its parameters simultaneously. The design of the topology of BBFNN is defined automatically at the higher level of the proposed algorithm, whereas in lower level, we optimize the free parameters of beta neural network. The second contribution of this work is to propose the multi-dimensional particle swarm optimization which represents the key point in determining the optimal number of beta basis function neural network. The experimental results on two problems demonstrate that HPSO is capable of evolving the BBFNN with good general generalization ability.

Acknowledgments. The authors would like to acknowledge the financial support of this work by grants from General Direction of Scientific Research (DGRST), Tunisia, under the ARUB program. Ajith Abraham acknowledges the support from the framework of the IT4Innovations Centre of Excellence project, reg. no. CZ.1.05/1.1.00/02.0070 supported by Operational Programme 'Research and Development for Innovations' funded by Structural Funds of the European Union and state budget of the Czech Republic.

References

Alimi, A.: The Beta System: Toward a Change in Our Use of Neuro-Fuzzy Systems. International Journal of Management, 15–19 (2000)

Chen, Y.Y.: Time series prediction using a local linear wavelet neural network. Neurocomputing 69(4-6), 449–465 (2006)

Chen, Y., Abraham, A.: Tree-Structure based Hybrid Computational Intelligence: Theoretical Foundations and Applications. Intelligent Systems Reference Library Series. S. Verlag, Germany (2009)

Chen, Y., Yang, B.D.: Time Series Forecasting Using Flexible Neural Tree Model. Information Sciences 174(3-4), 219–235 (2005)

Dhahri, H.A.: The modified particle swarm optimization for the design of the Beta Basis Function neural networks. In: Proc. Congress on Evolutionary Computation, Hong Kong, China, pp. 3874–3880 (2008)

Dhahri, H., Alimi, A.K.: Opposition-based particle swarm optimization for the design of beta basis function neural network. In: International Joint Conference on Neural Networks, IJCNN, Barcelona, Spain, pp. 18–23 (2010)

Dhahri, H., Alimi, A.K.: Opposition-based particle swarm optimization for the design of beta basis function neural network. In: International Joint Conference on Neural Networks, IJCNN, Barcelona, Spain, pp. 18–23 (2010)

Guang-Bin, H.S.: A generalized growing and pruning RBF (GGAP-RBF) neural network for function approximation. IEEE Transactions on Neural Networks 16(1), 57–67 (2005)

Juang, C.: Hierarchical Cluster-Based Multispecies Particle-Swarm optimization for Fuzzy-System Optimization. IEEE Transactions on Fuzzy Systems 18(1), 14–26 (2010)

Kennedy, J.E.: Particle swarm optimization. In: International Conference on Neural Networks, pp. 1942–1948 (1995)

Mackey, M.G.: Oscillation and chaos in physiological control systems. Science 197, 287–289 (1977)

Pox, G.E.: Time series analysis, forecasting and control. Holden day, San Francisco (1970)

Sexton, R.D.: Optimization of Neural Networks: A Comparative Analysis of the Genetic Algorithm and Simulated Annealing. European Journal of Operational Research 114, 589–601 (1999)

Stepniewski, S.W.: Pruning Back-propagation Neural Networks Using Modern Stochastic Optimization Techniques. Neural Computing & Applications 5, 76–98 (1997)

Subudhi, B.: A differential evolution based neural network approach to nonlinear system identification. Applied Soft Computing 11(1), 861–871 (2011)

Xu, X.L.: Comparison between Particle Swarm Optimization, Differential Evolution and Multi-Parents Crossover. In: International Conference on Computational Intelligence and Security, pp. 124–127 (2007)

Fuzzy Aided Ant Colony Optimization Algorithm to Solve Optimization Problem

Aloysius George and B.R. Rajakumar

Abstract. In ant colony optimization technique (ACO), the shortest path is identified based on the pheromones deposited on the way by the traveling ants and the pheromones evaporate with the passage of time. Because of this nature, the technique only provides possible solutions from the neighboring node and cannot provide the best solution. By considering this draw back, this paper introduces a fuzzy integrated ACO technique which reduces the iteration time and also identifies the best path. The proposed technique is tested for travelling sales man problem and the performance is observed from the test results.

Keywords: Fuzzy logic (FL), ACO, Travelling sales man problem, fuzzy rules, shortest path.

1 Introduction

Difficult combinatorial optimization problems can be solved by the nature-inspired technique called ACO [12] in a moderate amount of computation time [3]. ACO simulates the behavior of ant colonies in identifying the most efficient routes from their nests to food sources [7]. Dorgio initiated the idea of identifying good solutions to combinatorial optimization problems by imitating the behavior of ants [9]. Ants when searching for food initially search the region adjacent to their nest in a random manner. The ant estimates the quantity and the quality of the food as soon as the food source is identified and takes a portion of it back to the nest [10]. An aromatic essence termed as pheromone is used by real ants to communicate with each other [13]. A moving ant marks the path by a succession of pheromone by laying some of this substance on the ground [11]. Both the length of the paths and the quality of the located food source determine the quantity of the pheromone dropped on the paths [13]. The pheromones have to evaporate for a longer time if it takes more time for the ant to travel to its nest and return again [2]. The reason for the ants to select the shorter path has been found

Aloysius George · B.R. Rajakumar
Research & Development, Griantek, Bengaluru, India
aloysius_g@griantek.com, rajakumar@ieee.org

A. Abraham and S.M. Thampi (Eds.): Intelligent Informatics, AISC 182, pp. 207–215.
springerlink.com

to be due to the pheromone concentration deposited mostly on such paths [15]. Therefore, computational problems which can be downsized to determination of good paths can be solved by ACO through graphs by using the concept of "ants" [6]. By employing this concept, a population of artificial ants that searches for optimal solutions is created by the combinatorial optimization problem creates according to the constraints of the problem [16].

Combinatorial optimization problems such as Routing problem (e.g., Traveling Salesman Problem (TSP) and Vehicle Routing Problem (VRP)), Assignment problem (e.g., Quadratic Assignment Problem), Scheduling problem (e.g., Job Shop) and Subset problem (e.g., Multiple Knapsack, Max Independent Set) extensively employ the ACO algorithms [5,14, 19]. A majority of these applications require exponential time in the worst case to determine the optimal solution as they belong to NP-hard problems [21] [20]. In VRP the vehicle returns to the same city from where it started after visiting several cities [4].

2 Related Works

Some of the recent research works related to ant colony optimization are discussed below.

Chen *et al.* [22] have introduced a two-stage solution construction rule possessing two-stage ACO algorithm to solve the large scale vehicle routing problem. Bin *et al.* [23] have proposed an enhanced ACO to solve VRP by incorporating a new strategy called ant-weight strategy to correct the increased pheromone, and a mutation operation. Tao *et al.* [24] have proposed a fuzzy mechanism and a fuzzy probable mechanism incorporating unique fuzzy ACS for parameter determination.

Chang *et al.* [17] have proposed an advanced ACO algorithm to improve the execution of global optimum search. Berbaoui *et al.* [18] have proposed the optimization of FLC (Fuzzy Logic Controller) parameters of SAPF through the application of the ACO algorithm. Gasbaoui *et al.* [1] have proposed an intelligent fuzzy-ant approach for the identification of critical buses and optimal location and size of capacitor banks in electrical distribution systems.

Salehinejad *et al.* [8] have discussed a multi parameter route selection system employing an ACS, where the local pheromone has been updated using FL for detecting the optimum multi parameter direction between two desired points, origin and destination.

3 Shortest Path Identification Using ACO and Fuzzy Logic

In ACO, the shortest path is identified based on the pheromones deposited on the way by the traveling ants and the pheromones evaporate with the passage of time. Thus, based on the pheromones the ants identify the shortest path to reach the food from their home. There are some disadvantages in using this method because the local optimization technique only provides possible solutions from the neighboring node and cannot provide the best solution. Due to this draw back the iteration time is large and also the best solution is identified based on the selected best path in the first iteration. So here we propose a fuzzy integrated ACO technique to overcome this drawback. Here FL is used for selecting the shortest path weight between two places so that the number of iterations is less and this reduces the consumed time. Here, we have selected the TSP for demonstrating the optimization problem.

TSP is one of the most important problems in practical consideration. The main process is to identify the shortest path for the customer who is traveling from one place to another place. There may be different routes for reaching one place from another place. For selecting the shortest path two important conditions are considered, they are.

- The cost must be low
- Distance and time must also be low
- The customer starts from one place and ends at the same place itself with one condition that the customer should meet each cities once.

3.1 Fuzzy Integrated Ant Colony Optimization for TSP

FL is used for selecting shortest path for ants in ant colony optimization. By selecting the weightage of each path using fuzzy the time consumption is reduced considerably and also we get the correct shortest path. To accomplish this, firstly, the weight for each path between the two cities in forward and reverse directions is calculated and by using these values, FL rules are generated.

Let N be the number of cities to be traveled by the salesman, i be the source point of the salesman, j be the destination point of the salesman. The limit for i and j is $1 \le i \le N$ and $1 \le j \le N$.

The probability of moving from state i to state j is

$$P_{ij} = \frac{\tau_{ij} . \eta_{ij}}{\sum \tau_{ij} . \eta_{ij}} \tag{1}$$

where, τ_{ij} is the amount of pheromone deposited for transition from state i to state j and η_{ij} is the desirability of state transition ij.

By using the above equation the probability for each path between two cities are calculated. The iteration is said to be the number of iteration in which the ant is moving in the path i to j. And the contribution weight is the inverse of the cost value from the original data set.

3.1.1 Generating Fuzzy Rules

The most significant step of the proposed fuzzy integrated ACO is to generate fuzzy rules to feed fuzzy intelligence to the ants. Triangular membership function is used as the membership grade to train the FL. Probability, iteration and contribution weight are given as input to the FL and fuzzy weight matrix is obtained as the output. The inputs to the FL are fuzzified into three sets; they are high, medium and low. Similarly the output is fuzzified into two sets; they are high and low. By considering these input and output variables the fuzzy rules are generated, which are mentioned in Table 1.

Table 1 Fuzzy rules using AND logic

S.No	Fuzzy Rules
1	*if* $P = high$ *and* $I = low$ *and* $W = low$, *then* $F = low$
2	*if* $P = high$ *and* $I = low$ *and* $W = medium$, *then* $F = low$
3	*if* $P = high$ *and* $I = low$ *and* $W = high$, *then* $F = high$
4	*if* $P = high$ *and* $I = medium$ *and* $W = low$, *then* $F = low$
5	*if* $P = high$ *and* $I = medium$ *and* $W = medium$, *then* $F = high$
6	*if* $P = high$ *and* $I = medium$ *and* $W = high$, *then* $F = high$
7	*if* $P = high$ *and* $I = high$ *and* $W = low$, *then* $F = high$
8	*if* $P = high$ *and* $I = high$ *and* $W = medium$, *then* $F = high$
9	*if* $P = high$ *and* $I = high$ *and* $W = high$, *then* $F = high$
10	*if* $P = medium$ *and* $I = low$ *and* $W = low$, *then* $F = low$
11	*if* $P = medium$ *and* $I = low$ *and* $W = medium$, *then* $F = low$
12	*if* $P = medium$ *and* $I = low$ *and* $W = high$, *then* $F = high$
13	*if* $P = medium$ *and* $I = medium$ *and* $W = low$, *then* $F = high$
14	*if* $P = medium$ *and* $I = medium$ *and* $W = medium$, *then* $F = high$
15	*if* $P = medium$ *and* $I = medium$ *and* $W = high$, *then* $F = high$
16	*if* $P = medium$ *and* $I = high$ *and* $W = low$, *then* $F = high$
17	*if* $P = medium$ *and* $I = high$ *and* $W = medium$, *then* $F = high$
18	*if* $P = medium$ *and* $I = high$ *and* $W = high$, *then* $F = high$
19	*if* $P = low$ *and* $I = low$ *and* $W = low$, *then* $F = low$
20	*if* $P = low$ *and* $I = low$ *and* $W = medium$, *then* $F = high$
21	*if* $P = low$ *and* $I = low$ *and* $W = high$, *then* $F = high$
22	*if* $P = low$ *and* $I = medium$ *and* $W = low$, *then* $F = low$
23	*if* $P = low$ *and* $I = medium$ *and* $W = medium$, *then* $F = high$
24	*if* $P = low$ *and* $I = medium$ *and* $W = high$, *then* $F = high$
25	*if* $P = low$ *and* $I = high$ *and* $W = low$, *then* $F = high$
26	*if* $P = low$ *and* $I = high$ *and* $W = medium$, *then* $F = high$
27	*if* $P = low$ *and* $I = high$ *and* $W = high$, *then* $F = high$

3.2 Shortest Path Identification

ACO is used here to obtain the shortest path between the source and the destination point. The fuzzy weight matrix obtained from the FL is used to identify the shortest path. To calculate the shortest path, first the probability of moving from the source to destination i.e. i to j for all the possible combinations is calculated using the equation-1. The pheromone deposited for moving from one city to another city is to be updated during all iterations. The amount of pheromone to be deposited is calculated for each movement of ant from one city to another city.

The amount of pheromone to be deposited is calculated using the equation 2.

$$\Delta\tau_{ij} = \begin{cases} \dfrac{Q}{L_k}; \text{if ant k uses curve ij in its tour} \\ 0; otherwise \end{cases} \quad (2)$$

Where, $\Delta\tau_{ij}$ is the amount of pheromone deposited, Q is a constant and L_k is the constant for the k_{th} ant's tour.

After calculating the amount of pheromone to be deposited the next process is to calculate the pheromone to be updated. The pheromone to be updated mainly depends on the pheromone evaporation coefficient and the amount of pheromone deposited.

Pheromone update is calculated using the formula given below,

$$\tau_{ij}^{new} = (1-\rho).\tau_{ij} + \Delta\tau_{ij} \quad (3)$$

Where, ρ is the pheromone evaporation coefficient.

By updating the pheromone, the above process is repeated and from the result obtained by calculating the probability for every possible path, the shortest path for the corresponding route is identified. The two different paths calculated are the forward and the reverse path.

Forward path is said to be the path in which the ant moves from i to j. The cost function for the path is calculated based on the distance travelled by the salesman. For example we get one cost function by calculating the probability for city 1 to 2. The cost function for city 2 to 1 is also calculated for the returning path of the ant. This path is said to be a reverse path. The cost function is calculated for the path of city 2 to 1. The cost function of the path from city 1 to 2 and 2 to 1 are different. The forward and reverse cost functions are calculated for each path.

This fuzzy integrated ant colony optimization obtains the shortest path very quickly. The time required is very low because of the integration of FL with the ant colony optimization. The shortest path obtained is accurate because the weight matrix is obtained by considering the probability, iteration and contribution weight.

4 Results and Discussion

The proposed technique was implemented using MATLAB 7.10 and for testing we generated an asymmetric data that sales man traveling for 50 cities. After testing with the proposed fuzzy integrated ant colony optimization, the results are compared with the conventional ant colony optimization technique. The performance of proposed method is identified clearly from the comparison graph. Here, we selected paths by sending 30 and 20 ants for 50 tours respectively for every set of ants and evaluated the Total cost consumed by the proposed ACO and conventional ACO, which are tabulated in Table 2.

Table 2 Cost comparison between proposed and conventional ACO under different Test cases

Test Cases	Proposed ACO ($)	Conventional ACO ($)
30 Ants, 50 Tours	1211	1309
20 Ants, 50 Tours	1251	1313

4.1 30 Ants traveling for Different Tours and Identifying Shortest Paths

Here, using proposed technique we identified shortest path for 30 ants traveling in 50 tours is discussed. Initially we see the shortest path obtained for 30 ants travelling in 50 tours. Fig 2, shows the four different shortest paths identified using proposed method and Fig-3 shows the convergence comparison graph between the proposed method and conventional ACO for 30 ants traveling in 50 cities.

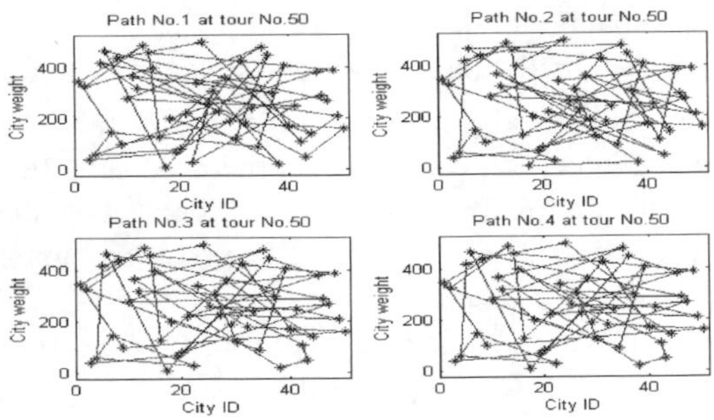

Fig. 1 Shortest path for traveling 50 cities identified using proposed method for 50 tours

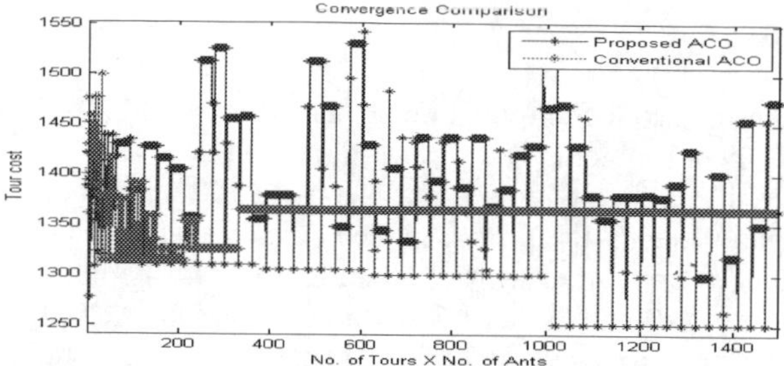

Fig. 2 Convergence comparison graph between the proposed ACO and conventional ACO

4.2 20 Ants Traveling for Different Tours and Identifying Shortest Paths

Here, from our proposed technique we identified shortest path for 20 ants traveling in 50 tours. Initially we see the shortest path obtained for 20 ants travelling in 50 tours. Fig 3 shows the four different shortest paths identified using proposed method for 20 ants travelling for 50 tours and Fig 4 shows convergence comparison graph between the proposed method and conventional ACO for 20 ants traveling in 50 cities.

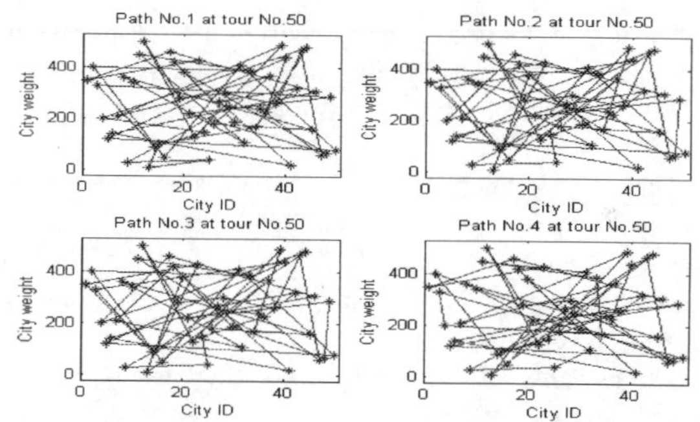

Fig. 3 Shortest path for traveling 50 cities identified using proposed method at tour number 50

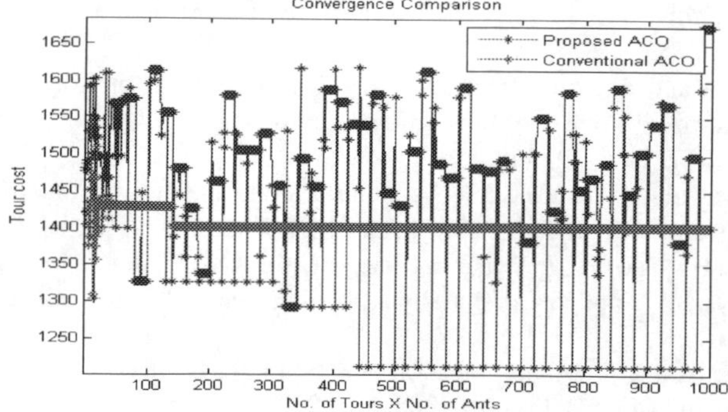

Fig. 4 Convergence comparison graph between the proposed ACO and conventional ACO

5 Conclusion

In this paper, the proposed technique is implemented in MATLAB 7.10 and tested for travelling salesman problem. The proposed method is tested for different ants, by sending different number of tours and identified different shortest path. The most important parameter in the proposed method is cost; the minimum tour cost of proposed method is compared with conventional ACO. The minimum tour cost for proposed ACO for 20 ants is 1251 $ & for conventional ACO 1313 $ and the minimum tour cost for proposed ACO for 30 ants is 1211 $ & for conventional ACO 1309 $. By using the shortest path identified using the proposed method; the total cost is low and also the time to meet all the cities also low. From the above results it is clear that the minimum tour cost of proposed method is very low when compared with conventional ACO method.

Acknowledgement. The authors would like to thank the anonymous reviewers and Editor-in-Chief for their valuable comments that helped to improve their manuscript. Also, the authors thank the International Institute of Applied Sciences and Technology (IIAST) for its partial support in providing the necessary supporting and learning articles and tools.

References

1. Gasbaoui, B., Chaker, A., Laoufi, A., Abderrahman, A., Allaoua, B.: Optimal Placement and Sizing of Capacitor Banks Using Fuzzy-Ant Approach in Electrical Distribution Systems. Leonardo Electronic Journal of Practices and Technologies 9(16), 75–88 (2010)
2. Baterina, A.V., Oppus, C.: Image Edge Detection Using Ant Colony Optimization. International Journal of Circuits, Systems and Signal Processing 4(2), 25–33 (2010)
3. Rezapour, O.M., Dehghani, A., Shui, L.T.: Review of Ant Colony Optimization Model for Suspended Sediment Estimation. Australian Journal of Basic and Applied Sciences 4(7), 2099–2108 (2010)
4. Nallusamy, R., Duraiswamy, K., Dhanalakshmi, R., Parthiban, P.: Optimization of Multiple Vehicle Routing Problems Using Approximation Algorithms. International Journal of Engineering Science and Technology 1(3), 129–135 (2009)
5. Chen, C., Tian, Y.X., Zou, X.Y., Cai, P.X., Jin, Y.M.: A Hybrid Ant Colony Optimization for the Prediction of Protein Secondary Structure. Chinese Chemical Letters 16(11), 1551–1554 (2005)
6. Darshni, P., Kaur, G.: Implementation of ACO Algorithm for Edge Detection and Sorting Salesman Problem. International Journal of Engineering, Science and Technology 2(6), 2304–2315 (2010)
7. Bella, J., McMullenb, P.: Ant colony optimization techniques for the vehicle routing problem. Advanced Engineering Informatics 18, 41–48 (2004)
8. Salehinejad, H., Talebi, S.: Dynamic Fuzzy Logic-Ant Colony System-Based Route Selection System. Applied Computational Intelligence and Soft Computing 2010, 1–13 (2010)
9. Toksari, D.: Ant colony optimization for finding the global minimum. Applied Mathematics and Computation 176, 308–316 (2006)

10. Dorigo, M., Blumb, C.: Ant colony optimization theory: A survey. Theoretical Computer Science 344, 243–278 (2005)
11. Dorigo, M., Maniezzo, V., Colorni, A.: The Ant System: Optimization by a colony of cooperating agents. IEEE Transactions on Systems, Man, and Cybernetics–Part B 26(1), 1–13 (1996)
12. Georgilakis, Vernados, Karytsas: An ant colony optimization solution to the integrated generation and transmission maintenance scheduling problem. Journal of Optoelectronics and Advanced Materials 10(5), 1246–1250 (2008)
13. Bouhafs, L., Hajjam, A., Koukam, A.: A Hybrid Heuristic Approach to Solve the Capacitated Vehicle Routing Problem. Journal of Artificial Intelligence: Theory and Application 1(1), 31–34 (2010)
14. Negulescu, S., Kifor, C., Oprean, C.: Ant Colony Solving Multiple Constraints Problem: Vehicle Route Allocation. Int. J. of Computers, Communications & Control 3(4), 366–373 (2008)
15. Saeheaw, T., Charoenchai, N., Chattinnawat, W.: Application of Ant colony optimization for Multi-objective Production Problems. World Academy of Science, Engineering and Technology 60, 654–659 (2009)
16. Thangavel, Karnan, Jeganathan, lakshmi, P., Sivakumar, Geetharamani: Ant Colony Algorithms in Diverse Combinational Optimization Problems - A Survey. ACSE Journal 6(1), 7–26 (2006)
17. Chang, Y.H., Chang, C.W., Lin, H.W., Tao, C.W.: Fuzzy Controller Design for Ball and Beam System with an Improved Ant Colony Optimization. World Academy of Science, Engineering and Technology 56, 616–621 (2009)
18. Berbaoui, B., Benachaiba, C., Dehini, R., Ferdi, B.: Optimization of Shunt Active Power Filter System Fuzzy Logic Controller Based On Ant Colony Algorithm. Journal of Theoretical and Applied Information Technology 14(2), 117–125 (2010)
19. Foundas, E., Vlachos, A.: New Approaches to Evaporation in Ant Colony Optimization Algorithms. Journal of Interdisciplinary Mathematics 9(1), 179–184 (2006)
20. Tripathi, M., Kuriger, G., Wan, H.D.: An Ant Based Simulation Optimization for Vehicle Routing Problem with Stochastic Demands. In: Proceedings of the Winter Simulation Conference, Austin, pp. 2476–2487 (2009)
21. Maria, L., Stanislav, P.: Parallel Posix Threads based Ant Colony Optimization using Asynchronous Communications. Journal of Applied Mathematics 2(2), 229–238 (2009)
22. Chen, C.H., Ting, C.H.: Applying Two-Stage Ant Colony Optimization to Solve the Large Scale Vehicle Routing Problem. Journal of the Eastern Asia Society for Transportation Studies 8, 761–776 (2010)
23. Bin, Y., Zhen, Y.Z., Baozhen, Y.: An improved ant colony optimization for vehicle routing problem. European Journal of Operational Research 196, 171–176 (2009)
24. Tao, C.W., Taur, J.S., Jeng, J.T., Wang, W.Y.: A Novel Fuzzy Ant Colony System for Parameter Determination of Fuzzy Controllers. International Journal of Fuzzy Systems 11(4), 298–307 (2009)

Self-adaptive Gesture Classifier Using Fuzzy Classifiers with Entropy Based Rule Pruning

Riidhei Malhotra*, Ritesh Srivastava, Ajeet Kumar Bhartee, and Mridula Verma

Abstract. Handwritten Gestures may vary from person to person. Moreover, they may vary for same person, if taken at different time and mood. Various rule-based automatic classifiers have been designed to recognize handwritten gestures. These classifiers generally include new rules in rule set for unseen inputs, and most of the times these new rules are distinguish from existing one. However, we get a huge set of rules which incurs problem of over fitting and rule base explosion. In this paper, we propose a self adaptive gesture fuzzy classifier which uses maximum entropy principle for preserving most promising rules and removing redundant rules from the rule set, based on interestingness. We present experimental results to demonstrate various comparisons from previous work and the reduction of error rates.

Keywords: Handwritten Gesture Classifier, Fuzzy Classifier, Fuzzy Logic, Entropy, Rule Pruning.

1 Introduction

Classification is a machine learning technique used to predict group membership for data instances. Classification techniques appear frequently in many applications areas, and have become the basic tool for almost any pattern recognition task.

Riidhei Malhotra
Department of Information Technology, Galgotias College of Engineering & Technology, Greater Noida (U.P.), India
e-mail: riidhei@gmail.com

Ritesh Srivastava · Ajeet Kumar Bhartee
Department of Computer Science & Engineering,
Galgotias College of Engineering & Technology, Greater Noida (U.P.), India
e-mail: {ritesh21july,ajeetkbharti}@gmail.com

Mridula Verma
Indian Institute of Technology Patna, India
e-mail: verma.mridula@gmail.com

* Corresponding author.

A. Abraham and S.M. Thampi (Eds.): Intelligent Informatics, AISC 182, pp. 217–223.
springerlink.com

In many applications, it is necessary to update an existing classifier in incremental fashion to accommodate new data, without compromising classification performance on old data, in handwritten gesture application, for instance. The advantage of online handwritten gesture classifier is that, it is built on-the-fly, from scratch and uses only few data, and incrementally adopts new unseen classes at any moment in the lifelong learning process. With the rapid development of gesture language, it nowadays has been applied in many fields such as human-computer interaction, visual surveillance.

Good feature selection is essential for gesture classification to make it tractable for machine learning, and to improve classification performance. *Shannon's entropy* and two *Bayesian* scores are available for columns that contain discrete and discretized data. In this work, we are using Shannon's Entropy Method. By comparing the features of an unknown gesture with the existing ones stored in the database, it is possible to identify the type of the gesture examined.

Abdullah Almaksour Eric Anquetil [14] proposed a hand written gesture classification system which is able to efficiently classify basic signs gestures for handwritten gesture recognition application. For gesture classification and recognition, different learning structures and algorithms as well as fuzzy logic sets are used. An automatic self adaptive Handwritten Gesture Classifier suffers from problems of rules over-fitting, because there is a huge possibility of addition of new rules in rule set for unseen inputs, most of the times these new rules are distinguish from existing one. Ultimately, we get a huge set of rules which incurs problem of over-fitting and rule base explosion. In this work, we have used a method ANFIS *(Adaptive-Network-based Fuzzy Inference System)* [2], [3] from Fuzzy Logic [4] for the task of classification.

A brief description of the related work is presented in section 2. Data collection is explained in section 3. Section 4 discusses proposed approach and results. Section 5 contains the conclusion and future work.

2 Related Work

The presented paper combines an incremental clustering algorithm with a fuzzy adaptation method in [1], in order to learn and maintain the model that is the handwritten gesture recognition system. The self-adaptive nature of this system allows it to start its learning process with few learning data, to continuously adapt and evolve according to any new data, and to remain robust when introducing a new unseen class at any moment in the life-long learning process.

Format of fuzzy implications and reasoning algorithm are used, for the method of identification of a system using its input-output data, as shown in [5]. The architecture and learning procedure underlying ANFIS is presented in [6], which is a fuzzy inference system implemented in the framework of adaptive networks. Takagi-Sugeno (TS) based on a novel learning algorithm [7] that recursively updates TS model structure and parameters, It applies new learning concept to the TS model called Evolving Takagi-Sugeno model (ETS). An online evolving fuzzy Model (efM) approach [8] to modeling non-linear dynamic systems, in which an incremental learning method is used to build up the rule-base, the rule-base is evolved when "new" information becomes available by creating a new rule or deleting an old rule depended upon the proximity and potential of the rules, and the maximum number of rules to be used in the rule-base. Various works have been done using maximum entropy approach [9], [10], [11].

Statistical modeling addresses the problem of constructing a stochastic model to predict the behavior of a random process. In constructing this model, we typically have a sample of output at our disposal from the process. Given this sample, which represents an incomplete state of knowledge about the process, the modeling problem is to parlay this knowledge into a representation of the process. After that, we can use this representation to make predictions about the future behavior about the process.

2.1 Maximum Entropy Modeling

We consider a random process that produces an output value y, a member of a finite set Y. In generating y, the process may be influenced by some contextual information x, a member of a finite set X.

Our task is to construct a stochastic model that accurately represents the behavior of the Random process. Such a model is the method of estimating the conditional probability, that a given context x, the process will output y. The probability that the model assigns to y in context to x is denoted by $p\ (y|x)$. The entire conditional probability distribution provided by the model is denoted by $p\ (y|x)$. We will denote by P the set of all conditional probability distributions. Thus a model, $p\ (y|x)$ is just an element of P. Here, we have considered y as class and x as feature i.e.

$$\mathbf{y} = \{Snail,\ Right\ Circle,\ Right\ Arrow\}$$
$$\mathbf{x} = \{Duration,\ Trace\ Points,\ Age\ Factor\}$$

2.2 Training Data

To study the process, we observe the behavior of the Random process for some time collecting a large number of samples $(x_1, y1), (x_2,y_2), (x_3,y_3)\ldots\ldots (x_N,y_N)$ [12]. We can summarize the Training sample in terms of its Empirical probability distribution $\tilde{p}$, defined by

$$\tilde{p}(x,y) \equiv (1/N) * number\ of\ times\ that\ (x,y)\ occurs\ in\ the\ sample \qquad (2.1)$$

2.3 Statistics, Features and Constraints

Our goal is to construct a statistical model of the process that generated the training sample $\tilde{p}(x,y)$. The building blocks of this model will be a set of statistics [13] of the Training sample. For example, Indicator function of a context feature f,

$$f(x,y) = \begin{cases} 1 \textit{ if } y{=}\textit{Snail and Duration follows Very High} \\ 0 \textit{ otherwise;} \end{cases} \tag{2.2}$$

The expected value of f with respect to the *empirical distribution*, is exactly the statistics we are interested in. we denote this expected value by

$$\tilde{p}(f) \equiv \sum_{x,y} \tilde{p}(x,y)\, f(x,y) \tag{2.3}$$

We call such a function, a Feature function or a Feature for short. When we discover any statistic to be useful, we can acknowledge its importance by constraining the expected value that the model assigns to the corresponding feature function f. The expected value of f with respect to the *conditional probability* $p(y|x)$

$$p(f) \equiv \sum_{x,y} \tilde{p}(x)\, p(y|x)\, f(x,y) \tag{2.4}$$

Where, $\tilde{p}(x)$ is the Empirical Distribution of x of the training sample. We constrain this expected value to be same as the expected value of f in the training sample. i.e.

$$\tilde{p}(f) = p(f) \tag{2.5}$$

We call the requirement (2.5) a *Constraint Equation* or *Constraint.* On combining (2.3), (2.4) and (2.5), we get the following equation

$$\sum_{x,y} \tilde{p}(x,y)\, f(x,y) = \sum_{x,y} \tilde{p}(x)\, p(y|x)\, f(x,y) \tag{2.6}$$

2.4 Maximum Entropy Principle

Given n feature functions f_i, we want $p(y|x)$ to maximize the entropy measure

$$H(p) \equiv - \sum_{x,y} \tilde{p}(x)\, p(y|x) \log p(y|x) \tag{2.7}$$

Where, p is chosen from;

$$C \equiv \{ p | p(fi) = \tilde{p}(fi)\ \forall\, i = 1,2,\ldots,n \} \forall \tag{2.8}$$

To select a model from a set C, of allowed probability distributions, choose the model $p^* \in C$ with Maximum Entropy $H(p)$:

$$P^* = \underset{p^* \in C}{\operatorname{argmax}} H(p) \tag{2.9}$$

Where, It can be shown that p^* is always well defined; that is there is always a unique model p^* with maximum entropy in any constrained set C.

3 Data Collection

We led the experiments on the SIGN-ON LINE DATABASE. The SIGN On-Line Database contains data acquired from 20 writers. The collection contains unistroke on-line handwritten gestures that were acquired on TabletPCs and whiteboards. The data collection sessions were performed at the Synchromedia laboratory (ETS, Montreal, Canada) and by the Imadoc team (Irisa laboratory, Rennes, France). In total, 17 classes of gestures were collected.

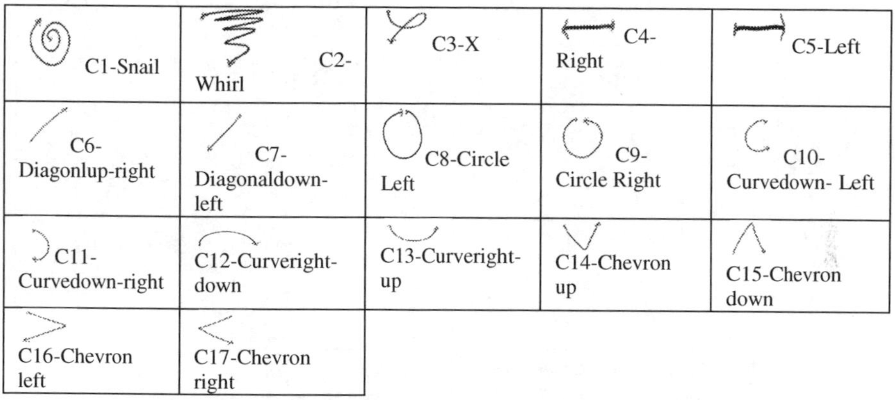

C1-Snail	C2-Whirl	C3-X	C4-Right	C5-Left
C6-Diagonlup-right	C7-Diagonaldown-left	C8-Circle Left	C9-Circle Right	C10-Curvedown- Left
C11-Curvedown-right	C12-Curveright-down	C13-Curveright-up	C14-Chevron up	C15-Chevron down
C16-Chevron left	C17-Chevron right			

Fig. 1 Seventeen Gestures Classes

Each gesture is described by a set of 21 features. The dataset and additional information on the data collection protocol can be found in [14].

4 Proposed Approach and Results

We propose an experimental setup for self adaptive gesture fuzzy classifier which uses maximum entropy principle for preserving most promising rules and removing redundant rules from the rule set, based on interestingness. The analysis of the result is done and shown by doing the comparison between the training and checking error, before pruning and after pruning of the rules and in addition to it using defuzzification method (Centroid and Lom–Som) and comparing the training error and checking error using both the defuzzification methods, after pruning the rules, and finding out which defuzzification method is most accurate among the two.

4.1 Evaluation

We have tested the whole database containing the 17 gestures, but here we have shown the result considering the 3 gestures,

- Class A = *Spiral*
- Class B = *Circle Right*
- Class C = *Right Arrow*

The dataset consist of the 21 features, but among all the 21 features, we have considered 3 features; *(i) Duration (ii) Trace points (iii) Age factor.*

After that, the main focus of our experiments is to find the performance before pruning of the rules in the beginning, on the stability and the recovery speed of the performance when introducing new unseen classes, and the performance after pruning of the rules to reduce the problem of over-fitting and limit the database.

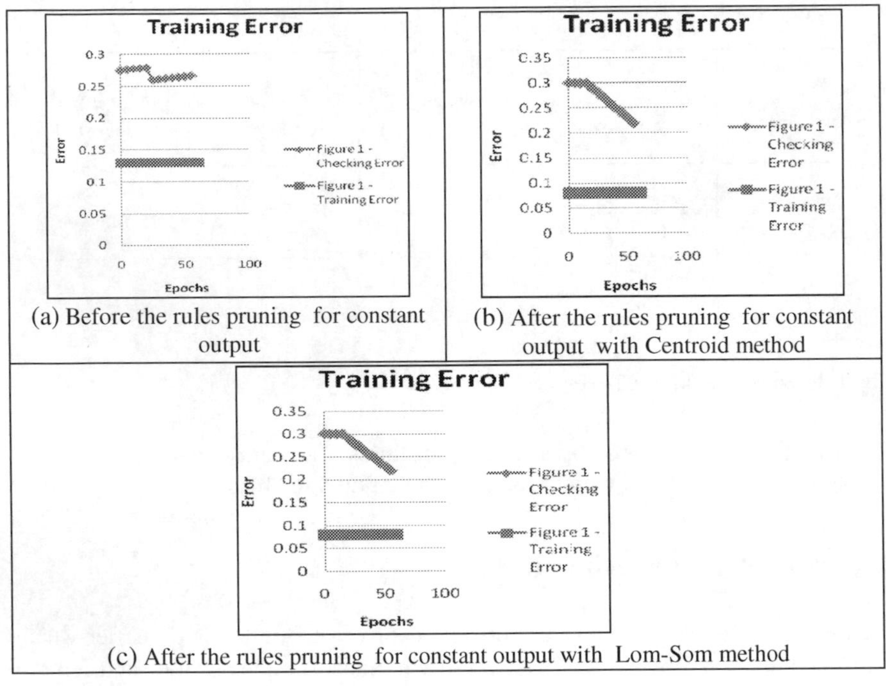

(a) Before the rules pruning for constant output

(b) After the rules pruning for constant output with Centroid method

(c) After the rules pruning for constant output with Lom-Som method

Fig. 2 Results

We have implemented the Result of the following experiments with the help of two defuzzification methods;

(i) *Centroid*
(ii) *Lom – Som*

By comparing the performance, of before pruning the rules and after pruning of the rules, we can see that the training error and checking error rate Decreases by 40%, for constant output and on comparing both the defuzzification methods for constant output, Lom - Som defuzzification method resulted better than Centroid method. In comparison to the previous work, after pruning of the rules, the over-fitting problem is reduced which further reduces misclassification error. It is also shown that the performance accuracy is maintained, Results have been shown in Fig. 2(a), 2(b) and 2(c) respectively.

5 Conclusion and Future Work

In the context of handwritten gesture recognition systems, we presented a self adaptive gesture fuzzy classifier which uses maximum entropy principle for preserving most promising rules and removing redundant rules from the rule set. Experiment results show the effectiveness of the approach in terms of reduction of over-fitting, which further reduces misclassification error, and thus limit the database. For future work, this approach can be applied to various datasets of other domains. Also, to compute interestingness in the fuzzy system, some other concepts may also be applied, in place of maximum entropy. More scalable methods can also be designed to apply this work on large datasets.

References

1. Almaksour, A., Anquetil, E., Quiniou, S., Cheriet, M.: Evolving Fuzzy Classifiers: Application to Incremental Learning of Handwritten Gesture Recognition System. In: International Conference on Pattern Recognition (2010)
2. Aik, L.E., Jayakumar, Y.: A Study of Neuro-fuzzy System in Approximation-based Problems. Matematika 24(2), 113–130 (2008)
3. Jang, J.-S.R.: ANFIS: Adaptive-Network-based Fuzzy Inference Systems. IEEE Transactions on Systems, Man, and Cybernetics 23(3), 665–685 (1993)
4. Bedregal, B.C., Costa, A.C.R., Dimuro, G.P.: Fuzzy rule-based hand gesture recognition. In: Artificial Intelligence in Theory and Practice, pp. 285–294. Springer (2009)
5. Takagi, T., Sugeno, M.: Fuzzy identification of systems and its applications to modeling and control. IEEE TSMC 15(1), 116–132 (1985)
6. Jang, J.-S.: Anfis: adaptive-network-based fuzzy inference system. IEEE Tr. on Systems, Man and Cybernetics (Part B) 23(3), 665–685 (1993)
7. Angelov, P., Filev, D.: An approach to online identification of takagi-sugeno fuzzy models. IEEE Tr. Systems, Man, and Cybernetics 34(1), 484–498 (2004)
8. de Barros, J.-C., Dexter, A.L.: On-line identification of computationally undemanding evolving fuzzy models. Fuzzy Sets and Systems 158(18), 1997–2012 (2007)
9. McCallum, A., Pereira, F.: Maximum entropy Markov models for information extraction and segmentation
10. Zellnerr, A., Highfiled, R.: Calculation of Maximum Entropy Distributions and Approximation of Marginal Posterior Distributions. Journal of Econometric 37, 195–209 (1988)
11. Berger, A.L., Pietra, S.A.D., Pietra, V.J.D.: A maximum entropy approach to natural language processing
12. Lazoand, V., Rathie, P.N.: On the Entropy of Continuous Probability Distributions. IEEE Trans. IT–24 (1978)
13. Jaynes: Papers on probability, statistics and statistical physics. Reidel Publishing Company, Dordrecht (1983)
14. http://www.synchromedia.ca/web/ets/gesturedataset

Speaker Independent Word Recognition Using Cepstral Distance Measurement

Arnab Pramanik and Rajorshee Raha

Abstract. Speech recognition has been developed from theoretical methods practical systems. Since 90's people have moved their interests to the difficult task of Large Vocabulary Continuous Speech Recognition (LVCSR) and indeed achieved a great progress. Meanwhile, many well-known research and commercial institutes have established their recognition systems including via Voice system IBM, Whisper system by Microsoft etc. In this paper we have developed a simple and efficient algorithm for the recognition of speech signal for speaker independent isolated word recognition system. We use Mel frequency cepstral coefficients (MFCCs) as features of the recorded speech. A decoding algorithm is proposed for recognizing the target speech computing the cepstral distance of the cepstral coefficients. Simulation experiments were carried using MATLAB here the method produced relatively good (85% word recognition accuracy) results.

1 Introduction

Speech recognition (also known as automatic speech recognition or computer speech recognition) converts spoken words to text. The term "voice recognition" is sometimes used to refer to recognition systems that must be trained to a particular speaker—as is the case for most desktop recognition software. Recognizing the speaker can simplify the task of translating speech. Speech recognition is a broader solution which refers to technology that can recognize speech without being targeted at single speaker—such as a call system that can recognize arbitrary voices.

Arnab Pramanik
Junior Project Officer, G S Sanyal School of Telecommunication,
Indian Institute of Technology, Kharagpur
e-mail: arnabpramanik.ece@gmail.com

Rajorshee Raha
Senior Project Officer, G S Sanyal School of Telecommunication,
Indian Institute of Technology, Kharagpur
e-mail: rajorshee87@gmail.com

A. Abraham and S.M. Thampi (Eds.): Intelligent Informatics, AISC 182, pp. 225–235.
springerlink.com

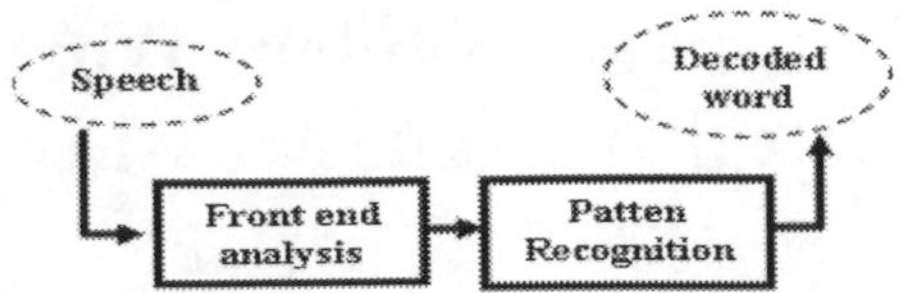

Fig. 1 Basic Speech Recognition System

In speech recognition cepstral distance measure is one of the most important aspect. A number of distance measures have been tried and still more have been proposed. In this paper we developed a simple low complexity speech detection algorithm which measures minimum cepstral distances of the ceptral coefficients to detect the speech. The cepstral distance approach which has been proposed is defined as z,

$$Z=\sqrt{(ct(i)^2-cr(i)^2)} \quad (1)$$

Where 'ct' and 'cr' are row vectors which are composed of the cepstral coefficients obtained from a test utterance and a reference template respectively.

1.1 Speech Recognition Implementation

A Speech Recognition system can be roughly devided into two parts namely front end analysis and pattern recognition as shown in the Fig 1. Speech recognition at its most elementary level, comprises a collection of algorithms drawn from a wide variety of disciplines including statistical pattern recognition, signal processing, communication theory etc. Although each of this areas is relied on to varying degrees in different recognizers perhaps the greatest common denominator of all recognition systems is the Front end analysis or Signal processing front end, which converts the speech waveform to some type of parametric representation (generally at a considerably lower information) for further analysis and processing.

1.2 Front End Analysis

Typically front end analysis comprises of data aquisation, noise removal, end point detection and feature excraction (see Fig 3). Data aquisition consists of recorded voice signal which is band limited to 300Hz to 3400Hz. Likewise sampling rate is selected 8000Hz. After that recorded signal is filtered using a band pass filter. After that a proper noise cancellation technique technique has taken to de-noise the filtered signal.

Denoised technique, used here is based on Wavelet Transform. After that uttered speech is selected from the recorded speech using end point detection technique. The end point detection determines the position of the speech signal in the time series. In the end-point detection method, it is often assumed that during several frames at the beginning of the incoming speech signal there the speaker has not said anything. So those frames give the silence or the background noise.

To detect the speech over the background noise concept of thresholding is used. This often is based on power threshold which is a function of time. Here the power frames are calculated and the threshold is set taking the noise frames. The frames above the calculated threshold are kept and other discarded leaving only high powered frames which consists of the speech. Proper end point detection requires proper calculation of threshold. After that the signal is pre-emphasized to avoid overlooking the high frequencies.

In this step the pre-emphasized speech signal is segmented into frames, which are spaced 10 msec apart, with 5 msec overlaps for short-time spectral analysis. Each frame is then multiplied by a fixed length window. Window functions are signals that are concentrated in time, often of limited duration N_0. While window functions such as triangular, Kaiser, Barlett, and prolate spheroidal occasionally appear in digital speech processing systems, Hamming and Hanning are the most widely used to taper the signals to quite small values (nearly zeros) at the beginning and end of each frame for minimizing the signal discontinuities at the edge of each frame.

Hamming window, which is defined as:

$$h[n] = \begin{cases} 0.54 - 0.46 \cdot \cos(2\pi n/N_0), 0 \leq \mathrm{n} \leq \mathrm{N}_0 \\ 0 \quad otherwise \end{cases} \tag{2}$$

The output speech signal of Hamming windowing can be described as:

$$\mathrm{Y[n] = h[n]*s[n],\ 0 \leq n \leq N_0} \tag{3}$$

After windowing the speech signal, Discrete Fourier Transform (DFT) is used to transfer these time-domain samples into frequency-domain ones. There is a family of fast algorithms to compute the DFT, which are called Fast Fourier Transforms (FFT)

$$X(k) = \sum_{n=0}^{N-1} x(n)\, e^{-j2\pi nk/N}, 0 \leq k < N \tag{4}$$

If the number of FFT points, N, is larger than the frame size $N0$, N-$N0$ zeros are usually inserted after the $N0$ speech samples.

After FFT the mel-frequency cepstral coefficients are calculated by passing the FFT coefficients through the mel filterbank frame by frame. In sound processing, the mel-frequency cepstrum (MFC) is a representation of the short-term power spectrum of a sound, based on a linear cosine transform of a log power spectrum on a nonlinear Mel scale of frequency. Mel-frequency cepstral coefficients (MFCCs) are coefficients that collectively make up an MFC. They are derived from a type of cepstral representation of the audio clip.

1.3 Mel Frequency Cepstral Coefficients

The difference between the cepstrum and the mel-frequency cepstrum is that in the MFC, the frequency bands are equally spaced on the Mel scale, which approximates the human auditory system's response more closely than the linearly-spaced frequency bands used in the normal cepstrum. In order to represent the static acoustic properties, the Mel-Frequency Cepstral Coefficients (MFCC) (see Fig 4) is used as the acoustic feature in the Cepstral domain. This is a fundamental concept which uses a set of non-linear filters to approximate the behaviour of the auditory system. It adopts the characteristic of human ears that human is assumed to hear only frequencies lying on the range between 300Hz to 3400Hz. Besides, human's ears are more sensitive and have higher resolution to low frequency compared to high frequency. The filter bank (see Fig 2) with M filters (m=1, 2,, M), where filter m is the triangular filter given by:

$$H_m(k) = \begin{cases} 0 & k < f(m-1) \\ \frac{k-f(m-1)}{(f(m)-f(m-1))} & f(m-1) \le k \le f(m) \\ \frac{f(m+1)-k}{(f(m+1)-f(m))} & f(m) \le k \le f(m+1) \\ 0 & k > f(m+1) \end{cases} \quad (5)$$

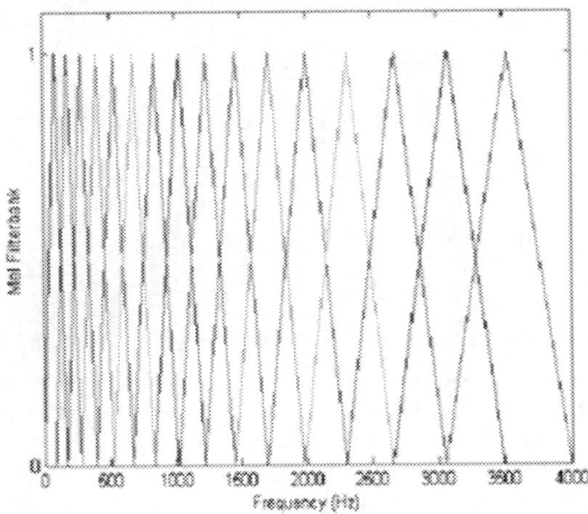

Fig. 2 Frequency response of Mel filter bank

If we define '*f'l* and '*fh*' be the lowest and highest frequencies of the filterbank in Hz, '*Fs*' the sampling frequency in Hz, '*M*' the number of filters and '*N*' the size of FFT, the cantering frequency '$f(m)$' of the mth filter bank is

$$f(m) = \left(\frac{N}{F_s}\right) Mel^{-1}\left(Mel(f_l) + m \cdot \frac{Mel(f_h) - Mel(f_l)}{M+1}\right) \quad (6)$$

1.4 Dynamic Featuring

In addition to the cepstral coefficients, the time derivative approximations are used as feature vectors to represent the dynamic characteristic of speech signal. To combine the dynamic properties of speech, the first and/or second order differences of these cepstral coefficients may be used which are called the delta differences of these cepstral coefficients may be used which are called the delta and these dynamic features have been shown to be beneficial to ASR performance. The first-order delta MFCC may be described as

$$\Delta C^{t}(k) = \frac{\sum_{l=-P}^{P} l \cdot \hat{C}^{t+l}(k)}{\sum_{l=-P}^{P} l^{2}} \tag{7}$$

Where $C^t(k)$ denotes the kth cepstral coefficient at frame t after liftering and P is typically set to the value 2 (i.e., five consecutive frames are involved).

1.5 Pattern Recognition

Speech patterns mainly differ in their cepstral domains. To distinguish one word from another their corresponding cepstral distances are noted and the minimum one is considered. This gives us fairly satisfactory result.

In case of our model the major compromise that had to be made to keep it simple and realizable in a short time was to have a small database. This constraint minimizes the versatility of our model but provides a satisfactory result in response to our algorithm. However this model can be furthur extented to a larger database and computation.

The database of this model comprises of four words. They had to be common and sensible yet acoustically distinguishable and not very long to be used as dataset in our model. For this purpose the common 4 colours were chosen to be our set of words stored in our database. These words are as follows: BLACK, BROWN, RED, WHITE.

The algorithm for pattern recognition using cenptral distance computation is as follows:

1. Algorithm for loading database and initilize matrix and vectors

```
Load Database   %MFCC of diffeneent words are loaded
initilization: Black=zeros(13,160)
               Brown=zeros(13,160)
               Red=zeros(13,160)
               White=zeros(13,160)
Black=[black1, black2, black3................black13]
Brown=[brown , brown2, brown3......... brown13]
Red=[red1, red 2, red3........................ red13]
White=[white1, white2, white3...........white13]
initilization:Cepstral_distance_black=zeros(13,160)
              Cepstral_distance_brown=zeros(13,160)
              Cepstral_distance_red=zeros(13,160)
              Cepstral_distance_white=zeros(13,160)
Load test vector file
Test←test_vector(1:160)
```

2. Algorithm for computing cepstral distances of test vectors

```
For i=1:13 do
    For j=1:160 do
        Cepstral_distance_black (i , j)← absolute[√((Black(i , j)²-test_vector(i , j)²)]
Cepstral_distance_brown(i , j)← absolute[√((Brown(i , j)²-test_vector(i , j)²)]
Cepstral_distance_red (i , j)← absolute[√((Red(i , j)²-test_vector(i , j)²)]
Cepstral_distance_white (i , j)← absolute[√((White(i , j)²-test_vector(i , j)²)]
    End for
End for
```

3. Algorithm for computing minimum spetral distance matrix

```
Initialization:min_distance_matrix=zeros(13,160)
For i=1:13
   For j=1:160
      min_distance_matrix=minimum of { Cepstral_distance_black (i , j),
Cepstral_distance_brown(i , j), Cepstral_distance_bred (i , j), Cepstral_distance_white(i , j)}
    End for
End for
```

4. Algorithm for minimum cepstral distance counting and decision making

```
initilization: counter1=0
               counter2=0
               counter3=0
               counter4=0
for i=1:13
    for=j=1:160
 if (min_distance_matri(i , j)==Cepstral_distace_black(i , j))
 then increment counter1 by 1
if (min_distance_matri(i , j)==Cepstral_distace_brown(i , j))
  then increment counter2 by 1
if (min_distance_matri(i , j)==Cepstral_distace_red(i , j))
  then increment counter3 by 1
if (min_distance_matri(i , j)==Cepstral_distace_black(i , j))
  then increment counter4 by 1
    End for
End for
initilization: x=0;
```

```
x= maximun of {counter1, counter2, counter3, counter4}
if counter1 is maximun then display "BLACK"
if counter2 is maximun then display "BROWN"
if counter3 is maximun then display "RED"
if counter4 is maximun then display "WHITE"
```

In our proposed algorithm database of four words namely BLACK, BROWN, RED, WHITE was taken. This datebase contains 13 samples of each words from different speak. Mel frequency coefficients (see Fig 4) of all 13 samples of each words are stored in matrix Black, Brown, Red and white respectively.

Then cepstral distance matrix (see Fig 7) is computed for each word using the proposed equation as shown in the algorithm. After that each element from the cepstral distance matrices of four words are taken and compared. After comparison the minimum cepstral distance among the four is stored in a matrix called min_distance_matrix. After that each element od the min_distance_matrix is searched from the four specral distance matrices. There are four counters acordingly. Whenever a element from min_distance matrix matches with a element of a specific cepstral distance matrix, the corosponding counter is incremented. Finally the values of all the counters is compared and the maximum value is selected. The counter having the higher value decides the detected word.

We developed this algorithm considering four words only. The basic aim of using a small database is to test our algorithm successfully. This algorithm can also be applicable using a larger database. Any word or any sound from any environment can be used to recognize using our speech recognition model.

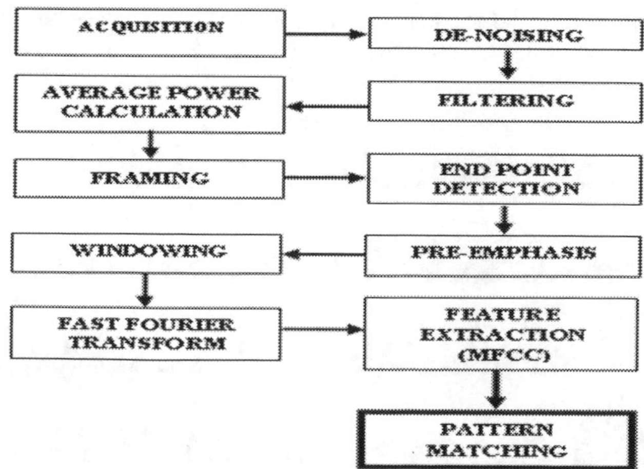

Fig. 3 Full Block Diagram representation of Pattern Recognition System

2 Results

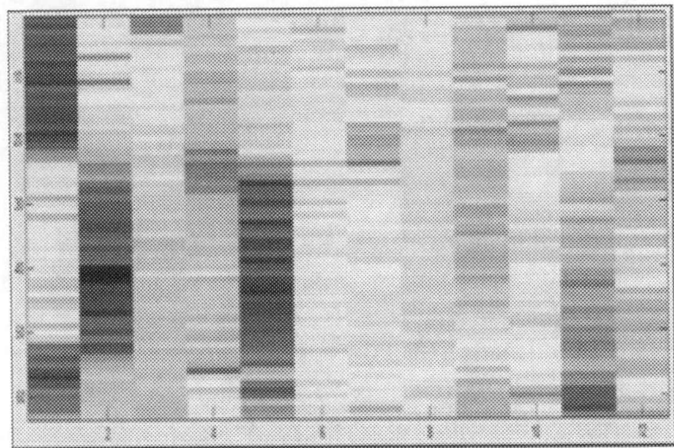

Fig. 4 Image plot of the Mel frequency cepstral coefficients the word "BLACK"

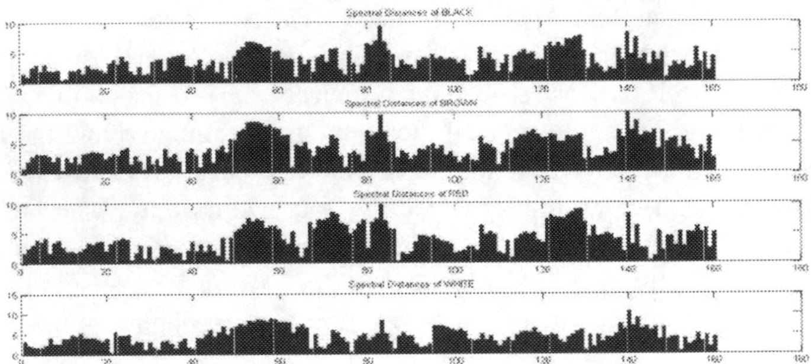

Fig. 5 Cepstral distance of the Mel frequency cepstral coefficients of the test speech sample (for the word "Black") with the respective words in the database

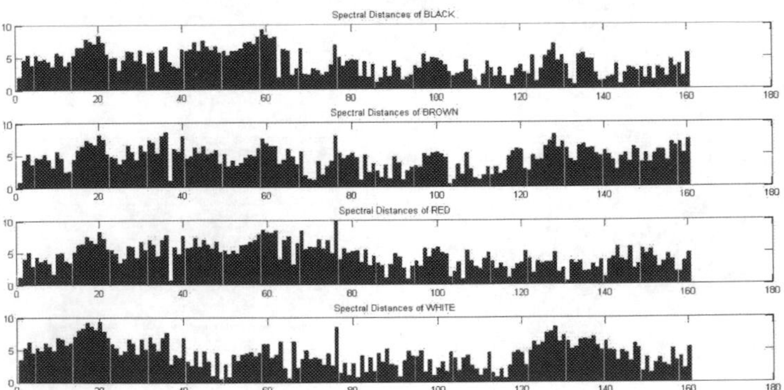

Fig. 6 Cepstral distance of the Mel frequency cepstral coefficients of the test speech sample (for the word "White") with the respective words in the database plot

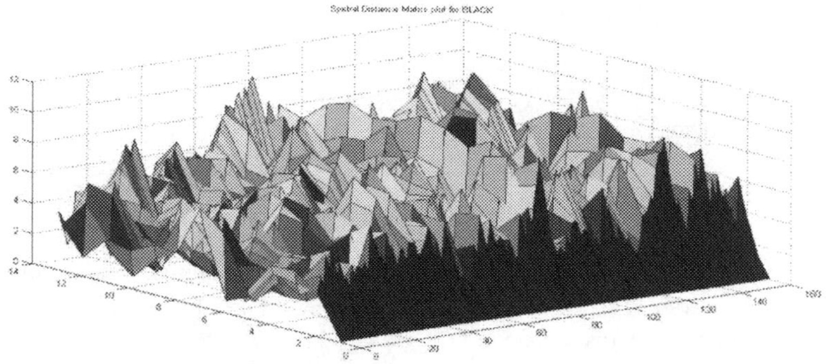

Fig. 7 Cepstral distance matrix plot for the word "Black"

- Following are the error matrices for some inputs:

Table 1 Input speech: Black

Word	Black	Brown	Red	White
Distance	9.7937121e+000	1.0452342e+001	1.1009596e+001	1.1020715e+001

Output response: Black Verdict: Hit

Table 2 Input speech: Brown

Word	Black	Brown	Red	White
Distance	9.1330997e+000	8.7554356e+000	9.1460794e+000	8.8259277e+000

Output response: Brown Verdict: Hit

Table 3 Input speech: Black

Word	Black	Brown	Red	White
Distance	8.8259277e+000	1.0452342e+001	8.1009596e+000	1.1020715e+001

Output response: Red Verdict: Miss

3 Application

In the health care domain, even in the wake of improving speech recognition technologies, medical transcriptionists have not yet become obsolete. The services provided may be redistributed rather than replaced. Speech recognition can be implemented in front-end or back-end of the medical documentation process. Application of ASR is in the military area also such as High-performance fighter aircraft, Helicopters; Battle management; Training air traffic controllers; Telephony and other domains.

4 Conclusion

This paper has presented simple and efficient pattern recognition methods using the cepstral separation of Mel frequency cepstral coefficients and has evaluated through four word vocabulary continuous speech recognition. For any speech recognition model database is the driving factor. In this model probably the greatest compromise made is the size of the database. We tested our algorithms with a smaller database because once it is implemented successfully then the same algorithms can be effective using larger database. As the algorithm development was the main area of concern for speech recognition system. We tested it with a database of sixty words. Out of these sixty words, fifty words were identified correctly. It is encouraged that for future improvements with larger database the better results will be generated and high accuracy will be delivered. The performance of the model can be further improved by using TMS320C6713 DSK i.e DSP toolkit. With its own set of function and modules computational load can be decreased and time-efficiency of the model can be considerably improved.

Acknowledgement. The authors are grateful to GSSST, Indian Institute of Technology, Kharagpur for allowing them to use the facilities in the hardware and project laboratory of GSSST.

References

[1] Huang, X.D., Lee, K.F.: Phonene classification using semicontinuous hidden markov models. IEEE Trans. on Signal Processessing 40(5), 1962–1067 (1992)

[2] Levinson, S.E., Rabiner, L.R., Juang, B.H., Sondhi, M.M.: Recognition of isolated digits using hidden markov models with continuous mixture densities. AT & T Technical Journal 64(6), 1211–1234 (1985)

[3] Acero, Acoustical and environmental robustness in automatic speech recognition. Kluwer Academic Pubs. (1993)

[4] Rabiner, L.R., Schafer, R.W.: Digital Processing of Speech Signals. Prentice Hall (1978)

[5] Jelinek, F.: Continuous Speech Recognition by Statisical Methods. IEEE Proceedings 64(4), 532–556 (1976)

[6] Young, S.: A Review of Large-Vocabulary Continuous Speech Recognition. IEEE Signal Processing Magazine, 45–57 (September 1996)

[7] Rabiner, L.R., Juang, B.-H.: Fundamentals of Speech Recognition. Prentice-Hall (1993)

[8] Mel Frequency Cepstral Coefficients: An Evaluation of Robustness of MP3 Encoded Music by Sigurdur Sigurdsson, Kaare Brandt Petersen and TueLehn-Schiøler

[9] Speech and speaker recognition: A tutorial by Samudravijaya, K., Young, S.J.: The general use of tying in phoneme-based hmm speech recognisers. In: Proceedings of ICASSP (1992)

[10] Nefian, A.V., Liang, L., Pi, X., Liu, X., Mao, C.: An coupled hidden Markov model for audio-visual speech recognition. In: International Conference on Acoustics, Speech and Signal Processing (2002)

[11] Neti, C., Potamianos, G., Luettin, J., Matthews, I., Vergyri, D., Sison, J., Mashari, A., Zhou, J.: Audio visual speech recognition. In: Final Workshop 2000 Report (2000)

[12] Oerder, M., Ney, H.: Word graphs: an efficient interface between continuous-speech recognition and language understanding. In: IEEE International Conference on Acoustics, Speech, and Signal Processing, vol. 2 (1993)

[13] Potamianos, G., Luettin, J., Neti, C.: Asynchronous stream modelling for large vocabulary audio-visual speech recognition. In: IEEE International Conference on Acoustics, Speech and Signal Processing, vol. 1, pp. 169–172 (2001)

[14] Dupont, S., Luettin, J.: Audio-visual speech modeling for continuous speech recognition. IEEE Transactions on Multimedia 151 (September 2000)

Wavelet Packet Based Mel Frequency Cepstral Features for Text Independent Speaker Identification

Smriti Srivastava, Saurabh Bhardwaj, Abhishek Bhandari, Krit Gupta, Hitesh Bahl, and J.R.P. Gupta

Abstract. The present research proposes a paradigm which combines the Wavelet Packet Transform (WPT) with the distinguished Mel Frequency Cepstral Coefficients (MFCC) for extraction of speech feature vectors in the task of text independent speaker identification. The proposed technique overcomes the single resolution limitation of MFCC by incorporating the multi resolution analysis offered by WPT. To check the accuracy of the proposed paradigm in the real life scenario, it is tested on the speaker database by using Hidden Markov Model (HMM) and Gaussian Mixture Model (GMM) as classifiers and their relative performance for identification purpose is compared. The identification results of the MFCC features and the Wavelet Packet based Mel Frequency Cepstral (WP-MFC) Features are compared to validate the efficiency of the proposed paradigm. Accuracy as high as 100% was achieved in some cases using WP-MFC Features.

Keywords: WPT, MFCC, HMM, GMM, Speaker Identification.

1 Introduction

Human beings possess several inherent characteristics that assist them distinguish from one another. Over the years, biometrics has emerged as the science which assimilates and tries to mimic the powers of the human brain by capturing unique personal features and consequently performing the task of human identification. Voice as a biometric tool has interested plethora of researchers as it can be easily intercepted, recorded and processed. Moreover, voice biometrics offers simple and secure mode of remote access transactions over telecommunication networks by authenticating the speaker first and then carrying out the required transactions. Hence, applications of speech processing technology are broadly classified into:

Smriti Srivastava · Saurabh Bhardwaj · Abhishek Bhandari · Krit Gupta ·
Hitesh Bahl · J.R.P. Gupta
Netaji Subhas Institute of Technology, New Delhi 110078, India

A. Abraham and S.M. Thampi (Eds.): Intelligent Informatics, AISC 182, pp. 237–247.
springerlink.com

Speech Recognition and *Speaker Recognition*. Speech recognition is the ability to identify the spoken words while speaker recognition is the ability to discriminate between people on the basis of their voice characteristics. Further the task of speaker recognition is dissected into two categories, *speaker identification* and *speaker verification*. Speaker identification is to classify that the test speech signal belongs to which one of the N- reference speakers whereas speaker verification is to validate whether identity claimed by an unknown speaker is true or not, consequently this type of decision is binary. Several recognition systems behave in a text-dependent way, i.e. the user utters a predefined key sentence. But, text dependent type of recognition process is only feasible with "cooperative speakers". Consider criminal investigation as an application (an unwilling speaker), here recognition can only be performed in text-independent mode. With increased applications of speech as a means of communication between the man and the machine, speaker identification has emerged as a powerful tool [1]. The phenomenon of speaker recognition has been in application since the 1970's [2]. Most of the state of art identification systems uses MFCC for front-end-processing as its performance is far superior compared to all other feature extraction mechanisms as described in [3]. The paper is organized as follows. Section 2 gives a description of the modules of speaker recognition. The proposed algorithm is described in section 3. Finally, the results are demonstrated in section 4.

2 Modules for Speaker Recognition

All speaker recognition systems contain two main modules, *feature extraction* and *feature or pattern matching*. Feature extraction is the process that extracts information from a voice signal of a speaker. Feature matching is the procedure to identify the unknown speaker by matching his features with those of known speakers. Sound pressure waves are acquired with the help of a microphone or some other voice recording device. This signal is then pre-processed. Speaker recognition using the pre-processed signal is accomplished in two stages, Enrollment or feature extraction and pattern matching or classification as depicted in fig.1.

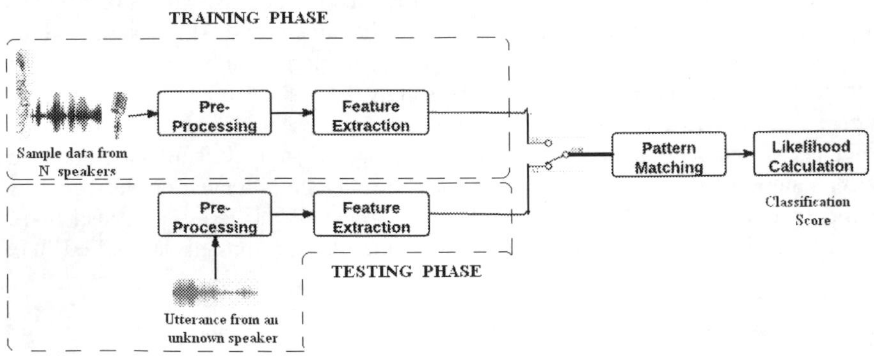

Fig. 1 Block Diagram of a Speaker Recognition system

During enrollment phase, speech sample from several speakers is recorded and a number of features are extracted using one out of the several methods available to produce individual's "voice model or template". During the next phase, pattern of an unknown utterance is compared with the previously recorded template. For speaker identification applications, speech utterance from an unknown speaker is compared with voice prints of all reference speakers. The unknown speaker is identified as that reference speaker whose voice model best matches with the model of unknown utterance. The performance of speaker identification system decreases with increasing population size. [1]

2.1 *Feature Extraction*

The mechanism of speech feature extraction reduces the dimensionality of the input signal by eliminating the redundant information while maintaining the discriminating capability of the signal [4]. Given the data of speech samples, a variety of auditory features are computed for each input set which constitute the feature vector. The present research proposes Wavelet Packet based Mel Frequency Cepstral feature extraction approach.

2.1.1 Mel Frequency Cepstral Coefficients

The advent of Mel Frequency Cepstral Coefficient (MFCC) technique for the task of feature extraction has over shadowed the existence of majority of its predecessor methods as it acknowledges human sound perception sensitivity with respect to frequency, providing better sound feature vectors. The most conspicuous difference between cepstral coefficients and MFCC is that the latter uses Mel filter banks to transform the frequency domain to Mel frequency domain [5]. The formula to convert f (Hz) into m (Mel) is as follows:

$$m = (2595 \log_{10} (1 + \frac{f}{700})) \quad (1)$$

The block diagram of MFCC feature extraction algorithm is as shown in fig.2.

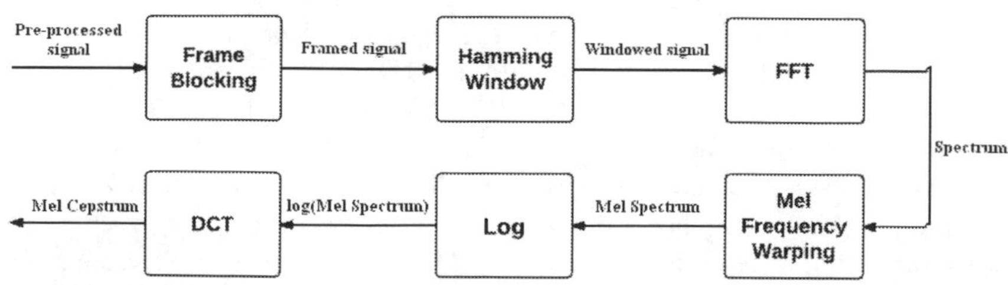

Fig. 2 Block Diagram implementation of the technique

Pre-processed speech signal is frame blocked with each frame having length of 25ms with an overlapping length of 15ms. The signal is then multiplied over short-time windows to avoid problems arising due to truncation of the signal. For our analysis, a hamming window is utilized. For each windowed frame, spectrum is computed using Fast Fourier Transform (FFT).Spectrum is passed through Mel filter bank to obtain the Mel spectrum. In the present work, 40 filters were used [6].Finally, Cepstral analysis is performed on the output of Mel filter banks using only 13 coefficients out of 40. The logarithm followed by the Discrete Cosine Transform (DCT) of the Mel spectrum produces a set of feature vectors (one vector corresponding to each frame) which are then termed as MFCC.

2.2 *Hidden Markov Model*

Hidden Markov Model (HMM) [7,8] springs forth from Markov Processes or Markov Chains. It is a canonical probabilistic model for the sequential or temporal data It depends upon the fundamental fact of real world, "Future is independent of the past and given by the present". HMM is a doubly embedded stochastic process, where final output of the system at a particular instant of time depends upon the state of the system and the output generated by that state. There are two types of HMMs: Discrete HMMs and Continuous Density HMMs. These are distinguished by the type of data that they operate upon. Discrete HMMs (DHMMs) operate on quantized data or symbols, on the other hand, Continuous Density HMMs (CDHMMs) operate on continuous data and their emission matrices are the distribution functions. HMM Consists of the following parameters

O {O1,O2…OT}	:	Observation Sequence
Z {Z1, Z2…ZT}	:	State Sequence
T	:	Transition Matrix
B	:	Emission Matrix/Function
π	:	Initialization Matrix
$\lambda(T, B, \pi)$	:	Model of the System
ρ	:	Space of all state sequence of length T
m $\{m_{q1}, m_{q2} \ldots m_{qT}\}$	:	Mixture component for each state at each time
$c_{il}, \mu_{il}, \sum_{il}$	:	Mixture component (i state and l component)

Single state HMM is known as GMM. For the purpose of text independent speaker identification, GMM has had a greater success over HMM [9]. There are three major design problems associated with an HMM outlined here. Given the Observation Sequence {O1, O2, O3,.., OT} and the Model $\lambda(T, B, \pi)$, the first problem is the computation of the probability of the observation sequence P $(O|\lambda)$.The second is to find the most probable state sequence Z {Z1, Z2,.., ZT}, the third problem is the choice of the model parameters λ (T, B,π), such that the probability of the Observation sequence, P $(O|\lambda)$ is the maximum. The solution to the above problems emerges from three algorithms: Forward, Viterbi and Baum-Welch [7].

2.2.1 Continuous Density HMM

Let O = {O1,O2...OT} be the observation sequence and Z {Z1, Z2…ZT} be the hidden state sequence. Now, we briefly define the Expectation Maximization

(EM) algorithm for finding the maximum-likelihood estimate of the parameters of a HMM given a set of observed feature vectors. EM algorithm is a method for approximately obtaining the maximum a posteriori when some of the data is missing, as in HMM in which the observation sequence is visible but the states are hidden or missing. The Q function is generally defined as

$$Q(\lambda,\lambda') = \Sigma_{q\varepsilon\rho} \log P(0,z \mid \lambda)P(0,z \mid \lambda') \tag{2}$$

To define the Q function for the Gaussian mixtures, we need the hidden variable for the mixture component along with the hidden state sequence. These are provided by both the E–step and the M-step of EM algorithm given

E Step:

$$\mathrm{Q}(\lambda,\lambda') = \Sigma_{Z\in\rho}\Sigma_{m\in M} \log P(O,z,m \mid \lambda)P(O,z,m \mid \lambda') \tag{3}$$

M Step:

$$\lambda' = \arg m_{\lambda} ax[Q(\lambda,\lambda')] + constrainment \tag{4}$$

The optimized equations for the parameters of the mixture density are:

$$\mu_{il} = \frac{\Sigma_{t=1}^{T} O_t P(z_{t=1}, m_{z_t t} = 1 \mid O_t \lambda')}{\Sigma_{t=1}^{T} P(z_{t=1}, m_{z_t t} = 1 \mid O_t \lambda')} \tag{5}$$

$$\sum_{il} = \frac{\sum_{t=1}^{T} (O_t - \mu_{il})(O_t - \mu_{il})^T P(z_t = i, m_{z_t t} = 1 \mid O_t \lambda')}{\sum_{t=1}^{T} P(z_t = i, m_{z_t t} = 1 \mid O_t \lambda')} \tag{6}$$

$$c_{il} = \frac{\sum_{t=1}^{T} P(z_t = i, m_{z_t t} = 1 \mid O_t \lambda')}{\sum_{t=1}^{T}\sum_{l=1}^{M} P(z_t = i, m_{z_t t} = 1 \mid O_t \lambda')} \tag{7}$$

3 Proposed Method

3.1 Discrete Wavelet and Wavelet Packet Transform

For discrete wavelet transform we have:

$$F[n] = \frac{1}{\sqrt{M}} . \sum_k W_\phi[j_0,k]\phi_{j_0,k}[n] + \frac{1}{\sqrt{M}} . \sum_{j=j_0}^{\infty} \sum_k W_\psi[j_0,k]\psi_{j,k}[n] \tag{8}$$

Here F[n], $\phi_{j0,k}[n]$ and $\Psi_{j,k}[n]$ are discrete functions defined in [0,M-1], a total of M points. Now,

$$W_{\phi}[j_0,k] = \frac{1}{\sqrt{M}}.\sum_n f[n]\phi_{j_0,k}[n] \tag{9}$$

$$W_{\psi}[j_0,k] = \frac{1}{\sqrt{M}}.\sum_n f[n]\psi_{j,k}[n] j \geq j_0 \tag{10}$$

$W_{\Phi}[j_0,k]$ are called *approximation coefficients* while $W_{\Psi}[j_0,k]$ are called *detailed coefficients*. These coefficients are obtained by using *Mallat algorithm* proposed in [10].

3.1.1 Wavelet Packet Transform

In the DWT decomposition, to obtain the next level coefficients, scaling coefficients (low pass branch in the binary tree) of the current level are split by filtering and down sampling [10]. With the wavelet packet decomposition, the wavelet coefficients (high pass branch in binary tree) are also split by filtering and down sampling. The splitting of both the low and high frequency spectra results in a full binary tree shown in fig.3 and a completely evenly spaced frequency resolution (In the DWT analysis, the high frequency band was not split into smaller bands).

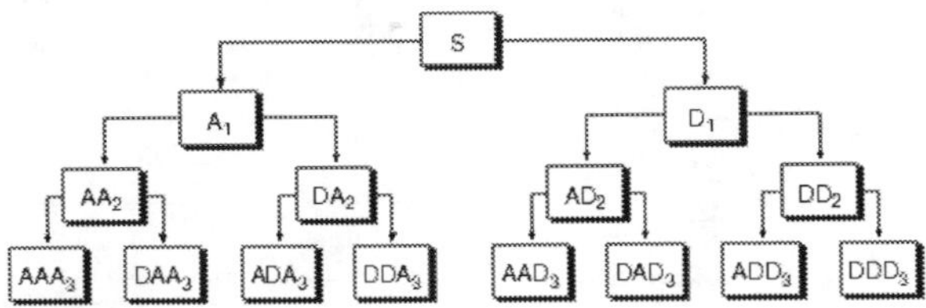

Fig. 3 Wavelet packet decomposition tree

3.2 *Motivation*

Speech is a “Quasi-stationary” signal. MFCC utilizes short time Fourier Transform (STFT) which provides information regarding the occurrence of a particular frequency at a time instant with a limited precision, with the resolution according to the Heisenberg Uncertainty principle dependent on the size of the analysis window.

$$Time * Frequency = \Delta t \Delta f \geq \frac{1}{4\pi} \tag{11}$$

Narrower windows provide better time resolution while wider ones provide better frequency resolution [11]. Even though STFT tries to strike a balance between the time and frequency resolution, it is admonished primarily as it keeps the length of the analysis window fixed for all frequencies resulting in uniform-partition of the time-frequency plane as shown in fig.4.

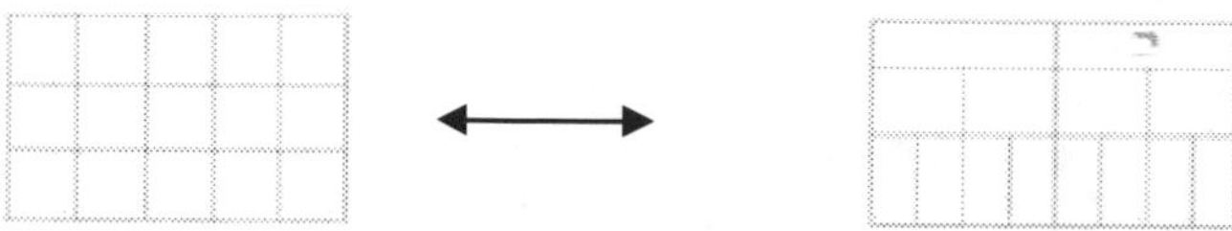

Fig. 4 Time-frequency plane uniformly partitioned in STFT

Fig. 5 Time-frequency plane non-uniformly spaced (constant area) in wavelet transform

Speech signals require a more flexible multi-resolution approach where window length can be varied according to the requirement to cater better time or frequency resolution. Wavelet Packet Transform (WPT) offers a remedy to this difficulty by providing well localized time and frequency resolution as shown in fig.5. Further, multi-resolution property of WPT makes it more robust in noisy environment as compared to single-resolution techniques and has better time-frequency characteristics. But, WPT increases the computational burden and is time consuming. Conventional wavelet packet transform mechanisms do not warp the frequencies according to the human auditory perception system. So, in this work an attempt is made for utilizing the advantages of the Mel Scale and multi-resolution wavelet packet transform to generate feature vector for the task of speaker identification.

3.3 *Proposed Paradigm*

3.3.1 Wavelet Packet Based Mel Frequency Cepstral Features

The block diagram for proposed approach is as shown in fig.6

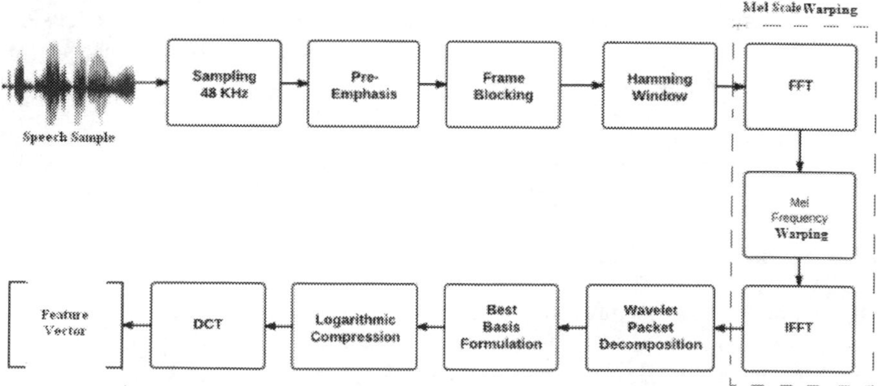

Fig. 6 Block diagram representation of proposed method

The analytical steps followed for feature extraction are as stated:

- The raw speech signal was primarily sampled at 48 kHz in order to further process it.

- Next, a framing window was utilized. The frame size was kept fixed to 25 milliseconds, a skip rate of 10 milliseconds was selected to accommodate for the best continuity.
- A pre-emphasis filter as described by equation (12) was next exercised in order to improve the overall signal-to-noise ratio. A rectangular Hamming window was deployed for framing.

$$H(z) = 1 - 0.97z^{-1} \tag{12}$$

- The resultant signal was transformed from time domain to frequency domain by applying Fast Fourier Transform (FFT). Then the signal in frequency domain was Mel-Warped using Triangular Mel Filter Banks. Afterwards, signal was again transformed to time domain by applying Inverse Fast Fourier Transform for further processing of signal.
- Next, wavelet packet decomposition was applied using daubechies4 (D4) wavelet. For a full j=7 level decomposition, the WPT corresponds to a maximum frequency of 31.25 Hz giving 128 sub-bands.
- Out of 128 frequency sub-bands 35 frequency sub-bands were used for further processing since higher frequency coefficient contained paltry amount of energy and first 35 coefficients represented 99.99%. The energy in each band was evaluated, and was then divided by the total number of coefficients present in that particular band. In particular, the sub band signal energies were computed for each frame as,

$$E_j = \frac{\sum_{j=1}^{N_j} [W_j^p f(i)]^2}{N_j}, j = 1,....,35 \tag{13}$$

- Lastly, a logarithmic compression was performed and a Discrete Cosine Transform (DCT) was applied on the logarithmic sub-band energies to reduce dimensionality:

$$F(i) = \sum_{n=1}^{B} \log_{10} E_n Cos(\frac{i(n-1/2)}{B}), i = 1,...., r. \tag{14}$$

3.3.2 Speaker Identification

After extracting the features we have used HMM or single state HMM called Gaussian Mixture Model (GMM) for the identification. The whole procedure is as explained in fig.7. Having the WP-MFC Feature from the speech signals, CDHMMs are trained for each speaker using Baum Welch (BM) algorithm which gives the parameters of the corresponding CDHMMs. Now the identification process can be described as follows: Given a test vector 'X' the log-likelihood of the trained batches with respect to their HMM models 'λ' is computed as

$\log P(X \mid \lambda)$. From 'N' HMMs $\{\lambda_1, \lambda_2, ...\lambda_N\}$ corresponding to 'N' speakers, the speaker can be identified with a test sequence using:

$$P(X \mid \lambda_{required}) = F[P(X \mid \lambda_1), ...P(X \mid \lambda_N) \tag{15}$$

Where $F()$ is the maximum of the likelihood values of the model $(\lambda_1, \lambda_2, ...\lambda_N)$. The model corresponding to the highest Log-Likelihood value is selected as the identified speaker.

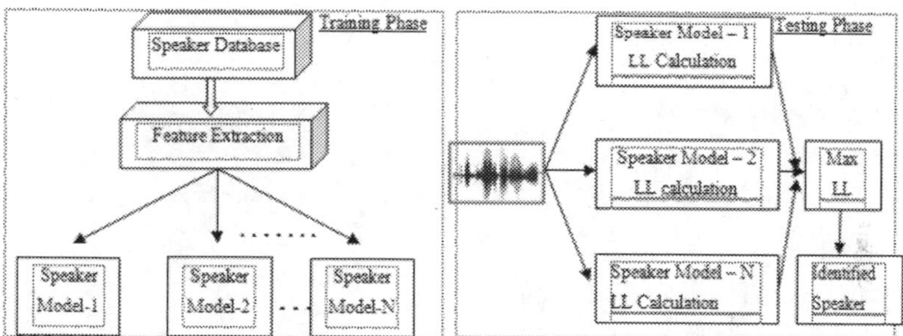

Fig. 7 Procedure for the classification algorithm

4 Experimental Results

Having acquired the appropriate test samples from the free online English speech database site [12], the database was created containing speech samples of 30 distinct speakers with 10 non-identical utterances each. Speaker models were created using 8 samples per speaker and testing was done using 2 samples of each speaker. The results of the identification process using GMM and HMM are displayed in table1, fig.8(a) and table 2, fig.8(b) respectively. The number of

Table 1 No. of states (Q) = 1 (GMM)

S.No.	No. of Gaussian Mixtures (M)	No. of States (Q)	No. of Speakers Recognized	
			WP-MFC Features	MFCC Features
1.	11	1	30	28
2.	12	1	30	28
3.	13	1	30	28
4.	14	1	30	28
5.	15	1	30	28

Table 2 No. of states (Q) = 2 (HMM)

S.No.	No. of Gaussian Mixtures (M)	No. of States (Q)	No. of Speakers Recognized	
			WP-MFC Features	MFCC Features
1.	11	1	27	25
2.	12	1	29	28
3.	13	1	30	27
4.	14	1	27	28
5.	15	1	27	29

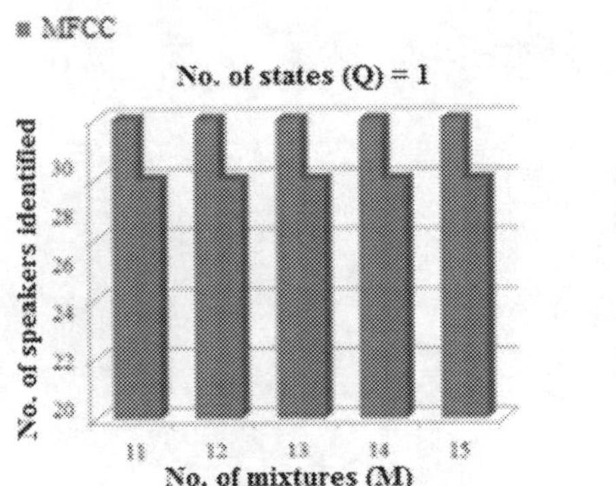

Fig. 8(a) Output Results with GMM

Fig. 8(b) Output Results with HMM

states (Q) was kept constant whereas the number of mixtures (M) was varied in each case.

5 Conclusion

Speaker Recognition is the use of machine to recognize a speaker from the spoken words. In this paper, we introduced a robust feature extraction technique for deployment with speaker identification system. These new feature vectors termed as Wavelet Packet based Mel frequency Cepstral (WP-MFC) Coefficients offer better time and frequency resolution. HMM and GMM were used to classify the acoustic data. Experimental results of the comparison between the performance of the proposed feature vectors and MFCC reveal the real life effectiveness of the proposed method. Also, better performance of GMM over HMM for speaker identification was confirmed.

References

[1] Reynolds, D.A.: Speaker Identification and Verification Using Gaussian Mixture Speaker Models. Speech Communication 17 (1995)

[2] Bolt Richard, H., Cooper Franklin, S., David Edward Jr., E., Denes Peter, B., Pickett James, M., Stevens Kenneth, N.: Speaker Identification by Speech Spectograms: A Scientists' View of its Reliability for Legal Purposes. The Acoustic Society of America 47 (1970)

[3] Reynolds Douglas, A.: Identification, Experimental Evaluation of Features for Robust Speaker. IEEE Transactions on Speech and Audio Processing 77, 257–285 (1994)

[4] Gaikwad Santosh, K., Gawali Bharti, W., Pravin, Y.: A Review on Speech Recognition Technique. International Journal of Computer Applications 10 (2010)

[5] Sirko, M., Michael, P., Ralf, S., Hermann, N.: Computing Mel-frequency coefficients on Power Spectrum. IEEE Proceedings of IEEE 1, 73–76 (2001)

[6] Chen, S.-H., Luo, Y.-R.: Speaker Verification Using MFCC and Support. In: Proceedings of the International MultiConference of Engineers and Computer Scientists (2009)

[7] Rabiner, L.: A tutorial on hidden Markov models and selected applications in speech recognition, pp. 257–286 (1989)

[8] Blimes, J.A.: A gentle tutorial of the EM algorithm and its application to parameter estimation for gaussian mixture and hidden markov models. International Computer Science Institute (1998)

[9] Reynolds, D.A., Campbell, W.M.: Springer Handbook of Speech Processing. Text Independent Speaker Recognition. Springer (2008)

[10] Mallat, S.G.: A theory for multiresolution signal decomposition: the wavelet representation. IEEE 111, 674–693 (1989)

[11] Robi, P.: The Engineers Ultimate Guide to Wavelet Analysis (2012), http://users.rowan.edu/~polikar/wavelets/wttutorial.html (accessed March 20, 2012)

[12] VoxForge (2012), http://www.voxforge.org/home/downloads/speech/english (accessed February 20, 2012)

Optimised Computational Visual Attention Model for Robotic Cognition

J. Amudha, Ravi Kiran Chadalawada, V. Subashini, and B. Barath Kumar

Abstract. The goal of research in computer vision is to impart and improvise the visual intelligence in a machine i.e. to facilitate a machine to see, perceive, and respond in human-like fashion(though with reduced complexity) using multitudinal sensors and actuators. The major challenge in dealing with these kinds of machines is in making them perceive and learn from huge amount of visual information received through their sensors. Mimicking human like visual perception is an area of research that grabs attention of many researchers. To achieve this complex task of visual perception and learning, Visual Attention model is developed. A visual attention model enables the robot to selectively (and autonomously) choose a "behaviourally relevant" segment of visual information for further processing while relative exclusion of others (Visual Attention for Robotic Cognition: A Survey, March 2011).The aim of this paper is to suggest an improvised visual attention model with reduced complexity while determining the potential region of interest in a scenario.

1 Introduction

The visual cognition in primates is due to a custom built attentional circuit that helps them respond to what they perceive through vision. This process is simple, yet robust. The endeavour of robotics research to design a bio-inspired visual attention model for the cognitive robot has strong connectivity with the research in cognitive psychology, computational neuroscience and computer vision as these are the three disciplines which cultivated the basic research on artificial modelling of human visual attention. The major motivation for developing computational models of visual attention was two-fold :1) creating a computational tool to test the validity of the theories/hypothesis of visual attention proposed in psychology and neuroscience; and 2)the potential applications of the principle of focused attention in computer vision, video surveillance, and robotics.

J. Amudha · Ravi Kiran Chadalawada · V. Subashini · B. Barath Kumar
Amrita School of Engg., Bangalore-35
e-mail: j_amudha@blr.amrita.edu,
{ravisrhyme,preetyramesh,bbmk2050}@gmail.com

A. Abraham and S.M. Thampi (Eds.): Intelligent Informatics, AISC 182, pp. 249–260.
springerlink.com

One of the most influential theories that served as a basis for many of the current research findings in the field of visual attention is the *feature integration theory (FIT)* [12] which was introduced in 1980. FIT states that different features are registered early, automatically and in parallel across the visual field, while objects are identified separately and only at a later stage, which require focussed attention. Visual features like colour, orientation, edge, intensity and others are processed and their saliency is captured in separate feature maps. All these feature maps are summed up to get master map. This map contains regions that attract our focus in order of decreasing saliency. FIT went through several stages of modifications to accommodate new findings on attention. One noticeable setback of early FIT is that it considers only the effect of visual strength of different features.

A model developed by Wolfe termed as the *guided search* model [5] of visual attention overcomes FIT's shortcomings by considering the effect of top-down information along with bottom-up saliency. Similar to FIT even *guided search* theory faced gradual modifications.

Next comes the CODE theory[4] which is a combination of two theories: 1) COntourDEtector theory for perception grouping[van Oeffelen and Vos 1982], 2) Theory of Visual Attention (TVA) [Bundesen 1990]. This theory is based on a *race model* of selection. Here parallel processing of a scene takes place and the element that finishes the processing first is selected. A major dissimilarity of CODE theory from previous two theories i.e. FIT and GST, is that CODE theory considers both space and object as the elemental unit for attention selection, whereas latter two only deal with space-based attention. Many other models for visual attention do exist that are not mentioned here. Refer [6] for further information on other models.

The first computational architecture of visual attention was introduced by Koch and Ullman [3] in 1985.An important contribution of their work is winner-take-all (WTA) network. WTA allows us to select one region at a time from the saliency map and if the selected region is not the region of interest, it inhibits the selected region and finds the next winner region. This is known as inhibition of return [8].

A saliency based visual attention model was developed by Itti[7] . The most influential computer vision model of visual attention is the neuromorphic vision toolkit (NVT) which has been extensively used in many other models of visual attention in computer vision and robotics. A number of concepts used in NVT are inspired by attention model of Koch and Ullman, e.g., feature map, saliency map, WTA along with IOR.NVT proposes a computationally better process for calculation of saliency map. Here, a multi-scale analysis of the input images is performed for calculation of individual feature maps which are combined together to create a centralised saliency map.

A new model *visual object detection with a computational attention system* (VOCUS) [10] proved to be efficient in overcoming the drawbacks of NVT. Functionality of VOCUS is similar to that of NVT, e.g. computation of saliency map, implementation of WTA and IOR. VOCUS, however, performs a number of improvements over NVT. Some of which are: 1) allows easy extension to more features, 2) in learning mode VOCUS computes most significant region and then learns merely these features whereas NVT considers whole region of object,3) in

VOCUS an algorithm is used to select training images while in NVT all images are used. Hence, this results in better accuracy in finding focus of attention in a given image. Fig. 2 shows the basic structure of VOCUS [10].

Models discussed above have proven to be the most influential models. There are many other models existing besides the above mentioned and these are considered to be more or less derived model. The objective of this paper is to propose an Improvised architecture of visual attention model to reduce the computational complexity by reducing the number of features considered in order to find the object of interest in a given image.

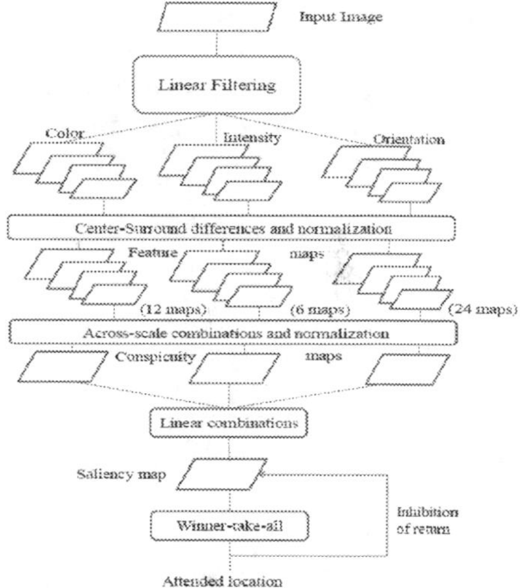

Fig. 1 Architecture of NVT [7]. A Bottom-up approach. It shows the architecture of NVT. The early version of NVT [7] performed only bottom-up analysis of attention but a later and newer version reported in [12] uses top-down cues during visual search.

2 General Characteristics of Computer Vision Based Visual Attention Models

Visual attention models in computer vision share some generic characteristics [7][12]. Some of them are discussed below.

2.1 Saliency Operator

Most of the existing vision models of visual attention use the concept of saliency map to guide attention toward different regions of a given image. A saliency map is also an image whose pixel intensity values indicate relative visual saliency of that corresponding pixel in the original input image. Higher value indicates higher saliency. Most salient pixels in the saliency map will be the current focus of attention. Inhibition of return is performed in order to prevent revisiting the same region by suppressing the saliency of current focus of attention.

2.2 Bottom-Up and Top-Down Analysis

The early attention models of computer vision used only bottom-up approach in attention selection, while the recent models impart top-down cues to get better results. Thus, the bottom-up cues guide visual exploration (focusing on most salient stimuli), while top-down cues guide visual search (an active scan of visual field in search of pre-specified object or stimuli).

2.3 Off-Line Training for Visual Search

Off-line training phase is required by the models prior to performing visual search. During this phase the model learns target specific features and uses this information to increase the saliency of the target-like features in the test images.

Since the way these characteristics are achieved is different, the performance also varies in every computational model.

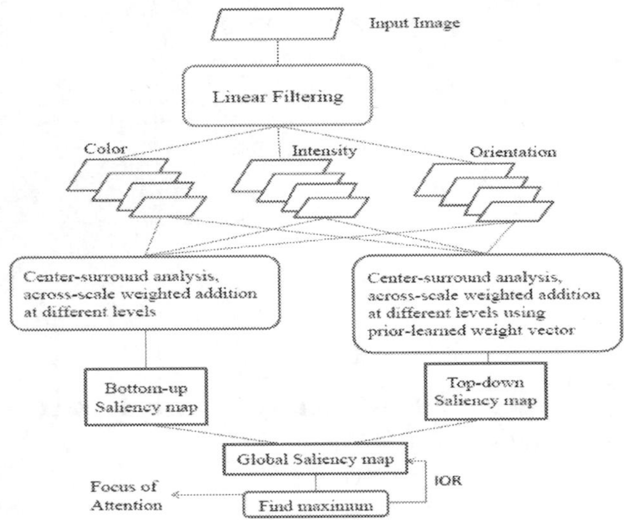

Fig. 2 Architecture of VOCUS[10]. A Top-down approach

Hence, the issues and challenges involved with robotic visual attention [7][12][10] are:

2.4 Optimal Learning Strategy

Information about the visual features of the target is necessary for a robot to perform visual search. Multitudinal sensors and actuators present in the robot facilitate the robots with higher degrees of freedom in their visual perception. As of now almost every feature varies with arbitrary affine transformation, change in viewing angle and lighting condition. Thus, the robot needs to learn many different views of the object in order to identify it in an arbitrary setting. A good strategy to learn the features of an object placed at different angles with minimum human supervision must be present in the attention model of a robot.

2.5 Generality

In most of the computer vision models of visual attention present today, the bottom-up (visual exploration) and top-down(visual search) phases are mutually exclusive. The programmer selects the required loop of attention. This decreases the generality of the attention model and makes it unfit for robotic systems. The selection of the mode of attention should be autonomous depending on the requirement in a robotic visual attention model.

2.6 Prior Training

In general, cognitive robots are expected to learn while working in a manner similar to human, but almost all of the visual attention models for the robot has a prior separate off-line training phase to improve the recognition performance.

Many of the above mentioned issues are interdependent on each other. For instance, if the two modes of visual attention are integrated in the same framework for the sake of generality, there will be no room for prior training. If learning is performed online, it must be ensured that an intelligent learning strategy should be used to enable the robot to obtain enough information about the target.

The proposed robotic cognition model developed looks at issues and challenges and operates in two modes: training (learning) and testing (searching).In learning mode the system analyses the scene and learns how to focus its attention towards the object of interest. The model learns the salient features and switches to searching mode to target itself to the object with less computational time (minimising the amount of inherent information it has to explore).

Conceptually the model shares similar behaviour to VOCUS [10]. The major limitation of VOCUS is that it learns the features from selected region, but doesn't identify the significant features used by the cognition models. In learning phase, a training algorithm selects the image to be used which adds restriction to the robustness of the system.The computation of bottom-up and top-down maps adds up to the complexity and is taken care in the proposed system.

In the proposed model, both bottom-up and top-down cues have been used. In bottom-up, three features (colour, intensity and orientation) are used to obtain saliency map. Features of target object are extracted by selecting the most salient region from the saliency map. Decision tree classifier is used to extract the important features among the selected features that help in classifying the object. In the top-down approach as prior information these important features are used to find the target object. Since, the number of features calculated for any test image are selective it will reduce the computational complexity.

3 System Architecture

Fig.3 represents the architecture of the proposed system for robotic cognition. It can be broadly divided in to two phases: A. Training, B. Testing. Initially, a three channel RGB image is given to the system. Each phase has the following modules.

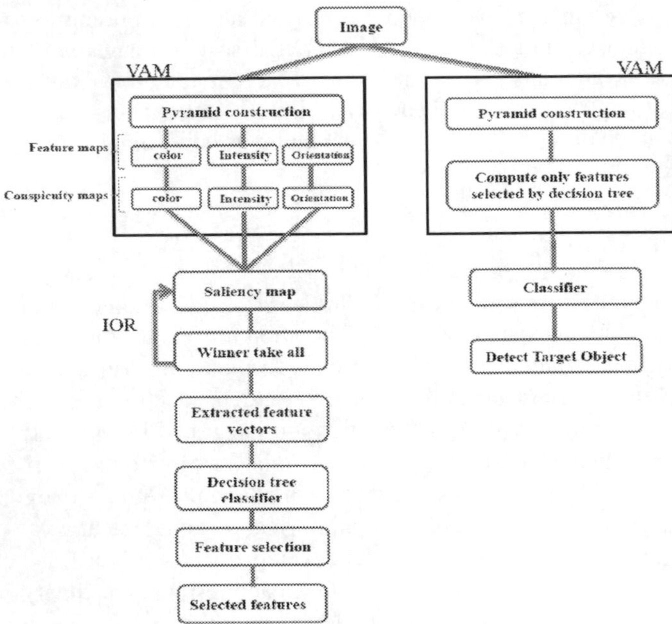

Fig. 3 Architecture of Proposed model for robotic cognition

3.1 Training

In training phase the system is made to learn the required features to find the class of target object.

3.1.1 Visual Attention Module (VAM)

The image is passed to the module of visual attention model. In this module, image is analysed with respect to three different features i.e. colour, intensity and orientation. Three channel Image which is given as input is split in to three images each of single channel. Each image is of one of red(r), green (g) or blue(b) channels. As images are analysed with respect to three different features image to be obtained for each feature is attained with following metrics. To get intensity image all the three channel intensities are linearly combined. It can be given as following.

$$\text{Intensity\_image} = (r+g+b)/3 \tag{1}$$

With respect to colour we find red(R), green (G), blue(B), yellow(Y) values using following equations.

$$R = r - (g + b)/2 \tag{2}$$

$$G = g - (r + b)/2 \tag{3}$$

$$B = b - (r + g)/2 \tag{4}$$

$$Y = r + g - 2(|r - g| + b) \tag{5}$$

(Negative Values are set as zéro)
To analyse the image with respect to orientation two dimensional gabor filtres are used. A 2D gabor kernel can be mathematically defined as

$$G(x,y) = e^{\left(\frac{x'^2+\gamma'^2 y'^2}{2\sigma^2}\right)} \cos\left(2\pi \frac{x'}{\lambda}\right) \tag{6}$$

Where

$$x' = x\cos\theta + y\sin\theta \tag{7}$$

$$y' = -x\sin\theta + y\cos\theta \tag{8}$$

The parameters involved in the construction of a 2D Gabor filter are: The variance σ of the Gaussian function. The wavelength λ of the sinusoidal function. The orientation θ of the normal to the parallel stripes of the Gabor function. The spatial aspect ratio γ specifies the ellipticity of the support of the Gabor function. For $\gamma=1$, the support is circular. For $\gamma< 1$ the support is elongated in the orientation of the parallel stripes of the function. Here Orientation is considered for four angles. i.e 0,45,90,135 degrees.

Pyramids are constructed next to this, from which centre and surround are fixed to get the absolute differences. It's done with following relation. S=C+L where S is surround, C is centre and L=1, 2; Once after getting centre and surround, centre surround differences are calculated to attain the feature maps. Intensity feature map can be obtained using following relation.

$$I(c,s) = |I(c) \ominus I(s)| \tag{9}$$

i.e. the absolute value of difference of intensities between centre and the surround. Similarly the colour feature maps are obtained by using the following equations.

$$RG(c, s) = |(R(c) - G(c)) \ominus (G(s) - R(s))| \tag{10}$$

$$BY(c, s) = |(B(c) - Y(c)) \ominus (Y(s) - B(s))| \tag{11}$$

Orientation feature maps are calculated as shown below.

$$O(c, s, \theta) = |(O(c, \theta) \ominus O(s, \theta)| \tag{12}$$

Where, 'c' indicates the center level of corresponding color pyramid which is 1. 's' indicates the surround levels of corresponding color pyramid which is 2,3.$\ominus$ denotes across scale difference.

3.1.2 Saliency Map Module

In this module conspicuity maps with respect to each feature are linearly combined to compute the saliency map. The feature maps of each feature are linearly combined to get corresponding conspicuity maps.

3.1.3 Winner-Take-All Module

In this module "winner-take-all algorithm" is applied on saliency map by dividing it in to a set of overlapping salient regions. It gets the most salient region in an image with respect to its intensity feature. If that region is not the area of interest or target object that region is inhibited, next salient region in the image is obtained and the process continues till the target object is attained.

3.1.4 Extracting-Features

Once the target object is attained then the features of that block are extracted from all the feature maps, conspicuity maps and saliency map.

3.1.5 Decision Tree Module

Extracted features of target region from different feature and conspicuity maps are passed as input to decision tree classifier, which in turn gives features necessary and sufficient to classify the target object.

3.2 Testing

In testing phase, the robotic cognition model detects the object of interest with mush less computational effort looking at information to which it has been trained. The cognitive model has been tuned in such a way that it automatically identifies the focus of attention as the object of interest. This approach is based on human perception whose attention is driven towards the object of interest and not towards any other scene information. Hence, we have modelled a driver assistance cognitive system whose object of interest is a signboard.

3.2.1 Computing Selected Features

In this module prior knowledge attained during training is used. Only the features selected by the decision tree during training were computed. As shown in fig.4., the features used by decision tree to classify the target object is RGFM1, o135FM1, ICONS, IFM2, o90FM2,o90FM1.Only these features were computed to classify the target object.

3.2.2 Classification Module

This module classifies the target object based on computed features.

3.2.3 Detect Target Object

This module detects the target object under one class based on the decision of classifier.

In this way the complexity of the system is reduced by reducing the number of features to be computed.

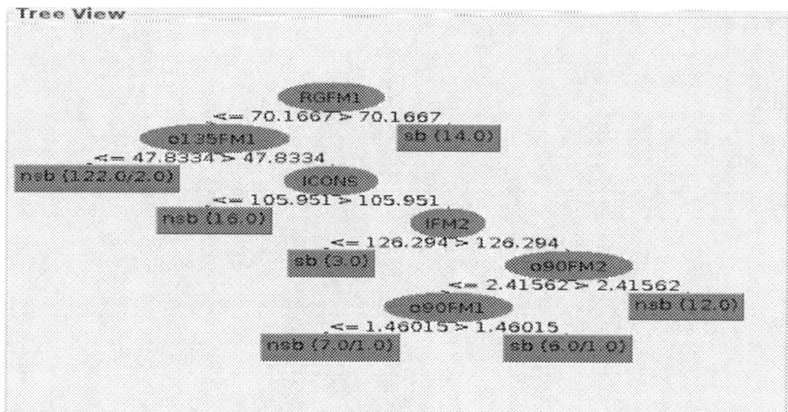

Fig. 4 Output of Decision tree classifier

4 Implementation and Platform

The proposed model of visual attention is being implemented using Opencv libraries with C++ bindings. Domain chosen for this implementation is detection of traffic sign boards. Fig.4 shows the number of training and testing images of different kinds used. All the objects detected will fall under one of the following two classes for this domain i.e. either sign board (Sb) or Non-sign board (Nsb). Every classification will fall under one of the following categories: True Positive, True Negative, False Positive and False Negative.

Parameters that evaluate the performance are:

4.1 Detection Rate

Ratio of total number of objects correctly detected to the total number of detections. It gives the efficiency of the system.

$$\text{Detection Rate} = \text{Number of true detections} / \text{total detections} \tag{13}$$

4.2 Computation Time

It is defined as the time taken to detect the target object in a given scenario. The Number of training images used are show in Table 1 and the number of testing images used were as shown in Table 2.

Table 1 Training images used for proposed model

SIGNBOARD	CLASS	NUMBER OF SAMPLES
Pedestrian	Sb	10
Bike	Sb	10
Crossing	Sb	10
Total		30

Table 2 Testing images used for proposed model

SIGNBOARD	CLASS	NUMBER OF SAMPLES
Pedestrian	Sb	6
Bike	Sb	6
Crossing	Sb	6
Total		18

Table 3 Results obtained in comparing detection rates in bottom-up and top-down approaches

SIGN-BOARD	NUMBER OF IMAGES	NUMBER OF SIGN BOARDS	NUMBER OF SIGNBOARDS DETECTED		DETECTION RATE	
			BOTTOM-UP	TOP-DOWN	BOTTOM-UP	TOP-DOWN
Bike	16	19	13	16	74%	84.21%
Crossing	16	20	17	18	85%	90%
Pedestrian	16	20	15	15	75%	75%
Total	48	59	45	49	78%	83%

The results are as given in Table 3. It shows the number of test images used and corresponding results obtained in both bottom-up and top-down approaches. The fact that can be noted here is detection rate will either improve or remains same in top-down when compared to bottom-up approach.

Table 4 shows the comparison of computation times to detect target objects, number of attempts to detect target objects and percentage reduction in computation times for each image. Fig.5. Shows the comparison of computation times for bottom-up and top-down approaches.

Table 4 Time taken to detect in bottom_up and top-down approaches and percentage reduction of time in top-down approach

Image	TIME IN BOTTOM-UP APPROACH(IN SECONDS)	NUMBER OF ATTEMTS IN BOTTOM-UP	TIME IN TOP-DOWN APPROACH(IN SECONDS)	NUMBER OF ATTEMPTS IN BOTTOM-UP	%REDUCTION IN TIME
Image1	1	1	0.5	1	50%
Image2	5	2	2	2	60%
Image3	7	3	3	3	57.3%
Image4	4	2	2	2	50%
Image5	6	3	3	3	50%
Image6	1	1	0.5	1	50%

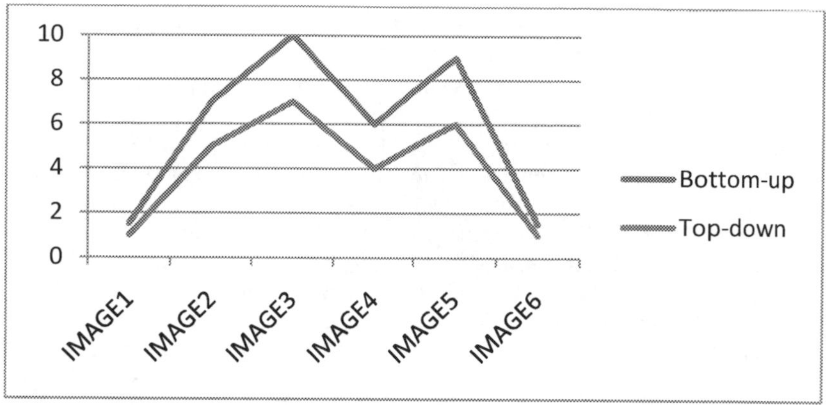

Fig. 5 Comparison of computation times of Bottom-up and Top-down approaches

5 Conclusion

In this paper, we have proposed an optimised computational visual attention model for robotic cognition. This method mimics the process of human attention towards known object of interest. In real environment, the proposed model successfully localizes a salient area and detects the road traffic sign regions with an average reduction of 57.3% time in processing the attention model when compared to other existing models. Detection rate can also be improved by considering more number of samples. And it is robust to illumination scale, view point and partial occlusion. In future the model can be extended to different objects or to object categorization to prove the robustness of the system.

References

[1] Amudha, J., Soman, K.P., Kiran, Y.: Feature Selection in Top-Down Visual Attention Model using WEKA. International Journal of Computer Applications (0975 – 8887) 24(4), 38–43 (2011)

[2] Triesman, A.M., Gelade, G.: A feature integration theory of attention. Cogn. Psychol. 12, 97–136 (1980)

[3] Koch, C., Ullman, S.: Shifts in selective visual attention: Toward the underlying neural circuitry. Human Neurobiol. 4, 219–227 (1985)

[4] Logan, G.D.: The CODE theory of visual attention: An integration of space-based and object-based attention. Psychol. Rev. 103, 603–649 (1996)

[5] Wolfe, J.M., Cave, K., Franzel, S.: Guided search:An alternative to the feature integration model for visual search. J. Exp. Pyschol.: Human Percept. Perform. 15, 419–433 (1989)

[6] Itti, L., Koch, C.: A saliency based search mechanism for overt and covert shift of visual attention. Vis. Res. 40, 1489–1506 (2000)

[7] Itti, L., Koch, C., Niebur, E.: A model of saliency based visual attention for rapid scene analysis. IEEE Trans. Pattern Anal. Mach. Intell. 20(11), 1254–1259 (1998)

[8] Posner, M.I., Cohen, Y.: Components of visual orienting, pp. 531–556. Erlbaum, Hillsdale (1984)

[9] Momotazbegam, FakhiriKarray: Visual Attention for Robotic Cognition: A Survey. IEEE Transactions on Autonomous Mental Development 3(1) (March 2011)

[10] Frintrop, S.: VOCUS: A Visual Attention System for Object Detection and Goal-Directed Search. LNCS (LNAI), vol. 3899. Springer, Heidelberg (2006)

[11] Frintrop, S., Rome, E., Christension, H.I.: Computatinal visual attention sys-tem and their cognitive foundation: A survey. ACM Journal Name 7(1), 1 (2010)

[12] Navalpakkamand, V., Itti, L.: Top down attention selection is fine grained. J. Vis. 6, 1180–1193 (2006)

A Rule-Based Approach for Extraction of Link-Context from Anchor-Text Structure

Suresh Kumar, Naresh Kumar, Manjeet Singh, and Asok De

Abstract. Most of the researchers have widely explored the use of link-context to determine the theme of target web-page. Link-context has been applied in areas such as search engines, focused crawlers, and automatic classification. Therefore, extraction of precise link-context may be considered as an important parameter for extracting more relevant information from the web-page. In this paper, we have proposed a rule-based approach for the extraction of the link-context from anchor-text (AT) structure using bottom-up simple LR (SLR) parser. Here, we have considered only named entity (NE) anchor-text. In order to validate our proposed approach, we have considered a sample of 4 ATs. The results have shown that, the proposed LCEA has extracted 100% actual link-context of each considered AT.

Keywords: Ontology, Augmented Context-Embedded grammar, SLR parser, Indexing, Focused-Crawling, Semantic-Web, NLP, Bare-Concept.

1 Introduction

World Wide Web (WWW) is collection of billions of pages that are linked together by hyperlinks. The hyperlink is often described in the following format: <A HREF= http://www.ai.ac.in/about.html/> Ambedkar Institute of Technology</A>. The first part indicates the target web-page location and the second part, i.e. Ambedkar Institute of Technology gives information about the content of the web-page, which is called anchor-text (AT). It has been very

Suresh Kumar · Asok De
Ambedkar Institute of Advanced Communication Technologies & Research, Delhi-31, India
sureshpoonia@yahoo.com, asok.de@gmail.com

Naresh Kumar
AIIT, Amity University, Noida, India
naresh.dhull@gmail.com

Manjeet Singh
YMCA University of Science & Technology, Faridabad, India
mstomer2000@yahoo.com

A. Abraham and S.M. Thampi (Eds.): Intelligent Informatics, AISC 182, pp. 261–271.
springerlink.com

challenging for the crawler and indexer to get relevant web-pages because of enormous size of www. Most of the researchers have widely explored the use of link-context to determine the theme of target web-page [1], [2], [3], [4], [5], [13]. Link-context has been applied in areas such as search engines [11], and focused crawlers [12] and automatic classification [14]. The extraction of precise link-context may be considered as an important parameter for extracting more relevant information from the web-page. This saves the effort of crawling and indexing of useless and irrelevant pages [6]. In most of the cases AT or text around AT was used to derive the context of a link. This motivates us to propose a rule-based approach for the extraction of the link-context.

Here, in this paper, we proposed a link-context extraction algorithm (LCEA) to derive context of a link from web-page. We have categorized AT into various bare-concepts (BC) such as named entity (NE), class name (CLN), technology (T), framework (F), entertainment (E), and sports (S) etc by manual analysis of 100 web pages from Wikipedia and Open Directory project (ODP). Further, we design ontology of these AT's. For illustration of our approach, we have considered only named entity (NE) ATs involving 14 terminals and 23 non-terminals. After that, we developed augmented context-embedded grammar which is used by SLR parser. Finally, the output of SLR parser is used by LCEA to extract precise link–context of web-page. We have validated our proposed approach by considering limited samples of ATs. The results have shown that, the proposed LCEA has extracted 100% actual link-context of each considered AT.

The rest of this paper is organized as follows: In section 2 related literature is discussed. In section 3, we present our proposed approach in detail, followed by conclusion and future work in section 4.

2 Related Literature

Since the search engines have come into existence, development of various techniques were witnessed in the literature, in order to get optimized result of the search engine. Some of the techniques were focused on the ranking of the search results while others are related to the crawling appropriate pages depending upon end user search trends and the focus area of the search engine [11], [12]. In order to achieve these objectives, both statistical and natural language processing [1], [2], [3], [4], [5], [13] based techniques have proposed. For example, in [1] a dependency analysis based link-context was extracted. The main idea in [1] is to simulate the browsing behavior of web readers. The author fractionize the behavior into four steps which were parsing, decomposing, grouping and selection. But in this technique as author itself made a statement that word variation between the link-context and the target web-page has made the quality of link-context derivation very low. In [2] authors have described an approach to generate automatic rich semantic annotations of text, which can be utilized by semantic-web. In [3] authors have given an idea that cohesive text and non-cohesive text surrounding the AT provide rich semantic cues about a target web-page. In [4] a scheme based on parsing of the text around anchor-link was proposed. In [4], it was tried to extract relevant sentence fragments in the sentence. But in this approach, the applicability of parser was confined to the single sentence only and things were arranged based on their semantic importance using ontology. However, they did not use the concept of AT to extract link-context. Technique in [5] was also related to deriving link-context from HTML tag tree, where firstly an HTML parser was used to find the hierarchical structure of the content arranged using HTML tags and then the actual AT is analyzed. In his approach two link-context derivation techniques were described. In first technique, context from aggregation nodes was derived and in second, context from text window was derived. But both of these techniques fail to capture the conceptual information of the AT, which result in finding poor quality of theme of the target web-page. In [13], a scheme related to link-context extraction based on semi-NLP approach was proposed and their results were not so encouraging with respect to precise extraction of link-context.

3 Proposed Method

In this section, we will first give the overall model of the system and then the issues related to the design of the system will be discussed in the subsequent sub-sections.

3.1 Model

In this model, first of all we have designed ontology OWL: thing of AT [8], [10]. Based on this ontology, we developed augmented context-embedded SLR grammar [7] of NE anchor-text only. Using this grammar the AT is parsed using SLR parser and the output of SLR parser is passed to LCEA as an argument. LCEA derive the context of AT. The proposed Model is shown in Fig.1.

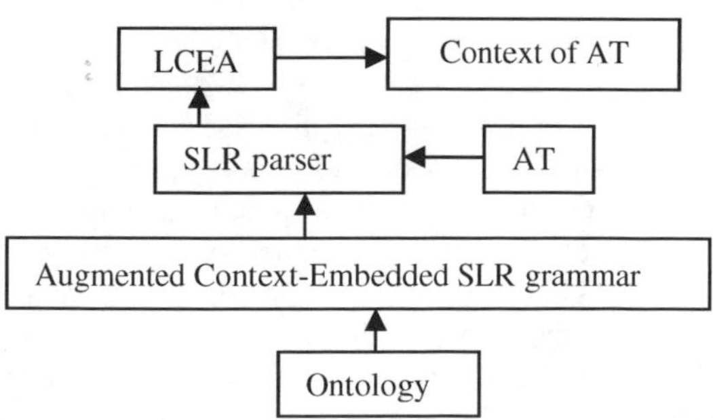

Fig. 1 Model for extraction of link-context

3.2 Types of Context Determination

In order to decide the types of context of AT, we analyzed 100 web pages from Wikipedia and Open Directory Project (ODP). Based on manual analysis of 100 web pages, we are able to categorize AT in various contexts or concepts such as named Entity (NE), technology (T), framework (F), phenomenon (P) and mechanism (M), sports (S) and entertainment (E). Ontology graph of these contexts/bare-concepts is shown in Fig. 2. In OWL [8], [10] there are two fundamental classes from which all other classes are derived, one is - OWL: Thing and second is - OWL: Nothing. The resource OWL: Thing is the class of all individuals, and every resource that is an instance of a class is implicitly a member of OWL: Thing. The resource OWL: Nothing represents the empty class, a class that has no members. In our ontology, AT is subclass of owl: thing. All of these concepts have rdf: is relationship with AT [9], [10].

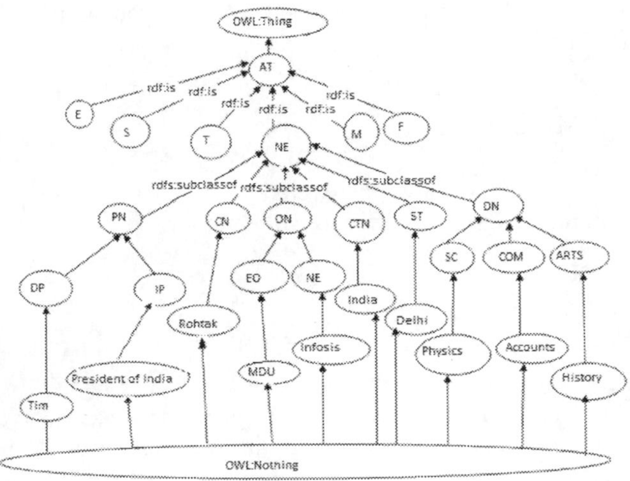

Fig. 2 Ontology graph of ATs

NE anchor-text has further divided into 6 subclasses named as person named entity (PN), organization named entity (ON), discipline named entity (DN), country name entity (CTN), state name entity (STN) and city name entity (CN). All these 6 subclasses have relationship rdfs: subclass-of with NE class [10]. PN entity has further subdivided into two classes named as direct person name (DP) and indirect person name (IP). "tim" is instance of DP and "president of India" is instance of IP class. These instances have rdf: is-a relationship with their respective base classes. ON is also further subdivided into education organization (EO) and non-education organization (NEO) subclasses. Both of these classes have rdfs: subclass-of relationship with ON class [10]. DN has sub classified into commerce (COM), science (SC), arts (ARTs). In this ontology graph, we have explored NE, anchor-text only.

3.3 Rule-Based Development

On the basis of above classification, we have constructed the context-embedded grammar as shown in table 2. Rule $AT^1 \rightarrow AT$ is added to make it an augmented grammar for the SLR parser. Initially, we have taken just 14 terminals and 23 non-terminals that are listed in abbreviation table 1 (used for the sake of simplicity). For example, for President, we have used "pre" as an abbreviation. In order to make discussion clear, we also used notational conventions like all terminals are represented using small case letters and non-terminals are represented using upper case letters.

Table 1 Abbreviation table

S. No.	Abbreviation	Semantic Description	S. No.	Abbreviation	Semantic Description
1	roh	Rohtak	22	SN	State Name Entity
2	raj	Rajasthan	23	CTN	Country Name Entity
3	md	MDU	24	CN	City Name Entity
4	uni	University	25	DP	Direct Person Name Entity
5	tim	Tim	26	IP	Indirect Person Name Entity
6	ber	Berner	27	EO	Education Organization
7	ind	India	28	NEO	Non-Education Organization
8	of	of	29	SC	Science discipline
9	pre	President	30	COM	Commerce discipline
10	acc	Accounts	31	ART	Art discipline
11	phy	Physics	32	FN	First Name
12	his	History	33	MN	Middle Name
13	lee	Lee	34	LN	Last Name
14	inf	Infosis	35	DES	Designation
15	AT	Anchor-Text	36	CNJ	Conjector
16	NE	Named Entity	37	UT	University Title
17	PN	Person Named Entity	38	CL	Class Name
18	ON	Organization Named Entity	39	COY	Company
19	DN	Discipline Name Entity			

Table 2 SLR Grammar

Rule No	Rule	Rule No	Rule
r0	$AT^1 \rightarrow AT$	r18	IP→DES CNJ CTN
r1	AT→NE	r19	EO→UT CN
r2	NE→PN	r20	EO→CL CNJ STN
r3	NE→ ON	r21	NEO→COY
r4	NE→DN	r22	STN→raj
r5	NE→ STN	r23	CN→roh

Table 2 (*continued*)

r6	NE→CTN	r24	C CTN → ind
r7	NE→CN	r25	CL→uni
r8	PN→DP	r26	FN→tim
r9	PN→IP	r27	MN→ber
r10	ON→EO	r28	LN→lee
r11	ON→NEO	r29	CNJ→of
r12	DN→SC	r30	DES→pre
r13	DN→COM	r31	COM→acc
r14	DN→ART	r32	SC→phy
r15	DP→FN MN LN	r33	ART→his
r16	DP→FN MN	r34	UT→ md
r17	DP→FN	r35	COY→inf

Using sets-of-items construction algorithm for SLR parser [7], the following table is constructed.

Table 3 Collection of set of items

State	Set of items	State	Set of items
I_0	AT1→.AT, AT→.NE, NE→.PN, NE→.ON, NE→.DN, NE→.STN, NE→.CTN, NE→.CN, PN→.DP, PN→.IP, ON→.EO, ON→.NEO , DN→.SC, DN→.COM, DN→.ART, DP→.FN MN LN, DP→.FN MN, DP→.FN, IP→.DES CNJ CTN, EO→.UT CN, EO→.CL CNJ STN, NEO→.COY, STN→.raj, CN→.roh, CTN→.ind, CL→.uni, FN→.tim, DES→.pre, COM→.acc, SC→.phy, ARTS→.his, UT→.md, COY→.inf	I_{21}	STN→raj.
I_1	AT1→AT.	I_{22}	CN→roh.
I_2	AT→NE.	I_{23}	CTN→ind.
I_3	NE→PN.	I_{24}	CL→uni.
I_4	NE→ON.	I_{25}	FN→tim.
I_5	NE→DN.	I_{26}	DES→pre.
I_6	NE→STN.	I_{27}	COM→acc.
I_7	NE→CTN.	I_{28}	SC→phy.
I_8	NE→CN.	I_{29}	ART→his.
I_9	PN→DP.	I_{30}	UT→md.
I_{10}	PN→IP.	I_{31}	COY→inf.
I_{11}	ON→EO.	I_{32}	DP→FN MN.LN DP→FN MN. LN→.lee
I_{12}	ON→NE.	I_{33}	MN→ber.

Table 3 (*continued*)

I_{13}	DN→SC.	I_{34}	IP→DES CNJ.CTN,CTN→.ind
I_{14}	DN→COM.	I_{35}	CNJ→of.
I_{15}	DN→ART.	I_{36}	EO→UT CN.
I_{16}	DP→FN. MN LN, DP→FN. MN, DP→FN., MN→.ber	I_{37}	EO→CL CNJ.STN, STN→.raj
I_{17}	IP→DESI.CNJ CTN, CNJ→.of	I_{38}	DP→FN MN LN.
I_{18}	EO→UT.CN, CN→.roh	I_{39}	LN→lee.
I_{19}	EO→CL.CNJ STN, CNJ→.of	I_{40}	IP→DES CNJ CTN.
I_{20}	NEO→COY.	I_{41}	EO→CL CNJ STN.

Thereafter using algorithm 6.1 [7] for the construction of SLR parsing table, the parsing table is constructed as indicated in table 4.

Table 4 Parsing table

State	Action																Goto																					
	roh	raj	md	uni	tim	ber	ind	of	pre	acc	phy	his	lee	inf	$	AT	NE	PN	ON	DN	STN	CTN	CN	DP	IP	EO	NEO	SC	COM	ART	FN	MN	LN	DES	CNJ	UT	CL	COY
0	s22	s21	s30	s24	s25		s23		s26	s27	s38	s29		s31		1	2	3	4	5	6	7	8	9	10	11	12	13	14	15	16			17		18	19	20
1															accept																							
2															r1																							
3															r2																							
4															r3																							
5															r4																							
6															r5																							
7															r6																							
8															r7																							
9															r8																							
10															r9																							
11															r10																							
12															r11																							
13															r12																							
14															r13																							
15															r14																							
16						s33									r17																	32						
17								s35																											34			
18	s22																						36															
19								s35																											37			
20															r21																							
21															r22																							
22															r23																							
23															r24																							
24								r25																														
25						r26									r26																							
26								r30																														
27															r31																							
28															r32																							
29															r33																							
30	r34																																					
31															r35																							
32													s39		r16																		38					
33													r27		r27																							
34							s23															40																
35		r29					r29																															
36															r19																							
37		s21																			41																	
38															r15																							
39															r28																							
40															r18																							
41															r20																							

3.4 Link-Context Extraction Algorithm (LCEA)

In this algorithm, we have taken three data structure as an input:

- AL [ ] array of strings of actions carried out during parsing (from SLR operation table).
- *n* is total number of moves to reach up to accept string in SLR operation table.
- AT is the input string of which context is to be determined.

```
LCEA (AL [ ], n, AT)
(1) Declare a string variable s, and set s=AL [n-3].
(2) Set NT= right hand side of arrow symbol in s
string.
(3) If NT = any of (STN, CTN, CN) then print NT as a
context of given AT, and stop.
(4) Otherwise Set s = AL [n-4].
(5) Set NT= right hand side of arrow symbol in s
string.
(6) Print NT as a context of given AT, and stop.
```

3.5 *Testing of LCEA*

In order to test LCEA, we have taken following set of test cases of AT's as input string:

- President of India
- Tim Berner Lee
- University of Rajasthan
- History

SLR operation of above AT's are shown in table 5 to table 8 respectively. The moves in these tables use the abbreviations which are already mentioned in table 1. The shift operation used in these tables, shifts the terminal lies on the front of the input string to the top of the stack and the reduce operation, pops twice the elements on the 'right hand side of the rule' and push 'left hand side of the rule' onto the stack.

Table 5 SLR Operation table of AT: "President of India"

Stack	**Input**	**Action**
0	pre of ind $	shift
0 pre 26	of ind $	reduce by DES→pre
0 DES 17	of ind $	shift
0 DES 17of 35	ind $	reduce by CNJ→of
0 DES 17 CNJ 34	ind $	shift
0 DES 17 CNJ 34 ind 23	$	reduce by CTN→ind
0 DES 17 CNJ 34 CTN 40	$	reduce by IPN→DES CNJ CTN
0 IP 10	$	reduce by PN→IP
0 PN 3	$	reduce by NE→PN
0 NE 2	$	reduce by AT→NE
0 AT 1	$	accept

Table 6 SLR Operation table of AT: "Tim Berner Lee"

Stack	Input	Action
0	tim ber lee $	shift
0 tim 25	ber lee $	reduce by FN→tim
0 FN 16	ber lee $	shift
0 FN 16 ber 33	lee $	reduce by MN→ber
0 FN 16 MN 32	lee $	shift
0 FN 16 MN 32 lee 39	$	reduce by LN→lee
0 FN 16 MN 32 LN 38	$	reduce by DP→FN MN LN
0 DP 9	$	reduce by PN→DP
0 PN 3	$	reduce by NE→PN
0 NE 2	$	reduce by AT→NE
0 AT 1	$	accept

Table 7 SLR Operation table of AT: "University of Rajasthan"

Stack	Input	Action
0	uni of raj $	shift
0 uni 24	of raj $	reduce by CL→uni
0 CL 19	of raj $	shift
0 CL 19 of 35	raj $	reduce by CNJ→of
0 CL 19 CNJ 37	raj $	shift
0 CL 19 CNJ 37 raj 21	$	reduce STN→raj
0 CL 19 CNJ 37 STN 41	$	reduce by EO→CL CNJ STN
0 EO 11	$	reduce by ON→EO
0 ON 4	$	reduce by NE→ON
0 NE 2	$	reduce by AT→NE
0 AT 1	$	Accept

Table 8 SLR Operation table of AT: "History"

Stack	Input	Action
0	his $	Shift
0 his 29	$	reduce by ART→his
0 ART 15	$	reduce by DN→ART
0 DN 5	$	reduce by NE→DN
0 NE 2	$	reduce by AT→NE
0 AT 1	$	Accept

We are now using above mentioned table 5 to table 8 as input to our proposed LCEA discussed in section 3.4. Result of manual execution of LCEA for these 4 ATs is shown in following table 9. Moreover, we have tested LCEA algorithm to compute the context of around 100 ATs and found 100% actual link-context in all cases. Due to the space constraint, in this paper, we have only shown the illustration of our LCEA considering above mentioned 4 ATs.

Table 9 Result of Link Context Extraction Algorithm

S. No.	Input string / AT	Execution Steps of LCEA
1	"pre of ind"	Here n=11 (1) s=AL [8] = NE→PN (2) NT = PN (3) Not Executed (4) s=AL [7] = PN→IP (5) NT=IP (6) Print IP (Indirect Person Name) as a context of AT="President of India" and stop
2	"tim ber lee"	Here n=11 (1) s=AL [8] = NE→PN (2) NT = PN (3) Not Executed (4) s=AL [7] = PN→DP (5) NT=IP (6) Print IP (Direct Person Name) as a context of AT= "Tim Berner Lee" and stop
3	"uni of raj"	Here n=11 (1) s=AL [8] = NE→ON (2) NT = ON (3) Not Executed (4) s=AL [7] = ON→EO (5) NT=EO (6) Print EO (Education Organization Name) as a context of AT= 'University of Rajasthan" and stop
4	"his'	Here n=6 (1) s=AL [3] = NE→DN (2) NT = DN (3) Not Executed (4) s=AL [2] = DN→ART (5) NT=ART (6) Print ART (Subject of Art Discipline) as a context of AT= "History" and stop

4 Conclusion and Future Work

In this paper, we have proposed a rule-based approach to derive link-context from anchor-text structure, where link-context is inherently embedded in rules. The approach has been successfully tested for new anchor-texts generated using the terminal symbols of the grammar. We have tested LCEA algorithm to compute the context of around 100 ATs and found 100% actual link-context in all cases. Due to the space constraint, in this paper, we have shown the illustration of our LCEA considering only 4 ATs. In this paper, we have taken a limited type of ATs ontology and hence the limited SLR grammar. In future, the type of ATs and their relations in our ontology will be explored. Finally, the performance of the LCEA would be tested. Subsequently, we would also try to extend terminal and non-terminal list, so that all categories of BC of ATs as shown in ontology graph could be covered.

References

1. Jing, T., Ping, T., Zuo, W.: Deriving Link Context through Dependency Analysis. In: IEEE International Conference on Education Technology and Computer (2009)
2. Java, A., et al.: Using a Natural Language Understanding System to Generate Semantic Web Content. International Journal on Semantic Web and Information Systems 3(4) (2007)
3. Chauhan, N., Sharma, A.K.: Analyzing Anchor- Links to Extract Semantic Inference of a Web page. In: 10th IEEE International Conference on Information Technology (2007)
4. Xu, Q., Zuo, W.: Extracting Precise Link Context Using NLP Parsing Technique. In: Proceeding of the IEEE/WIC/ACM International Conference on Web Intelligence, WI 2004 (2004)
5. Pant, G.: Deriving Link-context from HTML Tag Tree. In: Proceedings of 8th SIGMOD Workshop on Research Issues in Data Mining and Knowledge Discovery (2003)
6. Henzinger, M., et al.: Link Analysis in Web Information Retrieval. IEEE Data Engineering Bulletin 23(3), 3–8 (2000)
7. Aho, A.V., Ullman, J.D.: Principals of Compiler Design, pp. 197–214. Narosa Publishing House (25th reprint 2003)
8. Fensal, D., Van Harmelen, Horrocks, I., McGuinness, Patel-Scheider: OIL: An ontology Infrastructure for the Semantic Web. IEEE Intelligent Systems 16(2), 38–45 (2001)
9. Klein, M.: Tutorial: The Semantic Web- XML, RDF, and Relatives. IEEE Intelligent Systems 16(2), 26–28 (2001)
10. Hebeler, J., Fisher, M., Blace, R., Lopez, A.P.: Semantic Web Programming, pp. 63–139. Wiley Publication (2009)
11. Brin, S., Page, L.: The anatomy of a large-scale hypertextual web search engine. Computer Networks and ISDN Systems 30(1-7), 107–117 (1998)
12. Aggarwal, C.C., AI-Garawi, F., Yu, P.S.: Intelligent crawling on the World Wide Web with arbitrary predicates. In: WWW 10, Hong Kong (May 2001)
13. Chauhan, N., Sharma, A.K.: A framework to derive web page context from hyperlink structure. International Journal of Information and Communication Technology 1(3/4), 329–346
14. Attardi, G., Gulli, A., Sebastini, F.: Automatic Web page categorization by link and context analysis. In: Proceeding of THAI 1999, 1st European Symposium on Telematics, Hypermedia and Artificial Intelligence (1999)

Malayalam Offline Handwritten Recognition Using Probabilistic Simplified Fuzzy ARTMAP

V. Vidya, T.R. Indhu, V.K. Bhadran, and R. Ravindra Kumar

Abstract. One of the most important topics in pattern recognition is text recognition. Especially offline handwritten recognition is a most challenging job due to the varying writing style of each individual. Here we propose offline Malayalam handwritten character recognition using probabilistic simplified fuzzy ARTMAP (PSFAM). PSFAM is a combination of SFAM and PNN (Probabilistic Neural Network). After preprocessing stage, scanned image is segmented into line images. Each line image is further fragmented into words and characters. For each character glyph, extract features namely cross feature, fuzzy depth, distance and Zernike moment features. Then this feature vector is given to SFAM for training. The presentation order of training patterns is determined using particle swarm optimization to get improved classification performance. The Bayes classifier in PNN assigns the test vector to the class with the highest probability. Best n probabilities and its class labels from PSFAM are given to SSLM (Statistical Sub-character Language Model) in the post processing stage to get better word accuracy.

1 Introduction

Off-line handwriting recognition is the process of finding letters and words present in digital image of handwritten text. Due to various applications such as postal automation, form processing, storing large volume of manuscripts into digital form and reading aid for blind, offline handwritten recognition (OHR) become a more popular research area. Because of large variation of writing styles of individuals at different times and among different individuals OHR turn to be most interesting and challenging task. Various techniques used in each stage of OHR were reviewed by Nafiz Aric and Yarman-Vural [5]. Lots of works have been proposed by researchers in Chinese, Arabic, Japanese, Latin scripts [7, 12, 2, 8].

V. Vidya · T.R. Indhu · V.K. Bhadran · R. Ravindra Kumar
Centre for Development of Advanced Computing, Trivandrum, Kerala, India
e-mail: {vidyav,indhu,bhadran,ravi}@cdac.in

A. Abraham and S.M. Thampi (Eds.): Intelligent Informatics, AISC 182, pp. 273–283.
springerlink.com

Indian languages such as Tamil, Bangla, Devanagari and Oriya have a few numbers of studies in this area. In Tamil various approaches are used for classification such as Support Vector Machine (SVM) [25], Hidden Markov Model (HMM) [20], Back propagation networks [13]. Bangla Handwritten recognition using Super Imposed Matrices was proposed by Ahamad Shah [9]. Also SVM based [14] and Modified Quadratic Discriminant Function (MQDF) [16] based works were reported in Bangla. Quadratic classifier [15, 17] is suggested in Oriya and Devanagari languages. HMM based handwritten recognition was proposed by Shaw [23] in Devanagari script.

Malayalam is a prime language of Kerala, one of the South Indian state. Offline handwritten recognition of Malayalam language is in initial stage. G. Raju [11] proposed a wavelet based Malayalam handwritten recognition. Fuzzy zoning and class Modular Neural Network method is reported by Lajish V.L [18]. 1D Wavelet transform is applied on projection profile of character image is given in the paper [19]. Another study is based on HLH intensity patterns by Rahiman [27]. All these works are experimented on isolated Malayalam characters.

Here we propose the offline handwritten recognition for Malayalam document images. Probabilistic simplified fuzzy ARTMAP is used for classifying characters. Classified labels are mapped to Unicode values. Experimental result shows that our approach is very promising for future developments.

2 Handwritten Recognition System

As any other OCR, the proposed off-line handwritten character recognition system includes four stages: preprocessing, feature extraction, classification and post processing.

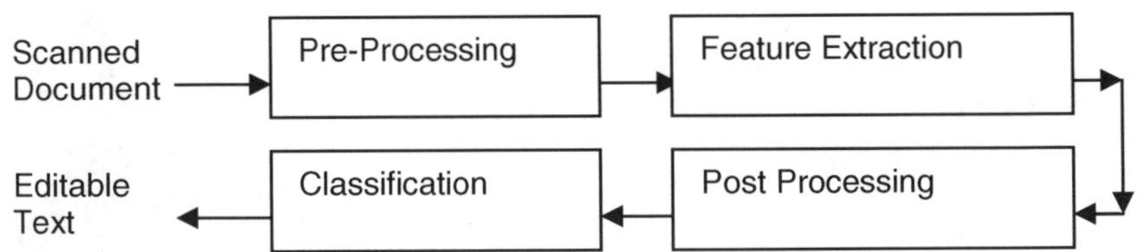

Fig. 1 Block diagram of the system

In our method, we classify an unknown character into one of the most probable known class by using a set of distinctive features. Scanned input image is refined in the preprocessing stage. After preprocessing, the image is segmented into lines, words and characters. In feature extraction, each character image glyph is converted into meaningful features. In classification stage the features are classified into one of the several classes using probabilistic classifier. Context information is used in post processing stage to improve the accuracy of the system. The block diagram of the proposed system is shown in Fig. 1.

2.1 Preprocessing

Objective of preprocessing is to remove distortions in the image, extract relevant text parts and make it suitable for next stage. Otsu' Technique [1] is used for binarization which converts gray scale image into binary image. Modified Directional Morphological Filter (MDMF) algorithm [4] is used to remove the noise present in the binarized image. Next step is to remove skew in the image; which may be introduced into the image while scanning or due to writing styles of human. Hough transform [24] estimates the skew angle in the image and rotates the image using the estimated skew angle to get the skew corrected image. Segmentation is an essential preprocessing stage, because the extent one can reach in separation of lines, words or characters directly affects the recognition rate of the script. The task of individual text-line segmentation from unconstrained handwritten documents is complex because the characters of two consecutive text-lines may touch or overlap. We have used piecewise projection profile method [22] combined with connected component labelling to segment the lines (Fig. 2). Skewed or moderately fluctuating text lines can be segmented properly by this method.

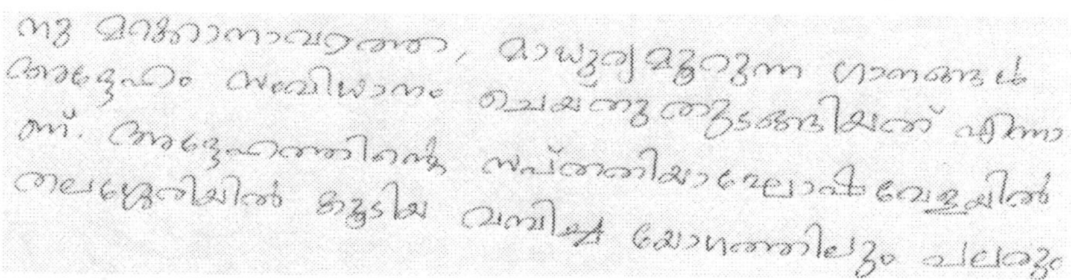

Fig. 2(a) Original image

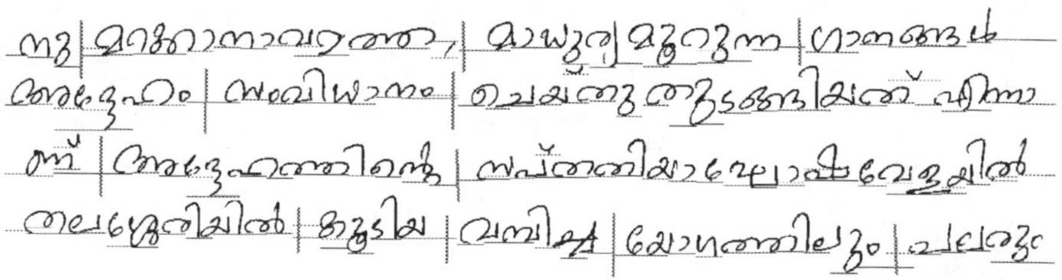

Fig. 2(b) Image segmented into lines and words after preprocessing

2.2 Feature Extraction

Feature extractor will extract most relevant information from the preprocessed character image thereby minimize the within-class variability at the same time enhance the between-class variability [21]. Fuzzy, geometrical, structural and reconstructive features are used in our system.

2.2.1 Fuzzy Depth Feature

Divide each character image into nine segments as shown in Fig. 3(a). Then calculate the distance from border of character block to line of the character. Segments 1,4,7 are left border view, 1,2,3 is the top border view, 3,6,9 is right border view and 7,8,9 is bottom border view. For each segment of the view, transform the distance into depth level by averaging it. Depth levels are converted to degree of membership in fuzzy set DEPTH. The fuzzy subsets of DEPTH are min, medium, and max and is shown in Fig. 3(b).

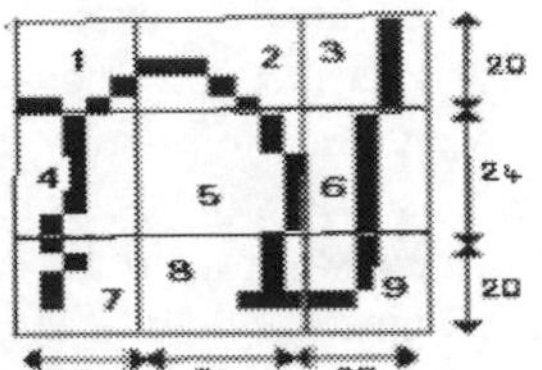

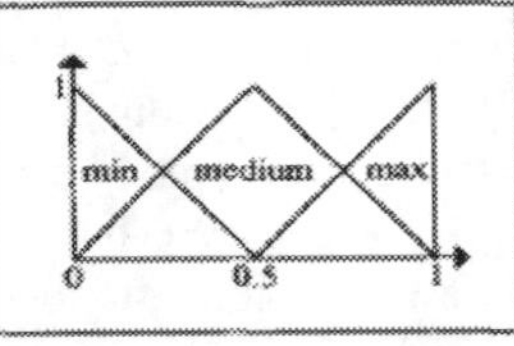

Fig. 3 (a) Division of an image 3(b) Fuzzy values of depth

2.2.2 Distance Feature

In this method character image is divided to 16 equal zones. Compute distance from character origin to each pixel present in the zone. Each distance is divided by $(h^2 + w^2)^{½}$ for normalization where h, w are width and height of the image. Then compute average distance for each zone.

2.2.3 Cross Feature

For each row and column count the number of crossing as shown in Fig. 4. Thus we get horizontal vertical cross features of the particular character image. It is normalized between 0 and 1.

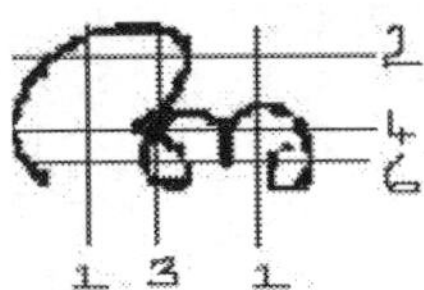

Fig. 4 Cross feature extraction

2.2.4 Zernike Moment Feature

Zernike moments capture global character shape information. Zernike moments based on a set of complex polynomials $\{V_{nm}(x,y)\}$ which form a complete orthogonal set over the unit disk of $x^2 + y^2 \leq 1$ in polar coordinates. The form of the polynomials is:

$$V_{nm}(x,y) = V_{nm}(\rho,\theta) = R_{nm}(\rho)e^{im\theta} \tag{1}$$

Where n is positive integer or zero; m is integers subject to constraints n-|m| is even, and |m| ≤ n; ρ is the length of the vector from the origin to the pixel (x, y); θ is the angle between the vector ρ and x axis in counter clockwise direction; $R_{nm}(\rho)$ is Radial Polynomial defined as:

$$R_{nm}(\rho)=\sum_{s=0}^{(n-|m|)/2}(-1)^s\frac{(n-s)!}{s!(\frac{n+|m|}{2}-s)!(\frac{n-|m|}{2}-s)!}\rho^{n-2s} \tag{2}$$

Zernike moment of a digital character image $f(x, y)$, of order n with repetition m is defined in equation (3). A relatively small set of Zernike moments can characterize the global shape of a pattern effectively. 12th order Zernike moment is used in our system.

$$A_{nm}=\frac{n+1}{\pi}\sum_x\sum_y f(x,y)V_{nm}^*(x,y), x^2+y^2\le 1 \quad \text{where} \quad V_{nm}^*(x,y)=V_{n,-m}(x,y) \tag{3}$$

2.3 *Classification*

Here we have used Probabilistic simplified fuzzy ARTMAP for training and classification of characters. Probabilistic simplified fuzzy ARTMAP (PSFAM) [3, 10] is an artificial neural network (ANN) classifier, which learns by supervised training, which can be trained with just one iteration of the training data, and which can be used in the incremental training mode without the need for retraining. It is a combination of simplified fuzzy ARTMAP (SFAM) and Probabilistic Neural Network. The architecture of PSFAM is shown in Fig. 5. Input layer, complement coded input layer, category layer are basic layers of SFAM. Summation layer and decision layer are part of PNN.

Compared to Multi-layer Perceptron (MLP) these networks have lower training time. Training of feature vector is done by SFAM [6]. The output class of test vector is determined as that of the highest estimated Bayes posterior probability using PNN.

All inputs are complemented in such a way that an input vector a is converted to an input vector $I = (a,a^c)$ where $a^c = 1- a$. w_j is the top down weight vector of the jth neuron in the output category layer : $w_j = [w_{j1}; w_{j2}; \ldots ; w_{j2d}]$.

Calculate the activation value for given input

$$T_j(I)=\frac{|I\wedge w_j|}{\alpha+|w_j|} for j=1,\ldots\ldots,N-1 \tag{4}$$

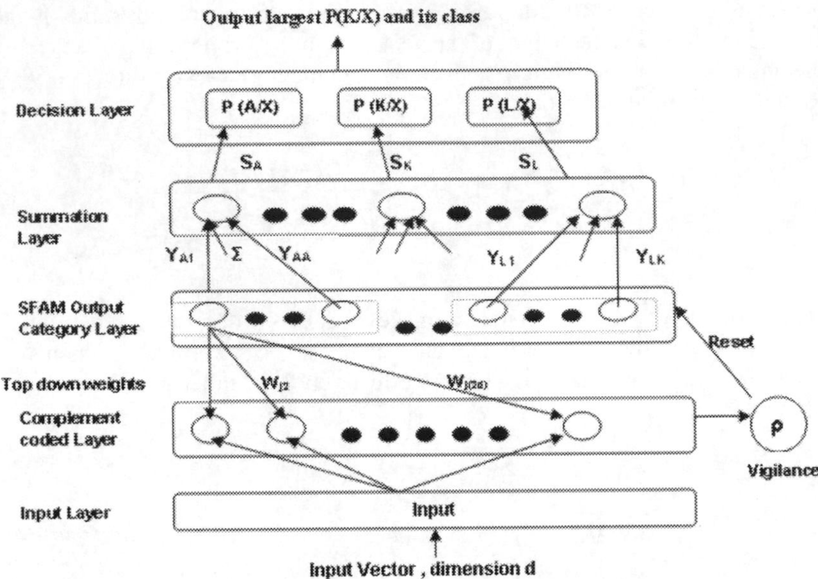

Fig. 5 Architecture probabilistic simplified Fuzzy ARTMAP

Find the winner by $J = arg[Max(Tj)]$. If winner neuron is uncommitted, create a new subclass and set the class label of the winner neuron to be as the class label of input pattern. If it is a committed neuron check the resonance condition or vigilance test i.e. the input is similar enough to the winner's prototype as

$$\frac{|I \wedge w_j|}{|I|} = \frac{|I \wedge w_j|}{2d} \geq \rho \tag{5}$$

If resonance condition is satisfied and also if the class label of the winner matches with the class label of input then update the top-down weight as given in Eq. (6)

$$w_j^{(new)} = \beta(I \wedge w_j^{(old)}) + (1-\beta)w_j^{(old)} \tag{6}$$

Where β is the learning factor $0 \leq \beta \leq 1$

Otherwise reset the winner $T_J = -1$, temporarily increase the vigilance factor.

Main disadvantage of SFAM is that the presentation order of training patterns will affect the classification performance. To avoid this we have used Particle Swarm Optimization (PSO) [26]. During testing, PNN is used to calculate the estimated posterior Baye's probabilities for each of the possible classes using the Parzen Window technique. The Baye's classifier is based on Bayes' theorem given in Eq. (7).

$$P(C_k/X)=\frac{P(X/C_k)P(C_k)}{P(X)}=\frac{P(X/C_k)P(C_k)}{\sum_k P(X/C_k)P(C_k)} \tag{7}$$

Where $P(C_k/X)$ is the posterior probability that vector X belongs to class C_k, $P(X/C_k)$ is the conditional probability that X is selected from the class C_k, and $P(C_k)$ is the prior probability that X belongs to class C_k. P(X/Ck) is determined from the Parzen Windows formula.

$$P(X/C_k)=\frac{1}{(2\pi)^{p/2}\sigma_k^p}\frac{1}{n_k}\sum_{l=1}^{nk}\exp\left(\frac{-(X-Y_{kl})^T(X-Y_{kl})}{2\sigma_k^2}\right) \tag{8}$$

In which p is the dimension of the vectors, σ_k is the smoothness parameter of class k (chosen either global to all classes or local to each class), n_k is the total number of nodes formed for class k. and Y_{kl} is the weight vector formed for l^{th} node of class k. Assume that a test vector X of class k is applied. X is complement coded. And find the exponential term for each node of class k. Then these exponential terms of all the nodes of class k is summed up together in summation layer using Eq. (8). Then Baye's posterior probability is estimated by Eq. (7). The process is repeated for each class. The highest estimated Baye's posterior probability and the associated class are determined by decision layer. Best m probabilities and classes are given to post processing.

2.4 *Post Processing*

Statistical Sub-character Language Model (SSLM) [28] is used to improve the recognition result at word level. Multi-Stage Graph (MSG) is used for getting ranked list of alternative words. MSG represents the probabilistic relationships from classifier as well as statistical language model. Here assumption taken is that neither broken characters nor touched characters are there in the word image. In MSG, node weights are taken from PSFAM and edge weights are computed from statistical bi-gram modeling. Traverse the graph stage by stage from the source node to the destination node, compute the product of node and edge weights on the path of partial sequences. Whenever number of partial sequence in the solution set exceeds a predefined value M, solution set is pruned. Thus get top M words that are sorted in descending order, based on their scores.

3 Result and Discussion

We have done our experiments in two phases. We have used 300 dpi digitized images in gray tone. In phase I, we have collected isolated samples from 26 informants. 2-dimensional array of rectangular boxes had been used for data collection purpose. Participants were requested to write one character per box with no other restriction imposed. Generally Malayalam character set includes 15 vowels and 36 consonants. Also the set includes five chillu characters, anuswaram, chandrakala and visargam. In addition to that numbers from 0 to 9, punctuations and 69 compound characters are also included in our set. Due to the writing style some characters look alike as shown in Fig. 6.

We divided collected data into 3 groups; (i) train set, which contains characters from 11 writers (ii) validation set, which contains characters from 8 writers and (iii) test set, which contains characters from 7 writers. Segmentation has no major role in this phase. Features are computed as mentioned in feature extraction stage. Train set is trained using SFAM. The parameters used are $\rho = 0.9$, $\alpha = 0.001$ and $\beta = 0.1$.With the aid of validation set PSO will minimize the generalization error according to which training patterns are arranged. Once optimization is done, the trained data is tested with validation set. Vectors which have not satisfied the match criteria i.e. resonance condition less than ρ, are also given for training in next iteration. The process will continue till it reaches the maximum iteration or it reaches the predefined accuracy. Accuracy obtained for each set is given in Table 1. Comparison of our method with other techniques for Malayalam mentioned in Section 1 is shown in Table 2.

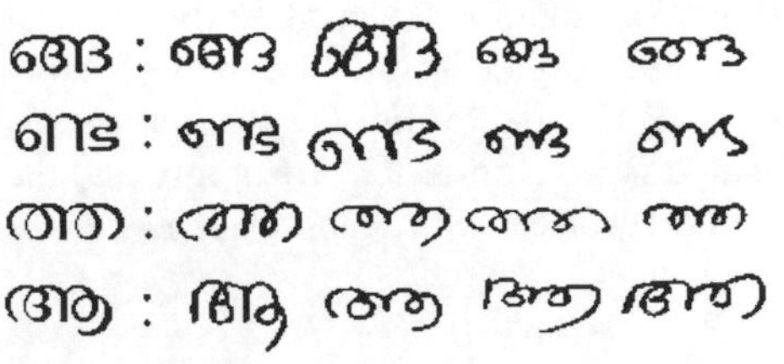

Fig. 6 Some samples of Malayalam characters

Table 1 Accuracy obtained for 3 sets

Set	Accuracy
Train	100%
Validation	98.28%
Test	87.81%

Table 2 Comparison of our method with other techniques

Method	Classifier	No. of Malayalam character used	Samples trained	Accuracy
[19]	MLP	33	4950	73.8
[18]	Class Modular NN	44	15,752	78.87
Our Method	PSFAM	145	2755	87.81

In phase II, 27 informants were asked to write contents from books or magazines in their own handwriting style. Trained network from first phase is given here for classification. Segmentation is the main task in this phase due to line proximity (touching or overlapping of ascenders or descenders due to small gaps between neighboring text lines) and line fluctuations (inter-line distance variability and inconsistent distance between the components due to writer movement).

After segmentation each character is given to feature extraction and classification. With the aid for language model SSLM will give the suggestions for error words. Classified output labels are mapped to Malayalam Unicode format. Recognized output and its original image are shown in Fig. 7. In this phase we get an average accuracy of 79.48%. In most of the images thicknesses of pen strokes are very less. Therefore most of the characters are broken down into two or more components. Touching of characters due to free style writing will be another cause for accuracy reduction.

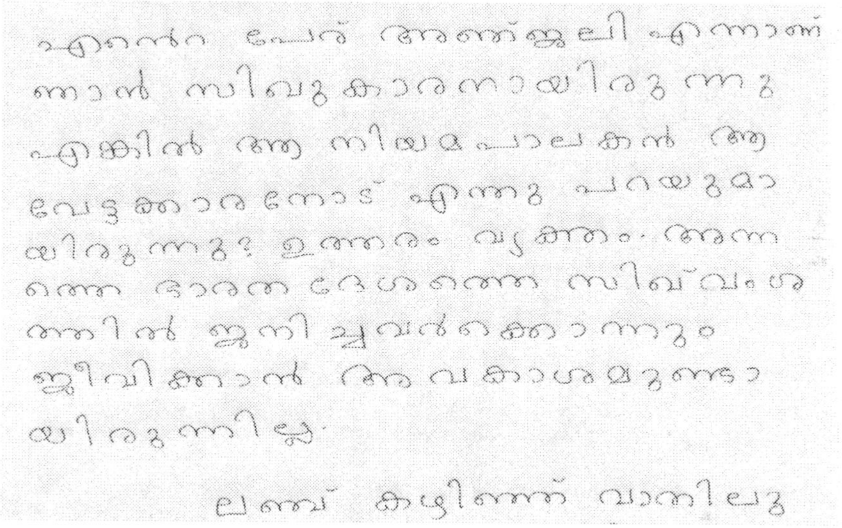

Fig. 7(a) Input Image

എൻെറ പേര് അണ്ജലി എന്നാണ്
ഞാൻ സിഖുകാരനായിരുന്നു
എങ്കിൽ ആ നിയമപാലകൻ ആ
വേട്ടക്കാരനോട് എന്തു പറയുമാ
യിരുന്നു9. ഉത്തരം വുക്തം അന്ന
ത്തെ ഭാരത ദേശത്തെ സിഖ്വംശ
ത്സിൽ ജനി ച്ചവർക്കൊന്നും
ജീവിക്കാൻ അവകാശമുണ്ടാ
യിരുന്നില്ല
ലഞ്ച് കഴിഞ്ഞ് വാതിലു

Fig. 7(b) Recognized output

4 Conclusion

In this paper we present a Probabilistic Simplified Fuzzy ARTMAP (PSFAM) classifier for the recognition of offline Malayalam handwritten characters. Incremental learning strategy of SFAM and probabilistic interpretation of predicated output using PNN will make this hybrid classifier more suitable for character recognition. Probabilistic output will be useful in the post-processing stage to generate suggestions. From the experiments we get an accuracy of 79.48%. Due to the presence of touching and broken characters in the test images, the recognition rate reduced from 87.81% (phase I) to 79.48% (phase II). Further research has to be done to find out how this can be handled during preprocessing and also to analyze the methods in feature extraction stage so that the recognition accuracy can be improved to the acceptable level.

References

1. Otsu, N.: A Threshold Selection Method from Gray-Level Histogram. IEEE Transaction on Systems, Man and Cybernetics (1979)
2. Saon, G., et al.: Off-Line Handwriting Recognition by Statistical Correlation. In: lAPR Workshop on Machine Vision Applications (1994)
3. Jervis, B.W., et al.: Probabilistic simplified fuzzy ARTMAP (PSFAM). In: IEE Proceedings of Science, Measurement and Technology (1999)
4. Zhang, P., et al.: Text document filters using morphological and geometrical features of characters. In: 5th International Conference on Signal Processing Proceedings (2000)
5. Arica, N., Yarman-Vural, F.T.: An Overview of Character Recognition Focused on Off-Line Handwriting. IEEE Transactions on System, Man and Cybernetics –Part C: Applications and Reviews (2001)
6. Taghi, M., et al.: A fast simplified Fuzzy ARTMAP Network. Neural Network Processing Letters 17 (2003)
7. Li, J., et al.: A New Approach for Off-line Handwritten Chinese Character Recognition using self- adaptive HMM. In: Proceedings of the 5th World Congress on Intelligent Control and Automation (2004)
8. Vinciarelli, A., et al.: Offline Recognition of Unconstrained Handwritten Texts Using HMMs and Statistical Language Models. IEEE Transactions on Pattern Analysis and Machine Intelligence (2004)
9. Shah, A., et al.: Bangla off-line Handwritten Character Recognition using Superimposed Matrices. In: 7th International Conference on Computer and Information Technology (2004)
10. Jervis, B.W., et al.: Integrated probabilistic simplified fuzzy ARTMAP. In: IEE Proceedings of Science, Measurement and Technology (2004)
11. Raju, G.: Recognition of Unconstrained Handwritten Malayalam Characters Using Zero-crossing of Wavelet coefficients. In: International Conference on Advanced Computing and Communications, ADCOM 2006 (2006)
12. Jannoud, I.A.: Automatic Arabic Hand Written Text Recognition System. American Journal of Applied Sciences (2007)
13. Sutha, J., Ramaraj, N.: Neural Network Based Offline Tamil Handwritten Character Recognition System. In: International Conference on Computational Intelligence and Multimedia Applications (2007)

14. Chaudhuri, B.B., Majumdar, A.: Curvelet–based Multi SVM Recognizer for Offline Handwritten Bangla: A Major Indian Script. In: Ninth International Conference on Document Analysis and Recognition (2007)
15. Pal, U., et al.: Off-Line Handwritten Character Recognition of Devnagari Script. In: 9th International Conference on Document Analysis and Recognition (2007)
16. Pal, U., et al.: Handwritten Bangla Compound Character Recognition Using Gradient Feature. In: 10th International Conference on Information Technology (2007)
17. Pal, U., et al.: A System for Off-line Oriya Handwritten Character Recognition using Curvature Feature. In: 10th International Conference on Information Technology (2007)
18. Lajish, V.L.: Handwritten Character Recognition using Perceptual Fuzzy-Zoning and Class Modular Neural Networks. In: 4th International Conference on Innovations in Information Technology (2007)
19. John, R., et al.: 1D Wavelet Transform of Projection Profiles for Isolated Handwritten Malayalam Character Recognition. In: International Conference on Computational Intelligence and Multimedia Applications (2007)
20. Kannan, R.J., et al.: Off-Line Cursive Handwritten Tamil Character Recognition. In: International Conference on Security Technology (2008)
21. Dalal, S., Malik, L.: A survey of methods and strategies for feature extraction in handwritten script identification. In: First International Conference on Emerging Trends in Engineering and Technology (2008)
22. Razak, Z., et al.: Off-line Handwriting Text Line Segmentation: A Review. International Journal of Computer Science and Network Security (2008)
23. Shaw, B., et al.: Offline handwritten Devanagari word recognition: A segmentation based approach. In:19th International Conference on Pattern Recognition, ICPR (2008)
24. Nandini, N., et al.: Estimation of Skew Angle in Binary Document Images Using Hough Transform. World Academy of Science, Engineering and Technology (2008)
25. Venkatesh, J., Sureshkumar, C.: Handwritten Tamil Character Recognition Using SVM. International Journal of Computer and Network Security (2009)
26. Keyarsalan, M., et al.: Font based Persian character recognition using Simplified Fuzzy ARTMAP neural network improved by fuzzy sets and Particle Swarm Optimization. IEEE Congress on Evolutionary Computation (2009)
27. Rahiman, M.A., et al.: Isolated Handwritten Malayalam Character Recognition using HLH Intensity Patterns. In: Second International Conference on Machine Learning and Computing (2010)
28. Mohan, K., Jawahar, C.V.: A Post-Processing Scheme for Malayalam using Statistical Sub-character Language Models. In: Proceedings of the 9th IAPR International Workshop on Document Analysis Systems, DAS 2010 (2010)

Development of a Bilingual Parallel Corpus of Arabic and Saudi Sign Language: Part I

Yahya O. Mohamed Elhadj, Zouhir Zemirli, and Kamel Ayyadi

Abstract. The advances in Science and Technology made it possible for people with hearing impairments and deaf to be more involved and get better chances of education, access to information, knowledge and interaction with the large society. Exploiting these advances to empower hearing-impaired and deaf persons with knowledge is a challenge as much as it is a need. Here, we present a part of our work in a national project to develop an environment for automatic translation from Arabic to Saudi Sign Language using 3D animations. One of the main objectives of this project is to develop a bilingual parallel corpus for automatic translation purposes; avatar-based 3D animations are also supposed to be built. These linguistic resources will be used for supporting development of ICT applications for deaf community. Due to the complexity of this task, the corpus is being developed progressively. In this paper, we present a first part of the corpus by working on a specific topic from the Islamic sphere.

Keywords: Sign Language, Saudi Sign Language, Arabic Sign Languages, Corpus, Linguistic Resources, 3D animation, Avatar, Virtual Reality, Multimedia Content, Parallel Corpus, Bilingual Corpus, Machine Translation.

1 Introduction

The importance of language corpora is obvious and already established in the Language-related studies as well as in the development of many sophisticated language processing tools. Corpora represent today an essential tool for both linguists and computer scientists, especially in the field of computational linguistics. They are used as infrastructure of many kinds of applications ranging from frequency counting systems, item search engines, text summarization

---

Yahya O. Mohamed Elhadj · Zouhir Zemirli · Kamel Ayyadi
Center for Arabic and Islamic Computing
Al-Imam Muhammad Ibn Saud Islamic University, Riyadh,
Kingdom of Saudi Arabia, P.O. Box 5701, Riyadh 11432, KSA
e-mail: yelhadj@ariscom.org, zouhir.zemirli@gmail.com,
ayadi_kamel1@yahoo.fr

A. Abraham and S.M. Thampi (Eds.): Intelligent Informatics, AISC 182, pp. 285–295.
springerlink.com

systems, annotation devices, information retrieval systems, automatic translation systems, question answering systems, etc. [1].

Since their introduction in the 1960, many specific- and/or general-monolingual corpora are developed for several languages around the world (e.g. the Corpus of Contemporary American English (COCA) [2], the American National Corpus (ANC) [3], the British National Corpus (BNC) [4], the Corpus for Spanish [5], the Corpus for Portugal [6], etc). Besides, bilingual corpora have also been developed for several languages [7] to support machine translation systems and other kinds of applications.

For the Arabic language, corpora are still limited in terms of size, coverage, and availability compared to European languages. However, some important Arabic corpora are now available (e.g. the CLARA Corpus, Al-Hayat Corpus, An-Nahar Corpus, the Arabic Gigaword Corpus, etc. [8]).

As far as the Saudi Sign Language (SaSL) is concerned, unfortunately there is no available signed content that can be used as a basis of any kind of processing tools to the best of our knowledge. Few existing contents are available for some other Arabic Sign Languages (ArSLs) in the form of movies, TV series, and news bulletins as indicated in [9]. This is may be due to the fact that almost all ArSLs are still in specification process.

In this paper, we present the task of building a well-designed bilingual parallel corpus, which is part of an ongoing funded project for automatic translation from Arabic to Saudi Sign Language (A2SaSL)[10]. The aim of this project is to convert an Arabic text to a 3D animated sequence of SaSL symbols in two main phases: translation phase and animation phase. The translation phase concerns the mapping between words (or morphemes) in the input text and their equivalent signs, which will rely on the sentence structure in both source and target languages; it represents a pure translation problem, and we want to base it on transformational grammars that can be deduced from the parallel corpus. The animation phase relates to the use of an avatar to render an animated sequence of signs; appropriate tools are used for both pre-creating animations and then signing them at the end of the conversion process. This phase is related to the graphical creation and animation of sign symbols for the Saudi Sign Language.

To build a comprehensive bilingual parallel corpus of Arabic and Saudi Sign Language, different topics in the Islamic field were selected. Here, we present a first part of the corpus related to one of these topics "the Prayer".

2 Overview of the Development of Saudi and Arabic Sign Languages

SaSL is a visual means of communication used by Saudi deaf community, not only to communicate between them, but also to communicate with the hearing society around them. SaSL shares many features with the other ArSLs due to the similarity of cultural values and gestural repertoire among the Arabic countries. In fact, the similarity of cultural traditions among the Arabic countries and the impact on the evolution of ArSLs to each other has led to believe that ArSLs are almost identical. Abdel-Fattah [9] argues that the existence of a standard Arabic spoken language has led to the expectation that there is a shared common sign language, especially in close Arabic regions such as gulf states, Magreb countries, etc. However, a lexical comparative study conducted by Kinda Al-Fityani and Carol Padden on four ArSLs shows that they manifest certain features of difference and therefore cannot be considered as just dialects of each other [11, 12]. This study has examined Jordanian Sign Language (LIU), Kuwaiti Sign Language (KSL), Libyan Sign Language (LSL), and Palestinian Sign Language (PSL).

Arabic Sign Languages are still in development and documentation processes. Indeed, the documentation process began in 1972 with the Egyptian dictionary published by the Egyptian Association for Deaf and Hearing Impaired [13]. Then, many attempts have been made in other Arab countries by trying to standardize and spread signs among members of their local deaf communities. In 1986, the Arabic sign alphabet (see fig 1) has been created and published around all Arabic countries. It is now used as a part of local dictionaries in all ArSLs. Since then, official sign dictionaries appeared in many Arabic countries, such as Morocco (in 1987) [14], Libya (in 1992) [15], Jordan (in 1993) [16], Tunisia (in 2007) [17], Kuwait (in 2007), Yemen (in 2009) [18], Sudan (in 2009) [19], Qatar (in 2010) [20], Palestine, UAE, and Iraq. Other countries are moving to this direction. **Saudi Arabia is one of the Arabic countries that still have no official sign dictionary**, but many efforts are now deployed to reach this goal.

An attempt of unifying ArSLs has been recently initiated by the League of Arab States (LAS) and the Arab League Educational, Cultural and Scientific Organization (ALECSO) in 1999. In 2000, a first part of what they called Unified Arabic Sign Dictionary (UASD) was published in Tunisia [21]. Some signs of UASD have been collected from local dictionaries in different Arab countries, while others are newly proposed and voted by deaf. This part contains more than 1600 words classified in categories: family, home, food, etc. A second part of the UASD appeared in 2006 with more signs and better items organization [22]. It was a result of strong collaboration between the LAS, the Arab Union for Deaf (AUD), and the ALECSO with support of the Arab deaf. In 2007, a DVD version with video-clips of signs of the two parts appeared [23].

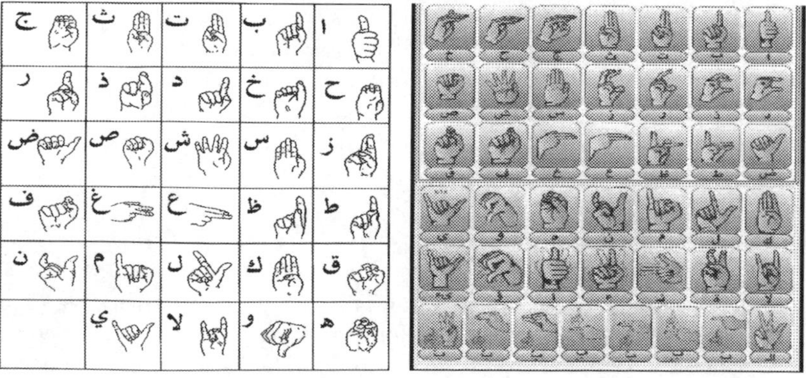

Fig. 1 Arabic Sign Alphabet (left) and Arabic Sign Alphabet with diacritics (right)

Unfortunately, the development of sign dictionaries was not accompanied by description of a linguistic structure of the Sign Language. We are aware only of a few studies that appeared recently in this field discussing some linguistic structures. In [9], Abdel-Fattah presents a linguistic comparative study between ArSLs. Based on certain ArSLs (Jordanian, Palestinian, Egyptian, Kuwaiti, and Libyan), he introduced some basic elements of ArSLs, such as the unity and diversity currently found in perceptions of signs across the Arab world. Hendriks discusses in his PhD [24] aspects of grammar from a cross-linguistic perspective based on Jordanian Sign Language (LIU). From a grammatical analysis point of view, he gives a broader description of linguistic structures in sign language across Jordan, Lebanon, Syria and Palestine. He showed that these SLs have quite uniform grammar. The Supreme Council for Family Affairs in Qatar recently published a book (in Arabic) on the Rules of Qatari Unified Sign Language [20]. It is a very interesting reference and is the first of its kind, to the best of our knowledge, for documenting ArSLs. It first discusses main features of Arabic sign language, and then focuses on Qatari Sign Language (QSL).

As a conclusion, ArSLs still need a lot of efforts for documentation and for specifying the linguistic structure. The Unified Arabic Sign Dictionary was a good step for documentation. However, the linguistic structure is still missing, not only in many individual ArSLs, but also in this new unification attempt. For SaSL, the situation is still not convenient despite the growing importance given officially to the Saudi deaf community. We hear, from time to time, about projects for documentation and specification in specialized centers, but there are still no available references that discuss neither a unified dictionary nor the linguistic structure.

3 Corpus Preparations

As evoked in the introduction, there is no available signed content in SaSL that can be used as a basis neither for our work nor for any kind of processing tools. In fact, we did find during our survey, a few sign-contents taken from religious conferences or lessons; but they are only available as video recordings with no alignment or tagging. Moreover, these contents are not in a structured form as they are generally oral discussions with all possible kinds of hesitations, sentence-breaking, etc. Thus, since, there is no explicit specification of grammar rules in SaSL, we need to have a well-structured parallel corpus that can be used for both deducing some lexical/linguistic features and extracting transformational rules between Arabic and SaSL.

3.1 Corpus Type

Since, this is the first corpus prepared for SaSL, we decided to focus on a specific field that might be of great importance for deaf to support them in their education and to allow developing applications to facilitate their integration in the community. Based on the importance of the religion in everyone's life especially in Saudi Arabia and guided by suggestions and feedback from deaf centers, the Islamic field is chosen to be a domain of work for our project. The choice of this field is driven by essential factors: 1) the fact that educational programs in Saudi Arabia are heavily based on religious backgrounds, 2) the importance of teaching and disseminating Islamic values to all segments of the community, 3) the potential for expansion, knowing that the lexicon and keywords of this field are not just common to Arabic, but they could easily be expanded to all Muslims around the world.

3.2 *Selection of Arabic Texts*

Five basic Islamic topics are selected and will be covered in terms of elements, functioning and provisions. They are Prayer, Pilgrimage, Fasting, Zakat, in addition to the topic of purity (cleanliness) as it is the basis for the correctness of much Islamic worships. These selected topics will be considered progressively to build our overall corpus. Here, we are focusing on the first topic "prayer"; we collected texts describing it in a comprehensive and sufficient manner for the deaf. It is worth to mention that the collected texts are not free literature texts. They have to be taken from authentic Islamic references and then be refined to suit deaf needs. So, after collecting them, they are reviewed, revised linguistically and

Table 1 Extract of the selected texts

ID(رقم)	Content (المحتوى)	ID(رقم)	Content (المحتوى)
1	تعريف الصلاة	1	أول وقتها هو زوال الشمس
2	الصلاة لغة الدعاء	2	وآخر وقتها إذا صار ظل كل شيء مثله
3	وشرعا قربة فعلية ذات أقوال وأفعال مخصوصة	3	مضافا إليه القدر الذي زالت الشمس عليه
4	حكم الصلاة	4	ويعرف زوال الشمس بطول الظل بعد تناهي قصره
5	الصلاة فرضت بالكتاب والسنة والاجماع	5	والأفضل تعجيلها في أول الوقت
6	فمن أنكرها فهو مرتد بلا خلاف	6	إلا في شدة الحر فيستحب الإبرادبها
7	وقد فرضت الصلاة بمكة ليلة الإسراء قبل الهجرة	7	**صلاة العصر**
8	باب أوقات الصلاة	8	ولها وقتان
9	للصلاة أوقات محددة لا تؤدى خارجها	9	وقت اختيار
10	**صلاة الظهر**	10	ويبدأ إذا صار ظل كل شيء مثله سوى ظل الزوال

legitimately, and then simplified inline with the educational capacity of the deaf. An appropriate format and structure were created to ease dealing with them. To give a simple idea of the corpus, we show in table 1 some extracts from the collected texts.

3.3 Analysis

Different kinds of statistics and processings have been conducted on the collected texts. We report in the following table (table 2) the result of statistics in terms of number of total words, number of unique words, and frequency of each word. Other kinds of detailed statistics have been also conducted.

Table 2 Number of total and unique words and their frequencies

N. of Total Words عدد الكلمات الاجمالي	N. of Unique Words عدد الكلمات بدون تكرار	Frequency معدل التكرار
4500	**1862**	**2.42**

It is worth to mention that texts of the corpus are being revised continuously to fit the educational capacity of the deaf. Words and sentences that appear difficult to understand by the deaf community are modified and/or simplified. So, statistics reported here might be slightly different (less or more) in the last version.

4 Translations to SaSL

Our aim in this task is to produce a highly accurate translation of the prepared Arabic content to SaSL. Two kinds of translations are performed: textual translation written as Arabic words and visual translation directly recorded from signers. The first part of the translation can be used to deduce some lexical/linguistic features and extracting transformational rules between Arabic and SaSL; it will be also a good contribution for building a signed content in a written form. The second part of the translation will serve in different purposes. It can be aligned with the Arabic content for a visual translation, for educational aims, etc.

4.1 Textual Translation

In this phase, we made considerable efforts to find deaf people that are able to read and write Arabic texts, even in simple structure, to work with them directly. However, it appeared difficult to find the required competencies due to the weak level of education for deaf people in the kingdom. It was necessary to deal with interpreters specialized in Saudi Sign Language from hearing people to help deaf understand some of our texts. So, we formed working groups of deaf (focus groups), each consisting of at least four people and supervised by an official SaSL interpreter. Deaf and interpreters have been chosen from different clubs and associations to obtain qualified people and to guarantee the recognition from all the deaf community. We asked each group to write the best translation, they agree on, aside of each sentence by the help of the interpreter. Each group is working on specific texts. Groups have been working in parallel to accelerate the translation process. The following table (table 3) shows an extract.

Table 3 Example of the textual translation

ID	الجملة العربية (Arabic Sentence)	الجملة الإشارية (Signed Sentence)
1	كتاب الصلاة	كتاب + صلاة
2	تعريف الصلاة	(تعريف + تعريف) + الصلاة
3	الصلاة لغة الدعاء	الصلاة + لغة + الدعاء
4	وشرعا قربة فعلية ذات أقوال وأفعال مخصوصة	شرعا + (قربة + قربة) + أقوال + افعال + صلاة + ركوع + مخصوصه
5	حكم الصلاة	حكم + الصلاة
6	الصلاة فرض بالكتاب والسنة والاجماع	الصلاة + فرض + (الكتاب + الكتاب)+ (السنة + السنة) + (الاجماع + الاجماع)
7	فمن أنكر ذلك فهو مرتد عن دين الإسلام بلا خلاف	دين + اسلام + عن + قاضي + شيخ + علماء + اختلاف + لا
8	وقد فرضت الصلاة بمكة ليلة الإسراء قبل الهجرة	مكه + الى + المدينة + قبل + ليل + الاسراء + فرض + صلاة + خمس
9	باب اوقات الصلاة	باب + صلاة + وقت
10	للصلاة أوقات محددة لا تؤدى خارجها، وهي	حدود + وقت + صلاة + حد + اذان + قبل + لا + حد + اقامة + بعد + لا
11	صلاة الظهر	صلاة + ظهر
12	أول وقتها هو زوال الشمس	وقت + أول + صلاة + ظهر + شمس عاموديه
13	وآخر وقتها إذا صار ظل كل شيء مثله	صلاة + ظهر+ حركة الشمس + ظل شي + مثل+ وقت + اخر
14	مضافا إليه القدر الذي زالت الشمس عليه	صلاة + ظهر + حركة الشمس + ظل + اضافة

As a strategy of validation, the translation from each group is given to at least one of the other groups for verification. If the translation of a sentence is not accepted by the revising group, a meeting is set up to discuss and agree on a translation. Now, all the Arabic content in the corpus is translated and mutually revised by the working groups.

4.2 Visual Translation

Difficulties in this phase are due to the fact that some religious terms do not have specific signs (unknown for deaf), because they might not be used in their daily life. Some other religious terms may also have different signs. To handle these problems, our working groups coordinate with other specialized clubs to see if these terms are known somewhere. The available signed religious contents, we evoked before, may also be searched for a possible occurrence. Finally, they suggest and vote for appropriate signs. After selecting and fixing signs for the religious terminology in the corpus, we focused on the recording of our textual translation to produce a visual translation. The same groups that worked on the textual translation are also performing its visual translation.

Note that, the necessary technical standards for the production of high quality videos have been taken in consideration to produce an accurate and clear signs. Visual translation of the Arabic content is already done for all texts. We are just performing a phase of verification and validation to ensure that all the video-clips of the sentences are good and well recorded; also that signs are homogeneous over all the content (e.g. if the same words have the same signs in different groups). Any translation that has some anomalies is completely deleted and recorded again. The following figures shows extracted parts of some signs form their video-clips.

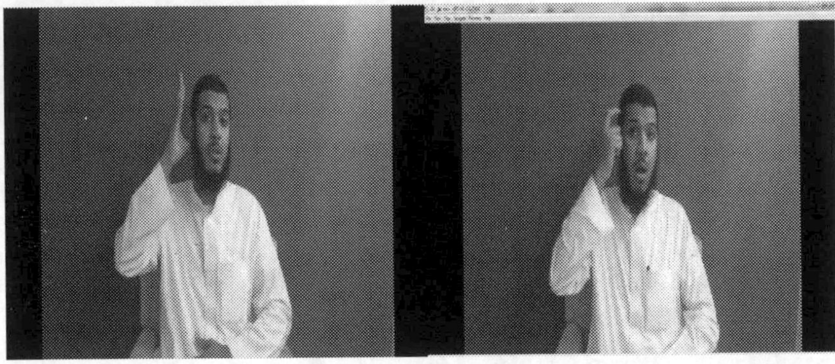

Fig. 2 The sign "prayer" (left side), the sign "mind" (right side)

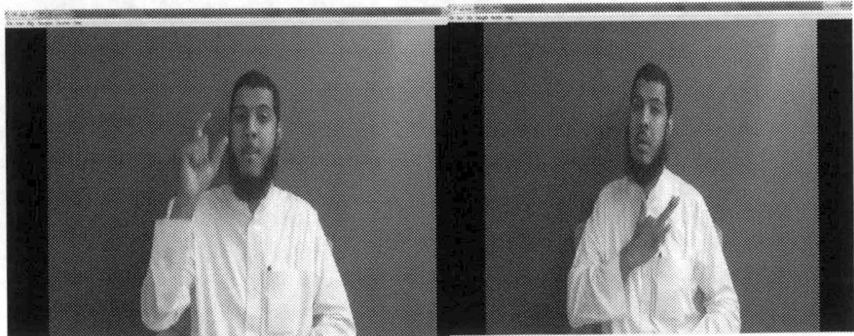

Fig. 3 The sign "Islam" (left side), the sign "faith" (right side)

5 Alignment of the Visual and Textual Translations: Segmentation

It is very important to have an aligned version of the translations, in which we can easily locate the sign and its caption in the video-clip of the sentence. This has many advantages in several situations and will be useful, particularly to train statistical model for machine translation. To realize this objective, we looked for available computer-tools that can be used for video segmentation. We selected an appropriate one (Aegisub [25]) and we are using it now for the subtitling. The Aegisub Project is a community-driven effort to write the BSDL licensed cross-platform subtitle editor Aegisub.

The alignment phase is almost finished, but the content is continuously updated when a video recording is replaced and/or corrected as indicated in the previous section. The following figures display extracted parts of some signs with their subtitles taken from the corresponding video-clips.

Fig. 4 Alignment of the sign "prayer" (only a part of the sign is visible in the window)

Fig. 5 Alignment of the sign "mind" (only a part of the sign is visible in the window)

6 Conclusions

In this paper, we presented the first part of our ongoing bilingual parallel corpus of Arabic texts and their translations into Saudi Sign Language; it is developed as part of a national project for translation from Arabic to Saudi Sign Language. Texts of this corpus are chosen from the Islamic religion field to help deaf Muslims (not only in Saudi Arabia but also in all Arabic and Islamic countries) improve their knowledge and become well religiously educated. They are revised linguistically and legitimately, and then organized in an appropriate format to ease dealing with them. Next, both textual and visual translations of these contents are performed as accurately as possible. New signs were proposed for unknown religious terms by choosing and/or voting among deaf community. Visual and textual translations are being aligned by segmenting the video clips to indicate borders of signs individually. This will allow to easily locate a sign and its caption in the video-clip of the sentence.

This bilingual parallel corpus is the first of its kind developed for Arabic Sign Languages. It will be very helpful for both studying the linguistic structure of SaSL and developing ICT applications for deaf. For the linguistic side, some lexical and/or linguistic features as well as transformational rules between Arabic and SaSL can be deduced and extracted from this corpus. For the ICT side, this corpus can be used to develop educational material for deaf and can also be used to build a translating machine, which may be employed for communication purposes in different kinds of applications.

The bilingual parallel corpus related to the selected topic (Prayer) is almost ready; we are just performing a last verification and validation of the recordings and their corresponding translations. We start working on the other selected topics (Pilgrimage, Fasting, Zakat, and Purity) with the same methodology to increase the size of the overall corpus and to enlarge its coverage. We invite other researchers to contribute by adding new contents from different topics either in the Islamic field or in other relevant ones.

Acknowledgments. This work is part of the A2SaSL project funded by Al-Imam University Unit of Sciences and Technology (grant number 08-INF432-8) under the framework of the National Plan for Sciences, Technology, and Innovation, Saudi Arabia. We thank the other project team members, especially our deaf groups, which actively participated in the elaboration of the work.

References

[1] Dash, N.S., Chaudhuri, B.B.: Why do we need to develop corpora in Indian languages? In: Proc. of the International Working Conference on Sharing Capability in Localization and Human Language Technologies (SCALLA 2001), Organized by ASI@ITC of the European Commission, and held at the National Centre of Software Technology, Bangalore, November 21-23 (2001)

[2] Davies, M.: Corpus of Contemporary American English: 425 million words, 1990-present (2008), http://corpus.byu.edu/coca

[3] Reppen, R., Ide, N., Suderman, K.: American National Corpus (ANC): Second Release. Linguistic Data Consortium, Philadelphia (2005), http://www.americannationalcorpus.org

[4] Davies, M.: British National Corpus from Oxford University Press (2004), http://corpus.byu.edu/bnc

[5] Davies, M.: Corpus del Español (2002), http://www.corpusdelespanol.org

[6] Davies, M., Michael, F.: Corpus do Português (2006), http://www.corpusdoportugues.org

[7] Canning, J.: Disability and residence abroad. Subject Centre for Languages, Linguistics and Area Studies Good Practice Guide (2008)

[8] Alansary, S., Magdi, N., Adly, N.: Building an international corpus of Arabic (ICA): Progress of Compilation Stage. In: 7th Int. Conf. on Language Eng., Cairo, Egypt, pp. 1–30 (2007)

[9] Abdel-Fattah, M.: Arabic sign language: A perspective. Deaf Studies and Deaf Education 10(2), 212–221 (2005)

[10] Elhadj, Y.O.M., Zemirli, Z.A.: Virtual Translator from Arabic text to Saudi Sign-Language (A2SaSL Proj). Technical Annual Report, Riyadh, Saudi Arabia (2011)

[11] Al-Fityani, K., Padden, C.: A lexical comparison of sign languages in the Arab world. In: Proceedings of the 9th Conference of Theoretical Issues in Sign Language Research, Florianopolis, Brazil (2008)

[12] Al-Fityani, K., Padden, C.: Sign language geography in the Arab world. In: Brentari, D. (ed.) Sign Languages: A Cambridge Survey. Cambridge University Press, New York (2010)

[13] Smreen, S.: Linguistic Culture of Deaf. In: International Workshop on "Deaf Communication", Tunisia (2004) (in Arabic)

[14] Wismann, L., Walsh, M.: Signs of Morocco. Peace Corps Morocco, Morocco (1987)

[15] Suwayd, A.: Al-qamus al-ishari. Dar Al-Madinah Al-Kadeemah Lil-kitab, Libya, Triploi (1992)

[16] Hamzeh, M.: Vocabularies of sign language. Jordan (1993)

[17] Institute for Promotion of the Handicapped. Tunisian Sign Dictionary. Ministry of Social Affairs and Solidarity, Tunisia (2007), http://www.iph.rnu.tn/handicap/signe_tunisienne/index.html

[18] 14 October Yemeni News Paper. Yemeni Sign Language Dictionary. Issue n. 14553 (August 7, 2009), http://www.14october.com/news.aspx?newsid=043bf351-638b-4ae6-ad22-a756bc72b34e

[19] Ahmed, A.K., Ajak, S.S.: Sudanian Sign Language Dictionary. Sudanese National Federation of the Deaf, Sudan (2009)

[20] Al-Benali, M., Smreen, S.: Rules of Qatari Unified Sign Language. The Supreme Council for Family Affairs, Qatar (2010), http://www.scfa.gov.qa

[21] LAS: First part of the Unified Arabic Sign Dictionary. The League of Arab States & the Arab League Educational, Cultural and Scientific Organization, Tunisia (2000) (in Arabic)

[22] LAS: Second part of the Unified Arabic Sign Dictionary. The League of Arab States & the Supreme Council for Family Affairs, Qatar (2006) (in Arabic)

[23] SCFA: The Unified Arabic Sign Dictionary (DVD). The Supreme Council for Family Affairs, Qatar (2007) (in Arabic)

[24] Hendriks, H.B.: Jordanian Sign Language: Aspects of grammar from a cross-linguistic perspective. LOT, Janskerkhof, Amsterdam, Netherlands (2008)

[25] The Aegisub Project, version is 2.1.9. It can be, downloaded from, http://www.aegisub.org

Software-Based Malaysian Sign Language Recognition

Farrah Wong*, G. Sainarayanan, Wan Mahani Abdullah, Ali Chekima, Faysal Ezwen Jupirin, and Yona Falinie Abdul Gaus

Abstract. This work presents the development of a software-based Malaysian Sign Language recognition system using Hidden Markov Model. Ninety different gestures are used and tested in this system. Skin segmentation based on YCbCr colour space is implemented in the sign gesture videos to separate the face and hands from the background. The feature vector of sign gesture is represented by chain code, distance between face and hands and tilting orientation of hands. This work has achieved recognition rate of 72.22%.

1 Introduction

Sign language is a very important language used by the deaf and mute in order to communicate to each other and to the general public. Sign language is a language that eliminates the audio limitation and makes it possible for the deaf and mute to communicate visually. Sign language consists of unique finger patterns and hand gestures and sometime face gestures to represent letters or words.

Malaysian Sign Language (MSL) is designed by the deaf naturally based on factors such as daily communication and lifestyle of the deaf and mute in Malaysia. MSL is not a spelled out of oral language or letters in gesture form. This is why hearing people with no knowledge in sign language had difficulties interpreting the message conveyed by the sign language user. This boundary of communication between the hearing and the deaf often is eliminated with the use of a sign language translator (person with hearing and knowledge of MSL). To solve this problem, it is proposed that a software-based sign language recognition system takes the task of translating sign language to verbal (voice). This system will involve image processing approach to track hand gesture of the sign language performed. In this write-up, the verbal (voice) part has been excluded.

Farrah Wong · G. Sainarayanan · Wan Mahani Abdullah · Ali Chekima ·
Faysal Ezwen Jupirin · Yona Falinie Abdul Gaus
Universiti Malaysia Sabah, School of Engineering and Information Technology,
Kota Kinabalu, Malaysia
e-mail: farrahwong@ieee.org

* Corresponding author.

A. Abraham and S.M. Thampi (Eds.): Intelligent Informatics, AISC 182, pp. 297–306.
springerlink.com

2 Preprocessing

2.1 Skin Colour Space

The first step taken is preparing a skin sample from various pictures. Skin region in these pictures are sampled at 10x10 pixels for each image. Two hundred samples are collected from 200 different images consisted of various skin tone at various lighting condition. The Y, Cb and Cr component is extracted from the images. The luminance factor, Y will not be used to map the skin colour. From the extracted Cb and Cr component, 2D-histogram of Cb and Cr element of colour pixel population is plotted.

It is observed that the skin colour distributions in both of the components are very close to each other and have small variances. This is a good indication because this shows that skin colour is suitable to be fitted into the Gaussian model. For a bivariate Gaussian distribution, the probability density function (PDF) is expressed with (1).

$$f(x_1,x_2)=\frac{1}{2\pi\sqrt{\det(\Sigma)}}\exp(-\frac{1}{2}[x_1-\mu_1,x_2-\mu_2]\Sigma^{-1}[x_1-\mu_1,x_2-\mu_2]^T) \quad (1)$$

where μ is the mean of distribution and Σ is the covariance matrix. The Cb and Cr values are fitted in into (1). Figure 1 shows the joint Gaussian distribution of Cb and Cr.

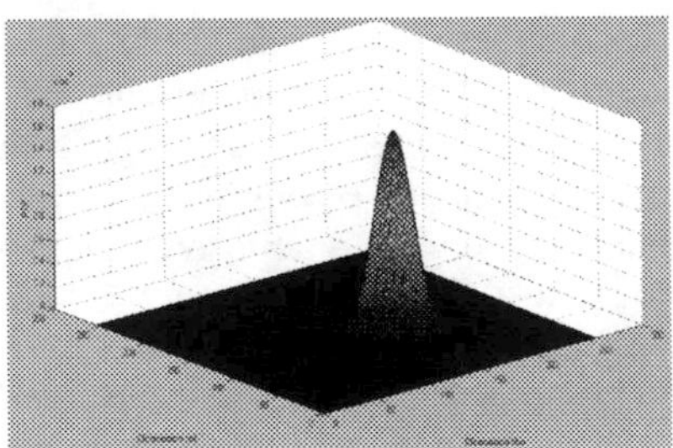

Fig. 1 Joint Gaussian distribution of Cb and Cr

The value of interest in this distribution are the mean of Cb colour value, mean of Cr colour value and the inverted covariance matrix of Cb and Cr and these values are shown in Table 1.

Table 1 Values of Interest

Mean Cb	107.6624
Mean Cr	154.8190
Inverted covariance matrix	$\begin{bmatrix} 0.0162 & 0.0092 \\ 0.0092 & 0.0130 \end{bmatrix}$

2.2 Grabbing Frames from Video

Video is built by frames of static image. To reduce memory and process time, only 30 frames will be taken from the sign language video. These 30 frames are chosen base on the degree of motion between consecutive frames. This means that only the 30 largest degree of motion is considered in the video. The degree of motion is calculated using (2).

$$\deg ree\ of\ motion = \max\left|\overline{frame_{t-1}} - \overline{frame_t}\right| \quad (2)$$

where, $i = 2,3,4,..., \overline{frame_t}$ is the vector representation of current frame, and $\overline{frame_{t-1}}$ is the vector representation of previous frame.

2.3 Skin Segmentation

Skin color segmentation has been applied for various image processing applications [1,2,3,4]. The selected frames are converted to YCbCr colour space. Each pixel value in Cb and Cr components is fitted into equation (1) using the parameter in Table 1 to produce the PDF representative in greyscale. The greyscale image produced is normalised to obtain better resolution. This greyscale image will be filtered by a low-pass filter to reduce the effect of noise. From the greyscale image, the skin can be segmented by applying thresholding to the image. The pixel that holds a higher value than the threshold value will be declared as skin. This process will produce a binary image where white pixel represents skin colour. Figure 2(b) shows the output after thresholding.

a) Original image

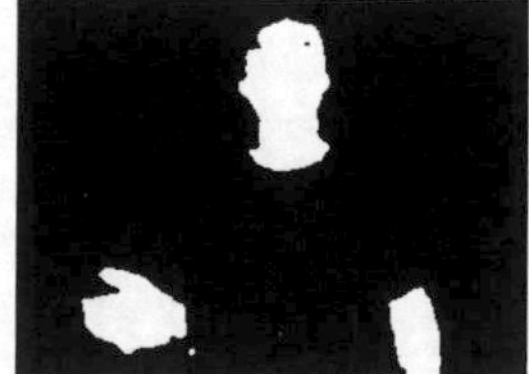

b) Skin segmentation by thresholding

Fig. 2 Skin segmentation

To improve the segmentation result, the 'noise" from false detection will be eliminated using blob area threshold. Segmented blob with an area smaller than the threshold value will be eliminated.

Another process involved to improve the segmentation result is to fill the "holes" in a blob with white pixel. Dilation followed by erosion is applied to connect separated blob cause by shadow or motion blur. The dilation and erosion processes used the same structuring element to conserve the original size of segmented blob. The final segmentation result shown in Figure 3.

2.4 Feature Vector

After the face and hands are correctly segmented, the feature vector is calculated. The feature vector used is the chain code of the hand trajectory, the distance between the head and the hand and the orientation of the hands. Figure 4, 5 and 6 shows the illustration of the feature vector respectively.

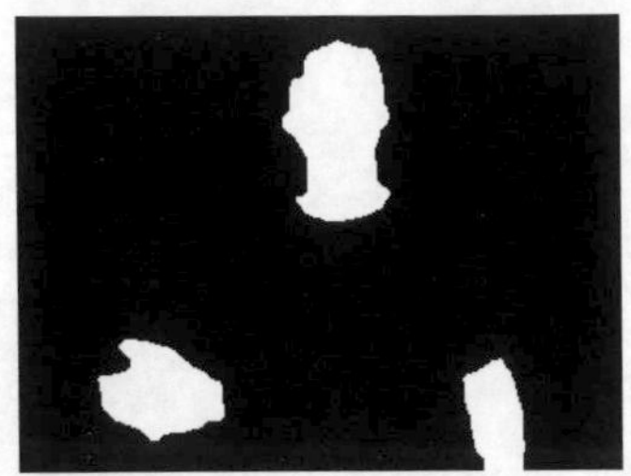

Fig. 3 Final Segmentation Result

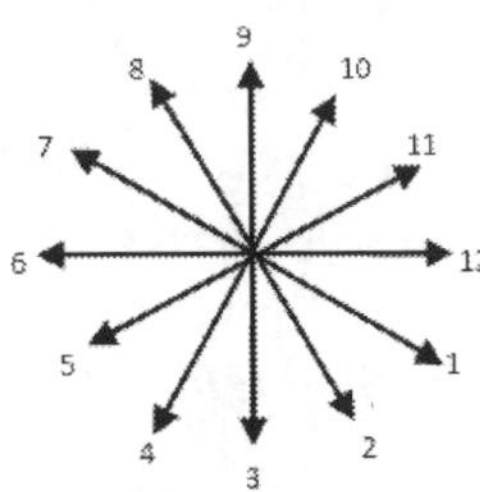

Fig. 4 Chain Code Direction Representation

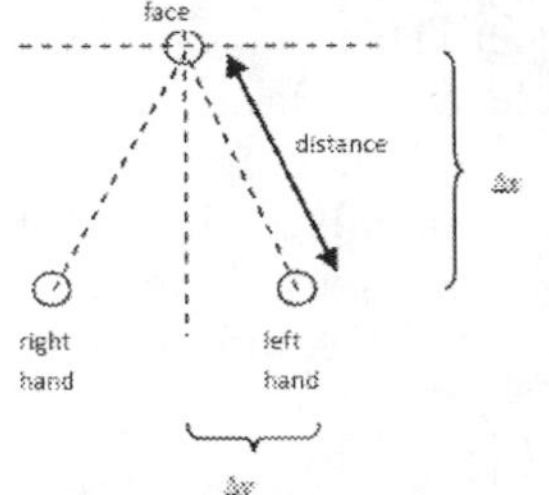

Fig. 5 Hand and Face Distance

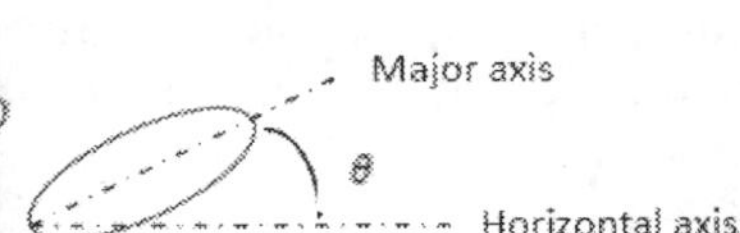

Fig. 6 Tilting Angle or Hand Orientation

To compute the feature vectors, center of mass or centroid of each blob (face and both hands) is to be obtained first. After that, each of the centroid is shifted with respect to three reference point correspond to the blob. This is to eliminate the issue of body posture and size that might affect the consistency of feature vector.

The shifting is done by calculating the difference of the centroids in the first frame and the reference point. The values obtained will be the shifting vector for all of the centroids throughout the whole 30 frames.

Sometimes, not all three centroids are able to be calculated. This is due to under-segmented of skin and overlapping problem (hand to hand or face and hand), which result in less than three blobs appear in the segmented image. To overcome this problem, Linear Kalman Filter is applied to estimate the coordinate of the centroids based on the previous coordinates.

The centroid obtained is used to calculate the feature vectors. To obtain the chain code, the angle of trajectory (Figure 7) is to be calculated first. The angle of trajectory is calculated using (3) and (4). Equation (5) is used to obtain the chain code. For static case, the chain code value is defined as "13".

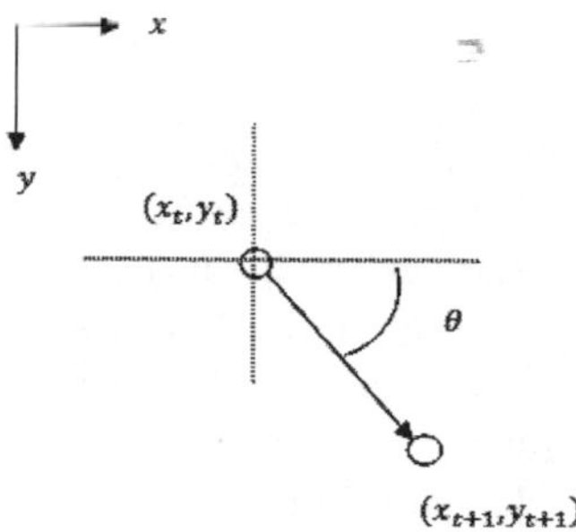

Fig. 7 Trajectory of centroid

$$\alpha = \tan^{-1}\frac{(y_t - y_{t+1})}{(x_t - x_{t+1})} \tag{3}$$

$$\theta = \begin{cases} \alpha & \text{if } y_t - y_{t+1} < 0 \text{ and } x_t - x_{t+1} < 0 \\ 180 - \alpha & \text{if } y_t - y_{t+1} < 0 \text{ and } x_t - x_{t+1} > 0 \\ 180 + \alpha & \text{if } y_t - y_{t+1} > 0 \text{ and } x_t - x_{t+1} > 0 \\ 360 - \alpha & \text{if } y_t - y_{t+1} > 0 \text{ and } x_t - x_{t+1} < 0 \\ 90 & \text{if } x_t - x_{t+1} = 0 \text{ and } y_t < y_{t+1} \\ 270 & \text{if } x_t - x_{t+1} = 0 \text{ and } y_t > y_{t+1} \end{cases} \tag{4}$$

$$\text{chain code} = \text{round}(\theta \div 30) \tag{5}$$

To obtain the distance between face and hands (Figure 5), the Pythagoras Theorem is used as shown in (6). The distance is measured in pixel.

$$\text{distance} = \sqrt{\Delta x^2 + \Delta y^2} \tag{6}$$

The angle of the hand orientation is measured in the range of -90° to 90° from the horizontal axis. The angle is determined by enclosing the blob in an ellipse. The angle between the major axis of the ellipse and the horizontal axis gives the hand tilting angle, representing the orientation of the hand. The orientation values range from 1 to 19, represent the angle -90° to 90° at the scale of 10°.

3 Recognition

3.1 Training

HMM [5,6] is used to perform recognition of sign language gestures. A compact notation for HMM can be expressed as:

$$\lambda = (A, B, \Pi) \tag{7}$$

where A is the state transition probability vector or also known as transition matrix that is represented by N x N matrix. Parameter B is the probability of the observation tied up with specific states, which is represented by N x M matrix.

The parameter Π represent the initial state distribution or also known as the state priority at time, t=0 [7].

The HMM training is performed towards three feature vectors, chain code, pixel distance between face and hand and the tilting orientation of the hand. Each of these vectors consists of two sets, right hand and left hand. In total, there will be six sets of data describing a sign gesture. These data will be trained separately, which means each set of data will have one HMM. The initial step to perform the training is to initialise the parameter.

The transition matrix, A selected is in the form of upper triangular matrix. This is to prepare for a more flexible model in case of state skipping due to faster performed gesture. Figure 8 visualises the transition pattern used in this project.

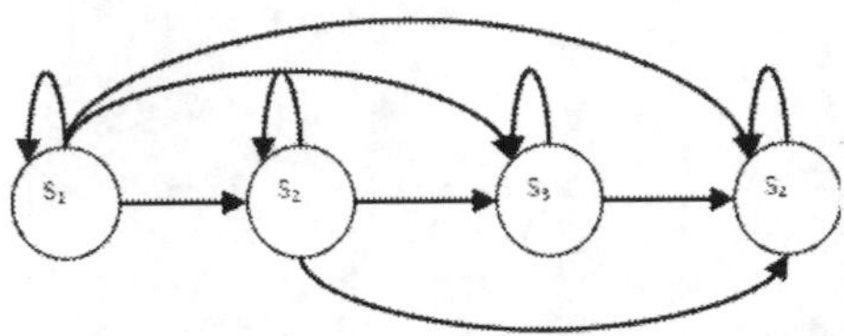

Fig. 8 Transition Pattern Used

The training is done with eight states, 10 states, 12 states, 14 states, 16 states, 18 states and 20 states. All these models will be compared in terms of recognition rate. The probability in the transition matrix is distributed uniformly row-wise.

In this project, 90 dynamic gestures are trained to obtain the HMM. For each gesture, eight set of sample (each sample six set of feature vector) are feed into the HMM training algorithm.

3.2 Evaluation

After each gesture has been trained, the models representing the gesture can be fitted in the recognition stage. Recognition stage involves decoding process by HMM towards the test data provided. The output is the logarithmic probability produced by each model. For a single data with multiple models, the decision of recognition is based on the highest value of probability.

However, this project involves separated data with multiple models. This means that each gesture will have six value of maximum logarithmic probability. This six values need to be combined in order to do classification based on highest logarithmic probability. In order to combine the logarithmic probability of each feature vector for the right and left hand, it is assumed that all the feature vectors are independent of each other. For an independent event, the probability for multiple occurrences to happen is equal to the product of the probability of each event (Figure 9).

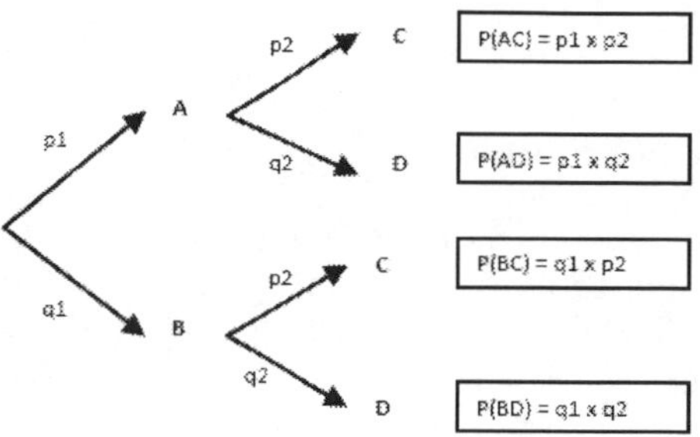

Fig. 9 Independent Event Probability

Because the evaluation output is in log likelihood, rather than multiplying, the values are summed up to represent the evaluation with a single value. This is possible because of the logarithmic identity of product value shown in (8). This method has also been used by J. E. Higgins et al. [8] in their work with multiple classifiers.

$$\log(a.b) = \log a + \log b \tag{8}$$

4 Result

After the training is completed, the models of each gesture are forwarded to the recognition stage as evaluation parameters. The HMM with eight states, 10 states, 12 states, 14 states, 16 states, 18 states and 20 states are tested with 2 sets of sample (90 gestures, 2 sets). The recognition rate for each of the different states is shown in Figure 10.

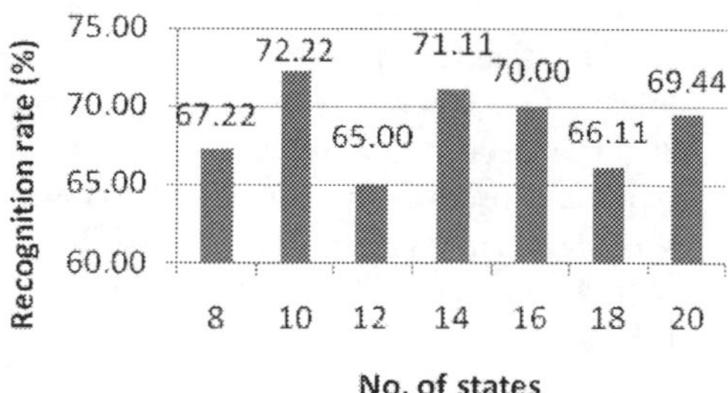

Fig. 10 Recognition Rate at Different Number of States

From the bar graph in Figure 10, the best recognition rate achieved is 72.22%, with 10 states. Thus 10 states HMM will be equipped to the sign language translator algorithm.

5 Discussion

5.1 Poor Skin Segmentation

Skin segmentation is an important procedure in this system in order to calculate the feature vector of sign gesture. Poor skin segmentation will affect the result of feature vector calculation. This is because the coordinate of centroid is dependent on the area of blob or segmented region. When this happens, the feature vector calculated may not be representing the trajectory of the gesture efficiently.

The presence of shadows brings two issues in segmentation. First, shadow that appears on the performer's hands or face will cause loss of information. Shadows create very dark spot that may jump out of the range of the skin colour. This sometimes caused the region of interest to become smaller. The worst case is the segmented skin may be eliminated because of the defined area threshold. Figure 11 below shows an example of this situation.

Skin-like background colour is an element that should be avoided when capturing videos of sign gesture. If this element is present in the video, the segmentation algorithm will declare this as a false segmentation and skip to the next frame. Skipping the frame causes loss of data, and may lead to false recognition. Yellow, brown and red are the common colours that cause this sort of problem. Figure 12 shows the example of this problem.

Fig. 11 Loss Segmentation Due to Shadow

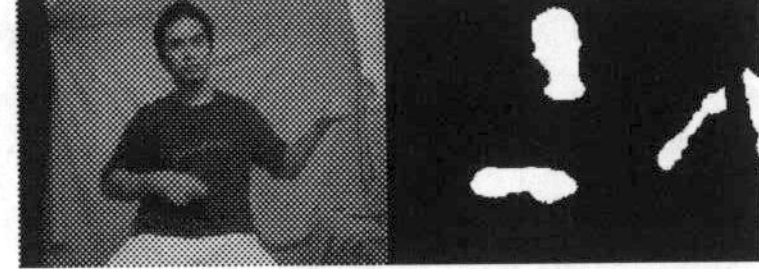

Fig. 12 False Segmentation for Pink Colour

Frame rate for a video is defined as the number of frames taken in one second. Frame rate limitation means that when a performer executes a sign too fast, the image in the frame will be blurred at the movement region. This is because the frame rate is not high enough to cope with the fast movement. This contributes problem to the skin segmentation algorithm. This blurred image usually will cause loss of information due to the segmented blob is too small, result in elimination by the blob area threshold (Figure 13).

5.2 Poor Centroid Estimation (Linear Kalman Filter)

The Linear Kalman Filter is triggered when overlapping occurs or when both or either one of the hands has failed to be segmented. Kalman filter estimates the centroid based on the previous centroid and its trajectory direction. However, it is observed that, in some cases, the estimated centroid does not represent the location of the hands. Figure 14 shows the example of this problem.

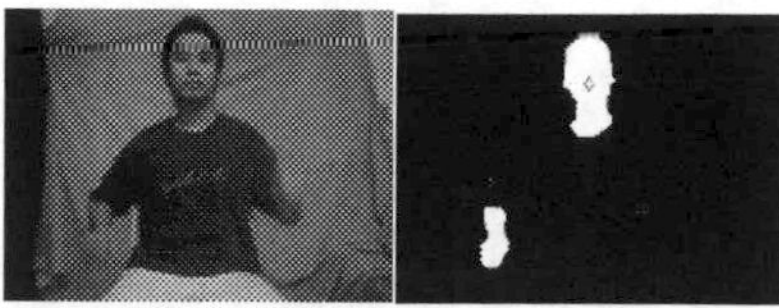

Fig. 13 Fast Motion Not Segmented

Fig. 14 Error in Linear Kalman Filter

This happens as the hands are still overlapping and moving down, the Kalman filter estimated that the centroids will still continue moving upwards, since the previous centroid mostly are moving upwards. Sign gestures that involve hand motion while overlapping also faced the same problem.

5.3 Feature Vector Issue

The sign language translator algorithm does not consider the finger pattern of the sign language performer. Thus, for some gesture with similar motion and position, false recognition will easily occur. Shown in Figure 15 are examples of gesture that has similar motion trajectory and hand position. These three gestures involve moving the right hand up to the level of the head (no overlapping) and shake it back and forward. The only different is the hand pattern.

Fig. 15 From left to right, *bagus*, *nakal* and *pensyarah*

6 Conclusion

In this project, 90 different Malaysian Sign Language gestures are intended to be recognised using HMM approach. The feature vector is based on the chain code, distance between face and hand and hand tilting orientation. Thirty frames are selected from the video. The HMM is built based on 10 states model. For each gesture, eight videos are used to train the model. For testing, two sets of video are used. The recognition rate achieve in this project is 72.22%, that is 130 out of 180 is recognised correctly.

Acknowledgments. We would like to express our deepest gratitude and appreciation to the Ministry of Higher Education (MoHE), Malaysia for the financial support under the Fundamental Research Grant Scheme (FRGS), FRG0026-TK-1/2006, received through Universiti Malaysia Sabah.

References

[1] Singh, S.K., Chauhan, D.S., Vatsa, M., Singh, R.: A robust skin color based face detection algorithm. Tamkang Journal of Science and Engineering 6(4), 227–234 (2003)

[2] Lin, C.: Face detection in complicated backgrounds and different illumination conditions by using YCbCr color space and neural network. In: IEEE International Conference on Multimedia and Expo., ICME, vol. 2, pp. 1187–1190 (2007)

[3] Tang, H.-K., Feng, Z.-Q.: Hand's skin detection based on ellipse clustering. In: International Symposium on Computer Science and Computational Technology, vol. 2, pp. 758–761 (2008)

[4] Butler, D., Sridharan, S., Chandran, V.: Chromatic colour spaces for skin detection using GMMS. In: IEEE International Conference on Acoustics, Speech, and Signal Processing, vol. 4, pp. 3620–3623 (2002)

[5] Maebatake, M., Suzuki, I., Nishida, M., Horiuchi, Y., Kuroiwa, S.: Sign language recognition based on position and movement using multi-stream HMM. In: Second International Symposium on Universal Communication, pp. 478–481 (2008)

[6] Matsuo, T., Shirai, Y., Shimada, N.: Automatic generation of HMM topology for sign language recognition. In: International Conference on Pattern Recognition, pp. 1–4 (2008)

[7] Rabiner, L.R.: A tutorial on Hidden Markov Models and selected applications in speech recognition. Proceedings of the IEEE 77(2), 257–286 (1989)

[8] Higgins, J.E., Damper, R.I., Dodd, T.J.: Information fusion for subband HMM speaker recognition. In: International Joint Conference on Neural Networks, vol. 2, pp. 1504–1509 (2001)

An Algorithm for Headline and Column Separation in Bangla Documents

Farjana Yeasmin Omee, Md. Shiam Shabbir Himel, and Md. Abu Naser Bikas

Abstract. With the progression of digitization it is very necessary to archive the Bangla newspaper as well as other Bangla documents. The first step of reading Bangla Newspaper is to detect headlines and column from multi column newspaper. But there is no such algorithm developed so far in Bangla OCR that can fully read Bangla Newspaper. In this paper we present an algorithmic approach for multi column & headline detection from Bangla newspaper as well as Bangla magazine. It can separate headlines from news and also can detect columns from multi column. This algorithm works based on empty space between headline- columns, column-column.

1 Introduction

OCR means Optical Character Recognition, which is the mechanical or electronic transformation of scanned images of handwritten, typewritten or printed text into machine-encoded text [1]. It is widely used for converting books and documents into electronic files, to computerize a record-keeping system in an office or to publish the text on a website. OCR has emerged a major research field for its usefulness since 1950. All over the world there are many widely spoken languages like English, Chinese, Arabic, Japanese, and Bangla etc. Bangla is ranked 5th as speaking language in the world [1]. With the digitization of every field, it is now necessary to digitized huge volume of old Bangla newspaper by using an efficient Bangla OCR. To do so, good page layout analyzer is necessary which is an important step for document analysis in advance OCR system. This step partition or separate several text block which can be used as input on basic OCR module. Moreover text block also contain several text format as headline, image, two or three column style. In this paper, we described an algorithmic process for separating headline from news block and column from column. The remainder of

Farjana Yeasmin Omee · Md. Shiam Shabbir Himel · Md. Abu Naser Bikas
Department of Computer Science and Engineering, Shahjalal University of Science and Technology, Sylhet, Bangladesh
e-mail: {fyomee,sabbirshiam}@gmail.com, bikasbd@yahoo.com

A. Abraham and S.M. Thampi (Eds.): Intelligent Informatics, AISC 182, pp. 307–315.
springerlink.com

the paper is organized as follow: In section 2 we discuss about the related works of page layout analysis as well as Bangla OCR. Section 3 describes the basic steps of Bangla OCR. The preprocessing step of basic bangle OCR needs attention because page layout analysis is one of the steps of it. So we point out common steps of The Preprocessing in section 4. Next in section 5 we described the page layout analysis with its importance in Bangla OCR. Our proposed algorithm describes in section 6. In section 7 we provided pseudo code of our algorithm. Section 8 contains our implementation. We calculate our performance in section 9. Limitations of our algorithm describes in section 10. We conclude our paper in section 11.

2 Related Work

On page layout section in Bangla OCR there are many works has been done. But none of them is error free [1]. Some existing text region separation technique is described in [2]. There is also a technique described in [2] for Bangla and for other multilingual Indian scripts. An algorithmic approach is described in [3] with its limitation. An open source based Bangla OCR has been launched by Md. Abul Hasnat and et al. [4]. The Center for Research on Bangla Language Processing (CRBLP) released *Bangla OCR* – the first open source OCR software for Bangla – in 2007 [5]. *Bangla OCR* is a complete OCR framework, and has a high recognition rate (in limited domains) but it also have many limitations. Another approach described in [6] for Bangla OCR. The open Source OCR landscape got dramatically better recently when Google released the Tesseract OCR engine as open source software. Tesseract is considered one of the most accurate free software OCR engines currently available. Currently research & development of OCR lay upon the Tesseract. Tesseract is a powerful OCR engine which can reduce steps like feature extraction and classifiers. The Tesseract OCR engine was one of the top 3 engines in the 1995 UNLV Accuracy Test. Between 1995 and 2006 however, there was very little activity in Tesseract, until it was open-sourced by HP and UNLV in 2005; it was again re-released to the open source community in August of 2006 by Google. A complete overview of Tesseract OCR engine can be found in [7] and a complete workflow for developing Bangla OCR is described in [1].

3 Basic Steps of Bangla OCR

For different languages the development procedure of an OCR can be different but the most basic steps of an Bangla OCR has the following particular processing steps[8] which also described in figure 1.

- Scanning.
- Preprocessing.
- Feature extraction or pattern recognition.
- Recognition using one or more classifier.
- Contextual verification or post processing

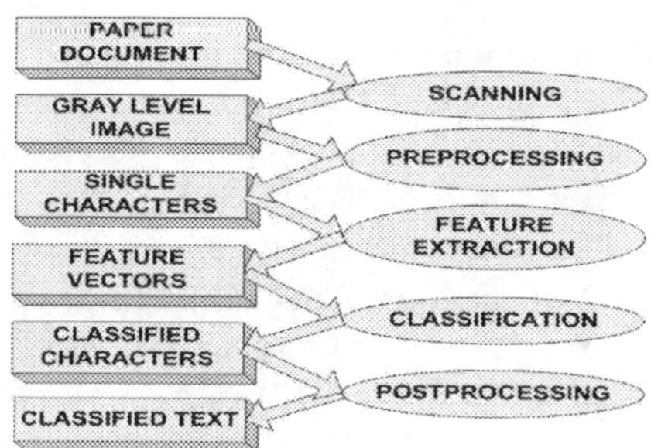

Fig. 1 Basic Steps of an OCR

4 Preprocessing

After scanning a document pre processing consists of a number of preliminary processing steps to make the raw data usable for the recognizer. The typical pre processing steps included the following process:

- Binarization
- Noise Detection & Reduction
- Skew detection & correction
- Page layout analysis
- Segmentation

5 Page Layout Analysis

Page Layout analysis is the most import part of pre processing and also the processing steps of OCR development. The overall goal of layout analysis is to take the raw input image and divide it into non-text regions and "text lines"–sub images of the original page image that each contains a linear arrangement of symbols in the target language. Some popular page layout algorithms are: RLSA [8], RAST [9] etc. An algorithmic approach is described in [3] with its limitation. Still this section is developing and thus become a major research field. Using page layout analysis in OCR helps to recognize text from newspaper, magazine, documents etc. None of these algorithms can be directly used for the Bangla OCR to detect the headline and multi column of Bangla newspaper. So we proposed an algorithm to detect headline and multi column and successfully implemented the algorithm on our ongoing Bangla OCR project.

6 Our Proposed Algorithm

The algorithm presented in this paper takes advantage of the "*Matra*" which is a distinctive feature of the *Bangla* Script. This algorithm presents a top-down approach by initially considering the complete document and trying to recursively extract text blocks from it. The processing steps of our proposed algorithm describe below:

- **Step 1:** First we load a noise eliminated binarized image.
- **Step 2:** Then we set a Fixed Space Width which means the minimum space between headlines and news body and Fixed Space Height which means the minimum space between columns. We get these two values from checking many different types of papers & magazines.
- **Step 3:** At first Space Width set as 0 and starting point is (0, 0).From starting point we check the pixel vertically and find the white line. If we find white line then we increase Space Width. At the point (x, y) where Space Width is greater than Fixed Space Width we cut the image in a rectangular shape from (x, y) to starting point. Then we save this image as a sub image and set starting point as (x, y). If we find a line which is not fully white then we set Space Width to 0 and continue this process for whole image.
- **Step 4:** If number of sub images from step 3 is greater than 1, we will consider the first sub image as headline image and other sub images as input for step 5.
- **Step 5:** We load each sub images for checking the pixel horizontally. To do so, set Space Height as 0 and starting point (0, 0). If we find the white line, we increase Space Height. At the point (x1, y1) where Space Height is greater than Fixed Space Height, we cut the image in a rectangular shape from stating point to (x1, y1). After that we save this image as a final child image and set starting point as (x1, y1). If we find a line which is not fully white, we set Space Height as 0 and continue this process for whole image.

After completing the above steps we find some sub images which consist of separated columns or headlines.

7 Pseudo Code of Our Algorithm

The pseudo code of our implemented algorithm describes below:

```
1. Load a noise eliminated Binarize image
//set the width and height fixed from observing various newspaper
2. Set FixedSpaceWidth and FixedSpaceHeight
3. Set TotalSubImageNumber = 0
4. Set XSize = InputImage.height
5. Set YSize = InputImage.width
6. Set FixedSpace = FixedSpaceWidth
7. Set Space = 0
8. for each i from 0 to XSize do
       sum = 0
       for each j from 0 to YSize do
               //Check all the pixels
               if pixel is black then
                       sum = sum + 10
                       if pixel is white then
                               sum = sum + 0
       end for
```

```
    if sum == 0 then
        //calculate space value using white line values
        SpaceValue = CurrentWhiteLineValue – PreWhiteLineValue
        if SpaceValue == 1 then
            Space++
            if Space > FixedSpace then
                //cut the image in rectangular shape
                Cut the image from point (0, y) to point (0,i)
                Save rectangular images as SubImage
                TotalSubImageNumber++
                y = i
                Space = 0
            end if
        end if
    end if
  end for
9. for each k from 0 to TotalSubImageNumber
    Load the sub image
    if TotalSubImageNumber > 1 && k == 0
        //First sub image saved as headline image
        HeadlineImage = SubImage[0]
    else
        Set XSize = SubImage[k].width
        Set YSize = SubImage[k].height
        Set FixedSpace = FixedSpaceHight
        Set Space = 0
        Repeat Step 8
    end if
  end for
```

8 Our Implementation

For using our algorithm and to test it we scanned news papers in a Flat bed scanner and use the following algorithm for creating noise eliminated binarized input image of our proposed algorithm:

- Binarization – Otsu Algorithm[10]
- Noise detection & elimination- Median filter [11][12]
- Skew detection & correction – Radon transform [13]

We have implemented our proposed algorithm described in the previous sections. This section encloses the implementation of our algorithm. We have used C++ for coding and integrated OPENCV library [14] which is for image manipulation. This section also encloses test cases going through by our application.

Figure 2 shows a sample inputs image of Bangla newspaper that we tested through our application. Figure 3 shows the main screen of our application.

Fig. 2 Newspaper cutting (original image) **Fig. 3** Main Screen of our application

Then we eliminate the noises of the input image.

Figure 4 shows the eliminated noise output in our application that will later go through our algorithm to detect the headline and columns.

Fig. 4 Output after Noise Elimination

Then after processing the noise eliminated image based on our algorithm we get the headline like the below images respectively where Figure 5 shows the detected output Headline and Figure 6 shows the detected Headline in our application.

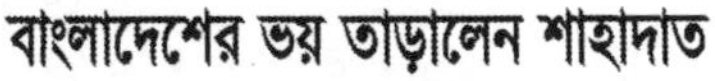

বাংলাদেশের ভয় তাড়ালেন শাহাদাত

Fig. 5 Output of Headline detection

Fig. 6 Our Application State after Headline Detection

The detected columns from the input image looks like Figure 7, Figure 9 and Figure 11. And Figure 8, 10, 12 are for the screenshot of each detected columns in our application respectively.

Fig. 7 1st column of the image

Fig. 8 Application State after 1st column detection

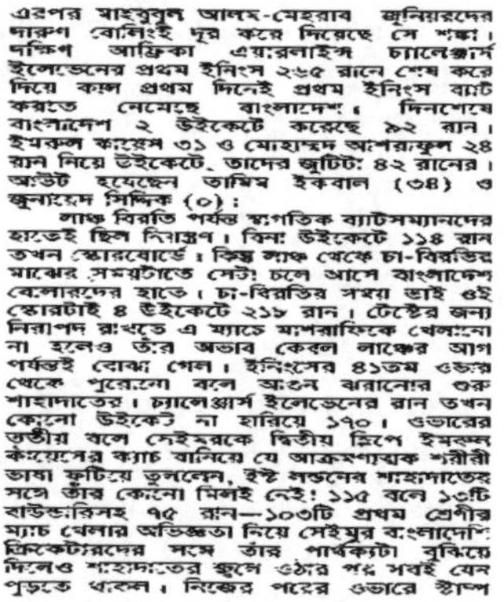

Fig. 9 2nd column of the image

Fig. 10 Application State after 2nd Column Detection

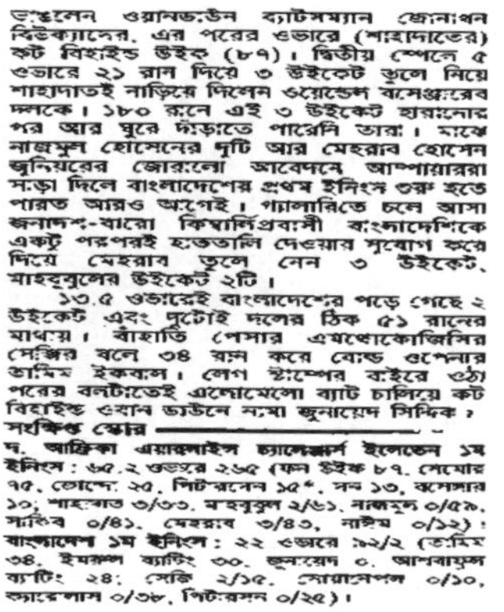

Fig. 11 3rd column of the image

Fig. 12 Application State after 3rd column Detection

9 Performance Analysis

We have tested this algorithm with some of the major daily newspapers of Bangladesh and our algorithm works well in maximum cases. We have tested our proposed algorithm with the daily "Prothom Alo" and the daily "Jugantor" which two are the most popular Bangla newspaper in Bangladesh. We have tested our proposed algorithm for 47 test cases and in almost 33 cases we got desired output. So the performance of our proposed algorithm has an accuracy of approximately 70.2%. The ratio of recognition can be improved if we train the tesseract with recent Bangla fonts.

10 Limitations and Future Work

In most of the cases in Bangla newspaper our algorithm works without any problem. But our implemented algorithm will not work if there are any image exists in a page layout of scanned newspaper. Our proposed algorithm is not based on the dynamic pixel space between column-column and headline-column. We are currently working to solve these limitations. We are devising some modules that can separate image from text and dynamically allocate space size using machine learning approach. Also we are working to train our tesseract engine with some popular Bangla fonts.

11 Conclusion

In this paper we present an algorithm to detect headline and multi column from Bangla Newspaper. Here we present the status of an ongoing project on page layout analysis of Bangla documents. Though this algorithm has some limitations in terms of contain images inside the newspaper which our algorithm currently cannot detect but we think this proposed algorithm will shed some light in the progress of the development of the Bangla OCR.

References

1. Omee, F.Y., Himel, S.S., Bikas, A.N.: A Complete Workflow for Development of Bangla OCR. International Journal of Computer Applications (IJCA) 21(9), 1–6 (2011), doi:10.5120/2543-3483
2. Ray Chaudhuri, A., Mandal, A.K., Chaudhuri, B.B.: Page Layout Analyzer for Multilingual Indian Documents. In: Proceedings of the Language Engineering Conference, LEC. IEEE (2002)
3. Khedekar, S., Ramanaprasad, V., Setlur, S.: Text-Image Separation in Devanagari Documents. In: 7th International Conference on Document Analysis and Recognition, ICDAR. IEEE (2003)
4. Hasnat, A., Murtoza Habib, S.M., Khan, M.: Segmentation free Bangla OCR using HMM: Training and Recognition. In: Proceeding of 1st DCCA, Irbid, Jordan (2007)

5. Open_Source_Bangla_OCR, http://sourceforge.net/project/showfiles.php?group_id=158301&package_id=215908
6. Hasnat, A., Murtoza Habib, S.M., Khan, M.: A high performance domain specific OCR for Bangla script. In: International Joint Conference on Computer, Information, and Systems Sciences, and Engineering, CISSE (2007)
7. Smith, R.: An Overview of the Tesseract OCR Engine. In: Proceeding of ICDAR 2007, vol. 2, pp. 629–633 (2007)
8. Description_Of_RLSA_Algorithm, http://crblpocr.blogspot.com/2007/06/run-length-smoothing-algorithm-rlsa.html
9. Breuel, T.M.: The OCRopus Open Source OCR System. In: Proceedings of the Document and Retrival XV, IS&T/SPIE 20th Annual Symposium, San Jose, CA, United States, vol. 6815. SPIE (2008)
10. Patnaik, T., Gupta, S., Arya, D.: Comparison of Binarization Algorithmin Indian Language OCR. In: Annual Seminar of CDAC-Noida Technologies, ASCNT (2010)
11. Gonzalez, Woods: Digital image processing, 2nd edn., ch. 4 sec. 4.3, 4.4; ch. 5 sec. 5.1–5.3, pp. 167–184, 220–243. Prentice Hall (2002)
12. Median Filter, http://en.wikipedia.org/wiki/Median_filter
13. Murtoza Habib, S.M., Noor, N.A., Khan, M.: Skew Angle Detection of Bangla script using Radon Transform. In: Proceeding of 9th ICCIT (2006)
14. OpenCV, http://opencv.willowgarage.com/wiki/

A Probabilistic Model for Sign Language Translation Memory

Achraf Othman and Mohamed Jemni

Abstract. In this paper, we present an approach for building translation memory for American Sign Language (ASL) from parallel corpora between English and ASL, by identifying new alignment combinations of words from existing texts. Our primary contribution is the application of several models of alignments for Sign Language. The model represents probabilistic relationships between properties of words, and relates them to learned underlying causes of structural variability within the domain. We developed a statistical machine translation based on generated translation memory. The model was evaluated on a big parallel corpus containing more than 800 millions of words. IBM Models have been applied to align Sign Language Corpora then we have run experimentation on a big collection of paired data between English and American Sign Language. The result is useful to build a Statistical Machine Language or any related field.

Keywords: Sign Language, Translation Memory, Probabilistic Model.

1 Introduction

Nowadays, many researches are focused on translating written texts to Sign Languages [1,2]. There exist several kinds of machine translation: Rule-based, Example-based and Statistical. For Statistical Machine Translation for Sign Language [3], it requires parallel corpora between two languages toward collecting data and statistics of words. These statistics are useful to build automatically a translation memory for machine translation that will be able to translation any new input. In this paper, we present a probabilistic approach to build a translation memory for Statistical Machine Translation between English and American Sign Language Gloss. Glosses are a written form of Sign Language. The originality of this is the application of notions of probabilistic alignment model to a language that is not based to grammatical rules. In section 2, we

Achraf Othman · Mohamed Jemni
Research Laboratory LaTICE, University of Tunis, Tunisia
e-mail: achraf.othman@ieee.org, mohamed.jemni@fst.rnu.tn

A. Abraham and S.M. Thampi (Eds.): Intelligent Informatics, AISC 182, pp. 317–324.
springerlink.com

present steps for collecting and computing statistics from existing data. In section 3, we describe the generation process of translation memory. We finish with experimentation and a conclusion.

2 Word-Based Translation Memory

In this section, we design a translation memory based on lexical content from parallel corpora. This step requires a dictionary which for each source word (in English) we have a corresponding word in American Sign Language.

2.1 Lexical Translation for ASL Translation Memory

If we take an input word in English (source), for example 'your', we found multiple translation in ASL like 'X-YOUR', 'DESC-YOUR' and others words that can be a possible translation (multiple targets). Here, the most appropriate translation of 'your' is 'X-YOUR' in affirmative sentence and 'DESC-YOUR' if 'your' means a possession. For this reason, if we have a large numbers of words in a collection, we need to collect statistics of possible translation for each word. Next, we describe our approach to collect data and extract statistics.

2.2 Collecting Data and Extracting Statistics

The notion of Statistical Machine Translation involves using statistics extracted from words and texts [4]. What kind of statistics we need to translation a word? Assuming we have a large corpus of data collection in English paired with a data collection in American Sign Language (ASL), where each sentence in English possess a translation in ASL, we can count how many times has been translated the word 'your' in 'YOUR', 'X-YOU' or 'DESC-YOUR'. For example, in a test corpus, the possible outcomes of translation of the word 'your', knowing that the number of occurrences of the original word in English is 10 000 times, is the term ' X-YOUR' for 8000 times and the term ' DESC-YOUR' 1600 times and 400 other words that do not correspond to the word 'your '.

2.3 Estimating Probability Distribution

If we want to estimate or calculate the probability distribution of lexical translation based on statistics collected corpus [5] of parallel English-ASL, we need to answer the following question: What is the most likely translation of the word 'your' in ASL? We define the problem as: $P_f : e \rightarrow P_f(e)$. Given a word f , we have the probability for each possible translation e . This function returns a big value if e is a possible translation, else a small value where e is rarely used. In case of zero, we don't have any possible translation of f . This function must approve two properties: $\sum_e P_f(e) = 1$ and $\forall e : 0 \leq P_f(e) \leq 1$. To calculate the probability distribution of a word f, it suffices to determine the number of occurrences of this word in the corpus in English and calculate the occurrence of these possible translations in the corpus in ASL. Then, we calculate the ration for each output e. We have $P_f(e) = 0.8$ if the output word is 'X-YOU' and $P_f(e) = 0.16$ if the output is 'DESC-YOUR' and 0.04 else. To choose the appropriate target word, we take to most important value. This kind of estimation is called 'Maximum Likelihood Estimate' that maximizes the similarity between data, in our case, between words [6,7].

2.4 Alignment

From the probability distributions already calculated previously for lexical translations, we can set up the first translation memory based statistics that uses only the lexical translations. In the following table, we show the probability distributions of four English words in American Sign Language. We denote the probability to translate a word f in a word e in ASL with a conditional probability function $t(e \mid f)$. Table 1 contains lexical translation probability of four words and Figure 1 shows the alignment diagram.

Table 1 Lexical translation of four words

your		car		is		blue	
e	t(elf)	e	t(elf)	e	t(elf)	e	t(elf)
DESC-YOUR	0.90	CAR	0.68	BE	0.80	CL :BLUE	0.51
X-YOUR	0.10	VEHICLE	0.21	DESC-BE	0.20	DESC-BLUE	0.43
		MOTOR	0.11			#BLUE	0.06

(a) 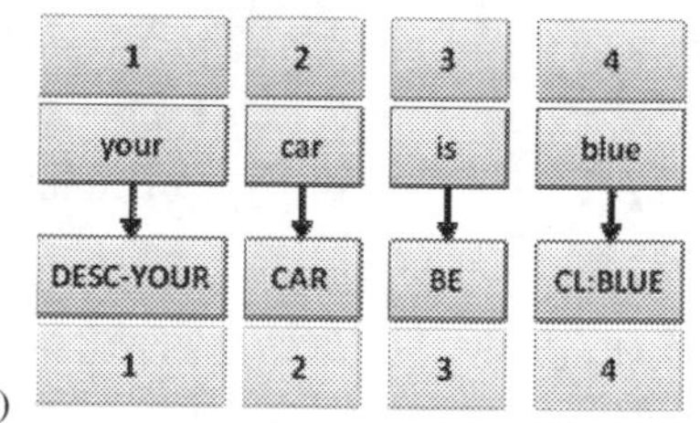

(b) 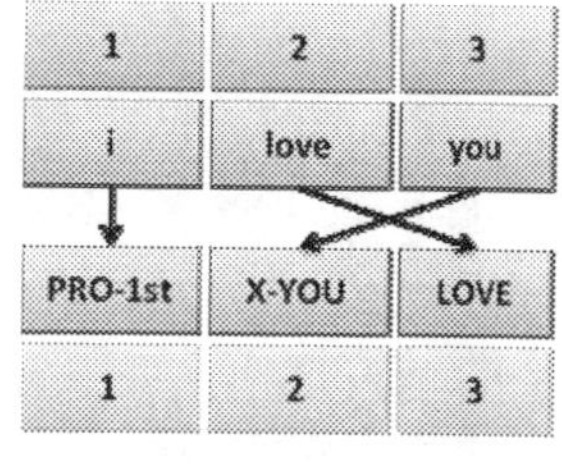

Fig. 1 (a) Alignment diagram of 'your car is blue' to 'DESC-YOUR CAR BE CL:BLUE' (b) Alignment diagram of 'PRO-1ST X-YOU LOVE'

The alignment between a source sentence f into a target sentence e is formalized by the alignment function a. This feature allows us to map the positions of words of f with the positions of words e. In the previous example, we have $a:\{1 \rightarrow 1; 2 \rightarrow 2; 3 \rightarrow 3; 4 \rightarrow 4\}$. To generalize, the alignment function is defined as $a:\{j \rightarrow i\}$. This is a simple alignment between two sentences having the same number of words and the same positions of words with its translation. Generally, the words which correspond do not have the same positions ranging from a source to a sentence target phrase in Figure 1(b). Alignment function will be $a:\{1 \rightarrow 1; 2 \rightarrow 3; \rightarrow 2\}$. Also, number of words in English is not the same in ASL (figure 3). Another case is a word in f does not have a translation in e (figure 4). Also, we can have a word in e that does not correspond to a word in f (figure 5). In this diagram, the target word 'HUH' does not match a word in the source sentence f. It is replaced by a NULL token.

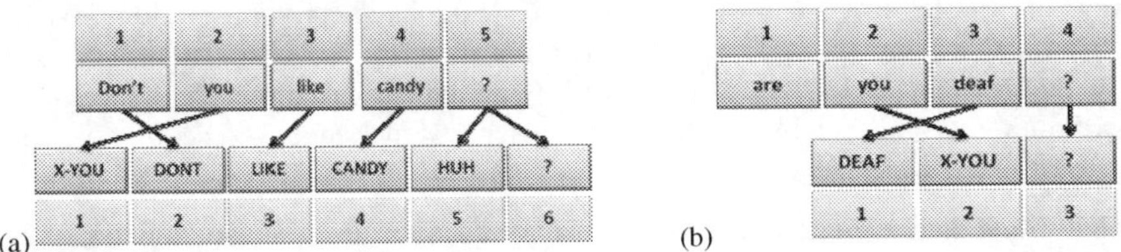

Fig. 2 (a) Alignment diagram of 'X-YOU DONT LIKE CANDY HUH ?' (b) Alignment diagram of 'DEAF X-YOU ?'

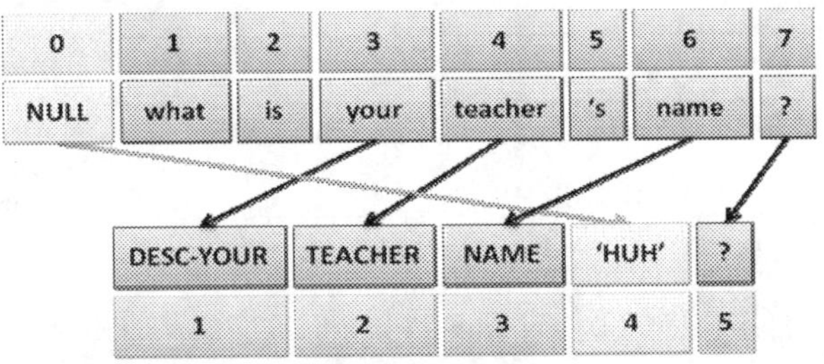

Fig. 3 Alignment Diagram of 'what is your teacher's name ?' to 'DESC-YOUR TEACHER NAME 'HUH' ?' en ASL

In this section, we presented the different possibilities of correspondence that can be found when translating a source sentence f into a target sentence e. The different models of possible alignment were formalized by a function that takes into account the deletion, addition and change of order in the translation. Given a parallel corpus English-ASL, how can determine the alignment between a source sentence and target sentence to generate a dictionary? This can be done by implementing the learning algorithms proposed by IBM Models. These models are called alignments IBM Model which are based on the probability of lexical translations using the alignment function a.

2.5 *IBM1 Alignment Model for American Sign Language*

Armed with an alignment function and lexical probability distributions, IBM Model 1 is used to calculate the probability of translating a sentence f into a sentence e. This probability is defined as $P(e \mid f)$. We define the probability to translation an English sentence $\left(f_1, f_2, \ldots, f_{l_f}\right)$ to an ASL sentence $\left(e_1, e_2, \ldots, e_{l_e}\right)$ with an alignment function a as $P(e, a \mid f) = \frac{\varepsilon}{\left(l_f + 1\right)^{l_e}} \prod_{i=1}^{l_e} t\left(e_j \mid f_{a(j)}\right)$. The probability of translation is calculated from the product of all probabilities of each word lexical e_j for j from 1 to l_e. At the quotient we added 1 to reflect the NULL TOKEN. So, we have $\left(l_f + 1\right)^{l_e}$ possible alignments to map $\left(l_f + 1\right)$ words from f with all words in e. ε is a normalization constant of $P(e, a \mid f)$ that ensure $\sum_{e,a} P(e, a \mid f) = 1$. Applying this formula to the previous table, we have: $P(e, a \mid f) = \frac{\varepsilon}{5^4} \times 0.9 \times 0.68 \times 0.8 \times 0.51 = 0.2497\varepsilon$.

3 Machine Learning for Lexical Translation

We introduced a model for translating an English sentence to ASL sentence base on distribution probability of lexical translations that are calculated in advance. In this section, we present methods for learning these distributions from a parallel corpus. We use the Expectation Maximization algorithm abbreviated as EM. This algorithm allows finding the maximum likelihood model parameters when we have a problem of incomplete data. Our goal is to estimate for each input word f from a big collection of data, what are the possible translations from existing data without using a dictionary. We are facing an incomplete data problem which is the alignment model.

3.1 *Estimation-Maximization Likelihood Algorithm (EM)*

The EM algorithm is directed to the problems of incomplete data. This is an EM-iterative method of learning works as follows:

i. Initialize the model with a uniform probability distribution.
ii. Apply the model to existing data (Estimation step).
iii. Learn the model from the data (Maximization step).
iv. Iterate steps 2 and 3 until convergence (usually around 1).

3.2 *EM for IBM 1 Model*

Starting from the previous example where we calculated the probability of translation based on the probability of lexical translations and adding the chain rule function, we have: $P(a \mid e,f)=\dfrac{P(e,a \mid f)}{P(e \mid f)}$ (I)

and $P(e \mid f)=\sum_a P(e,a \mid f)$. So, we have (II):

$$P(e \mid f)=\sum_{a(1)}^{l_f}\dots\sum_{a(l_e)}^{l_f}\frac{\varepsilon}{\left(l_f+1\right)^{l_e}}\prod_{j=1}^{l_e}t\left(e_j \mid f_{a(j)}\right)=\frac{\varepsilon}{\left(l_f+1\right)^{l_e}}\prod_{j=1}^{l_e}\sum_{i=1}^{l_f}t\left(e_j \mid f_i\right)$$

From (I) and (II), we conclude: $P(a \mid e,f)=\prod_{j=1}^{l_e}\dfrac{t\left(e_j \mid f_{a(j)}\right)}{\sum_{i=0}^{l_f}t(e_j \mid f_i)}$.

For the quotient $\dfrac{\varepsilon}{\left(l_f+1\right)^{l_e}}$, we suppose that is equal to 1 if $l_f \approx l_e$ to reduce the complexity of the algorithm. For the maximization step, we compute the possible translations from alignment probabilities. In this context, we define a function c applied to a pair of sentences (e,f) as follow:

$$c(e \mid f;e,f)=\sum_a P(a \mid e,f)\sum_{j=1}^{l_e}\delta(e,e_j)\delta(f,f_{a(j)}).$$

Kronecker delta function $\delta(x,y)$ is equal to if $x=y$ else zero. We have:

$$c(e \mid f;e,f)=\frac{t(e_j \mid f_{a(j)})}{\sum_{i=0}^{l_f}t(e_j \mid f_i)}\sum_{j=1}^{l_e}\delta(e,e_j)\sum_{i=0}^{l_f}\delta(f,f_{a(i)})$$

3.3 *EM for IBM 2 Model*

IBM Model 2 is characterized by an alignment function that supports changing the order of a word in the source sentence to a word in the target sentence. The first step is to compute the probability distribution of translation t. The second step is to align target words according to the lexical translation. Starting from previous function of translation probability and alignment function we have: $P(e,a \mid f)=\varepsilon\prod_{j=1}^{l_e}t(e_j \mid f_{a(j)})a(a(j) \mid j,l_e,l_f)$. So, $P(e \mid f)$ become: $P(e \mid f)=\varepsilon\sum_{a(1)=0}^{l_f}\dots\sum_{a(l_e)=0}^{l_f}\prod_{j=1}^{l_e}t(e_j \mid f_{a(j)})a(a(j) \mid j,l_e,l_f)$. IBM2 Model Algorithm is similar to IBM1 Model Algorithm. In addition to initializing initially by a uniform distribution, they are initialized by the resulting values of the first phase of learning. Then, we iterate until convergence.

3.4 EM for IBM 3 Model

Until now, we did not model the number of words resulting from the translation for each input word. Generally, every word in English is translated as one word in ASL, but you can find cases where a word in English is translated into two words at a time as the symbol '?' Can be translated as 'HUH ?', and even if we delete a word in the source sentence. This concept is modeled using $n(\varnothing \mid f)$. For each word in English f, the probability distribution indicates how many words will be translated $\varnothing = 1, 2, \ldots$. The model must also ensure the removal of an input word when $\emptyset = 0$. In addition, we can add a token to represent NULL. IBM3 Model illustrates four steps as shown in this example:

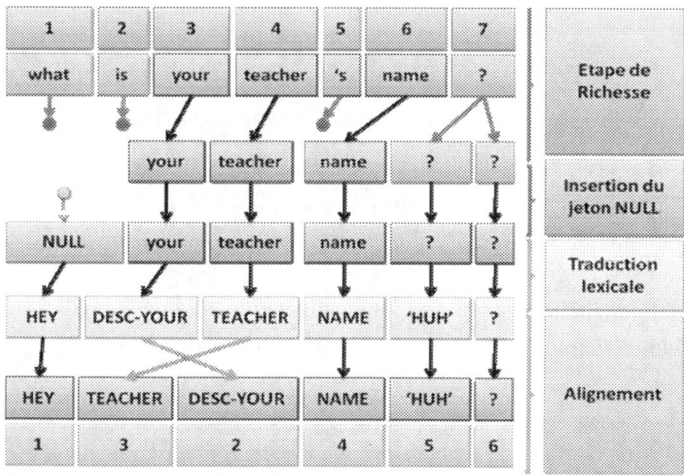

Fig. 4 Different steps in IBM3 Model

In this model, we improve the quality of alignment models. We support changing words order, removing and adding new other word in the target sentence.

4 Experimentations and Evaluation

In the experimentation, we applied this algorithm to a big parallel corpora [4] having more than 800 millions words in English. The result has been integrated in a Statistical Machine Translation aiming to translate an English text to American Sign Language Gloss [3].

5 Conclusion and Discussion

In this paper, we presented a probabilistic model for building translation memory between English and American Sign Language Gloss. It is based on Maximum Likelihood Estimation algorithm. The results can be used to build Statistical Machine Translation between written English text and ASL. As a future work, we are planning to add more features about Sign Language as mentioned in [8,9]. Many applications have been developed using these results like ASL-SMT [10] and others [11, 12].

References

1. Morrissey, S., Way, A.: Joining hands: Developing a sign language machine translation system with and for the deaf community. In: Proceeding CVHI Conference, Workshop Assistive Technol. People with Vision and Hearing Impairments, Granada, Spain (2007)
2. Morrissey, S.: Assistive technology for deaf people: Translating into and animating Irish sign language. In: Proceeding Young Researchers Consortium, ICCHP, Linz, Austria (2008)
3. American Sign Language Gloss Parallel Corpus 2012, ASLG-PC12 (2012), http://www.achrafothman.net/aslsmt/
4. Gutenberg Project (2012), http://www.gutenberg.org/
5. Koehn, P., Och, F., Marcu, D.: Statistical phrase-based translation. In: Proceedings of the 2003 Conference of the North American Chapter of the Association for Computational Linguistics on Human Language Technology, NAACL 2003 (2003)
6. Koehn, P.: Statistical Machine Translation. Cambridge University Press (2009)
7. Brown, P., Pietra, V., Pietra, S., Mercer, R.: The mathematics of statistical machine translation: parameter estimation. Computational Linguistics, Special issue on using large corpora: II (1993)
8. Huenerfauth, M., Zhou, L., Gu, E., Allbeck, J.: Evaluation of American sign language generation by native ASL signers. ACM Transaction Accessible Computing (2008)
9. Marshall, I., Sáfár, E.: Sign language generation using HPSG. In: Proceeding 9th International Conference Theoretical Methodological Issues Machine Translation, TMI 2002, Keihanna, Japan (2002)
10. Othman, A., Jemni, M.: English-ASL Gloss Parallel Corpus 2012: ASLG-PC12. In: 5th Workshop on the Representation and Processing of Sign Languages: Interactions between Corpus and Lexicon LREC 2012, Istanbul, Turkey (2012)
11. Jemni, M., Elghoul, O., Makhlouf, S.: A Web-Based Tool to Create Online Courses for Deaf Pupils. In: International Conference on Interactive Mobile and Computer Aided Learning, Amman, Jordan (2007)
12. Jemni, M., Elghoul, O.: An avatar based approach for automatic interpretation of text to Sign language. In: 9th European Conference for the Advancement of the Assistive Technologies in Europe, San Sebastián, Spain (2007)

Selective Parameters Based Image Denoising Method

Mantosh Biswas and Hari Om

Abstract. In this paper, we propose a Selective Parameters based Image Denoising method that uses a shrinkage parameter for each coefficient in the subband at the corresponding decomposition level. Image decomposition is done using the wavelet transform. VisuShrink, SureShrink, and BayesShrink define good thresholds for removing the noise from an image. SureShrink and BayesShrink denoising methods depend on subband to evaluate the threshold value whereas the VisuShrink is a global thresholding method. These methods remove too many coefficients and do not provide good visual quality of the image. Our proposed method not only keeps more noiseless coefficients but also modifies the noisy coefficients using the threshold value. We experimentally show that our method provides better performance in terms of objective and subjective criteria i.e. visual quality of image than the VisuShrink, SureShrink, and BayesShrink.

Keywords: Image denoising, Wavelet coefficient, Thresholding, Peak-Signal-to-Noise Ratio (PSNR).

1 Introduction

Image denoising has been one of the important research activities among the researcher working in the image processing area. Its main goal is to remove the additive noise while retaining the maximum important image characteristics. Degradation of an image by noise during its acquisition, processing, preservation, and transmission is unavoidable. This noise affects different processing tasks of the image such as edge detection, segmentation, feature extraction, and texture analysis. In order to analyze an image in a better way, it is reasonable to remove noise from the image and keep the information about its edge and texture [1-2]. Image denoising can also be achieved using the non-wavelet based technique such

Mantosh Biswas · Hari Om
Department of Computer Science & Engineering,
Indian School of Mines,
Dhanbad, Jharkand-826004, India
e-mail: {mantoshb,hariom4india}@gmail.com

A. Abraham and S.M. Thampi (Eds.): Intelligent Informatics, AISC 182, pp. 325–332.
springerlink.com

as Weiner filter. However, the non-wavelet based techniques tend to blur the sharp edges, destroy lines and others fine image details. Thus, these techniques fail to produce satisfactory results for a broad range of low contrast images and also are computationally expensive. To overcome these problems, the researchers use the wavelet based techniques such as VisuShrink [3-4], SureShrink [5-6], and BayesShrink [7-8]. These techniques have superior performance due to good energy compaction, sparsity and multiresolution structures. The wavelet transforms convert an image into wavelet coefficients in which the small and large coefficients, respectively, occur due to the noise and important signal features such as edges. Mathematically, the filter description using the additive white Gaussian noise (AWGN) via thresholding wavelet coefficients was developed by Donoho and Johnstone [3-6]. One of their important approaches is VisuShrink in which all the coefficients smaller than the Universal threshold are set to zero and the rest are preserved as such or shrunk by suitable threshold [3-4]. Furthermore, they have shown that the shrinkage approach is nearly optimal in the minimax sense. The VisuShrink however always produces an over-smoothed image in which many of the image details are lost when it is applied to natural images. This weakness has been overcome in the SureShrink [5-6] and BayesShrink [7] using subband adaptive techniques such as local parameter estimation. The SureShrink method is based on Stein's Unbiased Risk Estimator (SURE) [5-6]. This method is a combination of the Universal and SURE thresholds. Chang et. al propose BayesShrink method that minimizes the Bayesian risk [7]. These works have motivated us to develop a new threshold that can provide better results. In this paper, we propose a Selective Parameters based Image Denoising method in which the threshold value is estimated using different combinations of local parameters in each subband and it performs better than VisuShrink, SureShrink, and BayesShrink methods. The rest of the paper is organized as follows. Section 2 discusses the related work. Section 3 explains the proposed denoising method. The experimental results and Conclusions are given in Section 4 and 5, respectively.

2 Related Work

The Weiner filtering method requires information about spectra of the noise and original signal. It works well only for smooth signal and Gaussian noise. The Wiener Filter in the Fourier Domain is given as follows:

$$G(u,v) = \frac{H^*(u,v) P_s(u,v)}{\left|H(u,v)\right|^2 P_s(u,v) + P_n(u,v)}$$

where H(u, v) denotes degradation function, $H^*(u, v)$ its complex conjugate and $P_n(u,v)$ and $P_s(u, v)$ denote power spectral density of noise and non-degraded image, respectively.

To overcome the weakness of the wiener filter, Donoho and Johnstone have proposed wavelet based image denoising method i.e. VisuShrink. The Universal threshold, T_{Visu} is proportional to the standard deviation of the noise and is defined as [3-4]: $T_{Visu} = \sigma\sqrt{2 \log M}$; here M represents the signal size or number of samples and σ^2 is the noise variance. SureShrink method may be considered as a combination of the Universal and SURE thresholds and its threshold, T_{Sure}, is defined as [5-6]: $T_{Sure}=\min(t_J, \sigma\sqrt{2\log M})$; here threshold value t_J is associated to the Jth decomposition level in the wavelet transform. In fact it is defined for each level and thus it is referred as level dependent thresholding in the image. BayesShrink method is subband-dependent which means that the thresholding is done at each subband in the wavelet decomposition. It is also known as smoothness adaptive. The Bayes threshold, T_{Bayes}, is given as [7-8]: $T_{Bayes} = (\sigma^2 / \hat{\sigma}_y^2)$; where $\hat{\sigma}_y^2$ is the noise free signal variance. Now we discuss our proposed denoising method that removes the noise significantly.

3 Proposed Denoising Method

The image corrupted by white Gaussian noise can be written as follows:

$$Y_{i,j} = X_{i,j} + n_{i,j} \tag{1}$$

where $Y_{i,j}$ and $X_{i,j}$, $1 \le i, j \le M$, represent the corrupted image and original image, respectively, of size M×M each. $n_{i,j}$ is the independent identically distributed (i.i.d) zero mean additive white Gaussian Noise that is characterized by the noise level σ. We use W and W^{-1} to denote 2-D orthogonal discrete wavelet transform (DWT) and its inverse (IDWT), respectively [9-10]. On applying the wavelet transform W on the noisy image $Y=\{Y_{i,j}\}$, $1 \le i, j \le M$, we get the wavelet coefficients Y_w of the corrupted image as given below:

$$Y_w = W * Y \tag{2}$$

The wavelet transform divides the image into four partitions, called LL, HL, LH and HH, as shown in Fig. 1. After applying the threshold function on Y_w, we get an image, denoted by $\overline{X}$. The denoised estimate, denoted by $\tilde{X}$, of the original image X is obtained by applying the inverse wavelet transform on $\overline{X}$, i.e.

$$\tilde{X} = W^{-1} * \overline{X} \tag{3}$$

LL_1	HL_1
LH_1	HH_1

Fig. 1 2D-DWT with 1-Level decomposition

In our method, first parameter estimation is done then the proposed denoising algorithm.

3.1 Parameter's Estimation

This section describes computation of various parameters that are required to calculate the new threshold value (T_{NEW}) required in our proposed method. This new threshold is adaptive to different subband characteristics of the decomposition levels. Our method exploits the adaptive threshold using the combination of various parameters such as local wavelet coefficient, noise variance, decomposition level, and size of the subband. The adaptive threshold is defined as follows:

$$T_{NEW} = (1- e^{-\frac{Y_{i,j}^2}{t_l}}) \tag{4}$$

where the parameter t_l is computed once for each subband at each decomposition level using the following expression:

$$t_l = \sigma * \left(\frac{\sum Y_{i,j} - \sigma}{\sum Y_{i,j} + \sigma} \right) * \left(\sqrt{\frac{\log \hat{M}}{l}} \right) \tag{5}$$

Here $l = 1, 2, \ldots, J$; J signifies number of decomposition levels, $\hat{M} = M/2^l$ and σ^2 is noise variance, which is defined based on the median absolute deviation as:

$$\sigma^2 = [(\text{median}|\ Z_{ij}\ |) /0.6745]^2 \tag{6}$$

where $Z_{ij} \in HH_1$ subband (refer Fig. 1).

We shrink the wavelet coefficients as given below:

$$\overline{X}_{i,j} = X_{i,j} * \beta_{i,j} \tag{7}$$

where $\beta_{i,j}$ represents the shrinkage factor that is defined as

$$\beta_{i,j} = \max(0, T_{NEW})_+ \tag{8}$$

here + sign refers to keep the positive value and set it to zero for negative.

(a) (b) (c) (d)

Fig. 2 Test images: (a) Lena (b) Mandrill (c) Cameraman (d) Goldhill with size 512×512 pixels

3.2 Image Denoising Algorithm

The proposed procedure is given as follows.

Input: *noising image corrupted with Gaussian noise*

Output: *denoised estimate of original image*

Begin

i. *Perform 2-D Discrete Wavelet transform on noisy image up to J^{th} decomposition level.*

ii. *For each decomposition level of the details subband (i.e. HH, HL, and LH) with the wavelet coefficients do*

- *Estimate the noise variance σ^2 using (6).*
- *Calculate the new threshold, T_{NEW} using (4).*
- *Apply the shrinkage factor given in (7) to obtain the noiseless wavelet coefficients.*

  End {do}

iii. *Repeat steps (i) and (ii) for all decomposition levels.*

iv. *Reconstruct the denoised estimate image after apply the inverse wavelet transformation to the modified coefficients*

End

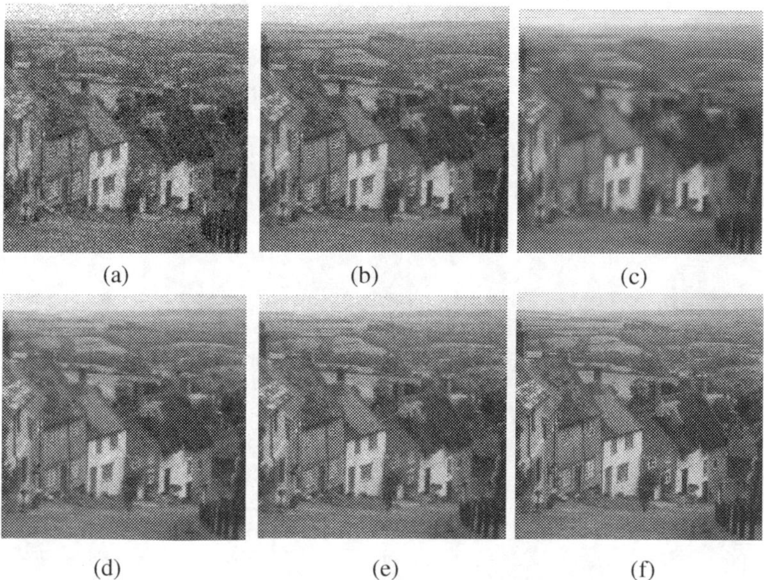

Fig. 3 (a) Noisy image with noise level 20 (b) Denoised image using Wiener Filter (c)Denoised image using Visu Shrink (d) Denoised image using SureShrink (e) Denoised image using BayesShrink (f) Denoised image using Proposed methods for Goldhill

4 Experimental Results and Discussion

The experiments are conducted on several test images that include Lena, Mandrill, Cameraman, and Goldhill of sizes 512×512 (refer Figs. 2) with various noise levels: 10, 20, 30, 50, 75, and 100. The wavelet transform that is employed is parametric multi wavelets compactly supported Symlet wavelet with eight vanishing moments at four scales of decomposition. To assess the performance of our proposed method, we compare PSNR (dB) result with that of the Weiner Filter, VisuShrink, SureShrink, and BayesShrink at various noise levels: 10, 20, 30, 50, 75, and 100 for all four test images. We observe that our method has higher PSNR than the Weiner Filter, VisuShrink, SureShrink, and BayesShrink methods for all noise levels and for test images (refer Table 1).

Table 1 PSNRs (in db) of various methods: Weiner Filter, VisuShrink, SureShrink, BayesShrink, and Proposed methods with noise levels: 10, 20, 30, 50, 75, 100 for the test images: Lena, Mandrill, Cameraman, and Goldhill

Image Name	Noise levels	Weiner Filter	VisuShrink	SureShrink	BayesShrink	Proposed
Lena	10	32.55	29.34	30.96	31.34	33.57
	20	28.99	26.40	26.43	29.09	30.49
	30	25.70	24.83	25.21	27.73	28.84
	50	21.40	23.00	23.84	25.91	26.59
	75	17.95	21.66	22.95	24.57	24.43
	100	15.49	20.76	22.38	23.67	23.82
Mandrill	10	26.50	23.74	27.76	27.75	29.80
	20	24.79	21.21	24.43	24.54	25.67
	30	23.08	20.20	21.57	22.79	23.27
	50	20.13	19.34	19.62	21.02	21.55
	75	17.30	18.84	19.29	20.07	20.34
	100	15.10	18.51	19.09	19.54	19.80
Camera man	10	32.77	27.96	31.13	30.78	32.83
	20	28.73	24.93	27.71	27.87	29.29
	30	25.52	23.39	24.10	26.04	27.56
	50	21.23	21.62	22.20	24.04	25.35
	75	17.82	20.32	21.18	22.78	23.53
	100	15.39	19.44	20.55	21.93	22.09
Goldhill	10	31.78	27.67	31.20	31.17	32.22
	20	28.26	25.25	27.86	28.33	29.09
	30	25.35	23.99	25.12	26.85	27.54
	50	21.27	22.51	23.32	25.21	25.69
	75	17.89	21.41	22.58	24.01	24.25
	100	15.45	20.64	22.10	23.18	23.46

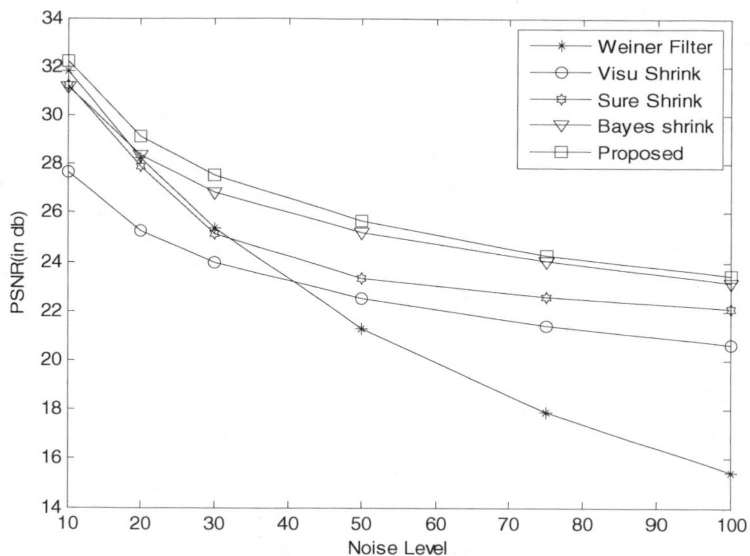

Fig. 4 PSNR vs. noise level of various methods: VisuShrink, SureShrink, BayesShrink, and Proposed for Goldhill

For the purpose of visual quality, we have taken original (noiseless) and noisy images with noise level 20 of Goldhill (refer Figs. 2(d) and 3(a)). We see that the denoised image using the proposed method has much less noise as it reduces the noise in a significant manner in comparison to the Weiner Filter, VisuShrink, SureShrink, and BayesShrink methods (refer Figs. 3(b)-(f)). Similar results have been obtained for other images. The PSNR gain curves for various denoising methods: Weiner Filter, VisuShrink, SureShrink, BayesShrink, and proposed methods are shown in Fig. 4 for Goldhill image. We have obtained similar types of PSNR curves for other images. Because of the repetitive nature of results, we have not shown their graphs. Our proposed method is having higher PSNR gain curve as compared to the Weiner Filter, VisuShrink, SureShrink, and BayesShrink methods for all noise levels (refer Fig. 4).

5 Conclusion

In this paper, we have discussed a new denoising method - Selective Parameters based Image Denoising Method that has improved the visual quality of the noisy image remarkably. Furthermore, our proposed method gives better performance in terms of PSNR and removes the noise significantly for all the test images taken and all noise levels under discussion than the Weiner Filter, VisuShrink, SureShrink, and BayesShrink methods.

Acknowledgements. The authors express their sincere thanks to Prof. S. Chand for his invaluable comments and suggestions.

References

1. Jansen, M.: Noise Reduction by Wavelet Thresholding. Springer, New York (2001)
2. Xie, J.C.: Overview on Wavelet Image Denoising. Journal of Image and Graphic 7(3), 209–217 (2002)
3. Donoho, D.L., Johnstone, I.M.: Ideal spatial adaptation via wavelet shrinkage. Biometrika 81, 425–455 (1994)
4. Donoho, D.L., Johnstone, I.M.: Wavelet shrinkage: Asymptotic? J. R. Stat. Soc. B 57(2), 301–369 (1995)
5. Donoho, D.L., Johnstone, I.M.: Adapting to Unknown Smoothness via Wavelet Shrinkage. Journal of American Statistical Association 90(432), 1200–1224 (1995)
6. Donoho, D.L.: De-Noising by Soft Thresholding. IEEE Trans. Info. Theory 41(3), 613–627 (1995)
7. Chang, S.G., Yu, B., Vetterli, M.: Adaptive Wavelet Thresholding for Image Denoising and Compression. IEEE Trans. Image Processing 9(9), 1532–1546 (2000)
8. Elyasi, I., Zarmehi, S.: Elimination Noise by Adaptive Wavelet Threshold. World Academy of Science, Engineering and Technology, 462–466 (2009)
9. Weeks, M., Bayoumi, M.: Discrete Wavelet Transform: Architectures, Design and Performance Issues. Journal of VLSI Signal Processing 35(2), 155–178 (2003)
10. Daubechies, I.: The Wavelet Transform, Time-Frequency Localization and Signal Analysis. IEEE Transaction on Information Theory 36(5), 961–1005 (1990)
11. Yang, Y., Wei, Y.: Neighboring Coefficients Preservation for Signal Denoising. Circuits, Systems, and Signal Processing 31(2), 827–832 (2012)
12. Om, H., Biswas, M.: An Improved Image Denoising Method based on Wavelet Thresholding. Journal of Signal and Information Processing 3(1), 109–116 (2012)
13. Chen, G., Zhu, W.-P.: Image Denoising Using Neighbouring Contourlet Coefficients. In: Sun, F., Zhang, J., Tan, Y., Cao, J., Yu, W. (eds.) ISNN 2008, Part II. LNCS, vol. 5264, pp. 384–391. Springer, Heidelberg (2008)

A Novel Approach to Build Image Ontology Using Texton

R.I. Minu and K.K. Thyagarajan

Abstract. The mere existence of natural living thing can be studied and analyzed efficiently only by Ontology, where each and every existence are concern as entities and they are grouped hierarchically via their relationship. This paper deals the way of how an image can be represented by its feature Ontology though which it would be easier to analyze and study the image automatically by a machine, so that a machine can visualize an image as human. Here we used the selected MPEG 7 visual feature descriptor and Texton parameter as entity for representing different categories of images. Once the image Ontology for different categories of images is provided image retrieval would be an efficient process as through ontology the semantic of image is been defined.

Keywords: Ontology, OWL, RDFS, MPEG7, Texton.

1 Introduction

Image Processing and analyzing is one of the active research areas from past decades. The main reason behind this is that there are different types of image format if one kind of technique is suitable for tiff image will not give same result for jpeg image. As the way human analyze and visualize an image is different from that of a machine. The main aim of this paper is that why can't we provide a way, so that a machine can also visualize an image as human so that the analyses would be an easier task. Thus, what if, we provide semantic for each categories of images though ontology? To provide Ontology for images the primary need is, image low level feature. Here we use MPEG 7 [13] descriptor and Texton [16]. As Ontology can be specified[5][6] by XML based RDFS and OWL the low level feature of image has to be in same format so that the integration of domain ontology with the feature ontology would not be cumbersome.

R.I. Minu
Anna University of Technology, Trichirappali, Tamil Nadu, India

K.K. Thyagarajan
Dept. of Information & Technology, RMK College of Engineering & Technology, Chennai, Tamil Nadu, India

A. Abraham and S.M. Thampi (Eds.): Intelligent Informatics, AISC 182, pp. 333–339.
springerlink.com

This paper is organized in such way that in next session the MPEG 7 feature descriptor is explained briefly, next the concept of Texton, then the formation of ontology through these feature and next at last with this ontology where we are applying the framework of it is been explained briefly.

2 Related Work

In [7] they uses the 16-bin color histogram, 62-D texture feature form Gabor filter and SIFT interesting point as visual feature for classification and they use multiple kernel (i.e.) each different kernel for different feature SVM classifier. They haven't explain the concept of ontology creation. In[8] general domain feature such as color and shape in used for creating image ontology which will not be so efficient for different domain images. In [9] they user the Region of interest concept for feature extraction and created a full ontology for Breast Cancer. In[10] [11] only the general local feature such as shape, size, intensity and position is used to define object ontology. In all those paper either they choice a single domain for creating image ontology or they have taken feature which won't give promising result regarding image classification.

3 Feature Descriptor

MPEG 7 is an ISO/IEC [13] standard for describing the multimedia content using different standard audio/video descriptor. Table 1 shows the list Visual Descriptor specified by the MPEG 7 standards. In most of the related work listed they used only the primary low level image feature such as color, shape and texture. In [7] only they specified how the feature is been extracted such as for color they use 16-bin color histogram, for texture 62-D of Gabor output. To provide a semantic for an image we need more specific details regarding the image feature. So, in MPEG 7 they have different feature descriptor for each primary local feature, in that we have chosen the promising feature descriptor such as Scalable Color Descriptor for providing color histogram details, Color layout Descriptor to provide the Spatial distribution of color, Dominant Color Descriptor provide possible 8 dominant color of an image and Edge Histogram Descriptor to provide the texture detail of an image. In most of the related work [12] they use any one of the feature here we use more than four so as to provide efficiency.

Table 1 MPEG 7 Visual Descriptor

VISUAL DESCRIPTOR			
COLOR	TEXTURE	SHAPE	MOTION
Histogram			
Scalable Color Descriptor	Texture Browsing Descriptor	Contour Shape Descriptor	Camera Motion Descriptor
Color Structure Descriptor			
Dominant color Descriptor	Homogenous Texture Descriptor	Region Shape Descriptor	Motion Trajectory Descriptor
Color Layout Descriptor	Edge Histogram Descriptor		Parametric Motion Descriptor

The concepts of said descriptor are all explained in our previous papers [1] [3]. So, in this paper we are going to explain in detail about the XML formation of each descriptor.

- **Scalable Color Descriptor**

The Color Space used in MPEG 7 is either RGB or HSV space. Here we convert the RGB space model images into HSV space model image. The reason behind is that all the images we may collected won't be an images taken by the expert photographer so due to dull luminance effect the RGB value will differ, so to avoided those problem we are converting to HSV model.

The Scalable color descriptor for the HSV model image in XML format is shown below:


```
<VisualDescriptor xsi:type="ScalableColorType" numOfBitplanesDiscarded="0" numOfCoeff="256">
<Coeff>-146 33 22 87 13 14 22 39 31 13 11 35 42 14 19 45 0 1 0 2 -6 5 0 10 -2 2 2 0 -15 5 1
19 0 0 0 1 0 0 1 2 4 1 1 3 1 2 4 9 1 0 2 2 2 3 3 0 15 0 0 -2 1 0 -3 -4 0 0 0 0 0 0 0 1 2 1 2
0 0 0 0 1 0 7</Coeff>
</VisualDescriptor>
```


Here the Scheme <xsi> is of type "ScalableColorType" which is the 256 coefficient value of the HSV space of the given image.

- **Color Layout Descriptor**

In Color Layout Descriptor [1][3][12] the image is partitioned in to 8 x 8 block and in each block's dominant color is determined. For each 8 x 8 block Discrete Cosine Transform for Y, C_r, and C_b color is determined and quantized for the required bit then using the Zigzag scanning the values are tabulated in matrix form. The concern XML is:


```
<VisualDescriptor xsi:type="ColorLayoutType">
    <YDCCoeff>13</YDCCoeff>
    <CbDCCoeff>6</CbDCCoeff>
    <CrDCCoeff>63</CrDCCoeff>
 <YACCoeff63>14 19 20 17 18 19 15 19 16 16 15 13 15 15 15 16 15 16 16 15 15 15 16 15 16 15
16 16 15 15 15 16</YACCoeff63>
 <CbACCoeff63>16 12 12 14 13 12 16 12 16 15 16 18 16 16 16 15 16 15 15 16 16 16 15 16 15 16
15 15 16 16 16 15</CbACCoeff63>
 <CrACCoeff63>16 17 17 16 16 18 15 17 14 15 15 14 14 15 14 16 14 16 15 15 16 15 16 17 15 16
15 15 16 15 15 16</CrACCoeff63>
 </VisualDescriptor>
```


Here the first three value of element <YDCCoeff>,<cbDCCoeff> & <CrDCCoeff> provides the said DC coefficient value for Y, C_r, and C_b color, where the other element gives the detail about the AC coefficient value.

- **Dominant Color Descriptor**

Among the list of Color descriptors, Dominant color is best suitable for local image. For given images maximum of 8 dominant colors is identified and label with unique numbering. Below shows the coding of identified one such dominant color.


```
<VisualDescriptor xsi:type="DominantColorType">
               <ColorQuantization>
                 <Component>R</Component>
                 <NumOfBins>8</NumOfBins>
                 <Component>G</Component>
                 <NumOfBins>8</NumOfBins>
                 <Component>B</Component>
                 <NumOfBins>8</NumOfBins>
               </ColorQuantization>
```

```
            <SpatialCoherency>31</SpatialCoherency>
            <Value>
              <Percentage>31</Percentage>
              <Index>255 0 0</Index>
          <ColorVariance>0 0 0</ColorVariance>
        </Value>
  </VisualDescriptor>
```


For the given input image one of the dominant color is Red whose color index is (255 0 0) , the Spatial Coherency represent the spatial homogeneity of the red color in the given image and its percentage of presence is given by percentage element.

- **Edge Histogram Descriptor**

Edge Histogram descriptor used to determine the edges of the images. By default the image is divided into 4X4 sub images and the edge magnitude of the each sub image is determined by comparing the standard edge provided by the MPEG 7 which is Vertical edge, Horizontal edge, 45° edge, 135° edge and non-directional edge. The first 4 bit represents the edge magnitude of the (0,0) sub block image and so on.


```
<VisualDescriptor xsi:type="EdgeHistogramType">
<BinCounts>4 2 6 7 4 5 2 3 6 5 4 2 7 3 5 3 3 6 7 5 4 2 6 3 6 4 1 5 4 6 5 2 6 4 5 3 4 6 4 5 4
2 5 6 6 4 2 6 3 6 4 1 7 4 4 3 5 6 3 5 4 4 4 5 5 3 4 6 6 5 4 3 7 7 4 1 2 4 7 6</BinCounts>
 </VisualDescriptor>
```


- **Texton**

The main objective of our work is to make the machine to visualize and identify an image as a human do. So, to think and act like human a human way of approach is to be provided to give semantic to images. The human process the visual information by pre-attentive visual perception, where each and every visual object information is consider as a micro visual structure. Likewise, an image can be represent as a superposition [15] of number of image base. Where the image base is a kind of dictionary which consist of various filtered images. The Major filter banks used are Gabor, Laplacian of Gaussian, wavelet and etc.

For creating a Texton dictionary nearly 64 filter bank of such filter has to be used. The output of such image has to be quantized. This quantized value has to be included to the MPEG 7 XML


```
<VisualDescriptor xsi:type="TextonDescriptor" numOfCoeff="256">
<Coeff> 1 1 1 1 1 1 2 1 1 1 1 1 1 1 1 1 0 1 0 2 -6 5 0 10 -2 2 2 0 -15 5 1 19 0 0 0 1 0 0 1
2 4 1 1 3 1 2 4 9 1 0 2 2 2 3 3 0 15 0 0 -2 1 0 -3 -4 0 0 0 0 0 0 0 1 2 1 2 1 2 2 2 4 0 0 0
1 0 0 1 1 6 1 2 1 3 3 2 7 0 0 0 1 0 0 2 2 8 2 2 1 1 -2 -3 0 0 0 0 0 1 6 0 0 0 0 0 0 0 0 0 0
0 -1 0 0 1 -1 0 0 1 -1 0 1 1 -2 -3 0 0 0 0 1 0 7 -1 0 1 0 0 0 0 -1 0 0 1 0 0 0 0 -1 -1 0 0 0
0 1 0 1 -3 0 0 0 0 1 0 7</Coeff>
</VisualDescriptor>
```


4 Multi-modal Fusion

The integration of derived MPEG 7 feature XML to Ontology is explained in this session. Here the fusion of visual feature and high level domain ontology is determined as shown in Fig.1. For the sake of experiment in our work we have taken three different domain image such as cup, Flower and Car images.

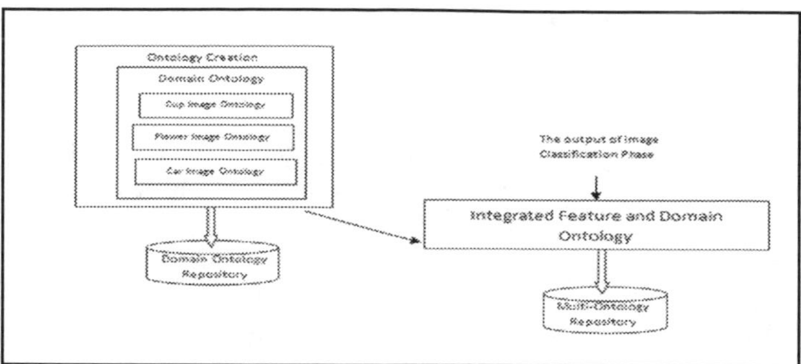

Fig. 1 Multimodal Ontology Fusion

- **MPEG 7 Ontology**

The specialty of MPEG 7 Descriptor is that it represents the image or video in XML format. In day-to-day advancement the performance of XML data is outperformed by OWL/RDFS data. Such kind of data representation is said to be Ontology data. The difference between XML and RDFS is that in XML the hierarchy data is represented in a well-formed way and to check its hierarchy the programmer has to provide schema or DTD thus again the machine doesn't understand why such hierarchy is needed. But in Ontology way of representation through RDFS/OWL we represent the reasons of such hierarchy representation. Thus for different image categories we can specify the cardinality of one kind of texture and color layout pattern is suitable only for one category of images. Thus we can provide restriction thus the image classification and categorization done in intelligent way through decision making technique rules. Fig.2 shows the ontology created with the said required feature.

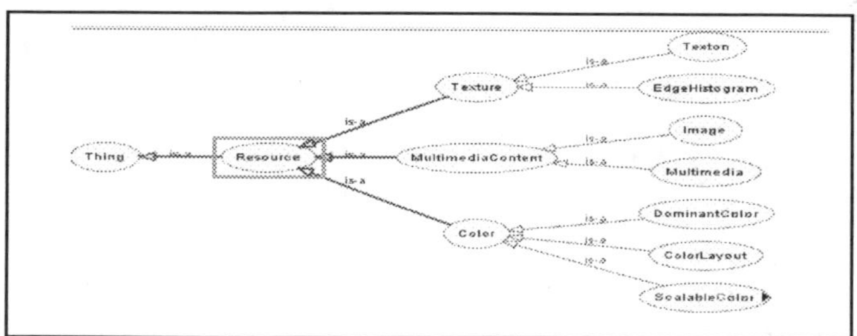

Fig. 2 MPEG 7 Ontology

Below show the code snippet of color ontology which has three sub-classes Dominantcolor, ColorLayout and ScalableColor

```
<owl:Class rdf:ID="Color">
                    <rdfs:label>Color</rdfs:label>
                    <rdfs:comment>Color of a visual resource</rdfs:comment>
<rdfs:subClassOf rdf:resource="http://www.w3.org/2000/01/rdf- schema#Resource"/>
</owl:Class>
<owl:Class rdf:ID="DominantColor">
```

```
                  <rdfs:label>DominantColor</rdfs:label>
<rdfs:comment>The set of dominant colors in an arbitrarily-shaped  region.</rdfs:comment>
                  <rdfs:subClassOf rdf:resource="#Color"/>
</owl:Class>
```

Now we needed a technique to populate the ontology with the MPEG 7 XML data. This mapping technique is explained in the forthcoming session.

- **XML to Ontology Mapping**

XML is a way of transferring the data in a tabular form in internet. Thus as the data has to be in a order we use either DTD or Schema to specify the format of the tabular data. OWL is also the same way of transferring the data but in semantic way. So, we can map those two language as said in [14]. The XML schema XSD element is converted to OWL element as shown in Table 2

Table 2 XML to OWL Conversion

XSD	OWL
Xsd:element with other elements	Owl:class
Xsd:element without any element	Owl:Datatype properties
Xsd:minoccurs, xsd:maxoccurs	Owl:mincardinality, owl:maxcardinality
Xsd:sequence, xsd:all	Owl:intersectionof

```
<owl:ObjectProperty rdf:ID=" EdgeHistogram ">
                  <rdfs:subPropertyOf rdf:resource="#visualDescriptor"/>
                  <rdfs:domain>
                            <owl:Datatypeproperties>
                                     <BinCounts>4 2 6 7 4 5 2 3 6 5 4 2 7 3 5 3 3 6 7 5 4
2 6 3 6 4 1 5 4 6 5 2 6 4 5 3 4 6 4 5 4 2 5 6 6 4 2 6 3 6 4 1 7 4 4 3 5 6 3 5 4 4 4 5 5 3 4
6 6 5 4 3 7 7 4 1 2 4 7 6</BinCounts>
                            </owl: Datatypeproperties >
                  </rdfs:domain>
                  <rdfs:range rdf:resource="#EdgeHistogram"/>
         </owl:ObjectProperty>
```

Above shown the code snippet of converted Edge histogram descriptor. By providing semantic for each category of images then image retrieval will be an efficient process such kind of approach is proposed in [1] my future work.

5 Conclusion

The Worldwide image repository is increasing enormously due to technology advancement. The literal data can be classified as integer, floating and character data type then they can be easy grouped and classified. Whereas Image don't have a specific data type for each category of image, so we need a technique equivalent to human visualization so that the image can be classified as human does, this is only possible by providing semantic for each category of images. So, in this paper we proposed a concept of Image ontology which paves an efficient way for semantic based image retrieval system.

References

1. Minu, R.I., Thyagharajan, K.K.: Multimodal Ontology Search for Semantic Image Retrieval. Submitted to International Journal of Computer System Science & Engineering for February Issue (2012)
2. Nagarajan, G., Thyagharajan, K.K.: A Novel Image Retrieval Approach for Semantic Web. International Journal of Computer Applications (January 2012)
3. Minu, R.I., Thyagharajan, K.K.: Automatic image classification using SVM Classifier. CiiT International Journal of Data Mining Knowledge Engineering (July 2011)
4. Minu, R.I., Thyagharajan, K.K.: Scrutinizing Video and Video Retrieval Concept. International Journal of Soft Computing & Engineering 1(5), 270–275 (2011)
5. Nagarajan, G., Thyagharajan, K.K.: A Survey on the Ethical Implications of Semantic Web Technology. Journal of Advanced Reasearch in Computer Engineering 4(1) (June 2010)
6. Minu, R.I., Thyagharajan, K.K.: Evolution of Semantic Web and Its Ontology. In: Second Conference on Digital Convergence (2009)
7. Fan, J., Gao, Y., Luo, H.: Integrating concept ontology and Multitask learning to achieve more effective classifier training for multilevel image annotation. IEEE Transaction on Image Processing 17(3) (2008)
8. Penta, A., Picariello, A., Tanca, L.: Towards a definition of an Image Ontology. In: 18th Int. Workshop on Database and Expert Systems Applications (2007)
9. Hu, B., Dasmahapatra, S., Lewis, P., Shabolt, N.: Ontology-based medical image annotation with description logics. In: 15th IEEE Int. Conf. on Tools with Artificial Intelligence (2003)
10. Mezaris, V., Kompatsiaris, I., Strintzis, M.G.: An Ontology approach to object-based image retrieval (2003)
11. Maillot, N., Thonnat, M., Hudelot, C.: Ontology based object learning and recognition: Application to image retrieval (2004)
12. Manjunath, B.S., Ma, W.Y.: Texture features for browsing and retrieval of image data. IEEE Transaction on Pattern Analysis and Machine Intelligence 18, 837–842 (1996)
13. ISO/IEC JTC1/SC29/WG11N6828 Palma de Mallorca, MPEG-7 Overview (version 10) (October 2004)
14. Bohring, H., Aure, S.: Mapping XML to OWL ontologies (2004)
15. Leung, T., Malik, J.: Representing and recognizing the visual appearance of materials using three-dimensional texton. IJCV (2001)

Cloud Extraction and Removal in Aerial and Satellite Images

Lizy Abraham and M. Sasikumar

Abstract. Aerial and satellite images are projected images where clouds and cloud-shadows cause interferences in them. Detecting the presence of clouds over a region is important to isolate cloud-free pixels used to retrieve atmospheric thermodynamic information and surface geophysical parameters. This paper describes an adaptive algorithm to reduce both effects of clouds and their shadows from remote sensed images. The proposed method is implemented and tested with remote sensed RGB and monochrome images and also for visible (VIS) satellite imagery and infrared (IR) imagery. The results show that this approach is effective in extracting infected pixels and their compensation.

Keywords: Adaptive segmentation, Average local luminance, Shadowing effect.

1 Introduction

A common and complex aspect in aerial and satellite image application is encountered when image is captured from above the clouds. This causes signal attenuation of the image acquisition above the cloud cover and cloud-shadows modifies the ground local luminance. Several methods were used to restore the cloud affected areas. But most of them are used for removing thin clouds [1]. Fusion techniques were also used to account for cloud and shadow defects [2]. These methods require cloud/shadow free images as reference images for processing and so not much reliable. Recent developments in the area include more efficient segmentation results but the performance is greatly influenced by the selection of the thresholds for various spectral tests [3]. T. B. Borchartt and R. H. C. de Melo [4]

Lizy Abraham
Department of Electronics & Communication Engineering,
LBS Institute of Technology for Women, Trivandrum, India

M. Sasikumar
Department of Electronics & Communication Engineering,
Marian Engineering College, Trivandrum, India

A. Abraham and S.M. Thampi (Eds.): Intelligent Informatics, AISC 182, pp. 341–347.
springerlink.com

suggested a method which works well but the algorithm fails for the compensation of cloud-shadows and the scaling factors have to be obtained experimentally for each part of the image.

In this paper, the cloud affected regions are detected and extracted by an adaptive segmentation algorithm. These cloudy regions, that were just extracted, are compensated for further processing. Then using a single program, later on, diminish the effects of both clouds and shadows. The output hence obtained, is the input image with reduced or diminished effects of clouds and shadows.

2 Methodology

Many of the techniques developed to enhance cloud-associated regions ignore the information in the shadow regions. Figure 1 presents the overall flow of our proposed system, considering both cloud and shadow regions which requires only a single image and single program for processing.

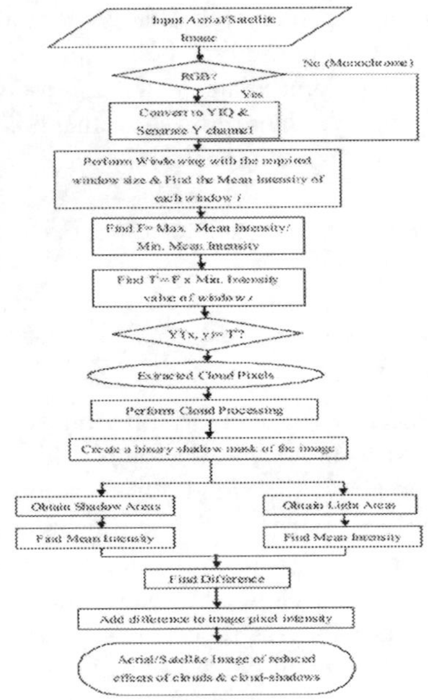

Fig. 1 Flow Chart of the Proposed Method

2.1 Extraction of Clouds

The clouds are detected by considering the hypothesis that, regions of the image covered by clouds present increased local luminance values (due to the direct reflection of the sunlight by the cloud). To automatically detect the presence of cloud in a region, the average of the local luminance (that is the mean image intensity) is used in the algorithm. For this the gray-scale image is divided to a number of small windows and finds the mean intensity μi of each window using the following equation (1):

$$\mu^i = \frac{1}{M}\sum_{k\in X^i}^{k=1...M} X_k^i \tag{1}$$

where M is the number of pixels in ith window and X_k^i denotes kth pixel in the ith window. For an RGB image, convert it to YIQ space and calculate the average intensity of Y channel alone. The window size can be varied depending on the cloud size. If clouds are of varied sizes and occupying only 1% of the larger image resolution, the best window size is 1.5% of image resolution. Next, cloud extraction is done by choosing a threshold depending on the statistical properties of the image. Here the threshold is depending on the parameter F which is the ratio of maximum mean intensity μmax and minimum mean intensity μmin for adaptive segmentation (2).

$$F = \frac{\mu_{\max}}{\mu_{\min}} \tag{2}$$

If xi(i,j) represents minimum intensity value of the image in the ith region, the threshold Ti is calculated by the following equation (3):

$$T^i = F \times x^i(i, j) \tag{3}$$

An image A having cloud infected pixels is obtained by (4):

$$\mathrm{A} = \{(\mathrm{x,y})|\ (\mathrm{x,y}) \subset X^i, X^i\ (\mathrm{x,y}) \geq \mathrm{T}^i\} \tag{4}$$

2.2 *Cloud Processing*

In cloud processing, every independent regions with cloud and without cloud are processed. Thus the luminance content in cloud affected regions is reduced, this, in turn reduces the effect of clouds in the image. Processing of the cloud affected pixels is done using (5):

$$I' = m_c + \frac{I - m_s}{\sigma_s}.\sigma_c \tag{5}$$

where I is pixel grey level value in cloud regions before processing, I' is pixel grey level value after processing. mS and σS are mean and variance of cloudy regions. mC and σC are average value and variance of regions without cloud.

2.3 Shadow Segmentation and Compensation

For detecting shadows, the RGB image is first converted to YCbCr space. Shadows are segmented based on mean values in each region of Y channel [5] using the equation (6):

$$T = \frac{1}{M} \sum_{k \in X^i}^{k=1...M} X_k^i \tag{6}$$

where pixel values less than this threshold is taken as shadow pixels. After this, compute the median filter for the image to reduce the noise. The result of the shadow detection is a binary shadow mask, which will be the input to the shadow removal algorithm.

We use a simple shadow model [6], where there are two types of light sources: direct and ambient light. Direct light comes directly from the source, while environment light is from reflections of surrounding surfaces. For shadow areas part or all of the direct light is occluded. The shadow model can be represented by the following formula (7):

$$l_i = (t_i \cos \Theta_i L_d + L_e) R_i \tag{7}$$

li represents the value for the i-th pixel in RGB space
Ld and Le represent the intensity of the direct light and environment light
Ri is the surface reflectance of that pixel
Θi is the angle between the direct lighting direction and the surface norm
ti is the attenuation factor of the direct light

If ki= ticos Θi is the shadow coefficient for the ith pixel and r denotes the ratio between direct light Ld and environment light Le, the shadow free pixel of Cb and Cr channels are computed using the equation (8):

$$l_i^{shadow\_free} = \frac{r+1}{k_i r+1} . l_i \tag{8}$$

For correcting the shadow infected pixel in the Y channel, find the average pixel intensities of shadow μshadow and light areas μlit. Then the corrected pixel is obtained by (9):

$$Y_i^{shadow\_free} = Y_i^{Shadow\_infected} + (\mu_{shadow} - \mu_{lit}) \tag{9}$$

Converting the YCbCr model back to RGB space we get the resultant image not having any of the cloud interferences and shadowing effects. For monochrome images process the gray level intensities using equations (6) and (9).

3 Results and Discussions

The proposed algorithm is evaluated with different types of images including visible and infrared (VIS & IR) satellite images, RGB and monochrome aerial images,

RGB satellite images having considerable amount of clouds, and in some cases, shadows in them. The experimental results are as shown in below figures (Fig.2 - Fig.7). The method first extracts shadows and clouds from the image and later on, reduce the effects of both clouds and shadows. The output, hence obtained, is the input image with reduced or diminished effects of clouds and shadows. All these processes are based on the values of parameters that are calculated from within the image, and so the algorithm is 'adaptive'.

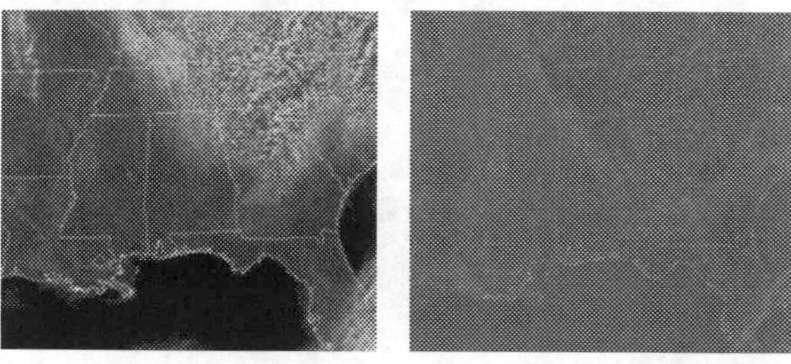

Fig. 2 (a) Visible Satellite Image of SE USA (b) Cloud processed image

Fig. 3 (a) Infrared Satellite Image of SE USA (b) Cloud processed image

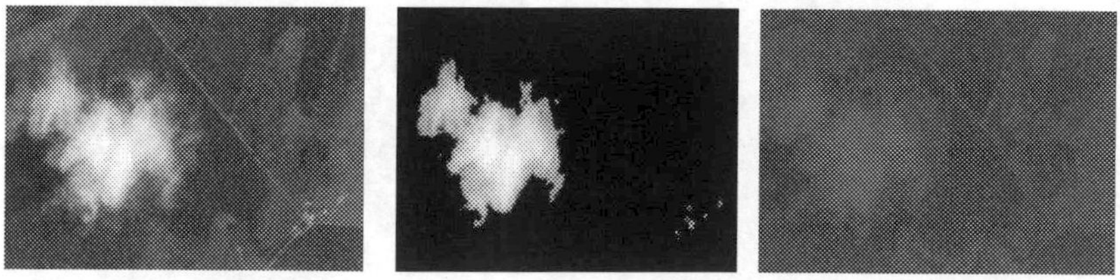

Fig. 4 (a) Monochrome Aerial Image (b) Cloud extracted image (c) Cloud processed image

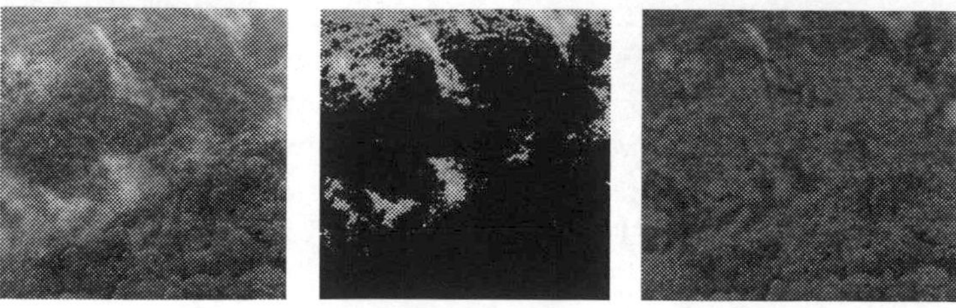

Fig. 5 (a) RGB Aerial Image (b) Cloud extracted image (c) Cloud processed image

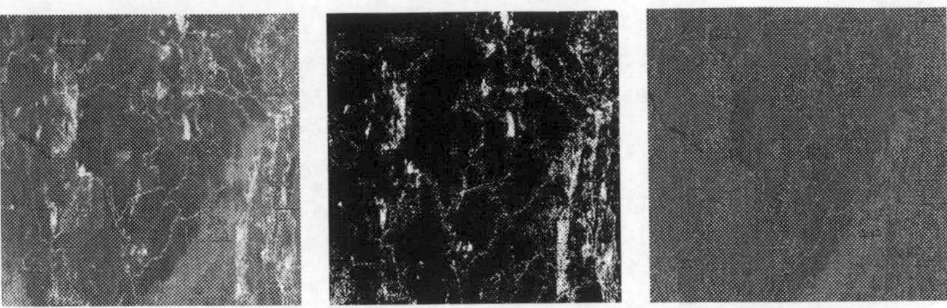

Fig. 6 (a) RGB Satellite Image (b) Cloud extracted image (c) Cloud processed image

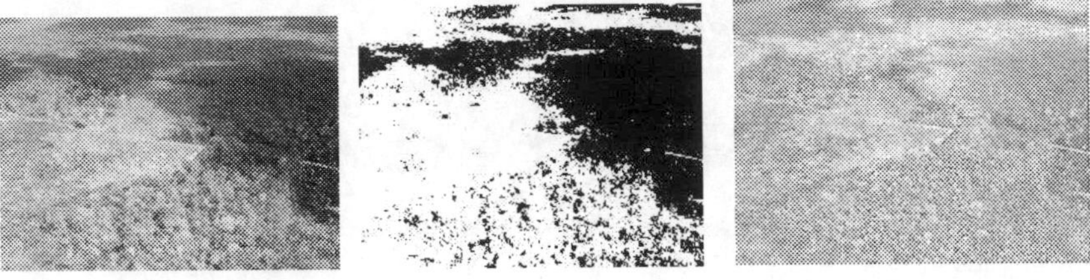

Fig. 7 (a) Aerial Image having cloud-shadows (b) Shadow extracted image (c) Shadow processed image

The results show that the algorithm can be used even for removing the shadow areas of dense urban area. The output image is free from shadow. Thus it is evident that the algorithm works well for both cloud removal and shadow removal. Such cloud and shadow free images can be used as inputs for feature extraction processes. In the case of image segmentation programs, where regions like vegetative areas are to be extracted, the clouds or shadows if present, do not pose a problem and they provide an output much reliable and accurate, than the same from an image where the effect of clouds and shadows are present. The algorithm can also be used for cartographic purposes to an extent, as it can be used to find areas of same visual characteristics.

4 Conclusion

The solution for signal attenuation and the shadowing effect caused by the clouds is designed in this paper. The system manages to diminish the effect of clouds and shadows and also estimate the region underlying it based on the pixel values of the surrounding regions. Unlike most previous methods which are suitable for image sequences, our method can extract and compensate clouds and shadows using only a single image. However, the algorithm can be modified to enhance the results for denser cloud images.

References

[1] Chanda, B., Majumder, D.D.: An iterative algorithm for removing the effects of thin cloud cover form Landsat imagery. Mathematical Geology 23(6), 853–860 (1991)

[2] Abd-Elrahman, A., Shaker, I.F., Abdel-Gawad, A.K., Abdel-Wahab, A.: Enhancement of Cloud-Associated Shadow Areas in Satellite Images Using Wavelet Image Fusion. World Applied Sciences Journal 4(3), 363–370 (2008)

[3] Reuter, M., Thomas, W., Albert, P., Lockhoff, M., Weber, R., Karlsson, K.G., Fischer, J.: The CM-SAF and FUB Cloud Detection Schemes for SEVIRI: Validation with Synoptic Data and Initial Comparison with MODIS and CALIPSO. J. Appl. Meteor. Climatol. 48, 301–316 (2009)

[4] Borchartt, T.B., de Melo, R.H.C., Gazolla, J.G.F.M., Resmini, R., Conci, A., Sanchez, A., de A. Vieira, E.: On the reduction of cloud influence in natural and aerial images. In: 18th International Conference on Systems, Signals and Image Processing, pp. 1–4 (June 2011)

[5] Tian, J., Sun, J., Tang, Y.: Tricolor attenuation model for shadow detection. IEEE Trans. Image Process. 18(10), 2355–2363 (2009)

[6] Guo, R., Dai, Q., Hoiem, D.: Single-Image Shadow Detection and Removal using Paired Regions. In: 24th IEEE Conference on Computer Vision and Pattern Recognition, USA, pp. 2033–2040 (2011)

3D360: Automated Construction of Navigable 3D Models from Surrounding Real Environments

Shreya Agarwal

Abstract. In this research paper, a system capable of taking as input multiple 2-dimensional images of the surrounding environment from a particular position and creating a 3-dimensional model from them, with navigation possible inside it, for the 360 degree view is developed. Existing approaches for image stitching, which use SIFT features, along with approaches for depth estimation, which use supervised learning to train a Markov Random Field (MRF), are modified in this paper in order to improve their efficiency. Also, an improvement in accuracy of depth estimation is suggested for the 3-dimensional model using matching SIFT features from the multiple input images. A method for enabling navigation in the 3D model through which we prevent motion in areas where movement is not possible is outlined, thus making the 3-dimensional model realistic and suitable for practical use. The proposed system is also an application of Neural Networks.

1 Introduction

A lot of work has been recently done to achieve the task of generating 3D models from 2-dimensional images of a scene. People have been working on developing virtual worlds and progress has been made in the field of virtual reality. This paper aims at developing 3D models of 'real environments' that the user can navigate inside. The proposed system takes as input multiple images of a scene. The output is a 3-dimensional model generated out of the images inside which the user can navigate, but only in areas where movement is possible.

Most systems which exist are computationally expensive, have limitations or are patented. This system aims at being fast and sufficiently accurate so that all users, whether or not technically sound, can use it with their own images.

Shreya Agarwal
S. V. National Institute of Technology, Surat, India
e-mail: a.shreya@coed.svnit.ac.in

A. Abraham and S.M. Thampi (Eds.): Intelligent Informatics, AISC 182, pp. 349–356.
springerlink.com

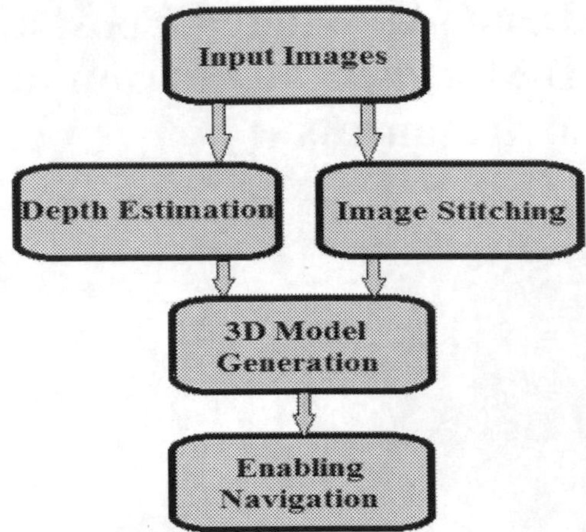

Fig. 1 3D360 system working flowchart

The first task is to estimate the depths of different points in the images. This is followed by stitching the images together and then generating the 3D model. The final step will be to enable navigation in the model. The methodology followed is shown in Fig. 1.

2 Previous Work

Estimating depth in a 2-dimensional image has been a longstanding problem in the field of computer vision. The most widely used technique has been estimating depth from binocular vision (stereo-vision) [1]. Other methods, like using defocus[2] and focus to find depth have also been used.

However, these methods are not used in this system because they require multiple images of the scene. Saxena, Chung & Ng [3] predict depths from single monocular images using supervised learning with a Markov Random Field (MRF). They utilize monocular cues from the 2D images in order to discriminately train their MRF.

The second task is image stitching. The major challenge here is image matching/ Methods for doing this can be divided into: feature-based[4] and direct. Many algorithms have been recently developed to use invariant features[5]. Lowe's Scale Invariant Feature Transform (SIFT) features are now widely used.

Make3D [6] is one system, developed at Stanford University, which lets the user create 3-dimensional model from a single image. However, the system does not render a 3D model for the complete 360 degree view of a scene.

3 Depth Estimation

The method suggested overlaps with the one suggested by Saxena, Chung & Ng [3]. Their method is modified to make it much faster while compromising little on the accuracy of depth estimation.

Two types of depths are looked for: *relative* (between points) and *absolute* (from the camera). Depth of a point will depend on the depths of points near it. Hence, in this case it is not enough to use local features to estimate the depth. The technique needs to use the *contextual information* around each point to find the depth.

Each image is viewed as a composition of multiple planes. These planes can be at any angle to the horizontal and vertical. A process called *super-pixellation*[6] is used to group pixels into such planes. These are also known as *superpixels*. Fig. 2(b) shows the formation of superpixels in an input image.

Fig. 2. Super-pixellation (a) Original image, (b) Super-pixelled image, (c) MRF graph over super-pixelled image (showing only limited number of nodes)

To find the depth from a single image, monocular cues are used. For example, texture gradients capture the distribution of the direction of edges while haze is produced as a result of scattering of atmospheric light. Three types of local cues are used to represent the local monocular cues: *haze*, *texture variations* and *texture gradients*. Lower frequency components of an image contain information about Haze and hence a local averaging filer is used with the color channels (with the image represented in the YCbCr color space). The texture information is contained in the intensity channel. Laws' mask is applied to the intensity channel to calculate the texture energy. Finally, the intensity channel is convolved with six oriented edge filters to get the texture gradient.

Hence, *17 values* for every pixel are found: 6 from the edge filters, 2 from convolving the color channels with first laws' mask, and 9 from convolving the intensity channel with Laws' masks. However, a feature vector for each superpixel has to be generated. For every superpixel *i,* each feature is considered and the sum and the sum-squared value of that feature over all the pixels (x,y) in this superpixel is found. Hence, out sum absolute energy formula is,

$$E_i(n) = (\sum (x, y)_n)^k$$

where k= ¶1,2◇give the sum absolute energy and sum squared energy respectively.

Now a feature vector of length 34 is obtained for every superpixel (since k has 2 values in the previous equation). For contextual information in the feature vector of a superpixel, the features of the neighboring 4 superpixels are included. This is as per the *Markov Random Field* (MRF) model. Fig. 2(c) shows the superpixellated image overlay-ed with an MRF graph.

A *feature vector of length 170 for every superpixel* is formed. This is used as a training set for a *neural network* having 3 hidden layers and 30 neurons in each layer.

4 Image Stitching

The system will stitch the images together to find the joining points of the images. It is done by utilizing *Scale Invariant Feature Transform (SIFT) features* [7]. This ensures that the same features are obtained for the scene at different zoom levels. The approach used for image stitching is summarized in Fig. 3(a).

Fig. 3 Image Stitching (a) working flowchart, (b) 3 original images with generated panorama

The image is convolved with Gaussian filters of different scales and then the difference of these Gaussian smoothed images is found. Once the DoG (Difference of Gaussian) images have been obtained, each pixel in each DoG image is checked, except in the first and last ones, with its 8 neighboring pixels and 18 (9+9) neighboring pixels in the two DoGs adjacent to it (before and after). If the pixel in consideration is the minima or maxima amongst all of these 26 pixels, it is considered as a SIFT feature [7]. Those candidate points in which the contrast is too low are removed, where contrast is the difference between its intensity and the intensities of its neighbors. This process is known as *keypoint localization*.

Orientation and *gradient magnitude* are assigned to each feature by looking at the contrast in the neighboring pixels. The gradient magnitude is calculated by calculating the horizontal and vertical gradient by taking difference in intensities of two pixels on each of the 4-connected sides of each feature pixel. The orientation,

$Orientation = tan^{-1}(V.G./H.G.)$
Where, V.G. = vertical gradient. H.G. = horizontal gradient

Matching of features in different images with similar orientations and gradients is done. It is found that modifying the method suggested in [7] leads to better accuracy and also makes the process efficient. The horizontal and vertical distance differences between matched features give us the joining position for two images. Example of panorama generation is showed in Fig. 3(b).

5 3D Model Generation

Once the depths of the superpixels have been determined and the images have been stitched together, a 3-dimensional model of the environment in consideration can be generated. However, a novel approach is suggested to improve the accuracy of depth plotting in the system.

Since a smaller feature vector has been used while training the neural network for estimating depths, it is possible that the network predicts slightly different depths for regions which are present in multiple images (overlapping regions). Overlapping regions are represented by matching SIFT features. These features are utilized to adjust the depth over the multiple images in such a manner that the depths at matching SIFT features in the images are the same.

For this purpose, the depths at corresponding final matches of SIFT features in two images are checked. The *average difference in depths* over all the matching SIFT features in two images is found. Next step is to apply this average difference over the depths of the second image. The method is repeated with rest of the images which have overlapping areas.

This results in a more uniform rendition of the environment in the 3-dimensional model. In this way, the accuracy of depth estimation of the whole process has been improved after utilizing matching SIFT features.

6 Restricting Navigation

There are many obstacles in an image through or beyond which movement is not possible in the real world. A method is now suggested to restrict navigation of the user beyond such points in the scene.

In every scene that is created, a *mesh* of points which have been plotted is created. Multiple convex and concave polygons can be drawn using these points. The system finds the *largest possible concave polygon* from this mesh that is furthest from the camera. It makes sure that the camera co-ordinates do not lie on the convex side of this polygon.

However, there will be obstacles that will lie a significant distance in front of this concave polygon. For restricting movement through these, a list of points which lie in a *non-navigable region* is maintained. These regions are identified by calculating the shortest distance of points that lie beyond and on the mesh from the concave polygon we determined earlier. If this distance exceeds a particular threshold and the point lies directly behind a node in the mesh, the point is added to the list of points in the non-navigable region.

7 Implementation

The images were pre-processed to find the super-pixels. Each superpixel was processed to find the feature vector. The true depths of the training set of 86 images were used as the target output for the neural network which had 3 hidden layers with 30 neurons each. The trained neural network was stored for predicting the depths for test images.

Fig. 4 3D model with multiple images. (left) Model zoomed in and panned right, (right) larger view of 3D model.

Table 1 Test results of depth estimation on images in terms of Average Absolute error percentage with varying number of superpixels in each image

Image #	% Average Absolute Error	No. of Superpixels
1	15.632852	20
2	30.702746	2176
3	8.816632	11
4	21.335124	125
5	19.716501	1111
6	13.28419	987
7	15.531753	298
8	11.561463	42
9	12.678880	335
10	12.426758	551
11	14.121634	20
12	23.971524	1036

The suggested algorithms for image stitching were implemented using C++ and the computer vision library OpenCV. The final 3-dimensional model was generated using the graphics library OpenGL. Fig. 4 shows an example of 3D models generated from multiple images.

The test results for depth estimation for 12 images are shown in Table 1. Fig. 5 shows 3D model from a single image. Navigation was enabled using the arrow keys and by allowing him to zoom in and out.

Fig. 5 Single image 3D model. (left) Original image, (right) Generated 3-dimensional model.

8 Applications

The proposed system has widespread applications in the modern world. The fact that the system is fast, sufficiently accurate for general purposes and can be operated by non-technical personnel with images taken by any amateur photographer guarantees its ease of access amongst the general public.

The biggest application is that for tourism. By using the proposed system, people can reconstruct places they visited if they have enough images to cover the scene. People can visit new places in the virtual world and move around.

Another major application of such a system is deciding on war strategies. High ranking officers need to form strategies for attacking and defending places and objects in a war.

The system can be used to generate 3D models which can be used in games and simulations thus bypassing the work of graphic rendering by using models based on real world environments.

9 Conclusion

This paper makes multiple modifications to depth estimation and image stitching techniques to make them efficient with respect to the proposed system. Depth estimation is implemented as an application of neural networks. The technique which uses matching SIFT features to regulate depth in the 3D models is a novel one suggested in this paper. The method suggested to restrict user movement in inaccessible areas gives good results. In conclusion, the 3D360 system is successfully and efficiently able to generate a 3-dimensional model navigable in a 360 degree space by using the suggested algorithms and can be used for multiple real world applications.

References

[1] Scharstein, D., Szeliski, R.: A taxonomy and evaluation of dense two-frame stereo correspondence algorithms. International Journal of Computer Vision 47, 7–42 (2002)
[2] Surya, G., Subbarao, M.: Depth from defocus: A spatial domain approach. International Journal of Computer Vision 13(3) (1994)
[3] Saxena, A., Ng, A.Y., Chung, S.H.: Learning Depth from Single Mo-nocular Images. Neural Information Processing Systems (NIPS) 18 (2005)
[4] Torr, P., Zisserman, A.: Feature Based Methods for Structure and Motion Estimation. In: Triggs, B., Zisserman, A., Szeliski, R. (eds.) ICCV-WS 1999. LNCS, vol. 1883, pp. 278–294. Springer, Heidelberg (2000)
[5] Schmid, C., Mohr, R.: Local Grayvalue Invariants for Image Retrieval. IEEE Transactions on Pattern Analysis and Machine Intelligence 19(5), 530–535 (1997)
[6] Saxena, A., Ng, A.Y., Sun, M.: Make3D: Learning 3-D Scene Structure from a Single Still Image. Transactions on Pattern Analysis and Machine Intelligence (2008)
[7] Ostiak, P.: Implementation of HDR panorama stitching algorithm. In: Proceedings of the Central European Seminar on Computer Graphic (2006)

Real Time Animated Map Viewer (AMV)

Neeraj Gangwal and P.K. Garg

Abstract. 3D game engines are originally developed for 3D games. In combination with developing technologies we can use game engines to develop a 3D graphics based navigation system or 3D Animated Map Viewer (AMV). Visualizing geospatial data (buildings, roads, rivers, etc) in 3D environment is more relevant for navigation systems or maps rather than using symbolic 2D maps. As 3D visualization provides real spatial information (colors and shapes) and the 3D models resembles the real world objects. So, 3D view provides high accuracy in navigation. This paper describes the development of human interactive 3D navigation system in virtual 3D world space. This kind of 3D system is very useful for the government organization, school bodies, and companies having large campuses, etc for their people or employers for navigation purposes.

1 Introduction

Maps are important source of primary information for navigation. Geographical maps are two-dimensional representation of a portion of the Earth's surface. We have developed many kinds of maps for our different needs. A political map for example shows territorial borders between the countries of the world. A physical map presents geographical features like mountains, lakes, rivers, soil type, etc. A map uses colors, symbols, and labels to represent features found on the ground. The ideal representation would be realized if every feature of the area being mapped could be shown in true shape. Road maps are perhaps the most widely used maps today, and form a subset of navigational maps, which also include aeronautical and nautical charts, railroad network maps, and bicycling maps. These maps can be prepared by municipalities, utilities, emergency services providers, and other local agencies.

Now this system of navigation can be made more interactive, handy or more useful by developing a 3D system where the user is able to see all parts of an area simultaneously and free to select his source and destination to view the complete

Neeraj Gangwal · P.K. Garg
Civil Engineering Department
Indian Institute of Technology Roorkee, Roorkee-247667, India
e-mail: {neerajgangwal2006,pkgiitr}@gmail.com

A. Abraham and S.M. Thampi (Eds.): Intelligent Informatics, AISC 182, pp. 357–363.
springerlink.com

shortest path between them. User can follow that path for his navigation purpose. Users can also interrupt the system at any point of time and move freely or walk-through in any direction where he wants.

2 Developing Real Time Animated Map Viewer

Here, we consider an example of university to explain the methodology of our 3D animated map viewer. Steps needed to make AMV are same for any kind of university campus, organization, mall, hospital, etc. The university has many teaching departments, sub-departments and various facilities like hostels, hospital, canteen, playgrounds, etc. so, there is need to make navigational system which helps the newcomers, faculty members or other persons to find any place or facility at any time.

For making the 3D map of the university we need basic details of all buildings and departments like their positions relative to each other and their sizes. Steps followed are -

2.1 Taking Images and Measurements of Various Parts and Departments

The measurements and position of departments, banks, etc can be taken manually or by consulting some legal and authorized body. This work can be done in modern way by using GPS which provides a high accuracy in all measurements (length, breadth, height).

2.2 Use of 3D Modeling Software to Design 3D of All Part

After measurements and picture capturing of all departments, roads and hospitals, etc we have to design the 3D model of all these. In market there are many 3D modeling software available freely and commercially. We have used Autodesk 3ds max software which is commercially available.

2.3 Conversion of Generated 3D Files to Usable Form for Game Engines

Geographic maps can be used as guide for placing the 3D designed objects in 3D field or 3D world space at their accurate place so that all designed field looks realistic and accurate. Single or multiple cameras can be used for making a good walk-through. In market various kind of game engine (Torque, Quest 3D software)are available which can understand the above generated 3D file and apply various kind of real time simulation or help in making walkthrough of 3D designed field.

2.4 Features Available for User

To make AMV more realistic we can use the real time fog, shaders, particle emitters etc. There are many options available for user interaction. Main options are as follows:

1. Input box available for User to enter Source and Destination.
2. When the source and destination are identified than a complete road track or shortest path, along with a complete guide of information is shown like distance, time and it also suggests which method is best to reach there by auto-rickshaw or bus.
3. An interrupt function is available by which user move freely to check any part of the campus (for this some direction keys are available).

2.5 Algorithms Used for Path Finding

There are many algorithms that can be used for shortest path findings. Mainly these algorithms are used in network routing for path finding. Here we give an example.

A road network can be considered as an undirected graph (since a road can be used in both directions for movement) with positive weights (Fig. 1). The nodes represent road junctions or different points in any campus and each edge of the graph is associated with a road segment between two junctions. The weight of an edge may correspond to the length of the associated road segment, the time needed to traverse the segment or the cost of traversing the segment.

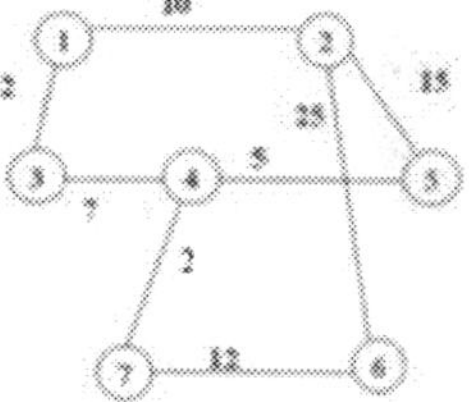

Fig. 1 Showing weight between different stations. A path with minimum cost is selected for navigation

Suppose we want to go from 1st to 6th position then the path followed is 1-3-4-7-6 as it has less cost on comparing with other paths. We solved our problem with Finite State machine (FSM) algorithm which gives shortest path between the two stations. Dijkstra's, algorithms can be used. These algorithms can be used in combination with the 3D programming language (OpenGL, 3D java, etc) or 3D game engines (as mentioned above) for rendering and simulation for path finding.

2.6 Installation of the AMV

There are many ways in which AMV can be used.

1. Developed system can be loaded or installed on a computer and computer is placed at different main points of the campus like main building, hospital, etc and used for navigational purpose.
2. This system can be developed in the form of information kiosk.
3. Program can be developed in the form of android application as android is an emerging filed today. The application can be given to each student or employer for navigation free of cost. Everyone is able to explore the 3D map in its mobile device. This is the most cost effective way.

In fig 2 to 8 we show some pictures from a university campus where some departments and residential colonies are shown. The user wants to go at location 'E' from location 'A'. There are many paths exist which can take user from location 'A' to location 'E', but the shortest and the main path is selected by AMV (path followed is A-B-C-D-E).

Fig. 2 At this point user enters its source and destination location. Here these locations are 'A' and 'E' respectively. User see every location in AMV as they appear in real world.

Fig. 3 Other departments are shown. 'B' is another location that comes in the midway

Fig. 4 'C' is another location in the path followed

Fig. 5 A complete path is shown from A to E from a top view camera

Fig. 6 Shows the final location 'E' which is the destination of the user. Now the user can follow this path to reach here and compare the real world objects with the objects shown by AMV while following his path.

Fig. 7 Free or walkthrough mode is selected by user. Now he can move in any direction and learn the campus area.

Fig. 8 Top view of the path selected area is shown. Various buildings, trees cars, are shown in this figure. We can make it realistic to any depth by adding other features.

3 Advantages

1. 3D animation and real time simulation is used in developing AMV so they are easy to use and self explanatory. No need of special training for the user.
2. AMV is not a collection of images. This is simply a 3D of whole area and you are moving through that area.
3. AMV shows the shortest or desired path.
4. These devices are very useful because of their user interactivity.
5. For purpose of security they are very good because we are designing the 3D of all parts by our own so it is the responsibility of the developer not to violates the security rules i.e. if a part or area is highly restricted then there is no need to design that area in 3D for navigation system. Hence, no violation of security rules.

4 Applications

1. Now a day, companies with large campuses are growing very fast. When the employers are newcomers then there is a big problem of navigation or path finding between different departments inside the campus. This type of situation can be avoided by using Animated Map viewer.
2. If the AMV is developed in the form of information-kiosk then they can be installed at complex malls, railway station where user or customer needs self navigation.
3. Sometimes people unfortunately skip the shortest path and take the large one and spend a lot of money and time in long run. Now if you have AMV installed nearby you, you can easily find the other shortest path by selecting a different or nearby source and save your lot of time.

5 Conclusion

AMV saves a lot of time and energy by providing us an accurate, attractive, easy to understand and similar to real world navigation. Sometimes 2D maps are hard to understand for a person as they use symbolic references which are quit confusing. AMVs have 3D of all areas with lot of effects, colors and lots of information about the area. AMVs are also the cost effective system because they need to manufacture only one time and can be used forever.

References

1. Bier, E.A., Stone, M.C., Pier, K., Buxton, W., De Rose, T.D.: Toolglass and magic lenses: The see-through interface. In: SIGGRAPH 1993: Proc. of the 20th Annual Conf. on Computer Graphics and Interactive Techniques, pp. 73–80 (1993)
2. Golledge, R.: Place recognition and wayfinding: Making sense of space. Geoforum 23(2), 199–214 (1992)

3. Schoning, J., Kruger, A., Müller, H.J.: Interaction of mobile camera devices with physical maps. In: Adjunct Proceedings of the 4th Intl. Conference on Pervasive Computing (Pervasive), pp. 121–124 (2006)
4. Siegel, A., White, S.: The development of spatial representations of large-scale environments. In: Reese, H. (ed.) Advances in Child Development and Behavior, vol. 10, pp. 9–55. Academic Press, New York (1975)
5. Urquhart, K., Cartwright, W., Miller, S., Mitchell, K., Quirion, C., Benda, P.: Cartographic representations in location-based services. In: Proc. 21st Intl. Cartographic Conference, pp. 2591–2602 (2003)
6. Montello, D.: A new framework for understanding the acquisition of spatial knowledge in large-scale environments. In: Egenhofer, M., Golledge, R. (eds.) Spatial and Temporal Reasoning in Geographic Information Systems. Spatial Information Systems, pp. 143–154. Oxford Univ. Press (1998)
7. Rekimoto, J.: The magnifying glass approach to augmented reality systems. In: International Conference on Artificial Reality and Tele-Existence/Conference on Virtual Reality Software and Technology (ICAT/VRST), pp. 123–132 (1995)
8. Rohs, M., Essl, G.: Which One is Better? Information Navigation Techniques for Spatially Aware Handheld Displays. In: Proc. of the Eighth International Conference on Multimodal Interfaces, pp. 100–107 (November 2006)
9. Yee, K.-P.: Peephole displays: Pen interaction on spatially aware handheld computers. In: Proceedings of the SIGCHI Conference on Human Factors in Computing Systems, pp. 1–8 (2003)
10. Mehra, S., Werkhoven, P., Worring, M.: Navigating on handheld displays: Dynamic versus static peephole navigation. ACM Trans. Comput.-Hum. Interact. 13(4), 448–457 (2006)
11. Jul, S., Furnas, G.W.: Critical zones in desert fog: aids to multiscale navigation. In: Proc. UIST 1998, pp. 97–106 (1998)
12. Rekimoto, J.: Tilting Operations for Small Screen Interfaces. In: Proc. UIST 1996, pp. 167–168 (1996)

Hyperlinks:

1. http://www.mapnavigation.net/
2. http://lumion3d.com/details/
3. http://quest3d.com/product-info/
4. http://www.opengl.org/documentation/current _version/
5. http://www.garagegames.com/products/torque-3d
6. http://en.wikipedia.org/wiki/Game_engine
7. http://www.blender.org/features-gallery/
8. http://usa.autodesk.com/3ds-max/
9. http://usa.autodesk.com/civil-3d/features/
10. http://www.vlabcivil-iitr.co.in/
11. http://www.instructables.com/id/How-to-Navigate-with-a-Map-and-Compass/
12 .http://help.arcgis.com/en/webapi/javascript/arcgis/index.html
13. http://www.mapmyindia.com/
14. http://www.satguide.in/#
15. http://www.abc-of-hiking.com/navigation-skills/
16. http://www.backpacking-lite.co.uk/map-navigation-skills.htm

A Novel Fuzzy Sensing Model for Sensor Nodes in Wireless Sensor Network

Suman Bhowmik and Chandan Giri

Abstract. To design an efficient Wireless Sensor Network application one need to understand the behavior of the sensor nodes deployed. To understand the behavior of a system we need a very good model, that can represent the system in a more realistic manner. There is a vagueness that we can identify in defining the sensing coverage of any sensor node. The human like reasoning that best suits for this vagueness is the fuzzy reasoning. Here in this paper, we are proposing a fuzzy based model and inference system to best way represent the sensing behavior of sensor nodes. Also, we propose a measure that can be used to check or compare the WSN system performance.

1 Introduction

The wireless sensor network (WSN) has become one of the favorite topic of research in recent years. WSN is designed to sense an environmental phenomenon. The term coverage in WSN has two aspect, one is called *sensing coverage* and the other is *linking coverage or communication coverage*. The former is the total area within the deployment field from where the sensors can detect the physical event occurred for which it is designed, and is defined [8] as the ratio of the sensible area to the entire deployment area. In the ideal situation these area must be equal. The success of WSN is highly dependent on efficient modeling of sensing capability of sensor nodes and good definition of different measures. When analyzing the sensors activity we can easily identify the inherent fuzziness in the sensing capability of

Suman Bhowmik
College of Engg. & Mgmt. Kolaghat, Dept. of CSE, India
e-mail: bhowmik.suman@gmail.com

Chandan Giri
Bengal Engg. & Sc. University, Dept. of IT, India
e-mail: chandangiri@gmail.com

A. Abraham and S.M. Thampi (Eds.): Intelligent Informatics, AISC 182, pp. 365–371.
springerlink.com

sensor nodes. In this paper we have used this fuzziness and proposed a unique fuzzy based system to model the sensing capability, also we have defined some quality of measure depending on this fuzzy model so that the performance of any WSN project can be qualified accordingly. The simplest, and widely used model for sensing and communication coverage is the circular disk model. In this model it is assumed that a sensor can sense or detect any physical event within a circular region of radius r_s around it with 100% certainty beyond which it can not sense any event. Parikh et. al. [4] used a constant value for the radius. Some researchers relates the sensing radius (r_s) with the communication range (r_c) to make certain that both the coverage will be maintained. Zhang et. al. [3] used $r_s >= \frac{r_c}{3}$, Tran-Quang et. al. used $r_s = \frac{r_c}{2}$ [7]. An improvement in [8, 9, 10] is to use extended range r_e beyond r_s, if the target is within r_s range, it will be sensed with probability 1, if the target is between r_s and r_e, it will be sensed with a finite probability p less than 1 but greater than 0, whereas if the target is out side of the range r_e, it is not at all sensed. According to the circular probabilistic model [11, 12, 13] if the target is within the radius r_e, the generated physical event will be sensed with probability p. The value of this probability p decreases with distance from the sensor, and if the target is out of the radius r_e range, it is not sensed. Soreanu et. al. [14] proposed an elliptical sensing area and sensors are capable to increase or decrease the area for variable power demand.

The rest of the paper is organized as follows: Sect. 2, discusses the proposed fuzzy sensing model, the inherent fuzziness, type of membership function used in defining the sensing profile and inference rules to be used. Sect. 3, defines some of the measures that can be taken, based on the proposed fuzzy model. Sect. 4, show the simulation result for some deployment pattern and judge their effectiveness by comparing the values obtained for the measures defined. Finally, Sect. 5 concludes the paper.

2 Proposed Fuzzy Sensing Model

There are two types of coverage problem in WSN. First is the sensing coverage, which deals with the fact that the whole deployment field should be covered by the sensor nodes or in other words every event occurring within the deployment field should be sensed by the sensors. Sensing coverage problem tries to find the optimal number of sensor nodes and the optimal pattern of deployment that are required to sense the whole deployment field. The second is the communication coverage, in our paper we will focus on the former coverage problem. The received signal power by the sensor is always less than the power of the signal emitted from the target. The sensing coverage is dependent on the received signal power. We can say qualitatively that the received signal strength is *"very good"* *"very near"* to the target, it then starts decreasing moving through *"good"* strength *"near"* to the target. When the sensor is at a *"moderate"* distance from target, received signal strength gradually decreases to *"moderate"* level. Then it falls through *"bad"* power level at *"far"* distance and decreases to *"very bad"* power level at a *"very large"* distance from the target. Thus we see that there is a fuzziness in both the received signal strength and the distance of sensor from the target, or in other terms the coverage is a fuzzy phenomena. It means that we can define a membership function for the sensing capability of a sensor node.

2.1 The Membership Function

The Wireless Sensor devices operates in a short range and a great many sensors are based on electromagnetic, acoustic or other types of mechanical wave propagation. The electromagnetic, acoustics and other mechanical signals obey the power law [2, 1] when propagating through free space. The fuzziness profile must be symmetric about the vertical axis drawn through the sensor and must have a gradual roll off to zero at the two sides of the vertical axis beyond a certain amount of flat top. Also the profile has to be normalized to be suitable to use as membership function i.e., we can express the received power as a ratio of actual power and the maximum possible power,. The above discussion shows that the most suitable mathematical function that models the normalized sensing capability of senors very closely is the generalized bell function. If the path loss exponent is $2b$, the target is at a distance c and the reference point is at distance a from target, then the generalized Bell function will look like the Fig. 1. The Bell function (parameters and their significances are shown in Fig. 1) $MF_{1D} = \frac{1}{1+|\frac{x-c}{a}|^{2b}}$ models the sensing capability in one dimension but we need a two dimensional model. Bell function is symmetric on the axis which is perpendicular to x axis and is passing through the point $(c,0)$ (as shown in Fig. 1). Since in both x and y direction we need the same shape, we can revolve the Bell function around the axis perpendicular to the $x-y$ plane at the co-ordinate (c,c). The resulting surface, of equation $MF_{2D} = \frac{1}{1+|\frac{(x-c)^2+(y-c)^2}{a}|^b}$ resembles with our model, which is shown in Fig. 2. One direct advantage of this fuzzy based sensing model is that the circular disk model can be shown as a special case of this model. This can be argued as follows: The 1D Bell function takes the shape of rectangle (red colored curve in Fig. 1) when the parameter $b = \infty$. In that case the width of the rectangle is

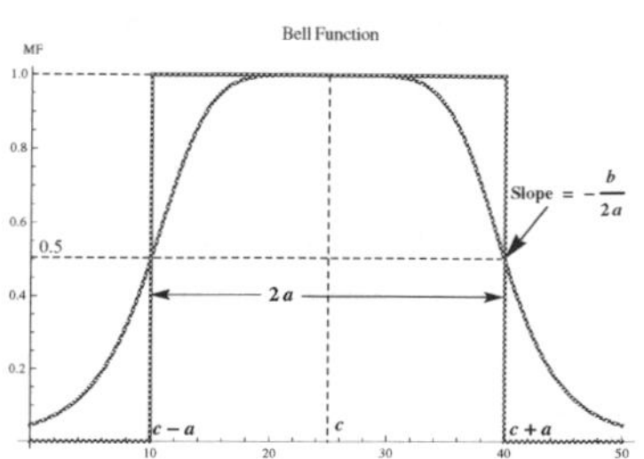

Fig. 1 Generalized Bell function

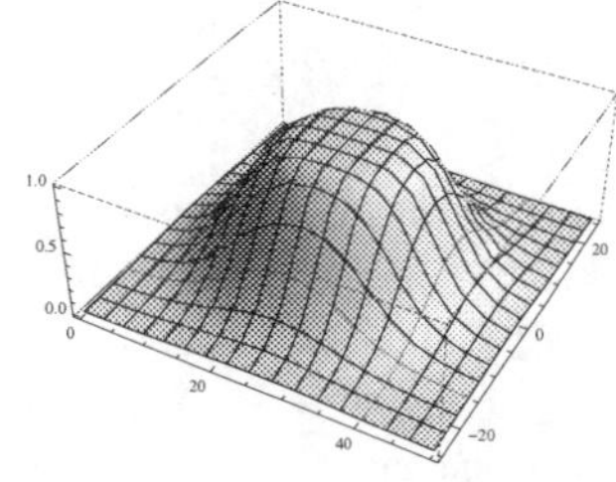

Fig. 2 Generalized Bell function rotated on its axis

$2a$. The corresponding 2D function will be a cylinder of radius $2a$, i.e., the sensing range will be circular disk of radius $2a$.

Theorem 1. *A generalized Bell function of expression* $\frac{1}{1+|\frac{x-c}{a}|^{2b}}$ *becomes a unit rectangular function of width* $2a$ *if* $b \to \infty$.

Proof. The 1D Bell function is $A_\alpha = \frac{1}{1+|\frac{x-c}{a}|^{2b}}$, where A_α is the α-level cut of membership function. The two roots of the equation are $x_1 = c - a(-1 + \frac{1}{A_\alpha})^{\frac{1}{2b}}$ and $x_2 = c + a(-1 + \frac{1}{A_\alpha})^{\frac{1}{2b}}$. So, the width of the Bell function is $Width(A_\alpha) = x_2 - x_1 = 2a(-1 + \frac{1}{A_\alpha})^{\frac{1}{2b}}$. Now $\lim_{b\to\infty} Width(A_\alpha) = \lim_{b\to\infty} 2a(-1 + \frac{1}{A_\alpha})^{\frac{1}{2b}} = 2a$ □

This fuzzy model is extending the sensing range of the sensor nodes, allowing these to accept signals from physical events coming from larger distance (than the discrete circular sensing range assumption) with a degradation membership function attached (roll off of the Bell function). The flat top is represented by the qualifier *"very good"* signal and the whole rolling off is divided by the qualifiers *"good"*, *"moderate"*, *"bad"*, *"very bad"* This extension of sensing range will clearly improve the sensing coverage over the circular disk model (as depicted in Fig. 3).

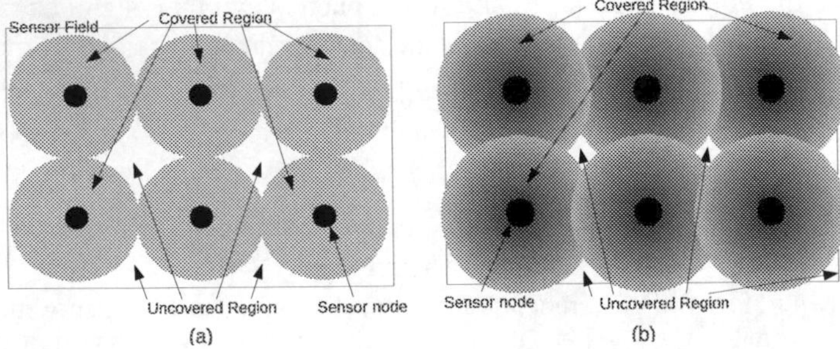

Fig. 3 Improvement of coverage using proposed fuzzy model: (a) coverage with circular model (b) coverage with fuzzy model

2.2 *Inference Rules*

The most important part of any fuzzy system is the inference rules defined, which controls the behavior of the system. In our case we have defined a threshold level (a crisp value) τ that discriminate between 'badly' and 'very badly' covered region, above which the signal from target is recognizable, but below it is non-recognizable. In sect. 3 we also used four other thresholds ($\tau_{vg}, \tau_g, \tau_m, \tau_b$) to identify four subspace for four different quality of coverage. The corresponding inference rules are as follows:

1. If sensor is *very near* to target then the received signal power is *very strong* i.e., *above* the threshold τ_{vg} where the coverage is *very good*.
2. If sensor is *near* to target then the received signal power is *strong* i.e., *above* the threshold τ_g and *below* the threshold τ_{vg}, where the coverage is *good*.
3. If sensor is at *moderate* distance from target then the received signal is at *moderate* power level i.e., *above* the threshold τ_m and *below* the threshold τ_g, where the coverage is *moderate*.
4. If sensor is *far* from target then the received signal power is *weak* i.e., *above* the threshold τ_b or τ and *below* the threshold τ_m, where the coverage is *bad* but still intelligible. This also defines the overall coverage.
5. If sensor is *very far* from target then the received signal power is *very weak* i.e., *below* the threshold τ_b or τ and is not at all intelligible. Here the coverage is *very bad*.

3 Sensing Coverage Measurement Model

There are several measures already defined by several researchers, but those are for existing sensing model. The simplest measure for sensing coverage for a particular deployment pattern is *Percentage of Coverage (PoC)* which is defined as: $PoC = \frac{A_{cov}}{A_{tot}} * 100$, Where A_{cov} is the total amount of area covered by all sensors and A_{tot} is the total area of the deployment field. Let, there are n number of sensor nodes deployed i.e., $S = \{s_1, s_2, \ldots, s_i, \ldots, s_n\}$. If the sensing area for i^{th} sensor (s_i) is A_{s_i} then $A_{cov} = \bigcup_i A_{s_i}$. For a particular situation the sensor can accept a certain level of signal, that may be in the range of very bad, bad or even moderate depending on the environmental conditions. This level of signal can be used as the threshold τ. From Fig. 1, we see that the threshold τ forms α-cut on the membership function. If the threshold τ or α-cut is A_α then from the proof of Theorem 1, we see that width of the membership function for A_α is $Width(A_\alpha) = x_2 - x_1 = 2a(\frac{1}{A_\alpha} - 1)^{\frac{1}{2b}}$, so the sensing radius is half of $Width(A_\alpha)$. Therefore, the radius of sensible region of sensor node s_i is $r_{s_i} = a(\frac{1}{A_\alpha} - 1)^{\frac{1}{2b}}$ (see Fig. 4). If the location of deployed sensor

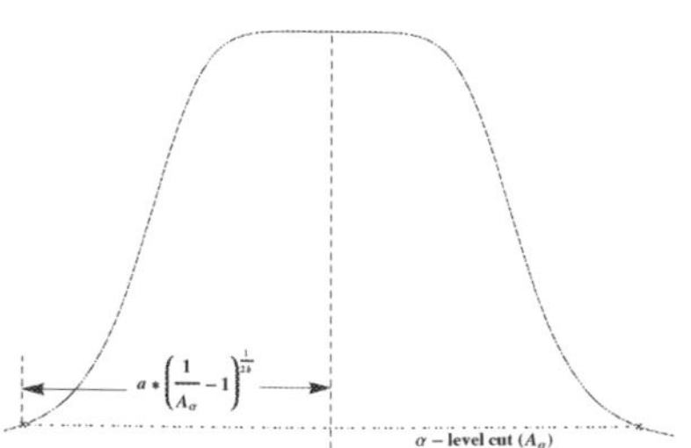

Fig. 4 Radius of the sensing region based on the threshold i.e., α-cut

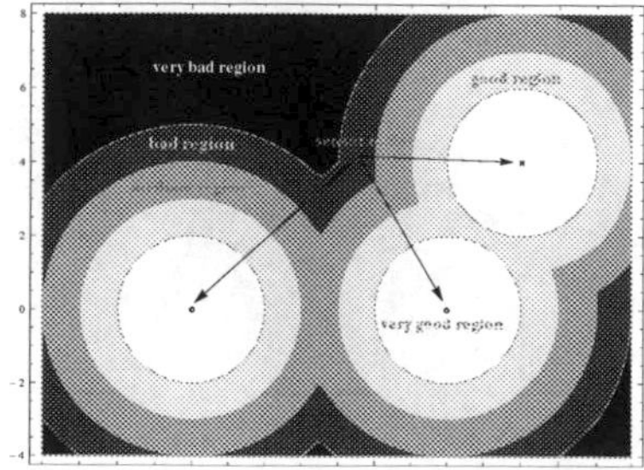

Fig. 5 Proximity of 3 sensors forms the fuzzy sub-spaces

s_i is $\{x_{s_i}, y_{s_i}\}$, then the circular sensible region (bad region in Fig. 5) for s_i is the region under the circular curve defined by the inequality $(x-x_{s_i})^2+(y-y_{s_i})^2 < r_{s_i}^2$ The total region for all sensors will be the space under the curve defined by the inequality $\bigcup_i (x-x_{s_i})^2+(y-y_{s_i})^2 < r_{s_i}^2$ Therefore, the area of the sensible region is the area under the above function, which is: $A_{cov} = \int_l \int_w (\bigcup_i (x-x_{s_i})^2 + (y-y_{s_i})^2 < r_{s_i}^2)\mathrm{d}x\mathrm{d}y = \int_l \int_w (\bigcup_i (x-x_{s_i})^2+(y-y_{s_i})^2 < a^2(\frac{1}{A\alpha}-1)^{\frac{1}{b}})\mathrm{d}x\mathrm{d}y$, where a rectangular deployment field of length l and width w is assumed. So, the percentage of coverage becomes: $PoC = \frac{\int_l \int_w (\bigcup_i (x-x_{s_i})^2+(y-y_{s_i})^2 < a^2(\frac{1}{A\alpha}-1)^{\frac{1}{b}})\mathrm{d}x\mathrm{d}y}{l*w} * 100 = \frac{\int_l \int_w (\bigcup_i (x-x_{s_i})^2+(y-y_{s_i})^2 < a^2(\frac{1}{\tau}-1)^{\frac{1}{b}})\mathrm{d}x\mathrm{d}y}{l*w} * 100.$

4 Simulation Results

For simulation purpose we have chosen a deployment field (rectangular) of size $200m \times 200m$ and deployed randomly a number of sensor nodes. For the deployment we have taken the following crisp value for those threshold such as $\tau_{vg} = 0.7$ (70%), the space where the coverage is very good, $\tau_g = 0.5$ (50%), the space where the coverage is good, $\tau_m = 0.3$ (30%), the space where the coverage is moderate and $\tau_b = \tau = 0.1$ (10%), the space where the coverage is bad. In the figure the gray region identifies "very good" covered subspace, the yellow colored region specifies the "good" covered space, the green colored region identifies the "moderate" covered subspace, blue colored region is "bad" covered subspace and the rest white region shows the "very bad" covered space. Figs. 6 shows the variations in the fuzzy covered regions for 100, 81, 64 and 49 number of nodes deployed with *random* deployment pattern.

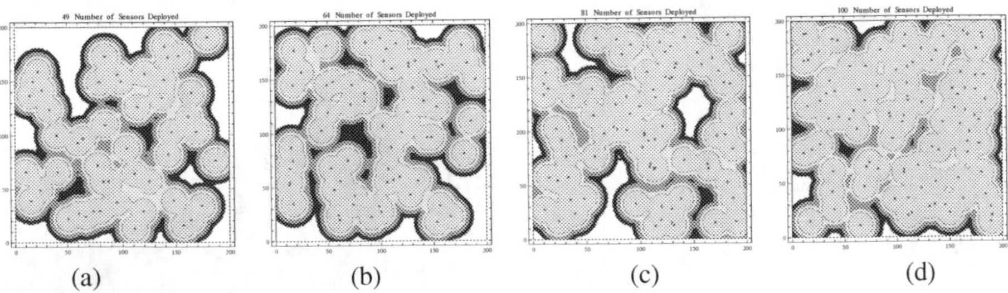

Fig. 6 Random deployment with: (a) 49 sensor (b) 64 sensors (c) 81 sensors and (d) 100 sensors deployed

5 Conclusion

Proper modeling of any system helps to improve the performance of the system and identifying good measure based on a particular model helps to evaluate the performance of the system. In our paper we have tried to address this two aspect of WSN system. We have proposed a new fuzzy based model for the sensing capability of sensor nodes which definitely improves the sensing coverage. We identified the membership function, and defined inference rules. We also defined some useful measure to quantify the system performance based on the proposed fuzzy based sensing model. We also shown the simulation results for some deployment pattern using the proposed model.

References

1. Pratt, T., Bostian, C.W.: Satellite Communications. John Wiley & Sons
2. Rappaport, T.S.: Wireless Communications: Principles and Practice. Prentice Hall (1996)
3. Zhang, H., Nixon, P., Dobson, S.: Partial Coverage in Homological Sensor Networks. In: IEEE Int. Conf. on Wireless and Mob. Computing, Networking and Comm., pp. 42–47 (2009)
4. Parikh, S., Vokkarane, V.M., Xing, L., Kasilingam, D.: Node-Replacement Policies to Maintain Threshold-Coverage in Wireless Sensor Networks. In: 16th Int. Conf. on Comp. Comm. and Net., pp. 760–765 (2007)
5. Song, P., Li, J., Li, K., Sui, L.: Researching on Optimal Distribution of Mobile Nodes in Wireless Sensor Networks being Deployed Randomly. In: Int. Conf. on Comp. Sc. and Info. Tech., pp. 322–326 (2008)
6. Mao, Y., Wang, Z., Liang, Y.: Energy Aware Partial Coverage Protocol in Wireless Sensor Networks. In: Int. Conf. on Wireless Comm., Net. and Mob. Computing, pp. 2535–2538 (2007)
7. Tran-Quang, V., Miyoshi, T.: A novel gossip-based sensing coverage algorithm for dense wireless sensor networks. Comp. Net. 53, 2275–2287 (2009)
8. Liu, T., Li, Z., Xia, X., Luo, S.: Shadowing Effects and Edge Effect on Sensing Coverage for Wireless Sensor Networks. In: 5th Int. Conf. on Wireless Communications, Networking & Mob. Computing, pp. 1–4 (2009)
9. Tsai, Y.R.: Sensing coverage for randomly distributed wireless sensor networks in shadowed environments. IEEE Trans. Veh. Tech. 57, 556–564 (2008)
10. Ghosh, A., Das, S.: Coverage and connectivity issues in wireless sensor networks: A survey. Pervasive Mob. Comp. 4, 303–334 (2008)
11. Wang, Q., Xu, K., Takahara, G., Hassanein, H.: WSN04-1: Deployment for Information Oriented Sensing Coverage in Wireless Sensor Networks. In: IEEE Globecom, pp. 1–5 (2006)
12. Xing, G., Tan, R., Liu, B., Wang, J., Jia, X., Yi, C.W.: Data Fusion Improves the Coverage of Wireless Sensor Networks. In: 15th Annual Int. Conf. on Mob. Computing and Networking, pp. 57–168 (2009)
13. Mao, Y., Zhou, X., Zhu, Y.: An Energy-Aware Coverage Control Protocol for Wireless Sensor Networks. In: Int. Conf. on Information and Automation, pp. 200–205 (2008)
14. Soreanu, P., Volkovich, Z.: Energy-Efficient Circular Sector Sensing Coverage Model for Wireless Sensor Networks. In: 3rd Int. Conf. on Sensor Tech. and Apps., pp. 229–233 (2009)

Retraining Mechanism for On-Line Peer-to-Peer Traffic Classification

Roozbeh Zarei, Alireza Monemi, and Muhammad Nadzir Marsono

Abstract. Peer-to-Peer (P2P) detection using machine learning (ML) classification is affected by its training quality and recency. In this paper, a practical retraining mechanism is proposed to retrain an on-line P2P ML classifier with the changes in network traffic behavior. This mechanism evaluates the accuracy of the on-line P2P ML classifier based on the training datasets containing flows labeled by a heuristic based training dataset generator. The on-line P2P ML classifier is retrained if its accuracy falls below a predefined threshold. The proposed system has been evaluated on traces captured from the Universiti Teknologi Malaysia (UTM) campus network between October and November 2011. The overall results shows that the training dataset generation can generate accurate training dataset by classifying P2P flows with high accuracy (98.47%) and low false positive (1.37%). The on-line P2P ML classifier which is built based on J48 algorithm which has been demonstrated to be capable of self-retraining over time.

Keywords: Peer-to-peer, machine learning, traffic classification, self-retraining.

1 Introduction

In recent decades, peer-to-peer (P2P) applications have become widespread among network users and consume a large proportion of total network bandwidth [2]. From the quality-of-service (QoS) point-of-view, accurate traffic classification can serve as a tool for network resources identification and QoS utilization for different network applications [15]. Several traffic classification based on machine learning

Roozbeh Zarei · Alireza Monemi · Muhammad Nadzir Marsono
Faculty of Electrical Engineering, Universiti Teknologi Malaysia,
81310 Johor Bahru, Malaysia
e-mail: {roozbeh.zarei,monemi}@fkegraduate.utm.my,
nadzir@fke.utm.my

A. Abraham and S.M. Thampi (Eds.): Intelligent Informatics, AISC 182, pp. 373–382.
springerlink.com

(ML) have been proposed to classify the Internet traffic based on traffic's statistical characteristics, e.g. [1, 17]. They are able to identify encrypted traffic and applications that use dynamic ports.

There are some limitations on applying statistical classification in on-line traffic classification. The classification accuracy becomes low over time as the traffic behavior changes [9, 16]. The network itself is dynamic and its traffic parameters may change over time as new trends of applications become dominant [16]. Therefore the classifier generated from a training dataset may be come outdated when used to classify traffic at different network segment and time. The work in [9] tried to address this problem by generating a new ML classifier when the current one becomes outdated. However, the accuracy achieved for the ML classifier was not high (around 88-97%). This is probably due to the inaccurate training dataset generation using the payload-based classification.

A practical retraining mechanism is proposed in this paper to maintain the accuracy of on-line P2P ML classifier above a certain threshold to adapt with the changes in traffic behavior. This allows the retraining of the on-line ML classifier with recent accurate training dataset generated through an off-line, but fast heuristic P2P flow classification. The accuracy of the on-line ML classifier is evaluated from time to time based on the flows which are labeled by the heuristic training dataset generation. The system is evaluated with different accuracy thresholds in order to assess the performance of the system. A synthetic dataset which consists of more than 25,000 flows is used for the evaluation. The results show that the on-line ML classifier can classify P2P traffic flow and maintain its accuracy above a predefined threshold over time by utilizing of the proposed retraining mechanism.

The remainder of this paper is organized as follows. In Section 2, the related works on P2P traffic classification and its limitations are discussed. Section 3 discusses the proposed practical on-line P2P ML classifier retraining mechanism. The experimental results and discussion are in Section 4. Section 5 concludes the research work and point out potential future directions for this work.

2 Related Works

There are several research works done in the field of P2P traffic classification, e.g., [5, 6, 11, 14]. Different approaches have been proposed by researchers to classify P2P traffic including the port-based method [13], payload-based method [14], heuristic method [5, 11], and statistical method [6, 12]. The heuristic method was proposed by Karagiannis et al. [5] to classify P2P flows at the transport layer based on the connection patterns of P2P networks instead of relying on P2P packet payloads. Two main heuristics classes to examine the behavior of two different types of flow pairs were proposed: TCP/UDP IP pairs and the <IP,Port> Pairs. Perényi et al. [11] proposed an improved traffic classification based on a six-step heuristic with a reported 95% accuracy. The heuristic used to detect nodes' port behavior provides the highest accuracy (99.97%) among all heuristics. One of the major issues of this method is the high false positive. John and Tafvelin [4] addressed this issue in their work on heuristic methods in the internet backbone.

Supervised ML techniques, in general, use a set of predefined instances (training dataset) to generate the classifier model. The performance of such classifiers is very sensitive to the quality of the training dataset [10]. Training datasets which are used for statistical classification are commonly generated by using payload-based classification in a controlled environment [9]. Hassan et al. [3] proposes a three-class heuristic technique to generate the training dataset for a P2P statistical classification. Heuristic classification was used to classify traffic as either P2P, non-P2P, or unknown traffic. The problem of this approach is the high false positive, since all flows classified as unknown were discarded from the generated training dataset. The system failed to classify most nonP2P traffic, i.e. around 75% of nonP2P flows were classified as unknown.

The traffic classification based on supervised ML needs to be retrained since the behavior of the traffic changes over time. Tian et al. [16] confirms the dynamic feature of real-world traffic from both overall traffic and application levels. They showed that network traffic has dynamic statistical features due to two reasons. First, the parameters of network traffic change over time. Second, new trends of applications may appear and become more dominant in different internet segments. They propose the Data Stream based Traffic Classification (DSTC) to detect the changes of traffic behavior in order to update the classifier. The authors compared the DSTC with different algorithms such as C4.5 and BayesNet [17]. They show that the proposed method can classify network traffic over time with 95% average accuracy. Over time, the accuracy was reported to be between 81% and 97%.

Oriol Mula-Valls [9] proposed a retraining mechanism for service-based techniques in order to generate new models as the current one becomes outdated. IP-based, service-based, and ML-based classification techniques were utilized in this system to identify 14 different applications. Payload-based classifications (e.g., L7-filter and OpenDPI) were used in preparing both training and test dataset. The retraining mechanism is divided in three phases. First, all flows which labeled by payload-based classification are classified by different classifiers. Second, the accuracy of the system is evaluated based on the set threshold value. Finally, system is updated if the accuracy falls below the set threshold. The results show that the ML-based classification accuracy is between 88%-97%. The low accuracy is probably due to using inaccurate training dataset to build the classifier, which was generated by payload-based classification.

3 Proposed On-Line ML P2P Classification with Retraining

This paper proposes a retraining mechanism to automatically retrain the on-line ML classifier when it becomes outdated. It consists of three phases as shown in Fig. 1. In the first phase, flows are extracted from the sampled incoming packets after being classified as either P2P or nonP2P flows by the training dataset generator.

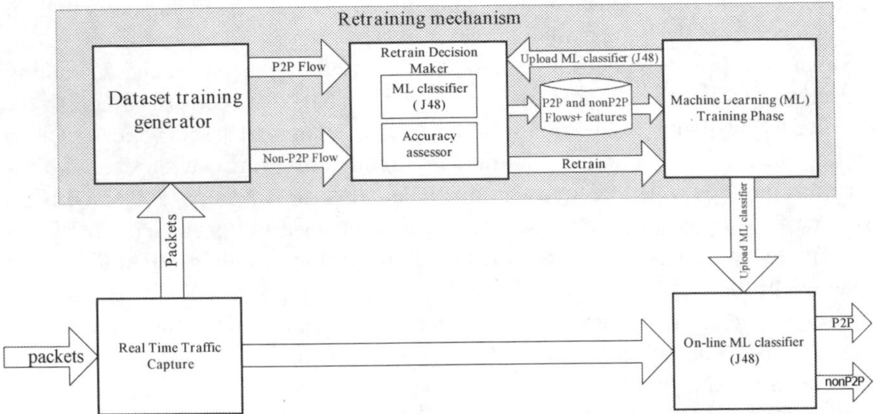

Fig. 1 Structure of Automatic Retraining Mechanism

In the second phase, the accuracy of on-line ML classifier is evaluated against the one generated by the training dataset generator. Finally, the on-line ML classifier is updated if the accuracy of current ML classifier falls below a predefined threshold.

3.1 Practical Training Dataset Generation

A training dataset generator is used for generating training datasets. P2P detection using heuristics has been proven to be able to differentiate P2P traffic from the background traffic. Some of the heuristics proposed in [4, 5, 11] are used to generate training dataset for the statistical ML classifier. This three-class heuristic classification involves several steps to label flows as P2P, nonP2P, and known flows. All heuristic are first applied independently to all flows. Then, flows are classified as P2P flow if they have been labeled by one or more of the P2P heuristics and at the same time not being classified by any of the nonP2P heuristics. On the other hand, flows are classified as nonP2P flows if they have been labeled by one or more of the nonP2P heuristics. The rest of flows are classified as unknown. Training dataset is built based on all P2P and nonP2P flows while the unknowns are discarded.

3.1.1 P2P Heuristics

Three main P2P heuristics types are used to classify P2P pairs. The first type examines the source-destination IP pairs that use both TCP and UDP to transfer data [5, 11]. The second is based on how P2P peers connect to each other by studying the connection characteristics of <IP,Port> pairs [5]. The third one relies on repeatedly use (identical) port numbers on a host within short time periods [11].

Many P2P applications use both TCP and UDP transport layers for communication. UDP connections are often used in P2P signaling phase where this phase involves high number of connections to peers. TCP connections require high overheads for sending queries as compared with UDP connection since it requires handshaking to establish the connection [7]. TCP connections are used to transfer data between hosts after a connection establishment. Six out of nine analyzed P2P protocols in [5] concurrently use both UDP and TCP connections.

The <Port.IP> pair is a type of heuristic to classify P2P applications. In hybrid P2P networks, each P2P client keeps a starting host cache which maintains the IP addresses of other clients or supernodes that connect to existing P2P networks. A new client with random port number uses the IP addresses of a supernode before finding other clients. Many clients start communicating with the new client by using the advertised <Port.IP> pair. All clients pick random source port number to connect to a new client. Therefore, seldom multiple clients choose the same source port number. In other words, for the advertised destination <Port.IP> pair of the client, the number of distinct IPs connected to the client will be equal to the number of distinct ports used [5].

Port usage heuristic was proposed by Perényi [11]. The basis of this heuristic is that the operating system allocates temporary port numbers to source ports at the beginning of the connection for normal applications. It is uncommon for a host to use the same port numbers within short time periods. In the case of P2P connections, both source and destination peers assign a fixed port for signaling traffic or data transfer. If an IP re-uses a source port repeatedly within t seconds, the <Port.IP> pair is classified as P2P and all flows to and from this pair are classified as P2P flows [4]. This P2P heuristics was not used in [3].

Simple port based classification is added in the three-class heuristic classification to classify P2P flows based on well-known P2P ports [5]. Although many P2P applications can dynamically choose arbitrary port numbers, the result in [8] shows that approximately one third of P2P traffic can still be classified by known P2P ports. At the same time, it is improbable for nonP2P applications to utilize known P2P port numbers. This heuristic was not included in [3].

3.1.2 NonP2P Heuristics

Six nonP2P heuristics are included in our training dataset generator to classify nonP2P traffic and to reduce false positive. The first nonP2P heuristic is used in our system to classify web access. A host usually uses multiple connections to a web server to improve the speed for downloading objects. Hence, in the case of web access, the ratio between the number of distinct ports and the number of distinct IPs connected to the web server is higher than the ratio in the case of P2P traffic [5]. To further identify web traffic, another heuristic is added based on the fact that web clients also use parallel connections to webservers. Meanwhile, data transmission between peers consists of one or more consecutive connections. Only a single connection can be active at any given time [11]. This nonP2P heuristics type was not used in [3].

The third nonP2P heuristic is used to identify traditional services such as NetBIOS and DNS based that use the same destination port and source port numbers. Hosts are classified as mail servers by observing of their port usage to connect to other hosts in the same time interval. This is the fourth nonP2P heuristic used in our system. Perényi [11] noted that some common applications can be suitably classified by known destination ports. Web ports are not included since some P2P applications like KaZaA also use these registered ports. The last nonP2P heuristic is utilized to classify attack pairs based on the *Sweep and Scan* in [4]. In the *Sweep* case, attack pairs are classified if one host uses a few different port numbers to connect to a lot of hosts. In the *Scan* case, attack pairs are classified if one host scans a few particular targets on a large number of distinctive ports. The last two nonP2P heuristics were not used in [3].

3.2 Retraining Mechanism

There are some limitations on applying statistical ML techniques for on-line traffic classification. The classification accuracy becomes low over time if the traffic behavior changes over time (i.e., concept drift) [9, 16]. We proposed a retraining mechanism which has the ability to retrain ML classification by creating a new classifier model when the classifier's accuracy falls below a set threshold. This is done by using the generated training dataset to retrain the classifier. It consists of two phases as shown in Fig. 1.

The retrain decision unit is responsible to evaluate the accuracy of the on-line ML classifier. Given an accuracy threshold, the retraining decision unit decides when the on-line ML classifier needs to be updated. This module consists of two parts as shown in Fig. 1. It consists an equivalent software ML classifier model (i.e., J48 in this paper) to the on-line J48 ML classifier (targeted as a hardware-based classifier). The software J48 model uses 25 features as shown in Table 1 to classify all flows which have been classified by the heuristic training dataset generation. The accuracy is calculated as the ratio of the correctly classified flows over the total number of flows. If the accuracy is below the set threshold, retraining process is applied.

The ML traning phase is responsible for generating a new on-line ML classifier. This module retrains the software J48 model and updates the online J48 classifier using the generated dataset by the training dataset generator. The training dataset is generated based on the last n classified flows by the training dataset generation in order to avoid large training time. Our empirical work shows that 2000 flows are sufficient to obtain a good accuracy.

Table 1 Set of features extracted from flow

Features (Flow)
<Src port, Dst Port, Protocol>
Total number of packets (in each direction and total for a flow)
Total number of bytes (in each direction)
Total bytes of payload (in each direction)
Flow duration(in each direction and total for flow)
Average payload per packet(in each direction)
Average length of packet(in each direction)
Average time interval (in each direction)
Number of packets per second(in each direction and total for flow)
Total number of bytes per second(in each direction and total for flow)

4 Results and Discussion

The results of the heuristics based training dataset generation and the retraining mechanism are discussed. Finally, the overall accuracy of the proposed on-line P2P ML classification over time is analyzed.

4.1 Dataset Preparation

Six datasets were captured from the Universiti Teknologi Malaysia academic network. Table 2 shows the captured datasets consisting of PPlive, uTorrent, FrostWire, eMule, Thunder, and BitTorrent P2P applications. We run each P2P application one application at a time to form different datasets. Each dataset contains one P2P application versus other nonP2P traffic. In total, all datasets consist of more than 53,000 flows (27,063 P2P/26,220 nonP2P).

Table 2 Dataset used to evaluate accuracy of ML classifier over time

dataset	P2P program	P2P / non-P2P (flows)
Dataset1	PPlive	4294/4157
Dataset2	uTorrent	4365/4325
Dataset3	FrostWire	4962/4407
Dataset4	eMule	3841/4501
Dataset5	Thunder	5283/4574
Dataset6	BitTorrent	4318/4256

4.2 Traffic Dataset Generation based on the Three-Class Heuristic

We used the heuristic classification in the first stage to generate the training dataset for the second stage of our proposed system. The work in [3] added a new class (unknown) to reduce false negatives. This added heuristics, which were not included in [3], improve the performance of classification in generating the training dataset. Based on our empirical study, our proposed generation of the training data set can process 250,000 packets in less than a minute.

Table 3 The accuracy of the heuristic classification. fp, tp, fn, tn denote false positive, true positive, false negative, and true negative, respectively.

Dataset Generation	$fp(\%)$	$tp(\%)$	$fn(\%)$	$tn(\%)$	Accuracy
Work in [3]	12.61	99.71	0.29	87.39	98.46
Proposed technique	1.37	98.38	1.62	98.63	98.47

Table 3 shows the comparison between our proposed dataset generation technique with the one proposed in [3]. As compared to work in [3] which has high fp (12.61%), our proposed three-class heuristic classification has less fp (1.37%). This is the result of adding better heuristics compared to [3]. Our three-class heuristic classification can classify more nonP2P flows (37% of the whole nonP2P flows) as compare to the result for heuristic classification in [3] which classified only 15% of the whole nonP2P flows. This improvement is result of adding more three nonP2P heuristics to our system.

4.3 Retraining Mechanism Evaluation

We generated two synthetic datasets for training and testing ML classifier by mixing flows of different datasets in Table 2. The first synthetic dataset contains 2,000 flows (800 nonP2P flows and 1,200 P2P flows) is used to generate the on-line ML classifier. Of the 1200 P2P flows, we select equally 600 P2P flows from two Dataset1 (PPLive) and Dataset3 (ForestFire). The second synthetic dataset which consists 25,000 flows is used for testing the on-line ML J48 classifier. This is to show the concept drift condition when ML classifier becomes outdated.

Fig. 2 shows the results of our experiment when the retraining mechanism is applied. We evaluate our system with two accuracy thresholds (i.e. 96% and 97%). The dashed line shows the accuracy of on-line traffic classifier decreases over time from 98.89% to 92.8% when retraining is disabled. By applying the retraining mechanism, the results show that the system is able to maintain the predefined accuracy over time. Setting a high threshold will results in more frequent retraining.

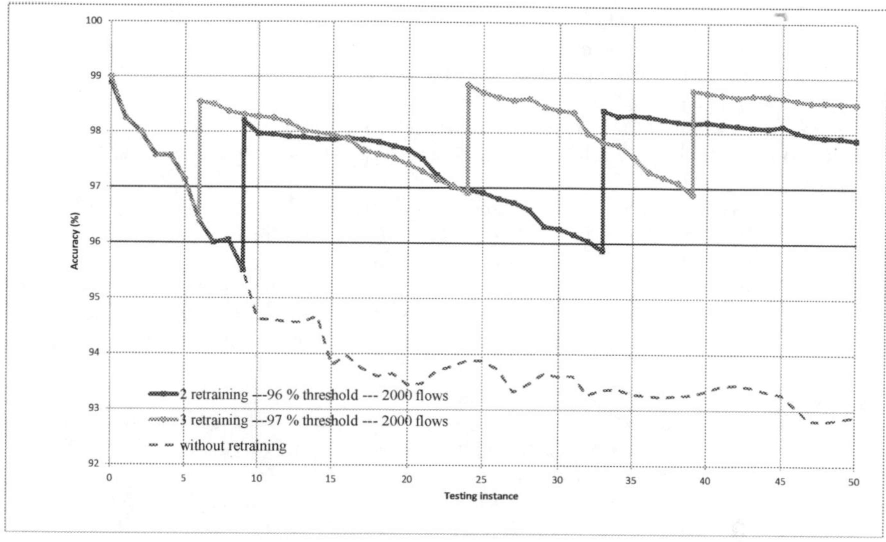

Fig. 2 The effect of Retraining mechanism on Accuracy of on-line ML classifier

5 Conclusion

We improved the training dataset generation based on three-class heuristic to generate the training datasets for training the on-line P2P ML classifier. Since the training dataset can be generated in a short period, frequent retraining of the on-line ML classifier is possible. We showed that the on-line P2P ML classifier suffers from the concept drift problem and low accuracy when frequent retraining is not applied. The on-line ML classifier was built based on the J48 algorithm and training dataset generated by the training dataset generator. The results of using automatic retraining mechanism shows that the online J48 ML classifier is able to maintain its accuracy above a predefined threshold (above 97%) over time.

Acknowledgements. This work was supported the Ministry of Higher Education of Malaysia Fundamental Research Grant, UTM Vote No 78577.

References

[1] Bernaille, L., Teixeira, R., Salamatian, K.: Early application identification. In: Proceedings of the 2006 ACM CoNEXT conference (CoNEXT 2006), Lisboa, Portugal, pp. 6:1–6:12 (2006)

[2] Chen, Z., Yang, B., Chen, Y., Abraham, A., Grosan, C., Peng, L.: Online hybrid traffic classifier for peer-to-peer systems based on network processors. Applied Soft Computing 9(2), 685–694 (2009)

[3] Hassan, M., Marsono, M.: A three-class heuristics technique: Generating training corpus for peer-to-peer traffic classification. In: Proceedings of the 2010 IEEE 4th International Conference on Internet Multimedia Services Architecture and Application (IMSAA 2010), pp. 1–5 (2010)
[4] John, W., Tafvelin, S.: Heuristics to classify internet backbone traffic based on connection patterns. In: ICOIN 2008: 22nd International Conference on Information Networking, pp. 1–5 (2008)
[5] Karagiannis, T., Broido, A., Faloutsos, M., Claffy, K.: Transport layer identification of P2P traffic. In: Proceedings of the 4th ACM SIGCOMM Conference on Internet Measurement, Taormina, Sicily, Italy, pp. 121–134 (2004)
[6] Li, W., Moore, A.W.: A machine learning approach for efficient traffic classification. In: Proceedings of 15th IEEE International Symposium on Modeling, Analysis, and Simulation of Computer and Telecommunication Systems, Washington, DC, USA, pp. 310–317 (2007)
[7] Madhukar, A., Williamson, C.: A longitudinal study of P2P traffic classification. In: Proceedings of the 2007 15th International Symposium on Modeling, Analysis, and Simulation of Computer and Telecommunication Systems (MASCOTS 2007), Washington, DC, USA, pp. 179–188 (2006)
[8] Moore, A.W., Papagiannaki, K.: Toward the Accurate Identification of Network Applications. In: Dovrolis, C. (ed.) PAM 2005. LNCS, vol. 3431, pp. 41–54. Springer, Heidelberg (2005)
[9] Mula-Valls, O.: A practical retraining mechanism for network traffic classification in operational environments. Master thesis, Universitat Politècnica de Catalunya (2011)
[10] Nguyen, T., Armitage, G.: A Survey of Techniques for Internet Traffic Classification using Machine Learning. IEEE Communications Surveys & Tutorials 10(4), 56–76 (2008)
[11] Perényi, M., Dang, T.D., Gefferth, A., Molnár, S.: Identification and analysis of peer-to-peer traffic. Journal of Communication 1(7), 36–46 (2006)
[12] Raahemi, B., Hayajneh, A., Rabinovitch, P.: Peer-to-peer IP traffic classification using decision tree and IP layer attributes. International Journal of Business Data Communications and Networking 3(4), 60–74 (2007)
[13] Sen, S., Wang, J.: Analyzing peer-to-peer traffic across large networks. IEEE/ACM Transaction on Networking 12, 219–232 (2004)
[14] Sen, S., Spatscheck, O., Wang, D.: Accurate, scalable in-network identification of P2P traffic using application signatures. In: Proceedings of the 13th International Conference on World Wide Web, WWW 2004, pp. 512–521. ACM, New York (2004)
[15] Soysal, M., Schmidt, E.G.: Machine learning algorithms for accurate flow-based network traffic classification: Evaluation and comparison. Performance Evaluation 67(6), 451–467 (2010)
[16] Tian, X., Sun, Q., Huang, X., Ma, Y.: A dynamic online traffic classification methodology based on data stream mining. In: 2009 WRI World Congress on Computer Science and Information Engineering, vol. 1, pp. 298–302 (2009)
[17] Williams, N., Zander, S., Armitage, G.: A preliminary performance comparison of five machine learning algorithms for practical IP traffic flow classification. SIGCOMM Computer Communication Review 36(5), 5–16 (2006)

Novel Monitoring Mechanism for Distributed System Software Using Mobile Agents

Rajwinder Singh and Mayank Dave

Abstract. As distributed system software gain complexity owing to increasing user needs, monitoring and adaptations are necessary to keep system fit and running. These distributed applications are difficult to manage due to changing interaction patterns, behaviors and faults resulting from varying conditions in the environment. Also the rapid growth in Internet users and diverse services has highlighted the need for intelligent tools that can assist users and applications in delivering the required quality of services. To address these complexities, we introduce mobile agent based monitoring for supporting the self healing capabilities of such distributed applications. We present the novel mobile agent based monitoring technique where the monitor agents constantly collect and update the global information of the system using antecedence graphs. Updating weights of these graphs further help in evaluating host dependence and failure vulnerability of these hosts. These graphs help monitoring mobile agents to detect undesirable behaviors and also provide support for restoring the system back to normalcy.

1 Introduction

Large scale distributed applications have become increasingly dynamic and complex. New requirements and flexible component utilization call for updates and extensions. In a ubiquitous network society, intelligent home appliances and other devices and systems in our everyday environment would provide services via cooperation with each other. Such complex system naturally becomes prone to faults thereby making constant monitoring necessary. In order to avoid unpredictable behaviors of such service components we present monitoring of

Rajwinder Singh
Department of CSE, Chandigarh Engineering College, Landran, Mohali, India
e-mail: rwsingh@yahoo.com

Mayank Dave
Department of CE, National Institute of Technology, Kurukshetra, India
e-mail: mayank@computer.org

A. Abraham and S.M. Thampi (Eds.): Intelligent Informatics, AISC 182, pp. 383–392.
springerlink.com

system components which is supported by a technology called multi-agent systems (MAS). Mobile agents are autonomous software programs that act on behalf of a user and moves through a network of heterogeneous machines [8]. These mobile agents form a group and monitor the distributed system service components by collecting global system information in form of antecedence graphs [9]. In this paper we propose a configuration that can easily improve the reliability of distributed applications.

The novelty of proposed scheme is that use of antecedence graphs for storing the component information by monitoring mobile agents. Antecedence graph is a directed acyclic graph which contains information of the events occurred before a state interval. When global information needs to be collected, all relevant monitoring mobile agents are identified and informed concurrently which dramatically decreases the time latency of tracing dependencies. Meanwhile, the proposed monitoring technique makes full use of the computation and storage ability of mobile agents and then reduces synchronization message transmission overhead. In this paper, we propose a new monitoring mechanism using antecedence graphs which exploits a new way to design cost-effective agent-driven software monitoring systems. Execution of weight updating also increases the system reliability.

The remaining paper is organized as follows: section 2 consists of discussion of existing work in monitoring methods. The proposed scheme is discussed in section 3 followed by performance analysis in section 4 and conclusions are drawn in section 5.

2 Related Work

Monitoring is a technique for designing systems with better reliability in case of fault or other unexpected events. Several authors have proposed monitoring methods. In [1], authors proposed that monitored values can be abstracted and related to the architectural properties of a model. This is the typical model representation scheme adopted by most ADLs. Authors in [2] proposed the use of probes and gauges report low level monitoring information that can be used to update an abstraction or to trigger alarm. Similarly, in [3], components like Event packager transform raw data collected from probes into smart, event compatible event streams which is collected by Event distillers. This may be supported by an architectural transformation tool that acts in response to the gauges that detect differences between running and actual system architecture. Combs et al. [4], propose external mirroring scheme that navigate probe data into the parallel space where services are executed, and return. Services are requested and translated into workflows that represent the service commitments of service providers. The parallel space can adaptively substitute services which may fail with eligible substitutes. Authors in [5] have created an architectural model which can be represented by plotting a graph of the interacting mechanism with the help of Acme, an Architecture Description Language (ADL). ADLs such as Acme, xADL, Darwin, etc. are being utilized to represent system architectures and facilitate the adaptation and reconfiguration of system components which are a necessary part of a healing system. In [6] authors proposed a method of diagnosis of distributed systems by using a fault model. In this method, performance is optimized locally but not globally, and therefore this method is not applicable to large-scale real-time distributed software operated in an open environment objects are added and deleted. Authors in [7] have used interdependency information in mobile agent based system, which is used for replication calculation. This approach is mainly for calculating number of replicas rather than exhaustive monitoring.

3 Monitoring Using Mobile Agents

In this section, we introduce the proposed scheme of using mobile agents for supporting monitoring capabilities in distributed applications. Further we explain the creation of antecedence graphs for storing the system information both at local as well as global level. There are many benefits [10] derived from applying mobile agents for using them for monitoring service components in distributed applications such autonomous and asynchronous execution , dynamic adaptation and robust and fault-tolerant behavior: Thus multi-agent systems can form the fundamental building blocks for distributed software systems, even if the software systems do not themselves require any agent-like behaviors. When a conventional software system is constructed with agents as its monitoring modules it increases the robustness and reliability of the system.

3.1 Monitoring Architecture

We consider simple scenario of distributed system software where each service host performs some service as per roles assigned. It performs interactions with other hosts in response to query or to complete the assigned task. If some host which may have higher importance (such as initiator of task) fails, then it may disorganize the whole system. Such dynamically changing distributed systems need constant monitoring. We propose distributed monitoring architecture to improve efficiency and robustness of distributed system software by means of employing various mobile agents. The above proposed multi agent monitoring model as shown in figure 1, uses the following mobile agents which improve the response of system to changes in behavior of the system.

- Monitor Agent: The lower-level components of the monitoring component module are Monitor Agents. These agents reside on service hosts and monitor various surrounding environment. These agents gather component information based on the host activity. This information is stored in form of antecedence graphs (discussed in detail in next section). Every state change is put as entry into the antecedence graph. Further in order to calculate dependence of system on service host, antecedence graph weights are also calculated and updated.

- Manager Agent: The manager is a higher-level component which maintains the overall multi-agent architecture. It is responsible for controlling the monitor agents in their monitoring process. Their main capability is to correlate different local events in time from different monitored distributed resources. They also store on the checkpointed antecedence graph. This serves two purposes, firstly vulnerable service hosts are recognized and it also keeps monitor agents light. These are designed as platform independent components and they accumulate and weigh the information gathered by the monitor agents.
- Alert Agent: An alert agent is built of several blocks including time-stamps corresponding to the creation time of the alert message, the detection time of threshold event, fault alert and other the alarm information. It carries out actions at command of Manager Agent.

  This multi-agent monitoring mechanism supports distributed application and responds adaptively to environmental changes. These agents involved in monitoring play the two roles: Supervision and control of service hosts and Generation of global information.

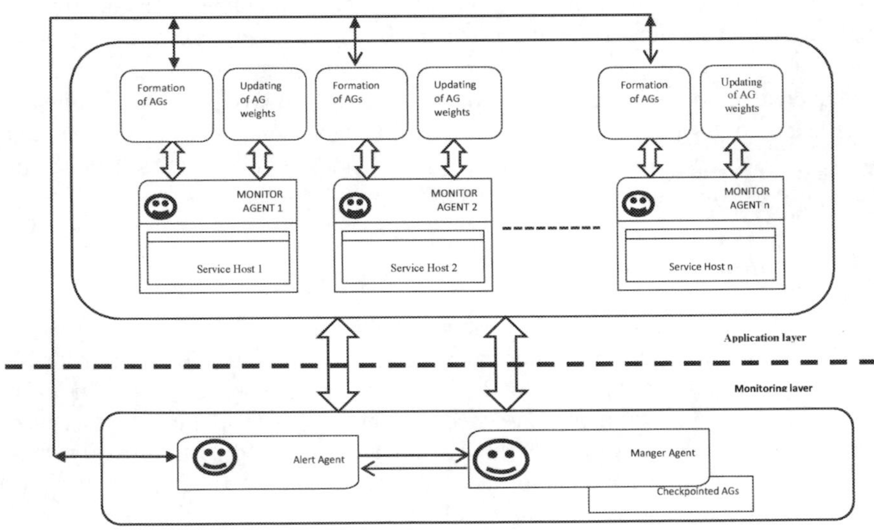

Fig. 1 Mobile Agent Based Monitoring Architecture

3.2 Antecedence Graphs

Antecedence graph is a directed acyclic graph which contains information of the events occurred before a state interval. Considering an example scenario of a distributed system with three service hosts monitored by three monitoring agents residing on three hosts of distributed systems, host A, host B and host C. Monitoring agents monitor the change of state of hosts. The inter host communication can be depicted in form of a graph as shown in figure 2.

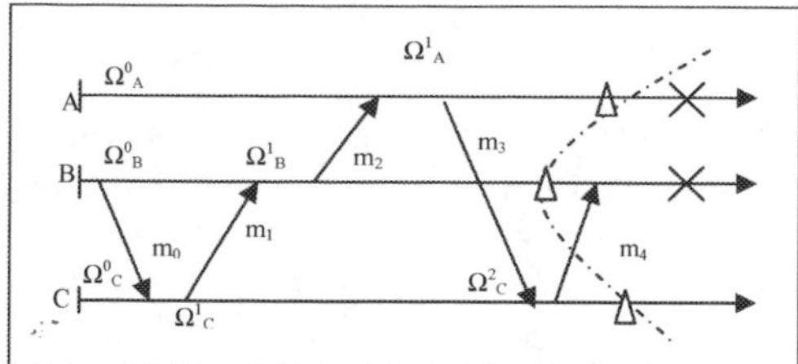

Fig. 2 An example of distributed system with three hosts

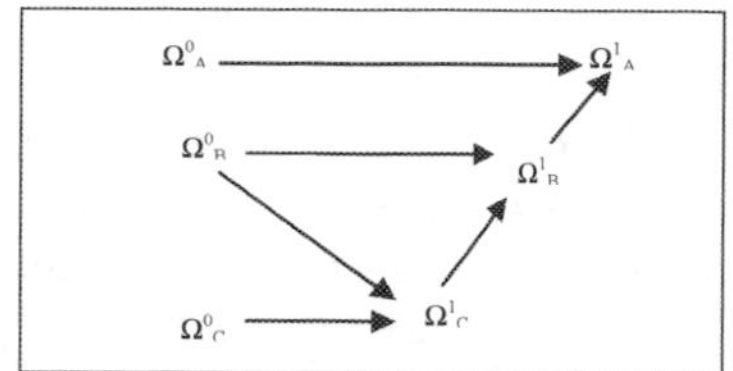

Fig. 3 AG for host A

- AG Formation at host A: The formation of antecedence graph for Host A takes the following steps: Message m_2 is received by Host A from Host B. A combines the antecedence graph received from B to its own graph for the formation of the event $\mathbf{\Omega^1_A}$. The resultant graph is illustrated in figure 3.
- AG Formation for Host B: Initially B sends a message m_0 to C. After this, the message m_1 is received by B from C; with the difference of antecedence graph of C (if a message from C has been previously received). B combines the antecedence graph (AG) received from C to its own graph for the formation of the event $\mathbf{\Omega^1_B}$. The resultant graph is illustrated in figure4.
- AG Formation for Host C: The formation of antecedence graph for host C takes the following steps: Initially C receives a message $\mathbf{m_0}$ from B for the formation of the event $\mathbf{\Omega^1_C}$. After this, the message $\mathbf{m_1}$ is sent to B from C. After this, C receives the message $\mathbf{m_3}$ from A, with the antecedence graph of A. C combines the antecedence graph received from A to its own graph for the formation of the event $\mathbf{\Omega^2_C}$. The resultant graph is illustrated in figure 5.

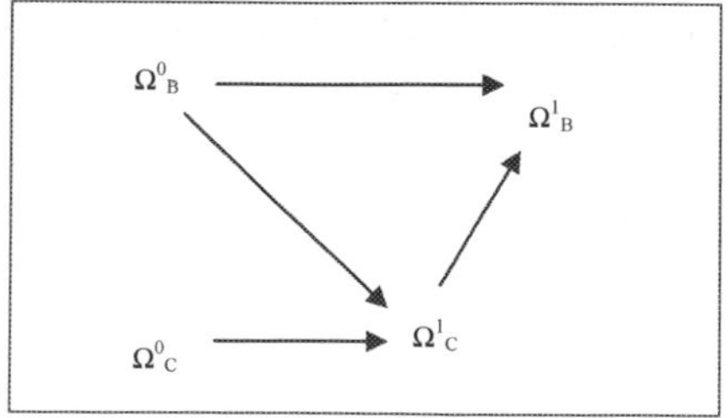

Fig. 4 AG for host B

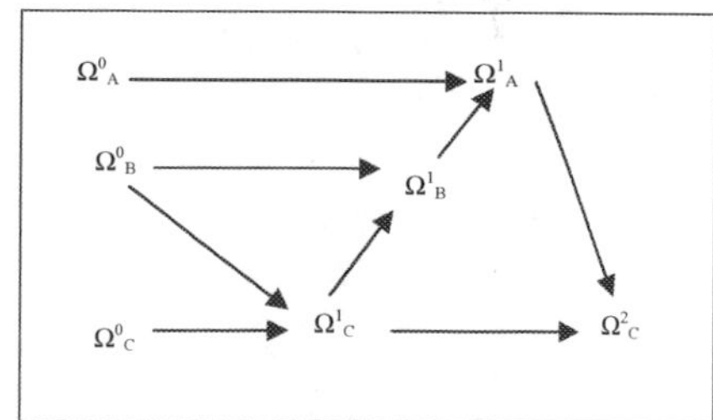

Fig. 5 AG for host C

Monitor Agents store these graphs and use them as catalogue of local information. They observe associated service hosts and update the local information in form of updating AGs). Beside they use this information to compute the dependence (or importance) of their respective service hosts and devise the next appropriate strategy.

3.3 Calculation of Host Dependence

Service hosts may need to communicate with other service hosts from time to time, either for required competencies or resources. These interactions are usually in form of message exchange. The Monitor Agent on each service host is in-charge of storing this information in terms of Antecedence graphs (as discussed above). In order to calculate dependence of other service hosts on their host, mobile agents use several parameters such as executed tasks and number of messages exchanged. This calculation is made by use of weights assigned to AGs. These weights may be useful in detecting which link becomes too heavy or if system is burdening few hosts. In this section we propose use of algorithm to compute the level of dependency on host (greater the dependency, higher is the importance role of host). At passage of Δt time threshold, the Monitor Agent executes the weight updating algorithm. The service hosts continue to act according to their goals and requirements.

3.3.1 Weight Updating Algorithm

Antecedence graph generated by Monitor Agent is a directed label oriented graph with n number of nodes and L set of links. Weights $W_{i,j}$ reflects the importance of the interdependence between the associated hosts ($Host_i$ and $Host_j$).The algorithm [11] assigns initial and updated weights to each link representing the exchange of messages, depending upon the class of messages. As per work in [12], there are six classes of messages which are given in table 1.

These classes have variations from 0 to 1 with value 0 corresponding to no influence and 1 corresponding to high influence. In the weight calculation algorithm we assign initial values to weights using the fuzzy granulation proposed by [13]. An influence is thus described by symbolic values (as shown in table 2) such as *low, medium, high* which correspond respectively to the intervals: [0, 0.35], [0.30, 0.65] and [0.60, 1]. The average value of each symbolic value is the median of its interval. It is used to define the weight of a message.

Table 1 Classes of performative

Class	Type of message
Class 1	request, request-whenever, query-if, query-ref, subscribe
Class 2	inform, inform-done, inform-ref
Class 3	cfp, propose
Class 4	reject-proposal, refuse, cancel
Class 5	accept-proposal, agree
Class 6	not-understood, failure

Table 2 Symbolic value of six message classes

Class	Symbolic values	Intervals
Class 4 , Class 6	Low	0 to 0.35
Class 2, Class 3, Class 5	Medium	0.30 to 0.65
Class 1	High	0.60 to 1.0

The weight of a message is defined by the median of the interval corresponding to the fuzzy value of its performative. The following algorithm is executed at occurrence of threshold time Δt to update the weights associated with links of AGs.

Algorithm: Basic calculations for updating weights

1: Let *threshold1* be weight threshold defined by designer.
2: At each occurrence of Δt do step 2 through
3: Let $NoM_{i,j}(\Delta t)$ be the number of messages sent by $host_i$ to $host_j$ where $i \neq j$ and $i, j <= n$ (where n is number of hosts)
4: Calculate

$$NoM_{max}(\Delta t) = \max_{1<=i,j<=n,\ i \neq j} (NoM_{i,j}(\Delta t))$$

$$NoM_{min}(\Delta t) = \max_{1<=i,j<=n,\ i \neq j} (NoM_{i,j}(\Delta t))$$

$$NoM_{avg}(\Delta t) = \sum_{1<=i,j<=n,\ i \neq j} (NoM_{i,j}(\Delta t)) / n * (n-1)$$

5: Let discount rate $\alpha(t) = 0.5$
6: **for** each $i \neq j$ **do**
7: **if** link between $host_i$ and $host_j$ exists then update

$$W_{i,j}(t+\Delta t) = W_{i,j}(\Delta t) + (NoM_{i,j}(\Delta t) - NoM_{avg}(\Delta t)) / (NoM_{max}(\Delta t) - NoM_{min}(\Delta t)) + \alpha(t)$$

8: end **if**
9: end **for**
10: **if** $W_{i,j}(t+\Delta t) > threshold1$ **then**
11. Checkpoint AGs at *Manager Agent* using *Alert Agents*
12. end **if**

Parameter α is the discount rate that dictates the degree to which existing weights are changed and updated. This parameter can be set high when weights need to be mane more sensitive to environmental changes and can be set low when empirical data needs to be given priority. After execution of above algorithm, the *Monitor Agents* are now equipped with dependence information of their respective hosts. The higher weights signify high dependence of system or other hosts on respective host. This is a measure of sensitivity of the host to failure which may lead to high vulnerability. Failure of hosts with higher vulnerability may cause total disorganization of system which may lead to domino effect on other hosts as well.

In order to prevent total loss of information in case of failure of such service hosts, *Monitor Agent* checkpoints and sends SOS message to *Manager Agent*. *Manager Agent* sends *Alert Agents* to collect AGs checkpointed at various hosts. These AGs from various *Monitor Agents* are then used by *Manager Agent* to collect global information in of form global AGs. In case of failure of its host, *Manager Agent* is informed by *Alert Agent*. *Manager Agent* provides global AGs to *Monitor Agents* of such recovering hosts. At time of recovery, it is necessary to bring hosts to its state prior to failure. These AGs can then be used to bring recovering host back to consistent state.

4 Experiments

In order to study the effectiveness of proposed mobile agent based monitoring technique, we used distributed ecommerce application. In this application, n hosts participate and offer bids for a product. The lowest bid is accepted as successful. Service hosts are designated several hosts in emarket such as Sales Host (offers bids), Purchase Host (calls for and accepts bids) and Manager Host (which manages these auctions). A collaborative multi-agent system consisting of *Manager Agent*, *Alert Agents* and *Manager Agents* is built as monitoring layer. The proposed monitoring system of multiple agents performing in collaboration in a group has been implemented on IBM Aglets[14] over a network of systems with configuration of 1 GB RAM and 3.2 GHz processor connected to 10/100 Mbps Ethernet. We conducted several experiments with varying number of service hosts from 5 to 30 in steps of 3 and 5.

- Execution Time : In first experimental run, we consider execution time as measure of effectiveness of proposed monitoring scheme. We consider execution time for two cases 1) without mobile agent based monitoring and 2) with use of mobile agent based monitoring using regular updating of weights of AGs. Figure 6shows that though execution time is more in case of mobile agent based monitoring yet it remains around constant and doesn't increase much with increase in number of agents. This can be attributed to the fact that in order to calculate the local and global information communications between agents of monitoring is now significantly high. Therefor the communication between agents in second case is optimized.
- Robustness : In second experiment round, we use simulation of faults. We chose host and stop its thread, which leads to failure of the host. The success rate of such failure simulation can be calculated as

$$\text{Success Rate} = (SS / TS) * 100$$

where SS is total number of successful simulations(i.e. the ones leading to failure) and TS is the total number of simulations.

Here we consider two cases 1) when host is chosen randomly 2) when choice of host s based is based on mobile agent based monitoring using weighted AGs. If killed host was of higher importance then its associated hosts may suffer as well

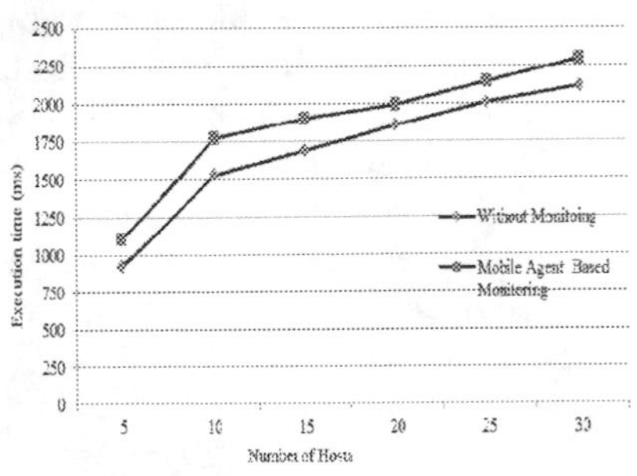

Fig. 6 Comparison of Monitoring Costs

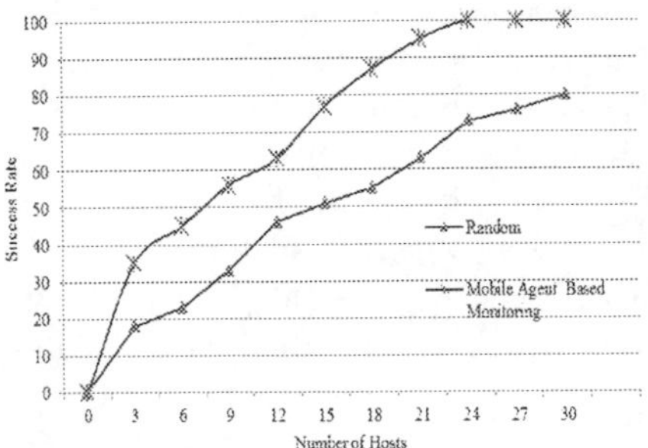

Fig. 7 Rate of succeeded runs for both cases

and may result in disorganization of whole marketplace (defined as failure). Note that in our experiment, we compute the number of failed negotiations instead of successful negotiations. Figure 7 shows that strategy using AGs and mobile agents to find dependence information for monitoring gives better results. This can be explained by the fact that mobile agents collaborate to update weights of AGs; thus improving reliability of application domains where roles of participating agents are quite crucial.

5 Conclusion

Distributed systems are often complex and change dynamically. Therefore constant monitoring is essential for such systems. To make these systems more reliable, we proposed mobile agents based monitoring approach. We introduced use of antecedence graphs which represent the interaction between various participant hosts. Updating of weights from time to time helped in calculating the dependence of hosts which in turn signifies the vulnerability of host to failure. From experimental results it can be safely inferred that he proposed monitoring technique for distributed application using mobile agents may effectively increase system tolerance beside effective recognition of vulnerabilities in system. In the future, we intend to work out a more formal model of the quantity of dependence and incorporate other parameters to gauge the efficiency of the model in accurately measuring host vulnerability.

References

[1] Garlan, D., Schmerl, B.: Model-based adaptation for self-healing systems. In: Proceedings of IEEE Conference on Future of Software Engineering (2007), doi:0-7695-2829-5/07

[2] Cheng, S.W., Garlan, D., Schmerl, B., Steenkiste, P., Hu, N.: Software architecture-based adaptation for grid computing. In: Proceedings of 11th IEEE Conference on High Performance Distributed Computing, Edinburgh, Scotland (2002)

[3] Valetto, G., Kaiser, G.E.: Case study in software adaptation. In: Proceedings of the ACM First Workshop on Self-Healing Systems (2002), doi:1-58113-000-0/00/0000

[4] Combs, N., Vagle, J.: Adaptive mirroring of system of systems architectures. In: Proceedings of ACM First Workshop on Self-Healing Systems, Charleston, SC, USA (2002), doi:1-58113-609-9/02/0011

[5] Dashofy, E.M., Hoek, A.V.D., Taylor, R.N.: Towards Architecture-based Self-Healing Systems. In: Proceedings of ACM First Workshop on Self-Healing Systems, Charleston, SC, USA (2002), doi:1-58113-609-9/02/0011

[6] Sims, M., Goldman, C.V.: Self Organization through Bottomup Coalition Formation. In: Proceedings of AAMAS 2003, Melbourne, Australia (2003), doi:1581136838/03/0007

[7] Tanimoto, A., Matsumoto, K., Mori, N.: An Adaptive and Reliable System Based on Interdependence Between Agents. Electrical Engineering in Japan 164(1) (2008)

[8] Lange, D.B., Mitsuru, O.: Programming and Deploying Java Mobile Agents Aglets. Book on Programming and Deploying Java Mobile Agents Aglets. Addison-Wesley Longman Publishing Co., Inc., Boston (1998) ISBN: 0201325829

[9] Khokhar, M.M., Nadeem, A., Paracha, O.M.: An antecedence graph approach for fault tolerance in a multi-agent system. In: Proceedings of 7th IEEE International Conference on Mobile Data Management (2006)

[10] Jansen, W.: Intrusion detection with mobile agents. Computer Communications 25(15) (2002)

[11] Guessoum, Z., Faci, N., Briot, J.P.: Adaptive Replication of Large-Scale Multi-Agent Systems – Towards a Fault-Tolerant Multi-Agent Platform. ACM Software Engineering Notes (2005), doi:1-59593-116-3/05/05

[12] Colombetti, M., Verdicchio, M.: An analysis of agent speech acts as institutional actions. In: Proceedings of the First International Joint Conference on Autonomous Agents and Multiagent Systems (2002), doi:1-58113-480-0/02/0007

[13] Zadeh, L.A.: A new direction in AI: Toward a computational theory of perceptions. AI Magazine 22(1), 73–84 (2002)

[14] `http://aglets.sourceforge.net/`

Investigation on Context-Aware Service Discovery in Pervasive Computing Environments

S. Sreethar and E. Baburaj

Abstract. The increasing transmission of portable devices with wireless connectivity enables new pervasive scenarios, where users require personalized service access according to their needs, position, and environment conditions (context-aware services). A fundamental requirement for the context-aware service discovery is the dynamic retrieval and interaction with local resources, i.e., resource discovery. The high degree of dynamicity and heterogeneity of mobile environments requires moving around and/or extending traditional discovery solutions to support more intelligent service search and retrieval, personalized to user context conditions. We have reviewed the research of context aware service discovery that based on semantic data representation and technologies; allow flexible matching between user requirements and service capabilities in open and dynamic deployment scenarios.

Keywords: pervasive computing, context aware service discovery.

1 Introduction

Recent advances in the computational capabilities of portable devices, such as cellular phones and palmtops, together with their increased wireless connectivity have favored the emergence of pervasive service provisioning scenarios. Mobile

S. Sreethar
Department of Computer and Information Technology,
Manonmaniam Sundaranar University, Tirunelveli, India
e-mail: sreethar78@yahoo.co.in

E. Baburaj
Department of Computer Science and Engineering,
Sun College of Engineering and Technology-Nagercoil, India
e-mail: alanchybabu@gmail.com

A. Abraham and S.M. Thampi (Eds.): Intelligent Informatics, AISC 182, pp. 393–403.
springerlink.com

users can access needed services from ubiquitous attachment points and even when changing physical locations, e.g., at their workplaces, at home or at publicly accessible places, such as airports and shopping malls. Moreover, users can increasingly benefit from services whose results adapt to the changing context, such as variations in user's position, preferences and requirements, and in locally available resources (context-aware services) [1]. Context is a complex notion that has many definitions. Here, we consider context as any key information characterizing the user, e.g., user preferences, needs, location, and any useful information about the environment where he operates, e.g., date, time, on-going activities and interactions with services and/or other users.

In these context-aware service provisioning scenarios, it is crucial to enable the dynamic retrieval of available services in the nearby of the user's current point of attachment, while minimizing user involvement in service selection, configuration and binding. Service discovery in pervasive environments, however, is a complex task as it requires to face several technical challenges at the state of the art, such as user/device mobility, variations (possibly unpredictable) in service availability and environment conditions, and terminal heterogeneity. Users might need to discover services whose names and specific implementation attributes cannot be known in advance, while service providers need to advertise services to clients whose technical capabilities and conditions at interaction time might be mostly unpredictable beforehand. In addition, service providers cannot exactly define and code at design time all possible configurations of devices accessing the service, e.g., by including any possible discovery protocol and data format.

2 Context-Aware Service Discovery

A context awareness system comprises three main components which we are concerned. To begin with, the system gathers context information available from user interface, pre-specified data or sensor and adds it to a repository. Furthermore, the system converts the gathered raw context information into a meaningful context which can be used. Finally, the system uses context and gives a reaction, and reveals the appropriate context to the user.

2.1 Service Discovery

In the past few years, industry and academia have investigated and proposed various discovery solutions and protocols, such as the Jini discovery architecture [5], the Bluetooth Discovery Protocol [2], the UPnP Simple Service Discovery Protocol [3], and the IETF Service Location Protocol [4]. Other discovery protocols have also emerged within the Web Services research community, such as ebXML [6] and UDDI [7]. All these protocols rely on exact syntactic based matching techniques based on unique identifiers, interfaces, names or XML-based keywords generally defined within fixed, standard taxonomies. However, the exact matching of patterns or keywords might be too restrictive when applied to dynamic and heterogeneous pervasive environments where little or no prior agreement exists on service description and user request format. Service discovery might possibly fail due to the syntactical mismatching between service names. For instance, if a user is looking for a ''News'' service, a service called ''Information'' would not match with the request, although their respective meanings are clearly compatible.

2.2 Context Based Service Discovery

Before introducing the design of context service discovery, let us define the terms to be used in the following sections.

- **Requester:** A device that requests contexts from remote devices for SA (Situation Awareness) application software running on the device.
- **Provider:** A device that provides requested contexts for Requesters
- **Request:** A beacon sent out by a Requester, requesting for contexts.
- **Result:** A beacon sent out by a Provider, containing requested context.
- **Neighbor:** A device in the locality of the network for another device.
- **Sensing Unit (SU):** A software or hardware unit, which can provide one or more contexts

The design of context aware service discovery is based on the following ideas:

Fig. 1(a) and (b) depict the different steps involved in the discovery process in Requester and Provider devices. Detailed descriptions of these steps will be presented below. In Fig. 1 (a), in the Requester device, an SA application registers with CDP (Context Discover protocol) by sending a list of required contexts for situation analysis. Then, CDP lookups the Context Repository (CR), which stores information of all contexts on the device, to see if the requested contexts are available locally. If a requested context C0 is available locally, C0 can be retrieved from the CR. Otherwise, the Requester sends out a Request beacon, advertising its need for C0, and waits for Result beacons. If CDP cannot receive any Result beacon for C0 within a time-out, the application is promptly de-registered.

In Fig. 1 (b), a new SU registers with CDP by sending the name and type of the contexts provided by the SU, the methods of acquiring contexts, and the range of supported frequency of context acquisition. On receiving an SU registration, R-CDP (Request-CDP) updates the CR. These contexts are indexed by their sources SUs, and a time-stamp to indicate when the contexts become invalid and should be removed from the CR. On receiving a Request for C0, the Provider searches the CR for C0. Depending on the availability of C0, a context match may occur, and trigger the sending of a Result beacon containing C0 to the Requester.

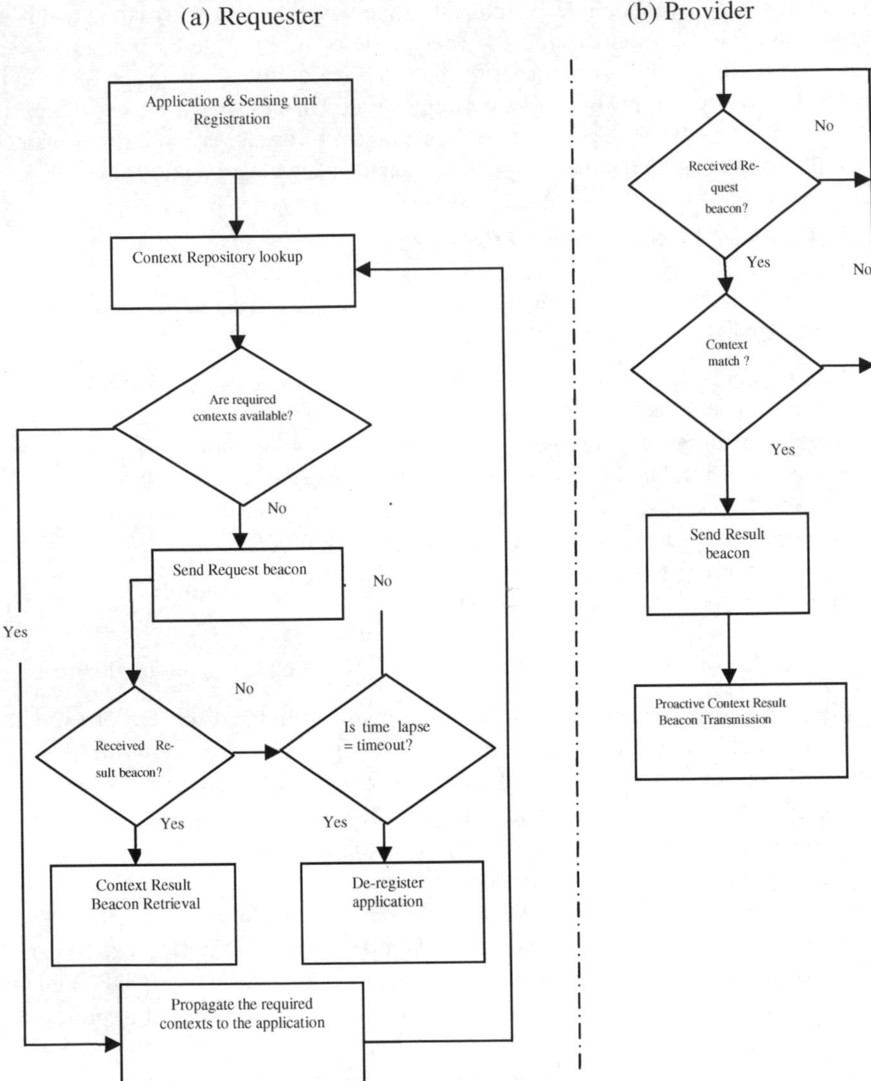

Fig. 1 Context service discovery

3 Literature Review

In this section we describe and explain various context aware service discovery that based on semantic data representation and technologies.

3.1 AIDAS: Adaptable Intelligent Discovery of Context-Aware Services

AIDAS frame work [8] have proposed semantic-based metadata (profiles) to describe properties and characteristics of the entities involved in the discovery process, i.e., services and clients. It associates each client with its context, which defines any client-related information relevant to the discovery process. In case of a user client, context might include personal information about the user, such as spoken languages, as well as technical features of the user's device.

AIDAS metadata model is specifically designed to support personalized user-centric discovery. In particular, the model focuses on the representation of service/user/device capabilities and requirements, whereas other parameters, e.g., service preconditions, post conditions, effects and assumptions taken into account in other solutions [9, 10], are not considered relevant since they mostly serve the purpose of supporting automatic service composition. As a key feature, AIDAS metadata model provides fine-grained profile modularization to improve precision and effectiveness in service selection. The comparison between service request and offer is performed by considering single capabilities and requirements of interest rather than the entire service profile. In addition, device profiles are exploited to refine the searching scope.

Finally, AIDAS allows users to further personalize service discovery by specifying preferences. In particular, users can define in advance a priority order amongst requested service capabilities, thus being relieved from the burden of manually selecting and ordering discovery results. Furthermore, users can specify which requirements to possibly relax in order to extend the set of potentially interesting services.

3.1.1 Management Services

The User Proxy is an application-independent middleware component that represents a portable device on the fixed network [1]. It adopts the general emerging design guideline of dynamically extending the Internet infrastructure with proxies acting possibly disconnected limited devices [1, 11, 12]. Proxies can handle several complex management tasks transparently to client devices by hiding network heterogeneity and managing network dynamicity. AIDAS associates one UP (User Proxy) with each portable device, with a 1-to-1 mapping.

Semantic languages seem to represent a suitable means to realize advanced discovery solutions that overcome the intrinsic limitations of traditional models. In [8] authors have proposed a middleware that exploits semantic techniques to perform context- aware discovery of services based on the requirements and preferences expressed by mobile users.

In [8] authors also aware that in open and dynamic scenarios it is necessary to deal with security issues. Therefore, they are enhancing the AIDAS framework with security features, e.g., authentication and access control features. It also have investigated the problem of resolution of conflicts that may arise between value or priority preferences. Another issue they have analyzed to plan to implement a new version of based on a network service discovery protocol instead of a registry, such as the Bluetooth discovery protocols or a P2P specific discovery protocol.

3.2 Novel Context-Aware Service Directory

In this approach, [13] authors focus on representation of context for web services and we choose a context representation formalism adopted from previous work on context-dependent representations [14]; context is represented by a set of dimensions that take one or more discrete values or range over a specific domain. Combining different dimensions of context that specify the conditions under which a service exists, results in a graph that represents a specific service category within the service directory. A set of such service categories form the service directory. Note that each service category may contain different contextual dimensions, based on the type of services it includes.

In [13] Authors define context in service discovery for mobile computing as the implicit information related both to the requesting user and service provider that can affect the usefulness of the returned results [15]. Two kinds of context are identified, namely service context and user context. Service context can be the location of the service, its version, the provider's identity, the type of the returned results and possibly its cost of use. On the other hand, each user is characterized by a user context that defines her current situation. This includes several parameters like location, time, temporal constraints, as well as device capabilities and user preferences. During service discovery user context is matched against service context in order to retrieve relevant services with respect to context-awareness. Context is very important in mobile service discovery, since it can play the role of a filtering mechanism returning only the subset of retrieved services that conform to the user's current context.

Limitations of this approach is both mobile devices and current wireless communication technology impose careful resource management. Applications running on mobile devices cannot afford to waste power nor consume available bandwidth by transmission of irrelevant data, not to mention unnecessary retransmissions.

In [13] authors have investigated the problem of screen size and, device capabilities impose restrictions on displayed data, for example presentation of high quality images is usually not supported. Returning long to mobile users leads to user annoyance, besides excessive bandwidth consumption. Finally, since timely responses are required for successful mobile applications, it is often necessary to exploit user location in order to provide resources from nearby locations rather than distinct ones. It also suggested focusing mainly in extending the context-aware service directory by means of contextual ontologies, in order to allow serving semantically enriched contextual requests for services, without the restriction of predefined contextual types or values.

3.3 Agent Based Middleware Infrastructure

In particular, components supporting ubiquitous computing can be classified according to their functionality, as illustrated in the following paragraphs.

Middleware architectures boost rapid application development in the scope of complex and heterogeneous network and computing infrastructures. The increasingly important role of middleware components is intensified, when it comes to addressing ubiquitous computing applications and services. Middleware infrastructures for such applications impose a need for balancing between transparency and context-awareness, while at the same time tackling with more sophisticated environments in terms of hardware and software. In addition, a ubiquitous computing application asks for a wide range of runtime services such as context-awareness, sensor streams capturing, transfer and processing, dynamic service discovery and invocation, as well as autonomic capabilities.

In supporting these features, In [16] authors developed an agent based middleware framework, which can ease the implementation of sophisticated context-aware services in appropriately configured in-door environments (called 'smart rooms'). Smart rooms comprise a rich set of video and acoustic sensors, enabling several perceptive interfaces to operate and provide elementary context cues. The introduced agent framework provides functionality for service access control, personalization, context modeling, as well as of dynamic control and management of sensors and actuating devices.

3.4 Context-aware Service Discovery in Multi-Protocol Pervasive Environments

The MSDA middleware [17] has been designed to support service discovery in heterogeneous and multi-protocol pervasive environments. We further built and integrated in MSDA the support for context awareness in order to accomplish more efficient service discovery. The key feature of our approach is that context information is used not only to select the most appropriate service instance, but also to improve the dissemination of service requests across heterogeneous pervasive environments, thus minimizing the resource consumption. As Authors[17] demonstrated by the evaluation, properly managing context information not only produces more accurate results, but also reduces the number of exchanged messages, at the cost of a limited overhead on the whole service discovery time. In future developments of MSDA, we plan to investigate how to support proactive provisioning of context information, how to optimize the exchange of such information among MSDA components (e.g., by aggregating network context information exchanged between MSDA Bridges), and how to guarantee the formulation of correct and non-conflicting context rules. Furthermore, we will investigate how to specialize the support offered by existing context ontologies to accomplish both efficient service selection and discovery request propagation.

3.5 Energy-Efficient Context Discovery Protocol

In [18] authors have presented an Adaptive, Lightweight and Energy-Efficient Context Discovery Protocol for discovering contexts for SA(Situation Awareness) application software in ubicomp environments. It has been implemented and evaluated in RCSM (Reconfigurable Context-Sensitive Middleware) [17-21]. It can achieve energy-efficiency with a small overhead in latency. The authors have analyzed as the numbers of Requesters and Providers vary also protocol can adapt to variations in network density. The design supports device mobility by dynamically discovering and continuously updating context Providers and Requesters. The design of Protocol is lightweight because it does not require the usage of resource-consuming mathematical computation and complex algorithms for collaborations among devices.

The [22] suggested on it includes real-time context delivery, inclusion of mechanisms to authenticate context data, Requesters, and Providers in order to enhance security, and investigation of other effective energy-efficient methods for coordinated context discovery among devices

4 Challenges and Open Research Issues

In order to develop a platform which provides context-aware service discovery in pervasive computing environments for many challenges and open research issues need to be faced. Many of them are common to all approaches but there are many others that arise from the built-in nature that these devices currently present. A list of the most challenging aspects of context-aware service discovery in pervasive computing environments is presented in this section. Table 1 also summarizes the challenges.

4.1 Resource-Limited Platform

Current Pervasive platforms have very limited resources because of the embedded of the devices, their small size, the dynamicity they present and also because of their autonomous nature. Sophisticated protocols are needed to implement complex context aware service discovery functionality in this kind of pervasive environment. The number of devices is expected to grow as time goes by, but not fast enough to satisfy the node demands. All protocols developed for these kinds of platforms need to be designed to be energy efficient.

As a result, memory consumption in the nodes of the networks will be unbalanced, which means, Request nodes near the Provider nodes will receive more packet traffic than the rest. Some techniques need to be developed to balance the network traffic to the extent possible.

4.2 Mobility

Request node and provider node mobility is allowed, the protocols are much more complex. Designing protocols that allow node mobility and effectively provide service discovery requirements is very challenging in pervasive environments. The fact that the location of a node can vary over time makes it hard to route packets under strict QoS requirements. Neighbor discovery techniques become much more complex especially if the node mobility range is not bounded.

4.3 Scalability

Many context-aware service discovery approaches have been presented in the literature but most of them have only been simulated. Simulation is important because it allows general information to be extracted about the behavior of the models and about its suitability in a current situation. However, current situation should be used whenever possible since performance results significantly vary among them. In addition, based on our experience, to obtain good scalability is much more difficult if the tests are done on real platforms than in a simulation. Even though further research will made it possible to all efforts made towards improving scalability.

4.4 Multiple Request

It is usual for some applications to share the same network, each of them having their own requirements. Thus, the platform needs to support different kinds of traffic and should adapt itself to the needs of the applications, which may be at some point being in conflict with each other. In case the application requirements cannot be met, the application should be notified.

Table 1 Summary of Challenges of context-aware service discovery in pervasive environments

Challenges	Descriptions
1. Resource-limited platform	Sophisticated protocols are needed to implement complex context aware service discovery function.
2. Mobility	Designing protocols that allow node mobility and effectively provide service discovery requirements.
3. Scalability	To provide scalability in pervasive environments.
4. Multiple Request	To support different kinds of traffic and should adapt itself to the needs of the applications.
5. Real-time and reliable	The delivery probability can be platforms raised up to acceptable levels.
6. Energy efficiency vs QoS	To adapt the network to dynamically choose the best configuration to provide the QoS requirements specified.

4.5 Real-Time and Reliable Platforms

The distributed nature of this technology makes it impossible to provide 100% reliability in the delivery of messages. However, by means of protocols, the delivery probability can be raised up to acceptable levels. To provide such features, complex protocols need to be used, which in turn have a negative influence on other aspects of the system such as energy efficiency or scalability. Most of the current solutions only provide soft real-time assurance. This is due to the limited real-time and reliable capabilities of the underlaying platform over which the application runs. In order for a protocol to provide hard real-time properties, the operating system of the platform where it runs needs to support it.

4.6 Energy efficiency vs. QoS

It is obvious that the more tasks need to be done, the more time they will consume. In this sense, providing QoS requires additional effort in terms of computation, memory and communication. All these things have an important influence on the energy consumption. It would be interesting to study under what circumstances this extra energy consumption is worth and moreover to adapt the network to dynamically choose the best configuration to provide the QoS requirements specified. More generically, the study of trade-offs is important in the way, that it allows the network to adapt itself to the pervasive environments, ensuring that no extra computation or communication is wasted.

5 Conclusions

Context-aware service discovery have been identified as a promising solution for pervasive environments. In order to do this, many challenges need to be faced and some open issues that hinder that goal are still unresolved. As the numbers of projects devoted to this goal are growing, future development in the field is expected. However, more effort is needed to develop middleware and to integrate existing protocols in a complete system in order to evaluate the feasibility of using this technology in real scenarios. We acknowledge the search for service discovery models that allows applications to declare their needs and also the development of context aware-protocols for pervasive environments as a significant factor that will allow the field to develop.

References

[1] Bellavista, P., Corradi, A., Montanari, R., Stefanelli, C.: Contextaware middleware for resource management in the wireless Internet. IEEE Transactions on Software Engineering (2003)

[2] Bluetooth white paper (2007), `http://www.bluetooth.com/Bluetooth/Learn/Technology/Specifications` (last visited)

[3] UPnP device architecture 1.0. UPnP forum, `http://upnp.org/resources/documents.asp`

[4] Guttman, E., et al.: IETF service location protocol v.2. IETF RFC 2608, `http://ietf.org/rfc/rfc2608.txt`

[5] Arnold, K., Wollrath, A., O'Sullivan, B., Scheifler, R., Waldo, J.: The Jini Specification. Addison-Wesley, Reading (1999)

[6] ebXML Registry Information Model v2.1, http://www.ebxml.org/specs/ebRIM.pdf
[7] The UDDI technical white paper (2007), http://uddi.org/pubs/uddi-techwp.pdf
[8] Toninelli, A., et al.: Semantic-based discovery to support mobile context aware service access. In: Proceedings Computer Communications (2008)
[9] Paolucci, M., Kawamura, T., Payne, T.R., Sycara, K.: Semantic Matching of Web Services Capabilities. In: Horrocks, I., Hendler, J. (eds.) ISWC 2002. LNCS, vol. 2342, pp. 333–347. Springer, Heidelberg (2002)
[10] Roman, D., Keller, U., Lausen, H., de Bruijn, J., Lara, R., Stollberg, M., Polleres, A., Feier, C., Bussler, C., Fensel, D.: Web service modeling ontology. Applied Ontology 1, 77–106 (2005)
[11] Karmouch, A. (ed.): Special section on mobile agents. IEEE Communications Magazine 36 (July 1998)
[12] Eliassen, F., et al.: Next generation middleware: requirements, architecture and prototypes. In: Proceedings of the 7th IEEE Workshop on Future Trends in Distributed Computing Systems (FTDCS 1999), pp. 60–65 (1999)
[13] Doulkeridis, C., et al.: A System Architecture for Context-Aware Service Discovery. Electronic notes in Theoretical Computer Science. Elsevier (2006)
[14] Stavrakas, Y., Gergatsoulis, M.: Multidimensional Semistructured Data: Representing Context-Dependent Information on the Web. In: Pidduck, A.B., Mylopoulos, J., Woo, C.C., Ozsu, M.T. (eds.) CAiSE 2002. LNCS, vol. 2348, p. 183. Springer, Heidelberg (2002)
[15] Doulkeridis, C., Vazirgiannis, M.: Querying and Updating a Context-Aware Service Directory in Mobile Environments. In: Proceedings of the 2004 IEEE/WIC/ACM Web Intelligence Conference (WI 2004), pp. 562–565 (2004)
[16] Soldatos, J., et al.: Agent based middleware infrastructure for autonomous context aware ubiquitous computing Services. Computer Communications (2007)
[17] Yau, S.S., Wang, Y., Karim, F.: Development of Situation-Aware Application Software for Ubiquitous Computing Environments. In: Proc. 26th IEEE Int'l Computer Software and Applications Conf (COMPSAC 2002), pp. 233–238 (August 2002)
[18] Yau, S.S., Karim, F., Wang, Y., Wang, B., Gupta, S.: Reconfigurable Context-Sensitive Middleware for Pervasive Computing. IEEE Pervasive Computing 1(3), 33–40 (2002)
[19] Yau, S.S., Karim, F.: An Energy-efficient Object Discovery Protocol for Context-Sensitive Middleware for Ubiquitous Computing. IEEE Trans. on Parallel and Distributed Systems 14(11), 1074–1084 (2003)
[20] Yau, S.S., Karim, F.: A Context-Sensitive Middleware-based Approach to Dynamically Integrating Mobile Devices into Computational Infrastructures. Jour. Parallel and Distributed Computing 64(2), 301–317 (2004)
[21] Yau, S.S., Karim, F.: An Adaptive Middleware for Context-Sensitive Communications for Real-Time Applications in Ubiquitous Computing Environments. Real-Time Systems 26(1), 29–61 (2004)
[22] Yau, S.S., et al.: An Adaptive, Lightweight and Energy-Efficient Context Discovery Protocol for Ubiquitous Computing Environments. In: 10th IEEE International Workshop on Future Trends of Distributed Computing Systems (2004)

A Fuzzy Logic System for Detecting Ping Pong Effect Attack in IEEE 802.15.4 Low Rate Wireless Personal Area Network

C. Balarengadurai and S. Saraswathi

Abstract. IEEE 802.15.4 is an emerging standard specifically designed for low-rate wireless personal area networks (LR-WPAN) with a focus on enabling the wireless sensor networks. It attempts to provide a low data rate, low power, and low cost wireless networking on the device-level communication. In low rate wireless personal area networks the position of each node changes over time. A network protocol that is able to dynamically update its links in order to maintain strong connectivity is said to be self-reconfiguring. In this paper, we propose a fuzzy logic system for detecting ping pong effect attack in low rate wireless personal area networks design method with self-reconfiguring protocol for power efficiency. The LR-WPAN is self-organized to clusters using an unsupervised clustering method, fuzzy clustering means (FCM). A fuzzy logic system is applied to master/controller selection for each cluster. A self-reconfiguring topology is proposed to manage the mobility and recursively update the network topology. We also modify the mobility management scheme with hysteresis to detect the ping-pong effect attack.

Keywords: FCM, FLS, IEEE 802.15.4, LR-WPAN, Ping-Pong effect.

1 Background and Related Work

IEEE 802 Working Group has developed a set of standards for short range wireless communications commonly referred as wireless personal networks (WPAN)

C. Balarengadurai
Department of Computer Science and Engineering,
Manonmaniam Sundarnar University, Thirunelveli, India
e-mail: balamtech@yahoo.co.in

S. Saraswathi
Department of Information Technology,
Pondicherry Engineering College, Puducherry, India
e-mail: swathi@pec.edu

A. Abraham and S.M. Thampi (Eds.): Intelligent Informatics, AISC 182, pp. 405–416.
springerlink.com

[1]. To address the need for low-power low cost wireless personal area networks, the IEEE New Standards Committee officially sanctioned a new task group in Dec 2003 to begin the development of low-rate WPAN (LR-WPAN) standard, called IEEE 802.15.4. The goal of Task Group 4, as defined in the Project Authorization Request is to provide a standard having ultra-low complexity, cost, and power for low-data rate wireless connectivity among inexpensive fixed, portable, and moving devices [2]. Generally an LR-WPAN network is organized as a star or peer to peer topology depending on the application. A network topology may considerably affect the overall power consumption of the system. Thus, it is important to design a topology for self-organizing LR-WPAN to reduce the power consumption. Conserving battery power is very significant because battery life is not expected to increase significantly in the coming years. In additions, for LR-WPAN, the position of each node changes overtime, the protocol must be able to dynamically update its links in order to maintain strong connectivity. A network protocol that achieves this is said to be "self reconfiguring".

Ping-pong effect attack defined as fast, repeated and undue handovers from a coordinator to the other, caused by wrong decisions. It produces loss of packets, loss of energy consumption, service interruption, reduced terminal performance and increased load in networks [3] [4]. A general mathematical theory for designing a minimum power topology within one cluster for a stationary network. Their approach only considers the immediate locality of a node [5]. A power control loop to control the transmitting and receiving power level in ad-hoc wireless network to reduce energy consumption in MAC layer, which selects the minimum, transmit energy needed to exchange message between any pair of neighboring nodes [6]. A power-efficient gathering in sensor information systems (PEGASIS) method is proposed, but no mobility of sensor nodes is assumed, which is not true for mobile ad hoc networks [7]. Power-aware routing and different metrics are discussed in power-aware routing [8]. A power aware virtual base stations (PA-VBS) protocol was proposed, which select a mobile node from a set of nominees to act as a base station [9]. A new power aware routing protocol was proposed to evenly distribute the power consumption rate of each node and minimize the overall transmission power for each connection request simultaneously [10]. In this paper, we propose a fuzzy logic system for detecting ping pong effect attack in low rate wireless personal area networks design method with self-reconfiguring protocol for power efficiency. The organization of the paper is as follows, power aware topology for LR-WPAN is discussed in section 2. Then, section 3 deals with master/controller selection in LR-WPAN using fuzzy logic systems. Next, self-reconfiguring topology for LR-WPAN in section 4. Finally, we conclude in section 5.

2 Power Aware Topology for LR-WPAN

2.1 Power Consumption Model and Cost Function

Three models are often used for wireless communications: path loss, large-scale variations, and small-scale variations [11]. Similar to minimum power energy for wireless sensor network[12], we concentrate only on path loss that has distance dependence which is well modelled by $1/d^p$, where d denotes the distance between the transmitter and receiver antennas, and the exponent p is determined by the field measurements for the particular system at hand [11], for example, $p = 2$ for free space, $p = 1.6 - 1.8$ for in building line-of-sight, and $p = 4 - 6$ for obstructed in building. Suppose there are c clusters in the LRWPAN, and m_i is the nodes in the ith cluster, we use the following cost function to minimize the power consumption.

$$j \triangleq \sum_{i=1}^{c} \sum_{k=1}^{m_i} (d_{ik})^p \tag{1}$$

Where d_{ik} is the degree of membership of x_k, p is a constant for a fixed environment, and

$$d_{ik} = \| x_k - v_i \| \tag{2}$$

Where $\| . \|$ is the Euclidean distance between one node ($\boldsymbol{x}_k$) and its cluster center ($\boldsymbol{v}_i$), where $\boldsymbol{x}_k$ and $\boldsymbol{v}_i$ can be 2-D or 3-D geography information. We partition the network to clusters via minimizing the total power consumption using an unsupervised fuzzy c-means (FCM).

2.2 Network Partition Using an Unsupervised Clustering in FCM

FCM clustering is a data clustering technique where each data point belongs to a cluster to a degree specified by a membership grade. This technique was originally introduced by Bezdek [13] as an improvement on earlier clustering methods. Here we apply FCM clustering to LRWPAN partition. Our objective is to partition n nodes to c clusters which will consume minimum power.

Definition 1 (Fuzzy c-Partition for LRWPAN)**.** Let $\mathbf{X} = x_1, x_2, \cdots, x_n$ be n nodes, $\mathbf{V}_{cn}$ be the set of real $c \times n$ matrices, where $2 \le c < n$. The Fuzzy c-partition space for $\mathbf{X}$ is the set

$$M_{fc} = U \in V_{cn} \mid u_{ik} \in [0,1] \forall i,k; \tag{3}$$

where

$$\sum_{i=1}^{c} u_{ik} = 1 \forall k \quad \text{and} \quad 0 < \sum_{k=1}^{n} u_{ik} < n \forall i$$

The row i of matrix $U \in M_{fc}$ contains values of the ith membership function, u_i, in the fuzzy c-partition U of $\mathbf{X}$.

Definition 2 (Fuzzy C Means Functionals)**.** we modify equation (1) to

$$J(U,v) = \sum_{i=1}^{c} \sum_{k=1}^{n} (u_{ik})^2 (d_{ik})^p \tag{4}$$

Where $u \in M_{fc}$ is a fuzzy c- partition of X; (v=v_1, $v2$,...v_c) where v_i is the cluster center of prototype u_i, $1 \leq i \leq c$; and, u_{ik} is the membership of x_k in fuzzy cluster u_i. J (U, v) represents the distance from any given data point to a cluster weighted by that point's membership grade. The solution of

$$U_{\in M_{fc},v}^{\min} J(U,v) \tag{5}$$

are least-squared error stationary points of J. The fuzzy clustering algorithm is obtained using the necessary conditions for solutions of equation (5), as summarized in the following:

Theorem 1: In equation (2) Assume || . || to be an inner product induced norm, let X have at least c < n distinct points, and define the sets $(\forall k)$

$$I_k = \{i \mid\leq i \leq c; d_{ik} = \| x_k - v_i \| = 0\} \tag{6}$$

$$\tilde{I}_k = \{1,2,...,c\} - I_k \tag{7}$$

Then (U, v) is globally minimal for J only if (ϕ denotes an empty sets)

$$I_k = \phi \Rightarrow u_{ik} = 1/\left[\sum_{j=1}^{c} (\frac{d_{ik}}{d_{jk}})^p\right] \tag{8}$$

or

$$I_k \neq \phi \Rightarrow u_{ik} = 0 \forall i \in \tilde{I}_k \; and \sum_{i \in I_k} u_{ik} = 0 \tag{9}$$

and

$$v_i = \sum_{k=1}^{n} (u_{ik})^2 x_k / \sum_{k=1}^{n} (u_{ik})^2 \forall i \tag{10}$$

The following iterative method is used to minimize J(U, v):

1. Initialize $U^{(0)} \in M_{fc}$ (e.g., choose its elements randomly from the values between 0 and 1). Then at step l ($l = 1, 2...$)
2. Calculate the c fuzzy cluster centers $v_i^{(l)}$ using (10) and $U^{(l)}$
3. Update $U^{(l)}$ using (8) or (9).
4. Compare $U^{(l)}$ to $U^{(l-1)}$ using a convenient matrix norm, i.e., if $\| U^{(l)} - U^{(l-1)} \| \leq \varepsilon_L$ stop; otherwise, return to step 2.
5. Each node has c membership degrees with respect to the c clusters. Determine which cluster this node belongs to based on the maximum membership. By

this means, every node is classified to one cluster and the network is partitioned to c clusters.

The master/controller for each cluster can be selected based on the centroid (center) of each cluster v_i ($i = 1, 2, \cdots, c$), and the remaining power of each node (a master/controller needs more power than a regular node). An ideal master/controller should be very close to the cluster centroid and has very high remaining battery capacity. But generally both conditions are not satisfied at the same time. To compromise this, we apply a fuzzy logic system to master/controller selection.

3 Master/Controller Selection in LR-WPAN Using Fuzzy Logic Systems

3.1 Overview of Fuzzy Logic System

Fuzzy logic uses fuzzy set theory, in which a variable is a member of one or more sets, with a specified degree of membership. Fuzzy logic allow us to emulate the human reasoning process in computers, quantify imprecise information, make decision based on vague and incomplete data, yet by applying a "defuzzification" process, arrive at definite conclusions. The block diagram representation of a fuzzy logic system (FLS) is shown Fig.1.

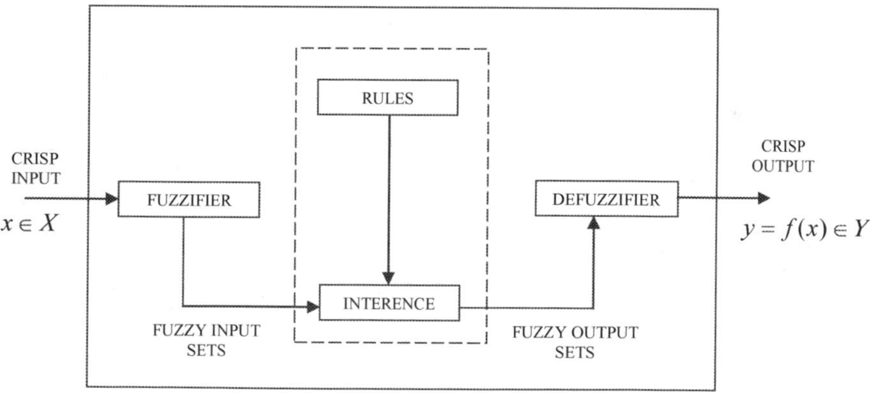

Fig. 1 Fuzzy Logic System

The FLC mainly consists of three blocks fuzzifier, inference and defuzzifier. The details of the above processes are given below.

Fuzzifier

The fuzzy logic controller requires that each input/output variable which define the control surface be expressed in fuzzy set notations using linguistic levels. The linguistic values of each input and output variables divide its universe of discourse into adjacent intervals to form the membership functions. The member value denotes the extent to which a variable belong to a particular level. The process of converting input/output variable to linguistic levels is termed as fuzzifier.

Inference
When an input is applied to a FLS, the inference engine computes the output set corresponding to each rule.The behaviour of the control surface which relates the input and output variables of the system is governed by a set of rules. A typical rule would be if x is A then y is B, when a set of input variables are read each of the rule that has any degree of truth in its premise is fired and contributes to the forming of the control surface by approximately modifying it. When all the rules are fired, the resulting control surface is expressed as a fuzzy set to represent the constraints output. This process is termed as inference.

Defuzzifier
Defuzzification is the process of conversion of fuzzy quantity into crisp quantity. The defuzzifier then computes a crisp output from these rule output sets. Consider a *p*-input 1- output FLS, using singleton fuzzification, center-of-sets defuzzification [14] and *"IF-THEN"* rules of the form

$$R^l: IF\ x_1\ is\ F_1^l\ and\ x_2\ is\ F_2^l\ and...and\ x_p\ is\ F_p^l, THEN\ y\ is\ G^l$$

Assuming singleton fuzzification, when an input $\{ x' = x'_1, ...x'_p \}$ is applied, the degree of firing corresponding to the lth rule is computed as

$$\mu_{F_1^l}(x'_1) * \mu_{F_2^l}(x'_2) * .... * \mu_{F_p^l}(x'_p) = \tau_{i=1}^p \mu_{F_i^l}(x'_i) \tag{11}$$

where * and τ both indicate the chosen *t*-norm. There are many kinds of defuzzifiers. In this paper, we focus, for illustrative purposes, on the center-of-sets defuzzifier [16]. It computes a crisp output for the FLS by first computing the centroid, C_{G^l} of every consequent set G^l, and then computing a weighted average of this centroid. The weight corresponding to the lth rule consequent centroid is the degree of firing associated with the lth rule, $\tau_{i=1}^p \mu_{F_i^l}(x'_i)$, so that

$$y_{cos}(x') = \frac{\sum_{l=1}^{M} c_{G^l} \tau_{i=1}^p \mu_{F_i^l}(x'_i)}{\sum_{l=1}^{M} \tau_{i=1}^p \mu_{F_i^l}(x'_i)} \tag{12}$$

where *M* is the number of rules in the FLS.

3.2 Fuzzy System in LR-WPAN

There are varied applications of intelligent techniques in wireless networks [15]. An energy efficient cluster formation for WSNs using subtractive and fuzzy c-mean clustering approach [16]. The increase in the growth of wireless application demands for the wireless network to have the capability to trace the locations of node user. Location updating scheme using fuzzy logic controls have been proposed in [17] that adaptively adjusts size of the location area for each user. Different approaches in improving the reliability and accuracy of measurement information from the sensor networks [18]. It offers a way of integrating sensor measurement results with association information, available or a priori, derived at aggregating nodes by using some optimization algorithm. They have considered both neuro-fuzzy and probabilistic models for sensor results and association information. The models carry out classification of the information sources, available in sensor systems and wireless networks. We have chosen a fuzzy logic in the base station for electing the cluster-heads. Several reasons support our use of fuzzy logic system in LRWPAN:

1. Representing the problem in mathematical (or probabilistic) model domain involves dealing with several variables and parameters at a time. Moreover these variables are to be defined separately for each scenario, in order to provide a collective output on the basis of the multiple input variables. Problem arises as the number of these variables increases. The mathematical model becomes too complex to handle so many parameters at a time, limited by the effective combination of different parameters together. Fuzzy logic systems on the other hand have got an inherent ability to integrate numeric ('fuzzy') and symbolic ('logic') aspects of reasoning. Therefore different parameters like concentration, energy, and centrality can be combined easily to give the desired result by defuzzifying the output fuzzy set.
2. Fuzzy logic is capable of making real-time decisions, even with incomplete information. Conventional control systems rely on an accurate representation of the environment, which generally does not exist in reality. Fuzzy logic systems, which can manipulate the linguistic rules in a natural way, are hence suitable in this respect. In addition, it can be used for context by blending different parameters - rules combined together to produce the suitable result.
3. Fuzzy logic offers a full range of operators to combine uncertain information in a better way than any other systems. Fuzzy logic control techniques can be used to design individual behaviour units. Fuzzy controllers incorporate heuristic control knowledge in the form of if-then rules. They have also demonstrated a good degree of robustness in face of large variability and uncertainty in the parameters. In fact considering only one parameter like energy is not suitable to select the cluster-head properly. This is because other conditions like centrality of the nodes with respect to the entire cluster, too gives a measure of the energy dissipation during transmission for all nodes. The more central the node is to a cluster the more is the energy efficiency for other nodes to transmit through that selected node. The concentration of the nodes in a given region too affects in some way for the cluster-head selection. It is more feasible to select a cluster-head in a region, where the node

concentration is high. In this paper, we design a FLS for master/controller selection initiation to detect the ping-pong effect.

3.3 Master/Controller Selection

The master/controller is selected based on two descriptors: distance of a node to the cluster centroid, and its remaining battery capacity. The linguistic variables used to represent the distance of a node to the cluster centroid were divided into three levels: near, moderate, and far; and those to represent its remaining battery capacity were divided into three levels: low, moderate, and high. The consequent – the possibility that this node will be selected as a cluster head – was divided into 5 levels, very strong, strong, medium, weak and very weak. We used trapezoidal membership functions (MFs) to represent near, low, far, high, very strong, and very weak; and triangle MFs to represent moderate, strong, medium, and weak. Fig. 2(a) &2(b) shows for MF_s for antecedents and MF_s for consequent.

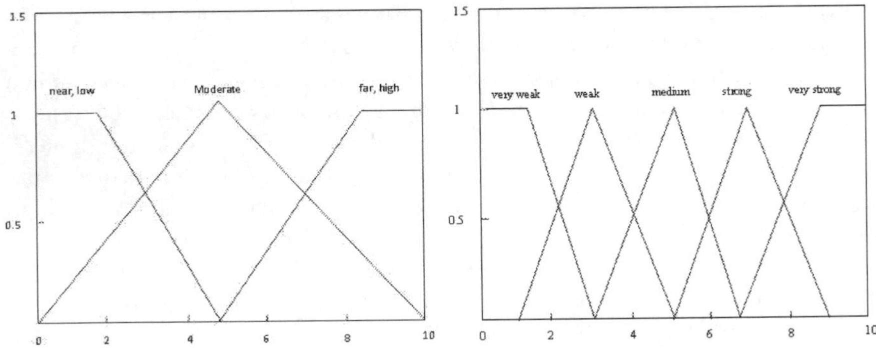

Fig. 2 The MFs used to represent the linguistic labels. (a) MFs for antecedents and (b) MFs for consequent.

Based on the fact that a master/controller should be very close to the cluster centroid and should have very high remaining battery capacity, we design a fuzzy logic system using rules such as R^l, IF distance of a node to the cluster centroid (x_1) is F_l^1, and its remaining battery capacity (x_2) is F_l^2, THEN the possibility that this node will be selected as a master/controller (y) is G^l, where $l = 1, \ldots, 9$ nodes. The rules for power aware packet routing control are Antecedent 1 is distance of a node to the cluster centroid, Antecedent 2 is its remaining battery capacity, and Consequent is the possibility that this node will be elected as a clusterhead. We summarize all the rules in Table 1. For every input (x_1, x_2), the output is computed using the below equation (13),

$$y(x_1,x_2)=\frac{\sum_{l=1}^{9}\mu_{F_l^1}(x_1)\mu_{F_l^2}(x_2)c_{avg}^l}{\sum_{l=1}^{9}\mu_{F_l^1}(x_1)\mu_{F_l^2}(x_2)} \tag{13}$$

As an example, we randomly generate 80 nodes (a cluster) within a square with 10 meters on each side. Each node has random battery capacity in [0, 10]. The distances of each node to the cluster centroid are normalized to [0, 10] scale. Each node is characterized by the two descriptors. We apply equation (13) to compute the selection possibility for each node, and pick the node having the highest selection possibility as the master/controller, as illustrated in Fig. 3. We also plotted the node having the maximum battery capacity and the node having the nearest distance to the cluster centroid in Fig. 3 shows the selected master/controller selection (distance: 3.75, battery capacity: 9.4265) which is denoted using "square box", the node with the maximum battery capacity (distance: 2.53, battery capacity: 9.9136) which is denoted using "oval box", and the node with the nearest distance (distance: 2.72, battery capacity: 1.4917) to the centroid is denoted using +.

Table 1 Fuzzy rules for power aware packet routing control

Rule #	Antecedent 1	Antecedent 2	Consequent
1.	near	low	medium
2.	near	moderate	strong
3.	near	high	very strong
4.	moderate	low	weak
5.	moderate	moderate	medium
6.	moderate	high	strong
7.	far	low	very weak
8.	far	moderate	weak
9.	far	high	medium

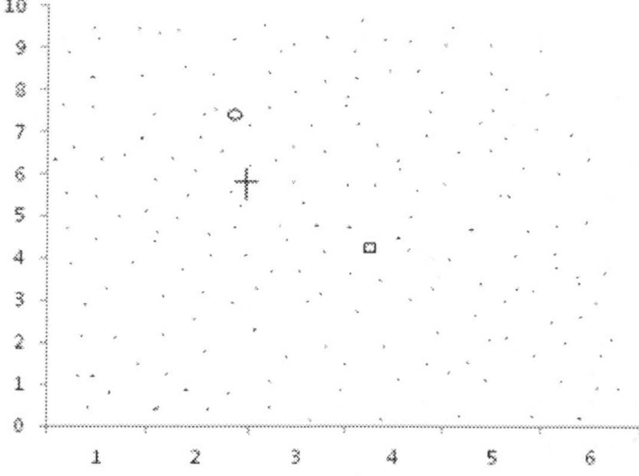

Fig. 3 Master/Controller Selection

4 Self-reconfiguring Topology for LR-WPAN

4.1 Mobility Management Scheme

A Network protocol that can update its links to maintain strong connectivity with the nodes is said to be "self reconfiguring". There exist different mobility patterns in a LRWPAN.

1. Nodes are moving in different directions with different speeds.
2. Some nodes die out while others are mobile.
3. New nodes join in while others are mobile.
4. Some nodes die out and some new nodes join in while other nodes are mobile.

In case 1, the total number of nodes doesn't change and in cases 2-4, the number of nodes may change. Without loss of generality, we assume that the number of nodes and their locations may change from time to time. We dynamically and recursively update the partition of clusters based on the assumption that the number of clusters is constant. This approach is possible because our approach is an iterative optimization method. We summarize the procedures for updating the connectivity among nodes:

1. Collect the status of each node including its geography information and its remaining battery capacity.
2. For every new node, randomly choose its membership degree to each cluster u^i and $\sum_{i=1}^{c} u^i = 1$. If a node dies out or leaves the network, delete its membership.
3. Update the total number of nodes n. Keep the existing c cluster centers $v_i^{(l)}$ as the initial values for the next iteration.
4. Calculate the c fuzzy cluster centers $v_i^{(l)}$ using (10)and $u^{(l)}$
5. Update $u^{(l)}$ using (8) or (9).
6. Compare $u^{(l)}$ to $u^{(l-1)}$ using a convenient matrix norm, i.e., if $\| u^{(l)} - u^{(l-1)} \|) \leq \in_L$ stop; otherwise, return to step 4.
7. Each node has c membership degrees with respect to the c clusters. Determine which cluster this node belongs to based on the maximum membership. By this means, every node is classified to one cluster and the network is partitioned to c clusters.
8. Select the master/controller for each cluster based on the scheme presented in Section 3.
9. Setup the star topology based on the partitioned clusters and selected master/controller for each cluster.

The above procedure can be used by a network periodically for every short period of time since every node is mobile and its remaining battery capacity is time varying.

4.2 *Mobility Management with Hysteresis*

In the network partition update, a node will be switched to another cluster if the membership degree to its current cluster is less than the membership degree to another cluster. Similarly, a master/controller will be switched if the selection possibility for the current master/controller is lower than one node in its cluster because of mobility and remaining battery capacity. Both schemes will have ping-pong effect, the repeated switch between two clusters caused by the rapid mobility. Motivated by the handoff scheme in cellular networks, we modify the mobility management scheme with hysteresis, which allows a new master/controller to be selected only if the selection possibility of a new master/controller candidate is sufficiently higher by a hysteresis margin. Similarly, the network partition with hysteresis will allow a node to switch to another cluster only if the membership degree to another cluster is higher enough by a hysteresis margin than the membership degree to the current cluster. This modification can detect the ping-pong effect attack in IEEE 802.15.4.

5 Conclusion

In this paper, we study the problem of ping-pong effect attack in wireless networks. We introduce a fuzzy system for detecting ping pong effect attack in low rate wireless personal area networks design method with power aware self reconfiguring topology. The LRWPAN is self organized to clusters using an unsupervised clustering method, fuzzy c-means. A fuzzy logic system is applied to master/controller selection for each cluster. A self-reconfiguring topology is proposed to manage the mobility and recursively update the network topology and minimizing the total energy of the system while distributing the load evenly to the nodes has a great impact on system lifetime. The mobility management schemes with hysteresis to detecting the ping-pong effect on the basis of ensure and improve the performance of the network.

References

[1] Siep, T.: Paving the way for personal area network standards: an overview of the IEEE P802.15 working group for wireless personal area networks. IEEE Pers. Communication 7(1), 37–43 (2007)

[2] Middleton, S.: IEEE 802.15 WPAN low rate study group PAR. doc no. IEEE P802.1500/248r3, http://grouper.ieee.org/groups/802/15/pub/2000/sep00/00248r3P802-15-LRSG-PAR.doc

[3] He, Q.: A novel vertical handoff decision algorithm in heterogeneous wireless networks. In: IEEE International Conference on Wireless Communications, Networking and Information Security (WCNIS), pp. 566–570 (June 2008)

[4] Sheng, J., Tang, L.: A Triangle module operator and Fuzzy Logic Based Handoff Algorithm for Heterogeneous Wireless Network. In: 12th IEEE International Conference on Communication Technology, pp. 448–451 (January 2011) ISBN: 978-1-4244-6868-3

[5] Meng, T., Rodoplu, V.: Minimum energy mobile wireless networks. IEEE J. Selected Areas in Communications 17(8), 1333–1344 (2008)

[6] Agarwal, S., et al.: Distributed power control in ad hoc wireless networks. In: IEEE 12th International Symposium on Personal Indoor Mobile Radio Communications (PIMRC), San Diego, CA (September 2008)

[7] Raghavendra, C.S., Lindsey, S.: PEGASIS: power efficient gathering in sensor information systems. Presented at IEEE Aerospace Conference (March 2006)

[8] Raghavendra, C.S., Singh, S., Woo, M.: Power aware routing in mobile ad hoc networks. In: Proc. of Annual ACM/IEEE International Conf. on Mobile Computing and Networking (MobiCom), Dallas, TX, pp. 181–190 (2008)

[9] Hassanein, H., Mouftah, H., Safwat, A.: Power-aware fair infrastructure formation for wireless mobile ad hoc communications. In: Proc. of Globecom 2001, San Antonio, TX, pp. 2832–2836 (September 2005)

[10] Toh, C.-K.: Maximum battery life routing to support ubiquitous mobile computing in wireless ad hoc networks. IEEE Communications Magazine 39(6), 138–147 (2009)

[11] Rappaport, T.S.: Wireless Communications. Prentice Hall, Upper Saddle River (2006)

[12] Rodoplu, V., Meng, T.: Minimum energy mobile wireless networks. IEEE J. Selected Areas in Communications 17(8), 1333–1344 (2008)

[13] Bezdek, J.C.: Pattern Recognition with Fuzzy Objective Function Algorithms. Plenum Press, New York (1999)

[14] Mendel, J.M.: Uncertain Rule-Based Fuzzy Logic Systems. Prentice-Hall, Upper Saddle River (2008)

[15] Tahon, C., Hammadi, S.: Special issue on intelligent techniques in flexible manufacturing systems. IEEE Transactions on Systems, Man and Cybernetics, 157–158 (May 2003)

[16] Guru, S.M., Halgamuge, M.N., Jennings, A.: Energy efficient cluster formation in wireless sensor networks. In: 10th International Conference on Telecommunications, pp. 1571–1576 (March 2003)

[17] Chou, L., Gen-Chung, L.: Location management using fuzzy logic control for wireless networks. In: International Conferences on Info-tech and Info-net, pp. 339–344 (October 2001)

[18] Kreinovich, V., Reznik, L.: Fuzzy and probabilistic models of association information in sensor networks. In: Proc. IEEE International Conference on Fuzzy Systems, pp. 185–189 (July 2004)

[19] Holtzman, J.M., Zhang, N.: Analysis of handoff algorithms using both absolute and relative measurements. IEEE Trans. Vehi. Technol. 45, 174–179 (2006)

[20] Agarwal, S., et al.: Distributed power control in ad hoc wireless networks. In: IEEE 12th International Symposium on Personal Indoor Mobile Radio Communications (PIMRC), San Diego, CA (September 2008)

[21] Gutierrez, J.A., et al.: IEEE 802.15.4: A developing standard for low power low cost wireless personal area networks. IEEE Networks, 12–19 (September/October 2008)

[22] Li, Q., Aslam, J., Rus, D.: Online power-aware routing in wireless ad-hoc networks. In: Proc. of Annual ACM/IEEE International Conf. on Mobile Computing and Networking (MobiCom), Rome, Italy, pp. 97–107 (2007)

Differentiated Service Based on Reinforcement Learning in Wireless Networks

Malika Bourenane

Abstract. In this paper, we propose a global quality of service management applied to DiffServ environments and IEEE 802.11e wireless networks. Especially, we evaluate how the IEEE 802.11e standard for Quality of Service in Wireless Local Area networks (WLANs) can interoperate with the Differentiated Services (DiffServ) architecture for end-to-end IP QoS. An Architecture for the integration of traffic conditioner is then proposed to manage the resources availability and regulate traffic in congestion situation. This traffic conditioner is modelled as an agent based on reinforcement learning.

Keywords: Wireless networks, IEEE 802.11e, DiffServ, end-to-end QoS, Traffic conditioner, Reinforcement learning.

1 Introduction

As wireless networks are being more and more widely deployed, the demands for QoS in the wireless networks are also increasing. The wireless access network will often be the bottleneck on the path between the source and the receiver, because of the nature of wireless communication. It is therefore important to integrate the WLAN QoS architecture with the end-to-end QoS concept adopted, for example DiffServ, to support applications with QoS requirements on wireless terminals.

The IEEE 802.11e is an emerging standard for QoS enabled MAC for the popular IEEE 802.11 WLAN. It was developed in order to meet the QoS requirements in 802.11 WLANs. In order to improve QoS provided by IEEE802.11e wireless network, we consider in our work an end-to-end QoS architecture which is based on dynamic mapping DiffServ to MAC differentiation. We will complete this architecture by adding at each station an adaptive traffic conditioner to control congestion from end to end and to regulate traffic entering the network. The next

Malika Bourenane
Computer Science Department
University of Es-Senia, Oran, Algeria
e-mail: mb_regina@yahoo.fr

Malika Bourenane
Industrial Computing and Networking Laboratory (LRIIR)

A. Abraham and S.M. Thampi (Eds.): Intelligent Informatics, AISC 182, pp. 417–424.
springerlink.com

section gives a brief overview on Diffserv architecture and the EDCA mechanism provided in 802.11e. In section 3, a state of art is provided. We present in the following section the proposed approach and the simulation. Finally, in section 6 a conclusion and some futures perspectives are given.

2 Background

2.1 *Differentiated Services (Diffserv)*

The DiffServ architecture consists of DiffServ domains that are divided into edge and core nodes [1]. The most complex part of the DiffServ functionality is implemented in the edge nodes, while the core nodes are kept as simple as possible. Packets from applications with similar QoS requirements are assigned the same service class at the edge of the DiffServ network and aggregated in the core network. DiffServ model meet the QoS requirements of different service classes by providing Per Hop Behaviors (PHB) that is the way in which a router will treat incoming packets (ie. the enqueuing plus managing congestion). The traffic classification and conditioning are conducted at the ingress network node while the PHB forwarding is conducted at the core nodes. Different traffic aggregates will then experience different forwarding treatment by the core nodes in the path depending on which PHB they belongs to, and three PHBs are currently supported that are the Expedited Forwarding (EF) PHB, the Assured Forwarding (AF) PHBs and the default PHB.

2.2 *IEEE 802.11e Norme*

The IEEE standard 802.11e provides QoS in two forms [2]. First it supports a priority based effort service similar to Diffserv and in the second it supports parameterized QoS for the benefit of applications requiring QoS for different flows. IEEE 802.11e enhanced version [2] introduced a single coordination function called hybrid coordination function (HCF) [3] used only in QoS enhance Basic Service Set (QBSS) and which combines the contention based EDCA (Enhanced Distributed Channel Access) and contention free medium access method HCCA (HCF Controled access Channel).

2.2.1 EDCA

The QoS [3] in a WLAN enhanced by EDCA, supports priority based best-effort service such as DiffServ. The standard defines four traffic classes (TC): Voice (VO), Video (VI), Best Effort (BE) and Background (BK). Prioritized QoS is realized through the introduction of four access categories (ACs) assigned to these TCs. The four access categories AC are such that AC_VO has the highest priority and is designed for voice traffic with strict latency, jitter and bandwidth requirements. AC_VI is meant for video traffic that has strict bandwidth requirements, but some looser latency and jitter requirements than voice does. AC_BK is meant for background traffic for which the lowest priority is given and AC_BE for best effort traffic. Each AC has its own transmit queue and its own set of AC parameters. The differentiation in priority between ACs is achieved by setting different values for those parameters.

3 Related Work

Only few contributions concerning the interoperability between 802.11e and DiffServ have been published, and very little work has been performed to evaluate the conformity between 802.11e and DiffServ PHBs. Skyrianoglou et. al. [4] have proposed a Wireless Adaptation Layer (WAL), between the link and IP layers. This WAL provides an interface to the upper and lower layer that is independent of the technology used and implements the DiffServ functionality in the nodes. Seyong Park et al. [5] have investigated the way the DiffServ PHBs can be translated into 802.11e priority classes. However, those architectures have some practical drawbacks. They require that QSTAs implement new features as DiffServ, which is not a trivial task because it requires changing all the QSTAs and the QAP.

4 Proposed Approach

We propose a solution for the implementation of QoS in wireless networks, based on the DiffServ model. The first part of this solution is to implement an autonomous agent that supports the dynamic setting of coupling between classes Diffserv and 802.11e ACs, the second part consists of introducing the traffic conditioner in order to ensure the availability of resources on the data path. This agent is placed in the QAP and its functionality is very simple. It only needs a table lookup, as well as reading and writing of the MAC and IP headers in a packet.

4.1 The Traffic Conditioner

The *traffic conditioner* is an important element of DiffServ, which consists of the meter, marker and shaper or dropper [2]. Traffic shaping can be done at the source prior to access into the network or within the network. Traffic shaping at the source is a means of self regulation in order to ensure conformity to the traffic contract. This Conformance is advantageous to minimize the amount of traffic discarded at the network ingress; this is why we define the traffic shaping mechanism at each QSTA. This traffic profile is to take into account the arrival rate of packets, so as not to exceed the maximum threshold of packet that can be sent over the network. Thus, a mechanism for traffic measurement indicates whether the stream of incoming packet matches the traffic profile negotiated. If this flow exceeds a certain threshold some packets are marked as low priority and will be automatically discarded when network congestion. In our solution, we implemented at each QSTA, a traffic conditioner as an agent to control and regulate the incoming traffic.

Due to the fact that current network management systems have limited ability to adapt to dynamic network conditions, providing intelligent control mechanisms for globally optimal performance is a necessary task. Therefore, we propose a network architecture that enables intelligent services to meet QoS requirements, by adding autonomous intelligence, based on reinforcement learning, to the traffic conditioner agent.

Reinforcement learning problems are typically modeled by means of *Markov Decision Processes* (MDPs) [6]. The model consists of the set *S* of potential environment states, the set *A* of possible actions, the designated reward function, R: S × A → $\Re$; and the policy which determines state transition, P: S × A → π(S) where π(S) represents the set of functions over *S*, which specify the actions required to transition by means of action a ∈ A from state *s* to state *s'*. The objective of the agent is to maximize the cumulative immediate rewards received. More specifically, the agent selects actions that maximize the expected discounted return:

$$\sum_{k=0}^{\infty} \gamma^k r_{t+k+1} \text{ where } \gamma,\ 0 \leq \gamma < 1, \text{ is the discount factor}$$

For any policy π, the action-value function, $Q^\pi(s, a)$, can be defined as the expected discounted future reward for a state s when action a is selected to be performed:

$$Q^\pi(s, a) = E_\pi \left\{ \sum_{k=0}^{\infty} \gamma^k r_{t+k+1} \middle| s_t = a, a_t = a \right\}. \quad (2)$$

The problem of solving an MDP is thus the same as the problem of finding the optimal Q-function Q*, which satisfies the Bellman equation [7]:

$$Q^*(s, a) = E\, \{r_{t+1} + \gamma \max_{a'} Q^*(s_{t+1}, a') \mid s_t = s, a_t = a\}$$

The optimal policy can be defined as $\pi^* = \arg\max_a Q^*(s, a)$. The particular form of reinforcement learning that our traffic conditioner agent employs is referred to as Q-Learning [6]. As a model-free reinforcement learning technique, Q-Learning is ideally suited for optimization in dynamic network environments.

5 Agent Model

We model the traffic shaping mechanism at each QSTA as a learning agent. This agent aims to avoid congestion at the entrance to the wired network, by performing three tasks iteratively: observation of the conditions of the environment, reasoning to interpret the observations, and then action. A Markov decision process characterizes this agent and the solution is approximated by reinforcement learning.

5.1 Action of Packet Handler

The objective of the control agent is to adjust either the transmission rate μ_i or packet generation rate r_i. The flow regulations are performed at different layers.

The first adjusts the local contention at the MAC layer, thus affecting the transmission rate. The second controls the admission of traffic from the application, influencing the packet generation rate.

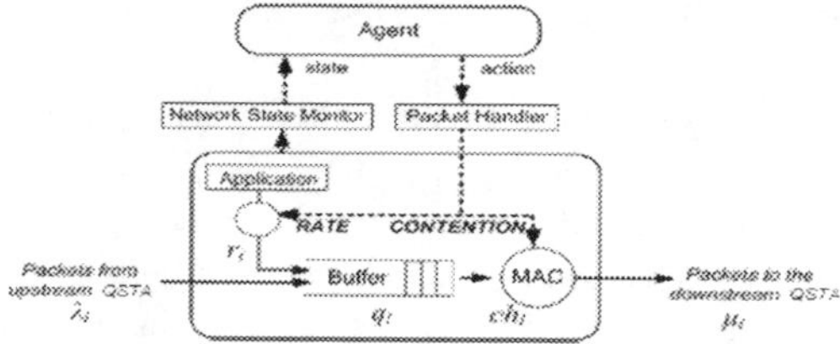

Fig. 1 Learning agent model

5.2 *Network State Monitor*

The state of the wired ingress node indicates the level of congestion. The buffer occupancy is simple to obtain and makes a reasonable indicator of local congestion. A link between a QSTA and the wired ingress router is considered as congested if the buffer is full.

6 Reinforcement Learning for Congestion Control

In this section, we establish the state, action spaces and rewards for the congestion control problem. As mentioned, the transmission rate can be controlled to mitigate congestion. This control is achieved by regulating the contention window when the channel is highly loaded.

6.1 *State and Action Spaces*

The state descriptor for a single agent is the queue length of the token bucket tb, then:

$$s = [tb]$$

The action is the choice of the contention window CW from a finite set,

$$a \in \{c_0, c_1, \ldots, c_i, \ldots\}$$

6.2 *The Reward*

The feedback signal is the cost related to the drop of packets when the wired buffer is bursting. Let D denotes the drops after a transmission attempt, then we calculate the reward r at state s and after executing action a as:

$$r(s,a) = -w \times D$$

where w is a weighting factor such that $0 \le w \le 1$.

The number of dropped packets reflects the state of network congestion, then a negative reward is assigned to this state.

7 Simulation and Evaluation of Results

The end to end QoS management implies the presence of specific mechanisms for supervising quality of service at each level (backbone and access networks). The bottleneck is often at the level of access networks. The introduction of QoS in these networks becomes a necessity, typically in a wireless environment. However, it must be done in two cases: for traffic entering in the access network and for the outgoing traffic. It is therefore a need for differentiation patterns not only in all areas that constitute the backbone network, but also in access networks which are often bottlenecks. To address the limitations posed by EDCA, namely the influence of flow with the highest priority on the lowest priority one and QoS constraints posed by the media stream (flow rate and time), we evaluate a coupling solution between EDCA and DiffServ as we complete by the integration of a traffic conditioner. This traffic conditioner is a learning agent that acts at each QSTA to regulate the incoming traffic and therefore reduce the congestion. We implemented the coupling solution in NS2. Our simulation model consists of three mobile stations that emit flows cbr0, cbr1 and cbr2, modeling respectively audio, video and data standard. We performed simulations for two scenarios: without mapping with mapping and traffic conditioner.

Analysis of the resulting curves shows that the EDCA can support the highest priority traffic. More traffic has priority, the more it is used. This curve confirms the hypothesis to be tested at this level which is as follows "the throughput of the lowest priority traffic degrades if the throughput of the higher priority traffic increases."

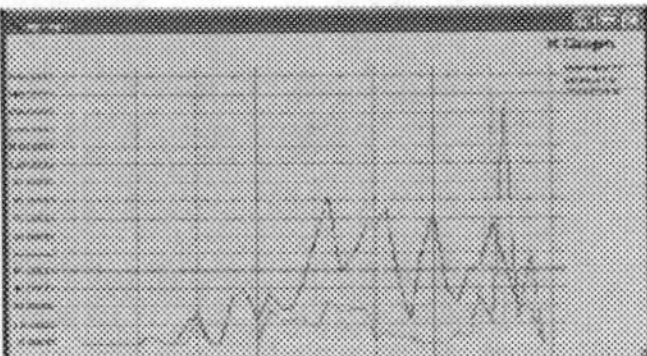

Fig. 2 Throughput versus time (EDCA)

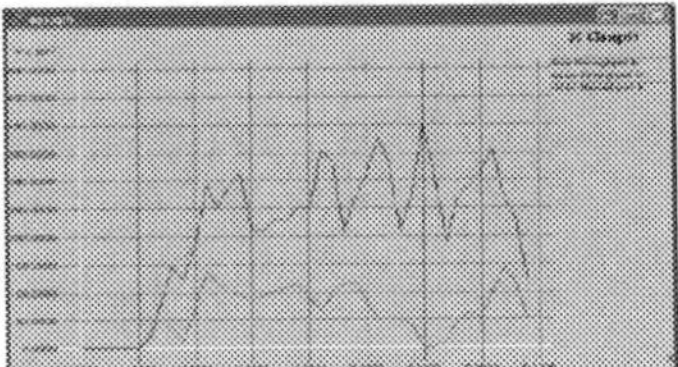

Fig. 3 Throughput versus time (proposed structure)

We observed, according to simulations conducted, that the proposed architecture performs well with respect to end to end service quality. The results show significant improvement in performance of data traffic (traffic lower priority). We can conclude from the curves obtained, that combining service differentiation and traffic conditioner provides better treatment for low priority packets in the queues, and increases its probability of channel access. We also evaluated the packet loss for each type of traffic, shown in Figure 5.

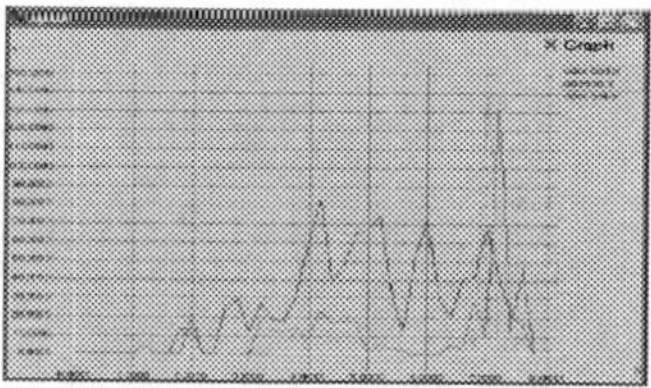

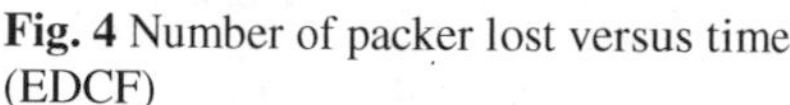

Fig. 4 Number of packer lost versus time (EDCF)

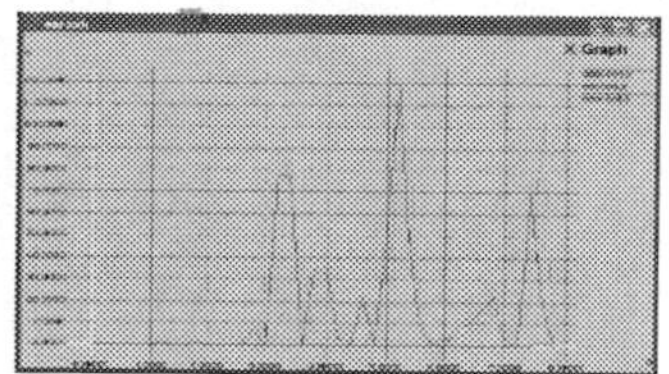

Fig. 5 Number of packet lost versus time (proposed structure)

Figure 6 depicts the time from end to end voice and video streams and Data. A time zero for voice traffic is observed indicating that the Voice packets are received immediately. This delay increases for other traffic at time t = 3s but it decreases until it reaches a very low period at the end of simulation.

In this first scenario, we can conclude that the coupling and the introduction of traffic conditioner helps to harmonize the quality of service in access networks and the backbone. Indeed, it allows a service class to observe the same behavior in the two levels of networks in order to improve the end to end QoS management of traffic classes.

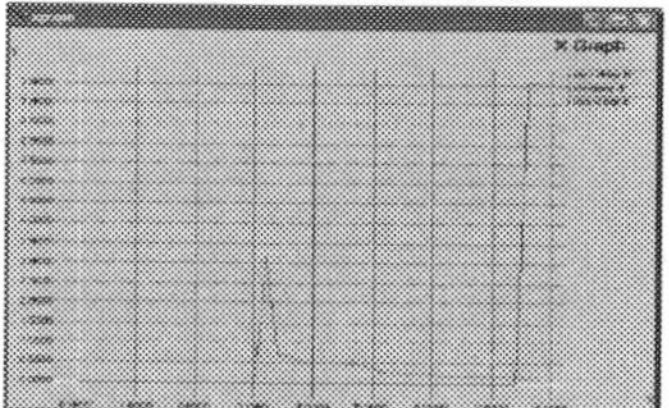

Fig. 6 End to end delay

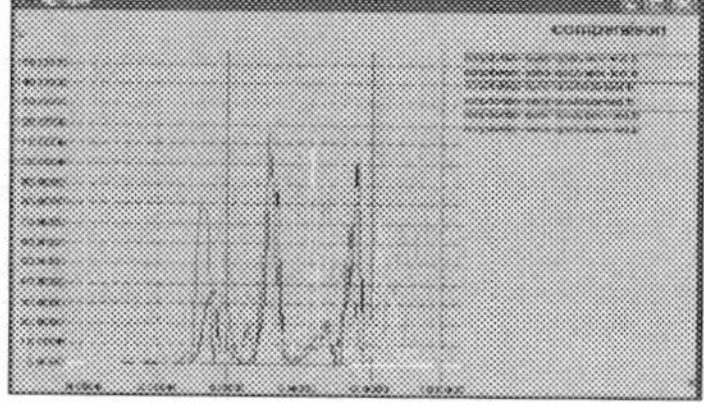

Fig. 7 Packet loss with/without mapping (congestioned network)

By observing the figure 7 on the losses, we note that the loss rate decreases through the EDCA in "mapping". This rate increased from 120 packets / second to 100 packets with mapping. This decrease amounts is due to the fact that data traffic is better served with mapping even in the presence of a large number of voice stations.

8 Conclusion and Future Work

In this paper we presented a method for managing quality of service based on the dynamic coupling between the Diffserv service classes and categories of access ACs of the IEEE 802.11e initially and integration of traffic conditioner in a second time. The subject discussed in this article can be further expanded and enhanced to contribute to the management of QoS in ad hoc networks.

References

[1] RFC Diffserv, http://www.ietf.org/rfc/rfc2475.txt
[2] Gavini, K.K., Apte, V., Iyer, S.: PLUS-DAC: A Distributed Admission Control Scheme for IEEE 802.11e WLANs. In: International Conference on Networking (ICON), Kuala Lumpur, Malaysia (November 2005)
[3] Acharya, R., Vityanathan, V., Chellaih, P.R.: WLAN QoS Issues and IEEE 802.11e QoS Enhancement. International Journal of Computer Theory and Engineering 2(1) (February 2010)
[4] Skyrinaoglou, D., Passas, N., Salkintzis, A., Zervas, E.: A Generic Adaptation Layer for Differentiated Services and Improved Performance in the Wireless Networks
[5] Park, S., Kim, K., Kim, D.C., Choi, S., Hong, S.: Collaborative QoS architecture between DiffServ and 802.11e Wireless LAN. IEEE (2003)
[6] Dayan, C.W.a.P.: Machine Learning, pp. 279–292 (1992)
[7] Sutton, R.S., Barto, A.G.: Reinforcement Learning: An Introduction. MIT Press, Cambridge (1998)

Multimodal Biometric Authentication Based on Score Normalization Technique

T. Sreenivasa Rao and E. Sreenivasa Reddy

Abstract. To achieve high reliability of biometric authentication using fusion of multimodal biometrics in authentication systems is a novel approach. In this paper we propose a method for the management of access control to ensure the desired level of security using the adaptive combination of multimodal matching scores. It uses a score normalization technique for multimodal biometric authentication using fingerprint, palmprint and voice. This technique is based on the individual scores obtained from each of the biometrics and then normalized to get a fused score. Training data sets are generated from genuine and impostor score distributions. Also this technique is compared with other score normalization techniques and the performance of the proposed system is analyzed. The proposed multimodal biometric authentication system overcomes the limitations of individual biometric systems and also meets the response time as well as the accuracy requirements.

Keywords: Palmprint, Fingerprint, Face recognition, Score level fusion, Individual score, Normalized score, Sum rule.

1 Introduction

An integration scheme to fuse the information obtained from the individual modalities is the main concept behind a multimodal biometric system [1]. Once the matcher of a biometric system is invoked, the amount of information available to the system drastically decreases. It is generally believed that a fusion scheme applied as early as possible in the recognition system is more effective. The amount of information available to the system gets compressed as one proceeds from the sensor module to the decision module.

Various levels of fusion can be classified into two broad categories: fusion before matching and fusion after matching, according to Sanderson and Paliwal [2]. A multimodal biometric system can work in three modes. They are serial

T. Sreenivasa Rao · E. Sreenivasa Reddy
Department of Computer Science & Engineering,
Acharya Nagarjuna University, A.P., India

A. Abraham and S.M. Thampi (Eds.): Intelligent Informatics, AISC 182, pp. 425–434.
springerlink.com

mode, parallel mode and hierarchical mode. In the serial mode the output of one biometric characteristic is used to reduce the no of possible identities before the next characteristic is used. So multiple source of information is not collected simultaneously. In parallel mode the information from multiple characteristics is taken together to perform recognition. In hierarchical mode individual classifiers are combined in a tree like structure. This mode is well suited where

Depending on the traits, sensors and feature sets multimodal systems are classified as:

1. Single biometric trait, multiple sensors**:** Multiple sensors are used to record the same biometric characteristic. The raw data taken from different sensors can then be combined at the feature level or matcher score level to improve the performance of the system.
2. Multiple biometrics: Multiple biometric traits such as fingerprints and face can be combined. Different sensors are used for each biometric characteristic. The interdependency of the traits ensures a significant improvement in the performance of the system.
3. Multiple units, single biometric traits: Two or more fingers of a single user can be used as a biometric trait. It is inexpensive way of improving system performance, as it doesn't require multiple sensors or incorporating additional feature extraction or matching modules. Iris can also be included in this category.
4. Multiple snapshots of single biometric: In this more than one instance of the same biometric is used for the recognition.
5. Multiple matching algorithms for the same biometric: In it different methods can be applied to feature extraction and matching of the biometric characteristic.

There are different issues when designing a multi biometric system. Some of them are, choice and number of biometric indicators, fusion Level, representation of features, matching score, decision support, fusion methodology etc.

The information of the multimodal system can be fused at any of the four levels.

a. Fusion at the sensor level: in this the raw data from different sensors are fused. In it we can either use samples of same biometric trait obtained from multiple compatible sensors or multiple instances of same biometric trait obtained using a single sensor. In it the data is fused at very early stage so it has a lot of information as compared to other fusion levels. Very less work has been done in this area.

b. Fusion at the Feature Extraction Level: The data or the feature set originating from multiple sensors or sources are fused together. Features extracted from each sensor form a feature vector. These features vectors are then concatenated to form a single new vector. In feature level fusion we can use same feature extraction algorithm or different feature extraction algorithm on different modalities whose features has to be fused. The feature level fusion is challenging because relationship between features is not known and structurally incompatible features are common and the curse of dimensionality. Because of these difficulties, only limited work is reported on feature level fusion of multimodal biometric system.

C. Matcher Score Level: Each system provides a matching score indicating the proximity of the feature vector with the template vector. These scores can be combined to assert the veracity of the claimed identity. The scores obtained from different matchers are not homogeneous, score normalization technique is followed to map the scores obtained from different matchers on to a same range. These scores contain the richest information about the input. Also it is quite easy to combine the scores of different biometrics so lot of work has been done in this field.

D. Fusion at the Decision Level: The final outputs of the multiple classifiers are combined. A majority vote scheme can be used to make final decision. Decision level fusion includes very abstract level of information so they are less preferred in designing multimodal biometric systems.

Fusion at the matching-score level is the most popular and frequently used method because of its good performance, intuitiveness and simplicity. In this paper we develop a fused palm-finger-face recognition system which overcomes a number of inherent difficulties of the individual biometric traits. All the three traits perform better individually but fail under certain conditions. The palmprint provides a larger surface area compared with the fingerprint, so that more features can be extracted for personal recognition. Thus the three recognizers (fingerprint, palmprint, voice) are combined at the match score level and final decision about the person's identity is made. Table 1 shows the summary of fusion techniques. Table 2 gives the summary of normalization techniques and Table 3 gives the summary of fusion techniques.

The rest of the paper is organized as follows. Section 2 addresses the literature study. In section 3 a brief overview is presented about the fingerprint, palmprint and speaker recognition. Section 4 describes the framework of the proposed system and fusion performed at the match score level. In section 5 we compare the results of the combined system with the results of the individual biometric traits. Finally, the summary and conclusions are given in last section.

Table 1 Summary of fusion Techniques

Simple Sum	$\sum_{i=1}^{N} S_i$
Minimum Score	min (S1, S2, … Sn)
Maximum Score	max (S1, S2, … Sn)
Sum of Probabilities	$\sum_{i=1}^{N} P(genuine \mid S_i)$
Product of Probabilities	$\prod_{(i=1)}^{N} [P(genuine \mid S_i)]$

Table 2 Summary of Normalization Techniques

Min – Max	S'= (s-min) /max – min)
Z-Score	S' = (s-mean) / standard deviation
MAD	S' = (s-median) / constant(median l s – medianl)
Tanh	S' =0.5[tanh(0.1(s-mean)/(standard deviation)) + 1]

s – output score, s' - normalized score.

Table 3 Summary of Fusion Techniques, GAR at 0.1 % FAR

Normalization Technique	Fusion Techniques				
	Simple Sum	Minimum Score	Maximum Score	Sum of Probabilities	Product of Probabilities
Min - Max	95.9 %	78.2 %	83.5 %	N/A	N/A
Z-Score	94.82 %	86.9 %	86.1 %	N/A	N/A
MAD	91.0 %	83.2 %	85.3 %	N/A	N/A
Tanh	95.24 %	887.5 %	86.1 %	N/A	N/A
None	88.5 %	82.0 %	83.6 %	87.3 %	87.2 %

2 Literature Review

The research on multi modal biometrics started in late 90's. Generally, FLF is achieved by concatenation of feature vectors [3,4,5]. A novel approach for fusion of ear and face profile features using Kernel Canonical Correlation Analysis (KCCA) by mapping the feature vectors into a higher dimensional space before applying correlation analysis is proposed by [6]. A number of research works on score-level fusion have been reported [7,8,9,10,11]. A detailed study on normalization techniques was proposed by Jain et al.[12]. Once the scores are normalized, simple arithmetic rules like sum, weighted sum, product, weighted product, product of probabilities, sum of probabilities, min–max score can be used [12,13,14,15]. Techniques like matcher weighting and user weighting were explored by Snelick et al [10]. Matcher weighting techniques assign weights to the matchers based on the accuracy of that biometric trait. While the user weighting techniques assign user-specific weights to each user [16]. These weights are determined empirically. Ross and Jain [17] combined face, fingerprint and hand geometry at the matching score level. In 2004, Toh et al. developed a multimodal biometric system using hand geometry, fingerprint, and voice at match-score-level fusion.

2.1 Fingerprint Recognition

Fingerprint is one of the most widely used biometric trait. The fingerprint is basically the combination of ridges and valleys on the surface of the finger. The lines that create fingerprint pattern are called ridges and the spaces between the ridges are called valleys or furrows. Once a high-quality image is captured, there are several steps required to convert its distinctive features into a compact template. The major steps involved in fingerprint recognition using minutiae matching approach after image acquisition are image enhancement, minutiae extraction, matching. The goal of fingerprint enhancement is to increase the clarity of ridge structure so that minutiae points can be easily and correctly extracted. The enhanced fingerprint image is binarized and submitted to the thinning algorithm which reduces the ridge thickness to one pixel wide for precise location of endings and bifurcations. This processed image is used to extract minutiae points which are the points of ridge endings and bifurcations. The location of minutiae points along with the orientation is extracted and stored to form a feature set. The minutiae based matching consists of finding alignment between the template and the input minutiae sets that result in the maximum number of minutiae pairings. This pairing generates a similarity score (MS_{finger}) which is used in score normalization module.

2.2 Palmprint Recognition

Palmprint is one of the physiological biometrics due to its stable and unique characteristics. The area of the palm is much larger than the area of a finger and as a result, palmprints are expected to be more distinctive than the fingerprints. Biometric palmprint recognizes a person based on the principal lines, wrinkles and ridges on the surface of the palm. These line structures are stable and remain unchanged throughout the life of an individual. Here PCA approach can be used which transforms palmprint images into specific transformation domains to find useful image representations in compressed subspace. It computes a set of basis vector from a set of palmprint images, and the images are projected into the compressed subspace to obtain a set of coefficients. New test images are then matched to these known coefficients by projecting them onto the basis vectors and finding the closest coefficients in the subspace. The basis vectors generated from a set of palmprint images are called eigenpalm. Recognition is performed by projecting a new image into the subspace spanned by the eigenpalms and then classifying the palm by comparing its position in palm space with the positions of known coefficients. The matching score (MS_{palm}) between two palmprint feature vectors can be calculated using the Euclidean distance.

2.3 Voice Recognition System

Speaker recognition is the task of recognizing speakers using their speech. Speaker recognition can be either identification or Verification depending on whether the goal is to identify the speaker among the group of speakers or verify the identity claim of the speaker. Further, speech from same text or arbitrary text may be used for recognizing speakers and accordingly we have text dependent and text independent modes of operation. The present work follows text dependent speaker identification approach. The fol1owing are the three phases in Automatic Speaker Recognition (ASR) task.

A. **Feature Extraction Phase:** The feature extraction phase involves estimation of Mel Frequency Cepstral Coefficients (MFCCs) from all training speech samples of all users. Estimation of MFCC involves steps such as windowing, calculation of Mel-Frequency bands, cepstral mean subtraction and filtering.

B. **Training Phase:** In training phase, we extract feature vectors from the speech signals of speaker #N. After finding the MFCC feature vectors, a small codebook that represents all the MFCC vectors in the minimum mean square sense has to be built. In the present work we have used Vector Quantization (VQ) method as the pattern matching method. K-Means Clustering has been used for code book generation.

C. **Testing Phase:** In this phase we try to find the spectral distance between testing utterance feature vectors and code vectors that were obtained in training phase, and classify the utterance to that speaker to whom it is nearer.

3 Proposed Multimodal System

Steps involved in the proposed system

1) Multiple biometric templates are acquired from an individual
2) Features are extracted individually by using the corresponding feature extraction method.
3) Compare these features with the features extracted from the samples stored in the database and matching score for each of the biometrics is estimated.
4) Using the min-max and z-score normalization techniques normalize the scores of biometrics individually.
5) Find the sum of the weighted sum of the normalized scores for each person individually.

The three resulting representations are then fed to the three corresponding matching modules. Here, they are matched with templates in the corresponding databases to find the similarity between the two feature sets. The matching scores generated from the individual recognizers are then passed to the fusion module. Finally, fused matching score ($MS_{final} = MS_{palm} + MS_{finger} + MS_{voice}$) is passed to the decision module where a person is declared as genuine or an imposter.

Scores generated from individual biometric traits are combined at matching score level using sum rule. MS_{palm}, MS_{finger} and MS_{voice} are the matching scores generated by palmprint, fingerprint and voice recognizers respectively. The first step involved in fusion is score normalization. Since the matching scores output by the three traits are heterogeneous because they are not on the same numerical range, so score normalization is done to transform these scores into a common domain prior to combining them. Min-max normalization transforms all the scores into a common range [0, 1]. The normalization of the three scores is done by min-max rule as follows:

$$N_{palm} = (Ms_{palm} - min_{palm}) / (Max_{palm} - Min_{palm}) \quad .... (1)$$

$$N_{finger} = (Ms_{finger} - min_{finger}) / (Max_{finger} - Min_{finger}) \quad .... (2)$$

$$N_{voice} = (Ms_{voice} - min_{voice}) / (Max_{voice} - Min_{voice}) \quad .... (3)$$

where $[min_{palm}\ ,\ max_{palm}]$, $[min_{finger}\ ,\ max_{finger}]$ and $[min_{voice}\ ,\ max_{voice}]$ are the minimum and maximum scores for palmprint recognition, fingerprint recognition and face recognition, $N_{palm,}$ N_{finger} and N_{voice} are the normalized matching scores of palmprint, fingerprint and voice respectively.

Fingerprints are represented using minutiae features, and the output of the fingerprint matcher is a similarity score. Palmprint images are also represented as eigenpalms and the matching score is generated as distance score. Voice signals are generated as code books and the matching score is generated as distance score. Prior to combining the normalized scores, it is necessary that all the three normalized scores are transformed as either similarity or dissimilarity measure.

Each biometric matcher produces a match score based on the comparison of input feature set with the template stored in the database. These scores are weighted according to the biometric traits used for increasing the influence of more reliable traits and reducing the importance of less reliable traits. Weights indicate the importance of individual biometric matchers in a multibiometric framework. The set of weights are determined for a specific user such that the total error rates corresponding to that user can be minimized.

4 Experimental Results and Discussions

It has already been proved that fusion of multiple biometrics improve the recognition performance as compared to the single biometrics. The performance of any biometric system is usually represented by the ROC (Receiver Operating Characteristic) curve. The ROC curve plots the probability of FAR (False Accept Rate) versus probability of FRR (False Reject Rate) for different values of the decision threshold (t). FAR is the percentage of imposter pairs whose matching score is greater than or equal to t and FRR is the percentage of genuine pairs whose matching score is less than t.

In order to show the effectiveness of the proposed method, we have plotted ROC of the individual biometrics (see Fig. 1) for Genuine Accept Rate (GAR) against False Accept Rate (FAR) where GAR (1–FRR) is the fraction of genuine scores exceeding the threshold.

The ROC curve of the proposed system is plotted for GAR versus FAR by putting appropriate weights and matching scores for different users at

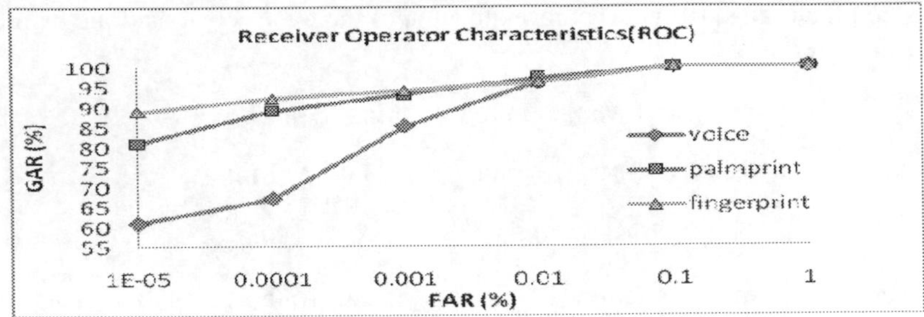

Fig. 1 ROC of the individual biometrics

Table 4 Adaptive weights of different biometrics for 10 users

USER	FINGERPRINT (W1)	PALMPRINT (W2)	VOICE (W3)
1	0.4	0.4	0.2
2	0.5	0.3	0.2
3	0.4	0.3	0.3
4	0.6	0.2	0.2
5	0.4	0.3	0.3
6	0.5	0.2	0.3
7	0.4	0.2	0.4
8	0.5	0.1	0.4
9	0.5	0.3	0.2
10	0.6	0.3	0.1

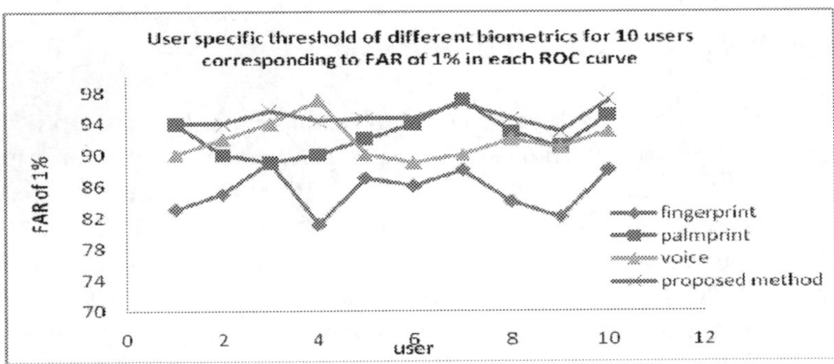

Fig. 2 User specific threshold of different biometrics for 10 users corresponding to FAR of 1% in each ROC curve

different thresholds. Table 4 shows the adaptive weights of 10 users for different biometrics. It is observed from the ROC curve that the performance gain obtained from the proposed system is higher as compared to the three individual traits (palmprint, fingerprint, voice) as it is evident from the ROC curves in Figure 2.

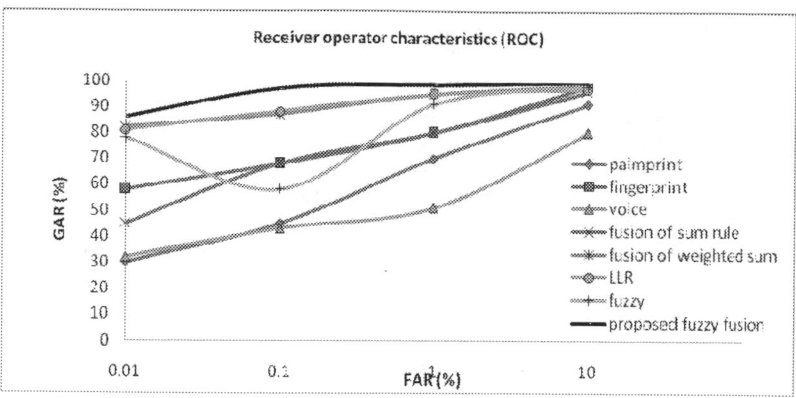

Fig. 3 Performance Comparison of proposed method with existing methods

5 Conclusion

Our experimental results shows that our system is not affected even when the biometric data are affected by noise and the performance speed is also considerably good. The present work includes fusing of the individual biometrics at the score level using weighted scores method. To overcome the problems faced by individual biometric recognizers of palmprint, fingerprint and face, a novel combination is proposed for the recognition system. The integrated system also provide anti spoofing measures by making it difficult for an intruder to spoof multiple biometric traits simultaneously.

It is well established that biometric features are unique to each individual and remain unaltered during a person's lifetime. In this paper, a multimodal biometric recognition system based on fusion of three biometric traits viz. palmprint, fingerprint, voice, has been proposed. Fusion of these three biometric traits is carried out at the matching score level. Our future work will be focused on integrating liveness detection with multimodal biometric systems and minimizing the complexity of the system.

References

1. Jain, A.K., Ross, A., Prabhakar, S.: An introduction to biometric recognition. IEEE Transactions on Circuits and Systems for Video Technology 14, 4–20 (2004)
2. Snelick, R., Indovina, M., Yen, J., Mink, A.: Multimodal biometrics: issues in design and testing. In: Proceedings of the 5th International Conference on Multimodal Interfaces, pp. 68–72 (2003)
3. Chetty, G., Wagner, M.: Audio-visual multimodal fusion for biometric person authentication and liveness verification. In: Chen, F., Epps, J. (eds.) NICTA-HCS Net Multimodal User Interaction Workshop, vol. 57, pp. 17–24 (2005)
4. Ross, A., Govindarajan, R.: Feature Level Fusion Using Hand and Face Biometrics. In: Proceedings of SPIE Conference on Biometric Technology for Human Identification II, vol. 5779, pp. 196–204 (2005)

5. Kumar, A., Zhang, D.: Personal recognition using hand shape and texture. IEEE Transactions on Image Processing 15, 2454–2461 (2006)
6. Xu, X., Mu, Z.: Feature fusion method based on KCCA for ear and profile face based multimodal recognition. In: IEEE International Conference on Automation and Logistics, pp. 620–623 (2007)
7. Ailisto, H., Lindholm, M., Makela, S.M., Vildjiounaite, E.: Unobtrusive user identification with light biometrics. In: Raisamo, R. (ed.) Nordic Conference on Human- Computer Interaction, pp. 327–330 (2004)
8. Zhang, W., Shan, S., Gao, W., Chang, Y., Cao, B., Yang, P.: Information fusion in face identification. In: Proceedings of the 17th International Conference on Pattern Recognition, vol. 3, pp. 950–953 (2004)
9. Dass, S.C., Nandakumar, K., Jain, A.K.: A principled approach to score level fusion in multimodal biometric systems. In: Proceedings of the 5th International Conference on Audio- and Video-Based Biometric Person Authentication, pp. 1049–1058 (2005)
10. Snelick, R., Uludag, U., Mink, A., Indovina, M., Jain, A.K.: Large-scale evaluation of multimodal biometric authentication using state-of-the-art systems. IEEE Transactions on Pattern Analysis and Machine Intelligence 27(3), 450–455 (2005)
11. Shu, C., Ding, X.: Multi-biometrics fusion for identity verification. In: 18th International Conference on Pattern Recognition, vol. 4, pp. 493–496 (2006)
12. Jain, A.K., Nandakumar, K., Ross, A.: Score normalization in multimodal biometric systems. Pattern Recognition 38, 2270–2285 (2005)
13. Ong, M.G.K., Connie, T., Jin, A.T.B., Ling, D.N.C.: A single-sensor hand geometry and palmprint verification system. In: WBMA 2003: Proceedings of the 2003 ACM SIGMM Workshop on Biometrics Methods and Applications, pp. 100–106 (2003)
14. Nakagawa, T., Nakanishi, I., Itoh, Y., Fukui, Y.: Multi-modal biometrics authentication using on-line signature and voice pitch. In: International Symposium on Intelligent Signal Processing and Communications, pp. 399–402 (2006)
15. TohK, A., Jiang, X., Yau, W.Y.: Exploiting global and local decisions for multimodal biometrics verification. IEEE Transactions on Signal Processing 52(10), 3059–3072 (2004)
16. TohK, A., Yau, W.Y.: Combination of hyperbolic functions for multi-modal biometrics data fusion. IEEE Transactions on Systems, Man, and Cybernetics 34(2), 1196–1209
17. Ross, A., Jain, A.K.: Information Fusion in Biometrics. Pattern Recognition Letters 24, 2115–2125 (2003)
18. Nandakumar, K., Chen, Y., Dass, S.C., Jain, A.K.: Likelihood ratio-based biometric score fusion. IEEE Transactions on Pattern Analysis and Machine Intelligence 30, 342–347 (2008)

Extracting Extended Web Logs to Identify the Origin of Visits and Search Keywords

Jeeva Jose and P. Sojan Lal

Abstract. Web Usage Mining is the extraction of information from web log data. The extended web log file contains information about the user traffic and behavior, the browser type, its version and operating system used. Mining these web logs provide the origin of visit or the referring website and popular keywords used to access a website. This paper proposes an indiscernibility approach in rough set theory to extract information from extended web logs to identify the origin of visits and the keywords used to visit a web site which will lead to better design of websites and search engine optimization.

Keywords: Web Usage Mining, Extended Web Log, Keyword Search.

1 Introduction

Web usage mining is the extraction of information from web log files generated when a user visits the web site [1][2][3]. We can study the behavior of a web surfer in a web site, extract navigational patterns about their favorite pages and check the reliability of web design and architecture with the analysis of such web log data. [4][5]. There are three main ways to access a website. 1) through a search engine request 2) through a link from another website and 3) through the website root by typing the URL of the website in a web browser [6]. Web logs are maintained by web servers and contain information about users accessing the site. Web usage analysis includes straightforward statistics such as page access frequency as well as more sophisticated forms of analysis like clustering, classification and

Jeeva Jose
BPC College, Piravom, 686664, Kerala, India
e-mail: vijojeeva@yahoo.co.in

P. Sojan Lal
School of Computer Science, Mahatma Gandhi University,
Kottayam, 686560, Kerala, India
e-mail: padikkakudy@gmail.com

A. Abraham and S.M. Thampi (Eds.): Intelligent Informatics, AISC 182, pp. 435–441.
springerlink.com

association rule mining. Access log files contain large amount of HTTP information. Every time a web browser downloads an HTML document from the internet, the images are also downloaded and stored in the log file [7]. A *hit* is any file from a web site that a user downloads and *accesses* are an entire page downloaded by users regardless of the number of images, sounds or movies [8]. Extended log file contains tremendous information about the referrer, search engines used, operating system, browser and its version for accessing a website.

2 Web Log Files

The most widely used log file formats are Common Log File Format and Extended Log File Format [9] [10]. Traditionally there are four types of server logs [11]. They are a) Transfer log b) Error log c) Agent log and d) Referrer log. There are three main sources to get the raw web log file [12]. They are the a) client log file, b) Proxy log file and c) Server log file. Client log files are most accurate and authentic to depict the user behavior but it is a difficult task to modify the browser for each client. In proxy servers, same IP address is used by many users and hence to identify users is difficult. Hence most researchers consider the web server log file as most reliable and accurate for web usage mining process. The extended web log contains the following information.

User's IP address- It is the visitor's hostname or IP address from where the visitor is making a connection. Hostname is the name of the machine where WWW or mail server is running.

Rfcname or User Authentication- This field returns user authentication. It operates by verifying specific TC/IP connections and returns the user identifier of the process who owns the connection. If the value is not present, it is indicated by a "-" character.

Logname- It is the user's login name in local directory. If the value is not present, it is represented by a "-" character.

The date-time stamp of the access - The access date defined by day(DD), month(MMM), year(YYYY), hours(HH),minutes(MM) and seconds(SS). The last symbol stands for the difference from Greenwich Mean Time.

The HTTP request method- This field contains the page/file access method. Usually there are four ways of HTTP requests namely PUT, GET, POST and HEAD. PUT is used by web site maintainers having administrator privileges. Access to this method by ordinary users is forbidden and this method rarely appears in web logs. GET transfers the content of the web document to the user. This is the most popular method. CGI functionality is served via POST or HEAD.

The URL requested or Path - It is the path and filename retrieved from the host. For example, /images/stories/home/index_05.gif

Protocol Version - It defines the version of the protocol used by the user to retrieve the information. The most commonly used are 1.0 and 1.1.

The response status - This is the status code returned by the server. The status code 200 is the successful status code. There are four class of status code [13].

Bytes or Size of the file sent-This field shows the number of bytes transferred from the web server to the user.

The referrer URL - It defines where the visitor came from. This is a source for identifying the various search engines used to access the website or other websites linking to a website.

The browser and its version- This field identifies the browser type and its version. This is needed because not all browsers can view all components of a web site and each of which have different viewing capabilities.

The operating system and its version - The operating system and its version is available in the extended log. This is useful for the website developers as the website may have different look on different platforms.

Cookie information - This is a token which defines the cookie sent to a visitor. These cookies can be used to track individual users. This is helpful in generating sessions. But cookies raise the concern of privacy and requires the cooperation of the users.

We have used the extended log file of a business organization NeST for 1 month ranging from January 1, 2011 to January 31, 2011. Fig. 1 shows a sample entry of the extended log file.

117.196.136.242 - - [01/Jan/2011:00:46:11 +051800] "GET/ templates/sfo_home/images/horiz_line.gif HTTP/1.1" 200 50 "http://www.nestgroup.net/" "Mozilla/5.0 (Windows; U; Windows NT 6.1)

Fig. 1 Example Entry of an Extended Web Log

3 Data Cleaning and Pre Processing

The first pre processing task is data cleaning. The process of data cleaning is to remove noise or irrelevant data. Web server access logs represent the raw data source. It is important to identify and discard the data recorded by web robots or web crawlers, the images, sounds, java scripts etc that is often redundant and irrelevant [14]. The problems identified in pre processing of log files are the need for a large storage space due to considerable volume of data saved on disk, the existence of large amount of data that is irrelevant for the web mining process containing images, sounds, movies, icons etc., the storage of requests performed by search engines and various automated scripts, the storage of data containing error messages such as 300(redirects), 400(bad request) and 500(server errors). The advantages of pre processing include the reduction in storage space and improved precision of web mining. Search engines can generate 90% of the traffic on web sites [16]. The robots contain specific codes that are able to capture a wider range of IP addresses. Such string examples can be 'Google' or 'googlebot' for Google search engine, 'Yahoo' or 'crawl' for Yahoo search engine, 'msnbot' for MSN, 'spider', and so on. The IPs containing any of the above mentioned substrings can be identified as a search engine visit and hence eliminated. Similarly the 'robots.txt' in the URL requested field and strings like 'bot', 'spider', 'crawl' etc in the browser field are also the requests from the search engine. The pre processing also includes the elimination of records containing PUT, HEAD and POST in the HTTP requested method field. All the records containing other than 200 in the status code field is also eliminated to filter the successful requests. The irrelevant entries like .html, .txt, .jpg, .wav, .ico, .png etc re removed as it may bias the web mining tasks to follow. Table 1 shows the results of pre processing. Each user is identified with a time out of 30 minutes [10].

Table 1 Results of pre processing

Total number of records	2,86,867
Number of records after removing search engine visits	2,58,594
Number of records after pre processing	20,934
Percentage in reduction	92.70%
Total number of users	8,677

4 Indiscernibility Relations in Rough Set Theory

Indiscernibility relations in rough set theory [15] can be used for grouping data with similar characteristic from web log files. Objects characterized by the same information are indiscernible (similar) in view of the available information about them. Any set of all indiscernible (similar) objects is called an elementary set and forms a basic granule of knowledge about the universe. Let a given pair S= (U,A) of non–empty finite sets U and A, where U is the Universe of objects and A is the set consisting of attributes. The function a: $U \rightarrow V_a$, where V_a is the set of values of attribute a, called the domain of a. The pair S=(U,A) is called an information system. Any information system can be represented by a data table with rows labeled by objects and columns labeled by attributes. Any pair (x, a) where $x \in U$ and $a \in A$ defines the table entry consisting of the value a(x). Any subset B of A determines a Binary relation I(B) on U, called an indiscernibility relation defined by xI(B)y if and only if a(x)=a(y) for every $a \in B$, where a(x) denotes the value of attribute a for object x. I(B) is an equivalence relation. The family of all equivalence classes of I(B) will be denoted by U/I(B) or simply U/B. Equivalence classes of the relation I(B) of the partition U/B are referred to as B-elementary sets or B-elementary granules. Let U represents the set of the set of all user sessions with the path traversed. Let A be the subset of U which represents the < HTTP request, URL requested and status code>. An indiscernibility relation I(B) is defined for every a(x)=a(y) [15].

5 Construction of Equivalence Classes from Web Logs

We consider the subset of web log file entry which includes <HTTP request, Referrer URL, status code> for indiscernibility relation and is denoted by B. Based on the values of these attributes, the family of all equivalence classes is generated and it is denoted as U(B) . Out of the equivalence classes generated, the indiscernibility relation having the attribute values<GET, *google*, 200> ,<GET, *yahoo*,200>,<GET, *bing*, 200> etc are the equivalence classes which is useful for identifying the search engines from which the users have made their request. Similarly equivalence classes of other websites referring the current website can also be identified. The advantage of this method is that the equivalence classes generated can be further used to perform statistical analysis. Each user is identified with a time out of 30 minutes [10]. Fig. 2 shows the graphical representation of the origin of visits. The top referring sites and percentage of users entered from those sites is shown in Table 2.

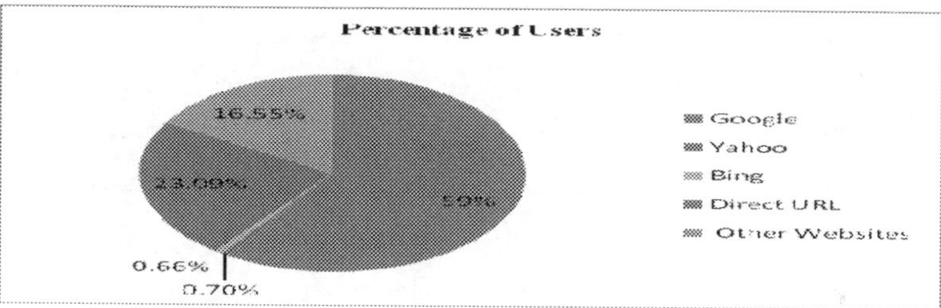

Fig. 2 Origin of visits and the percentage

Table 2 Top Referring Sites

Referring Site	Percentage of Users
www.nestsoftware.com	4.50%
www.linkedin.com	3.20%
www.opterna.com	2.90%
www.keralaindex.com	2.60%
en.wikipedia.org	1.50%
Others	1.85%

6 Extraction of Keywords and User Agents

The referrer field is further extracted to identify the popular search keywords used. It is helpful in identifying the most preferred content by the web user and properly analyzed results can even twist short and long term strategy of the organization typically in terms of products, services or intended research direction, in order to meet the dynamic expectation of the customer. Table 3 shows the most frequent keywords typed in various search engines to access the web site. Among the visitors through search engines, 11.33% of them used the keyword sfo technologies. Likewise the rest of the keywords were used. In addition to the keywords mentioned in Table 3, there were several other keywords which were less than 0.1 %. Hence these keywords were not considered relevant. The user agents are extracted to understand the browser and its version used. Equivalence class <browser, version> is generated. Table 4 shows the statistics of the different browsers and its version.

Table 3 Frequent Keywords and Percentage of Visits

Keywords	Percentage
sfo technologies	11.33%
nest group	7.97%
network systems	0.87%
nest power electronics	0.38%
nest photonics	0.38%
nest group Trivandrum	0.32%
nest information technology	0.28%

Table 4 Statistics of the browsers used

Browser	Percentage
Mozilla Version 5.0	50.91%
Microsoft Internet Explorer Version 8.0	20.94%
Microsoft Internet Explorer Version 6.0	14.15%
Microsoft Internet Explorer Version 7.0	10.78%
Opera Version 9.8	0.21%
Microsoft Internet Explorer Version 9.0	0.19%
BlackBerry	0.06%
Microsoft Internet Explorer Version 4.01	0.06%
Microsoft Internet Explorer Version 5.5	0.06%
Microsoft Internet Explorer Version 5.0	0.04%
Others	2.55%

7 Conclusion

The Extended Log File is an immense source of information to identify the user behavior, the origin of visits and sometimes even changes the long and short term strategy of the organization. It is important to know the origin of visitors for a web site developer, so that a better search engine optimization can be performed. Understanding the frequent keywords is important while creating meta tags. Search engines like Google uses several criteria for determining the PageRank. These keywords could be included as heading in a web page or if appeared in bold font may increase the PageRank. Thus the website may appear closer to the search results. The websites referring to a particular web site provides information about types of web sites linked to one's website. The information about the browser and its version is essential because the web site designers may develop sites that require viewing capabilities which may not be supported by certain browsers.

References

1. Facca, M.F., Lanzi, L.P.: Mining interesting knowledge from weblogs: a survey. Data & Knowledge Engineering 53, 225–241 (2005)
2. Cooley, R., Mobasher, B., Srivastava, J.: Web Mining: Information and Pattern Discovery on the World Wide Web. In: 9th IEEE International Conference on Tools with Artificial Intelligence (1997)
3. Pitkow, J.: In search of Reliable usage Data on WWW. In: Sixth International WWW Conference (1997)
4. Kohavi, R., Parekh, R.: Ten Supplementary Analyses to improve E-commerce Web Sites. In: Fifth KDD Workshop (2003)
5. Spiliopoulou, M.: Web Usage Mining for Web Site Evaluation. Communications of the ACM 43(8), 127–134 (2000)
6. Ortega, J.L., Aguillo, I.: Differences between web sessions according to the origin of their visits. Journal of Informetrics 4(1), 331–337 (2010)
7. Suresh, R.M., Padmajavalli, R.: An Overview of Data Pre processing in Data and Web Usage Mining. In: First International Conference on Digital Management, pp. 193–198 (2006)
8. Burton, M.C., Walther, B.J.: A Survey of Web Log Data and Their Application in Use-Based Design. In: 34th Hawaii International Conference on System Sciences, pp. 1–10 (2000)
9. Wahab, M.H.A., Mohd, M.N.H., Hanafi, H.F., Mohsin, M.F.M.: Data Pre-processing on Web Server Logs for Generalized Association Rules Mining Algorithm. In: Proceedings of the World Academy of Science, Engineering and Technology, pp. 190–197 (2008)
10. Pabarskaite, Z., Raudys, A.: A process of knowledge discovery from web log data: Systematization and critical review. Journal of Intelligent Information Systems 28, 79–104 (2007)
11. Bertot, J.C., Mcculure, C.R., Moen, W.E., Rubin, J.: Web Usage Statistics: Measurement Issues and Analytical Techniques. Government Information Quarterly 14, 373–395 (1997)
12. Hussain, T., Asghar, S., Masood, N.: Web Usage Mining: A Survey of Preprocessing of Web Log File. In: International Conference on Information and Emerging Technologies, pp. 1–6 (2010)
13. Internet: Hypertext Transfer Protocol Overview, `http://www.w3.org/protocols` (last retrieved October 2011)
14. Mican, D., Sitar-Taut, D.: Preprocessing and Content/Navigational Pages Identification as Premises for an Extended Web Usage Mining Model Development. Informatica Economica 13(4), 168–179 (2009)
15. Pawlak, Z., Skowron, A.: Rudiments of Rough Sets. Information Sciences 177(1), 3–27 (2007)

A Novel Community Detection Algorithm for Privacy Preservation in Social Networks

Fatemeh Amiri, Nasser Yazdani, Heshaam Faili, and Alireza Rezvanian

Abstract. Developed online social networks are recently being grown and popularized tremendously, influencing some life aspects of human. Therefore, privacy preservation is considered as an essential and crucial issue in sharing and propagation of information. There are several methods for privacy preservation in social networks such as limiting the information through community detection. Despite several algorithms proposed so far to detect the communities, numerous researches are still on the way in this area. In this paper, a novel method for community detection with the assumption of privacy preservation is proposed. In the proposed approach is like hierarchical clustering, nodes are divided alliteratively based on learning automata (LA). A set of LA can find min-cut of a graph as two communities for each iteration. Simulation results on standard datasets of social network have revealed a relative improvement in comparison with alternative methods.

Keywords: Social Networks, Privacy Preservation, Community Detection, Top-Down Hierarchical Clustering, Learning Automata.

1 Introduction

Nearly half of the internet users today are members of some online social network which results in change behaviors of users over the web. Indeed, users are needed to be both creators and managers of the content. For each piece of content, users who upload information must control what and with whom they share. When users post a status update, photo, wall and video, they must also select their

Fatemeh Amiri · Nasser Yazdani · Heshaam Faili
Department of Electrical and Computer Engineering, University of Tehran, Tehran, Iran
e-mail: f.amiri@ece.ut.ac.ir,
{yazdani,hfaili}@ut.ac.ir

Alireza Rezvanian
Computer & IT Engineering Department, Amirkabir University of Technology, Tehran, Iran
e-mail: a.rezvanian@aut.ac.ir

A. Abraham and S.M. Thampi (Eds.): Intelligent Informatics, AISC 182, pp. 443–450.
springerlink.com

audience at the time they post. According to [1] on average, each user has 130 friends and 80 groups and on average, 90 pieces of content are uploaded by him/her per month. So, the task of managing access to content has turned into a significant mental burden for many users. As a result, the issue of privacy on social networks has received important attention among numerous researches [19]. Privacy preferences are highly correlated with the community structure of the graph and tend to break down along the lines of the community structure [2]. The community can explore properties without releasing the individual privacy information. According to [3], communities are groups of vertices in a network, such that edges between vertices in the same community are dense though spare between different communities. Community addresses useful information about type of interpersonal relations, how to exchange content and the way of distribution within social networks.

In this work, it has been attempted to detect communities using graph theory in order to stop information leaks of communities and to ensure privacy. Conventionally, social network can be seen as a graph by representing users as vertices and relationship between the users as edges. Community detection is essential for identifying structure of the users according to the natural structure of users. Therefore, sharing the information alone within groups might have an important function of privacy preservation. A top-down hierarchical clustering algorithm is ad-dressed in this work to detect communities in graphs represented for the social network. We first assume that there is one community in the graph; the proposed method split this community into two communities during the next step. Min-Cut problem is employed in order to split a big community into smaller ones. In graph theory, a min-cut of graph is defined as a cut that has either the smallest number of elements (un-weighted case) or the smallest sum of possible weights. There are several algorithms to find the min-cut. This paper proposes an algorithm to compute min-cut based on Learning Automata. For this purpose, a learning automaton (LA) is assigned to each node of graph. Each LA chooses the optimal action at random, while the environment responds the taken action in turn with a reinforcement signal, in order to extracting strongly connected components as communities [4].

Several community detection algorithms have been presented in recent years. Two major methods recently adopted are shortest path betweenness [5] and network modularity [6-9]. In the former, the shortest paths between all pairs of nodes are calculated and the number of runs along each edge is counted. In the latter, it has been tried to minimize the number edges between groups based on a modularity function. In [10] by using a random walk, local community was composed followed by refining the nodes within communities. An application of community detection for security in the social networks were presented in [11], they have suggested a clustering method for analyzing the relations inside social networks with complex structures.

For the rest of the paper, Section 2 introduces learning automata briefly. Then proposed algorithm is described in Section 3. In Section 4, standard dataset is described then experimental results are discussed in Section 5. Finally, conclusion and future works are presented in Section 6.

2 Learning Automata

A learning automaton (LA) [12,13] is an adaptive decision-making unit that improves its performance by learning how to choose the optimal action from a finite set of possible actions through repeated interactions with a random environment. The action is chosen at random based on a probability distribution kept over the action-set and at each instant the given action is served as the input to the random environment. The environment responds the taken action in turn with a reinforcement signal. The action probability vector is updated based on the reinforcement feedback from the environment. The objective of a learning automaton is to find the optimal action from the action-set so that the average penalty received from the environment is minimized.

The environment can be described by a triple $E \equiv \{\alpha, \beta, c\}$, where $\alpha \equiv \{\alpha_1, \alpha_2, \ldots, \alpha_r\}$ represents the finite set of the inputs and $\beta \equiv \{\beta_1, \beta_2, \ldots, \beta_m\}$ denotes the set of the values can be taken by the reinforcement signal, and $c \equiv \{c_1, c_2, \ldots, c_r\}$ denotes the set of the penalty probabilities in which the element c_i is associated with the given action α_i. If the penalty probabilities are constant, the random environment is said to be a stationary random environment, and if they vary with time, the environment is called a non-stationary environment [14].

Variable structure learning automata is represented by a triple $<\beta, \alpha, T>$, where β is the set of inputs, α is the set of actions, and T is learning algorithm. The learning algorithm is a recurrence relation that is used to modify the action probability vector. Let $\alpha(k)$ and $p(k)$ denote the action chosen at instant k and the action probability vector on which the chosen action is based, respectively. The recurrence equation shown by equations (1) and (2) is a linear learning algorithm in which the action probability vector p is updated. Let $\alpha_i(k)$ be the action chosen by the automaton at instant k.

$$p_j(n+1) = \begin{cases} p_j(n) + a \times \lfloor 1 - p_j(n) \rfloor & i = j \\ (1-a)p_j(n) \quad \forall_j & i \neq j \end{cases} \tag{1}$$

When the taken action is rewarded by the environment (i.e. $\beta(n) = 0$) and

$$p_j(n+1) = \begin{cases} (1-b)p_j(n) & i = j \\ \left(\dfrac{b}{r-1}\right) + (1-b)p_j(n) \quad \forall_j & i \neq j \end{cases} \tag{2}$$

When the taken action is penalized by the environment (i.e. $\beta(n) = 1$). r is the number of actions which can be chosen by the automaton, *a(k)* and *b(k)* denote the reward and penalty parameters and determine the amount of increases and decreases of the action probabilities, respectively [16].

3 Proposed Algorithm

A top-down hierarchical algorithm is proposed for community detection. At first, all users as nodes inside a graph are considered in a big community. At each step, communities are formed iteratively. In order to stop splitting communities, it is suggested to apply modularity function, which is presented by Girvan-Newman to indicate the quality of the communities. Modularity is defined as follow [15]:

$$Q = \frac{1}{2m} \sum_{i,j} \left(A_{ij} - \frac{k_i k_j}{2m} \right) \delta\left(C_i, C_j\right) \tag{3}$$

Where, the sum of the runs over all pairs of vertices (*A*) is the adjacency matrix, *m* is total number of edges inside the graph, and K_i represents the degree of ith node. If vertices *i* and *j* are in the same community the δ function yields one and yields zero otherwise. A partitioning with maximum value of modularity measure is chosen. One pseudo-code of the proposed algorithm is depicted in Fig. 1 below.

Algorithm 1 proposed algorithm for community detection

in 1st step, graph is separated into 2 communities C^1_{11} and C^1_{12} by Community Splitting()
Do
Suppose the set of communities have been detected until i-1th step is $\{C^{i-1}\}$
a. In i^{th} step, computer Q metric for set of communities $\{C^{i-1}\}$: Q^{i-1}
b. for each member of $\{C^{i-1}\}$: C_j^{i-1} repeat:
b.1. split C_j^{i-1} into two small communities $\{C^i_{j1}, C^i_{j2}\}$ by **Community Splitting()**
b.2. compute Q measure for $\{\{C^{i-1}\} - C_j^{i-1}\} \cup \{C^i_{j1}, C^i_{j2}\}$:$Q^1$
b.3. If $Q^{i-1} - Q^1 < 0$ then add $\{C_{j1}{}^i, C_{j2}{}^i\}$ to communities discovered right now: $\{\{C^{i-1}\} - C^{ji-1}\} \cup \{\{C_{j1}{}^i, C_{j2}{}^i\}$
c. $\{C^i\} \leftarrow \{C^{i-1}\}$
d. compute Q metric for set of communities $\{C^i\}$: Q^i
f. $i \leftarrow i+1$
While $Q^{i-1} - Q^1 >$ threshold
Return $\{C^{i-1}\}$

Fig. 1 Pseudo-code of proposed algorithm

For dividing communities, min-cut problem is proposed for splitting the big communities into smaller ones using learning automata (Fig. 1 and line b.1 of algorithm 1). The proposed min-cut method creates two sets of nodes with the minimum number of edges between them. A set of learning automata is assigned to each node of the graph. Every node of learning automata could choose one of two actions, namely: "become a member of S1" or "become a member of S2", where, S1 and S2 are two disjoint communities. The pseudo-code for the community splitting algorithm is illustrated in Fig. 2.

Algorithm 2 Community Splitting()

Assign nodes into two set of S_1 and S_2 randomly
Initialize probability vector of every nodes of v_i for LA
Repeat N times
a. Select an action by LA for some vertex of S_1/S_2 and exchange assignment
b. Evaluate the performance of the selected action
c. Update the probability vector of LA of v_i

Fig. 2 Pseudo-code of community splitting algorithm

4 Social Networks Dataset

The performance of our approach is evaluated using four popular social networks. Zachary's karate (Fig. 3.a) is a social network of interactions between people in a karate club which has experienced 34 members of a karate club over a period of 2 years. During the study, a disagreement developed between the administrator of the club and the club's instructor, which caused the instructor to leave and start a new club. Another dataset is American College Football Teams (Fig. 3.b) with 115 vertices representing the teams with two vertices being connected when their teams play against each other. The teams are divided into 12 conferences. Games between teams in the same conference are more frequent than games between teams of different conferences, so one has a natural partition where the communities correspond to the conferences. The third network is Dolphin (Fig. 3.c), the social network of frequent associations between the dolphins. There are 62 dolphins and edges were set between animals which were seen together more often than expected by chance. The dolphins separated in two groups once a dolphin left the place for some time. Due to the natural classification, Lusseau's dolphin network is often used to test algorithms for community detection.

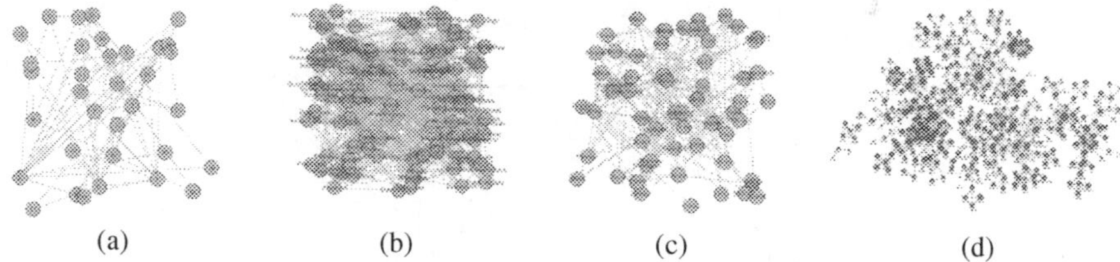

Fig. 3 Social network dataset

The last network is Netscience (Fig.3. d), a coauthorship network of scientists working on network theory and experiment, as compiled by M. Newman in May 2006. The network was compiled from the bibliographies of two review articles on networks. The network contains all components for a total number of 1589 scientists and not just the largest component of 379 scientists [15]. Description of the mentioned social network dataset is summarized in Table 1.

Table 1 Social networks descriptions

Dataset	Nodes	Edges
Zachary's karate	34	78
American college football teams	115	613
Dolphin	62	159
Coauthorships NetScience	379	914

5 Simulation Results

In experiment, L_{RI} is used for learning algorithm of learning automata having $\alpha = 0.01$. Different values of α have been examined to receive good results. Learning process for every learning automaton repeated 1000 times to determine the optimal min-cut and created two communities. All experiments were performed on a Windows platform having configuration Intel® core 2 Duo CPU 2.5 GHZ, 3 GB RAM and Windows XP. Our results on four networks are shown in Table 2. The experiments are repeated 30 times independently, the average values of experimental results are represented in Table 2. The run time and the value of modularity measure of our proposed method from min-cut and learning automata as *MCL* are given for each dataset in Table 2. Moreover, several experiments were conducted to compare our proposed method with CNM [6], PL [17], PBD [18]. The obtained results were compared with those of other community detection methods in terms of modularity and run-time by Tables 2 and 3, respectively. As can be seen in Tables 2 and 3, the recently proposed method is comparable and in some cases better than other methods.

Table 2 Comparison of modularity for our proposed algorithm and other methods

Data set	CNM[6]	PBD[18]	PL[17]	MCL
Zachary's karate	0.387	0.394	0.335	0.411
American college football teams	0.454	0.447	0.452	0.456
Dolphin	0.322	0.359	0.349	0.386
Coauthorships NetScience	0.837	0.837	0.828	0.837

Table 3 Comparison of run-time for our proposed algorithm and other methods

Data set	CNM[6]	PBD[18]	PL[17]	MCL
Zachary's karate	0.518	1.031	0.238	0.697
American college football teams	2.539	2.706	2.147	2.902
Dolphin	0.854	0.887	0.808	1.279
Coauthorships NetScience	9.134	10.342	8.609	12.932

For further improvements, one can apply a random walk to make initial sets, because it helps to make two partitions of neighborhood vertices. Moreover, for isolated nodes, one can randomly place them into one of the two subsets.

6 Conclusions

In this paper, a new community detection algorithm has been introduced for data privacy preservation, which is a top-down hierarchical algorithm based on learning automata. At each step, one community is divided into two smaller communities using the learning automata. The process of division continues until it satisfies a stop measure. The modularity measure was applied to terminate community division and also to reach better results. The results of experiments on Karate, Football, Dolphin and Netscience datasets are compared with other methods cited in relevant literatures in terms of modularity measure and run time. Experiments have demonstrated that our proposed method is comparable with other methods or even outperforms them. As the future work, one can apply a random walk to make initial set in this algorithm for the purpose of additional improvement. This would contribute to produce two partitions of neighborhood vertices. Moreover, they can be randomly placed into one of the two subsets for the isolated nodes.

References

1. Facebook Statistics (2011), `http://www.facebook.com/press/info.php?statistics` (Accessed November 10, 2011)
2. Fang, L., LeFevre, K.: Privacy Wizards for Social Networking Sites. In: IW3C2, pp. 351–360 (2010)
3. Newman, M.E.J.: Networks: an introduction. Oxford University Press (2010)
4. Fortunato, S.: Community detection in graphs. Phys. Rep. 486, 75–174 (2010)
5. Girvan, M., Newman, M.: Community structure in social and biological networks. P. Natl. Acad. Sci. Usa 99, 7821 (2002)
6. Clauset, A., Newman, M., Moore, C.: Finding community structure in networks using the eigenvectors of matrices. Phys. Rev. E 70(6), 066111 (2004)
7. Blondel, V.D., Guillaume, J.L., Lambiotte, R., Lefebvre, E.: Fast unfolding of communities in large networks. J. Stat. Mech. Theory Exp., P10008 (2008)
8. Wakita, K., Tsurumi, T.: Finding community structure in mega-scale social networks. In: WWW/Internet (2007)
9. Yan, B., Gregory, S.: Detecting communities in networks by merging cliques. In: ICIS (2009)
10. Thakur, G.S., Tiwari, R., Thai, M.T., Chen, S.S., Dress, A.W.M.: Detection of local community structures in complex dynamic networks with random walks. IET Syst. Biol. 3(4), 266–278 (2009)
11. Yang, B., Sato, I., Nakagawa, H.: Secure Clustering in Private Networks. In: ICDM (2011)
12. Rezvanian, A., Meybodi, M.R.: An adaptive mutation operator for artificial immune network using learning automata in dynamic environments. In: NaBic, pp. 479–483 (2010)
13. Rezvanian, A., Meybodi, M.R.: LACAIS: Learning Automata Based Cooperative Artificial Immune System for Function Optimization. In: Ranka, S., Banerjee, A., Biswas, K.K., Dua, S., Mishra, P., Moona, R., Poon, S.-H., Wang, C.-L. (eds.) IC3 2010. CCIS, vol. 94, pp. 64–75. Springer, Heidelberg (2010)
14. Thathachar, M., Sastry, P.S.: Varieties of learning automata: an overview. IEEE T. Syst. Man Cy. B 32(6), 711–722 (2002)
15. Newman, M.E.J.: Finding community structure in networks using the eigenvectors of matrices. Phys. Rev. E 74(3), 036104 (2006)

16. Narendra, K.S., Thathachar, M.A.L.: Learning automata: an introduction. Prentice-Hall (1989)
17. Pons, P., Latapy, M.: Computing communities in large networks using random walks. In: ISCIS (2005)
18. Pujol, J.M., Béjar, J., Delgado, J.: Clustering algorithm for determining community structure in large networks. Phys. Rev. E 74(1), 016107 (2006)
19. Amiri, F., Yousefi, M.M.R., Lucas, C., Shakery, A., Yazdani, N.: Mutual information-based feature selection for intrusion detection systems. J. Netw. Comput. Appl. 34(4), 1184–1199 (2011)

Provenance Based Web Search

Ajitha Robert and S. Sendhilkumar

Abstract. During web search, we often end up with untrusted, duplicates and near duplicate search results which dilutes the focus of search query. Factors that may influence the trust of web search results shall be referred to as 'Provenance'. Provenance is basically the information about the history of data. In this paper, we propose a provenance model which uses both content based and trust based factors in identifying trusted search results. The novelty of our idea lies in attempting to construct a provenance matrix which encompasses 6 factors (who, where, when, what, why, how) related to the search results. Inferences performed over the provenance matrix leads to trust score which is then utilized to remove near-duplicates and retrieve trusted search results.

Keywords: Web search, Provenance Mining, Provenance Matrix, Near-Duplicates, Trust, Semantics, Document Clustering, Ontology.

1 Introduction

With the growing number of Web pages on the Internet, it has become increasingly difficult for users to find desired information. The approaches have been made and still researches are going on to optimize the Web search [1]. Most users just view top 5-10 search results and therefore might miss relevant information because search results may contain unrelated, near-duplicate and untrustworthy results. So, Trustable and relevant web findability is a main issue in Web search.

A recent study shows that one of the main factors that influence the trust of users in Web content is provenance. Provenance information about a data item is information about the history of the item, starting from its creation, including information about its origins [2]. Information about provenance determines the quality, reliability and amount of trust. So based on the provenance information it can be ensured that the users are provided with the most trustworthy results at the

Ajitha Robert · S. Sendhilkumar
Department of Information Science & Technology,
College of Engineering Guindy, Anna University
e-mail: ajitharobert01@gmail.com,
ssk_pdy@yahoo.co.in

A. Abraham and S.M. Thampi (Eds.): Intelligent Informatics, AISC 182, pp. 451–458.
springerlink.com

top of the search results and also effectively eliminating the not so trustworthy duplicate documents. Provenance information is calculated based on Who (has authored a document), What (is the content of the document), When (it has been made available), Where (it has been published), Why (the purpose of the document), How (it is linked).

2 Related Works

Qiang Ma et. al. [3] stated a Content coverage based Search, where the topic sketch (overview) of the documents is compared with Wikipedia content for the same query. Xinye Li et. al. [4] used cosine similarity to remove the duplicate documents. In Danushka Bollegala et. al. [5] approach semantic similarity between words is found to eliminate the near duplicates. Query Expansion is stated in Nicole Anderson [6] work .(i.e) Contextual information such as Location, Personal preferences, Personal vocabularies (eg pc for personal computer) is added to queries which helps to find the relevant documents but it doesn't help effectively in finding trustworthy documents. Sunil et. al. [7] stated an approach to eliminate similar Web pages using Crawler which uses Filtering agent to remove pages with similar url's. Duygu Taylan et. al. [8] stated a focused crawler approach which calculates link score based on page relevancy, and crawler determines which links have to be added in URL frontier based on the link score.

Katsumi Tanaka et. al. [9] proposed knowledge and trust oriented search i.e Topic, coordinate and association terms are found. But topic is focused much rather than the content. Chuan Huang et. al. [10] proposed a trust and popularity based search which uses link analysis of author reputation for improved trustworthy results. Olaf Hartig [2] describes the provenance of a specific data item from the Web by a provenance graph. Ivan Vasquez et. al. [11] describes about data provenance which is the evolution of data, including the source and authority of data creation, changes to the data along the life history. Y. Syed Mudhasir et. al. [12] stated about near duplicate detection and elimination. Only four of the Provenance factors are used to detect and eliminate near duplicates.

3 Provenance Based Web Search

Overall architecture diagram is depicted in figure 1. Web documents are collected in the area of datamining using a specialized browser which downloads top 30 results for a given research article title query. Preprocessing should be done to those collected web documents for constructing Document Term Matrix (DTM). Since constructed DTM is large in size, SVD technique is used to reduce its dimension to compute the similarity with cosine query. After mining provenance values, Provenance Matrix is constructed for the six provenance values and its weight calculation is done. From provenance weight, Trust value is calculated. Cluster those documents based on its similarity. Duplicate documents are detected from each cluster using trust value and they are eliminated, Finally re-ranking is done for the original documents.

The collected Web documents were of varied formats with extensions.html, .pdf, .doc and .txt. Preprocessing to those documents involves: (i) removing html tags and scripting elements (parsing), (ii) pdftotext conversion (iii) tokenizing, (iv) removal of stop words, (v) Domain specific stemming. (a domain specific dictionary, (a list of domain specific words are maintained) is used to check for domain specific words. eg Web mining should not be stemmed as Web mine).

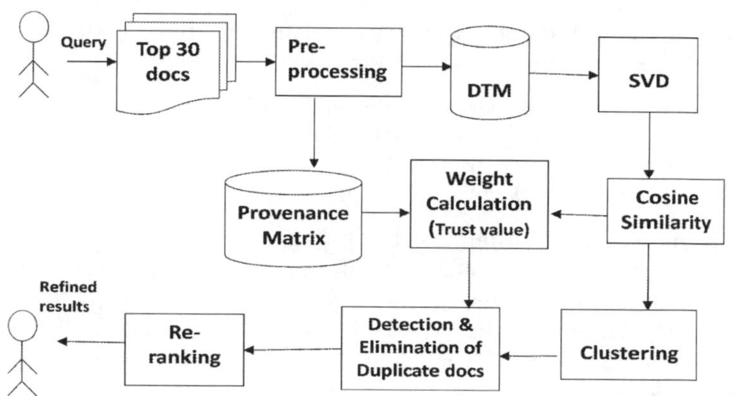

Fig. 1 Overall System Architecture

Document-term matrix(DTM) describes the frequency of terms that occur in a collection of documents. Original DTM is of dimensions 18936 rows and 250 columns i.e.,27936 terms and 250 documents. The matrix after k reduced SVD is 130 rows and 250 columns. Cosine similarity(Equation 1) is computed for document-query comparison and the resultant value is utilized in calculating the trust score. The resultant similarity value from Document Document comparison is used for document clustering.

$$SIM(D_i, D_j) = \cos\theta = \frac{(D_i \cdot D_j)}{|D_i||D_j|} \tag{1}$$

Web pages are clustered using k-means clustering algorithm[16]. The documents with the comparable similarity values are clustered.

3.1 Provenance Mining and Its Weightage

Provenance mining is extracting all the provenance values from document. Since Domain Specific Search is focused, search results mostly contain published papers. So search results are splitted into i) Ordinary web documents & ii) Published papers (Includes pdf and html version).

WHO (Author analysis): Author name is extracted from web documents. Author weightage is assigned based on the number of publications provided in DBLP[14]. If number of authors is more than one, then average is calculated.

WHERE (Location Analysis): For ordinary webpages Some website holders publish their websites in their own Web servers and others will publish in some other web servers in rental manner. Owners of web server is given more weightage. For Published papers Weightage is given based on the following priority.

Transactions>Journal>International conference>National Conference

WHEN (Temporal analysis) Extract the created date and last updated date(if applicable) from web documents. Recent created date and updated date has more weightage and the summation of both is taken as final weightage.

WHAT (Syntactic Analysis) The syntactic similarity (cosine similarity) value between document and query is used for the purpose of What factor.

WHY (Semantic analysis) The semantic similarity value between document and query is analyzed for the purpose of finding weight of Why factor for all the documents.Why depicts the purpose of the document. So the document level (Abstract intermediary or detailed) is analyze

Input: Query, DTM, **Output:** Document level

1. Mark the level of query domain specific terms in the domain ontology.
2. From dtm, get the domain specific terms which has count more than 0.
3. Check for domain specific terms in dtm, which falls inside the marked level
4. If more words match, (i.e above the threshold value), found inside the level of domain specific query term then document is in depth.
5. Use equation (2) to find the depth of the document

$$\boldsymbol{whywt} = \left(\frac{\boldsymbol{MDW}}{\boldsymbol{TDW}}\right) \tag{2}$$

TDW=Total domain specific words in a document & MDW=Matched domain specific words inside the level of domain specific query term

Ontology Construction[13]: Datamining ontology is built by referring the classification of database and Information retrieval of ACM taxonomy and Data Mining book of Jiawei Han & Micheline Kamber

HOW (Link Analysis)

Published paper:
Google scholar[15] is used for the purpose of finding **citation count**. Citation weight is assigned based on the citation count. For Title and reference relevancy, the similarity value itself is taken as its weight. Final weightage for how factor is computed using the equation 3.

$$How_{wt\_pub} = 0.5CW + 0.25TR + 0.25RR \tag{3}$$

CC-Citation Weight, TR-Title Relevancy, RR-Reference Relevancy

Ordinary Web pages
Inlink Weight: Google search special query is used for the purpose of finding inlink. Inlink weight is assigned based on inlink count

Quality Inlink Weight: Quality inlink is authoritative inlinks (i.e) inlinks from educational, organization and government sites. Google Page Rank is utilized for giving weightage for quality inlinks. Google gives page rank score for each webpage. It ranges from 0 to 10.

Knowledge Path Weightage (Webpage and its Outlinks)

Knowledge path weightage for given webpage and its outlinks is calculated using the Knowledge path count. For outlinks, average of outlinks is calculated. Final weightage of how factor for ordinary pages is calculated using the equation 4

$$How_{wt\_ord} = 0.3IW + 0.25KPW + 0.2KPOW + 0.25QIW \quad (4)$$

IW - Inlink weightage, KP - Knowlede path weightage of given web page, KPOW-Knowledge path weightage of outlinks for the given web page, QIW-Google Page Rank Weightage of Quality inlinked sites

Provenance Matrix is formed where, the rows represent the provenance factors and the columns represents the documents and the respective fields contain provenance weight. Table 3 show the weightage of provenance factors which is calculated from the extracted provenance values. The contents of provenance matrix is used to calculate the trust value for all the documents.

Table 1 and 2 summarizes the extracted provenance values. Since Link factor has subfactors. How factor values are shown separately in table 2.

Table 1 Provenance Matrix extracted values

Docs/ factrs	Doc 1	Doc 2	Doc 3	Doc 4
Who	Yan Wan	Katsumi Tanaka	QINGYUZ HANG	Yutaka Matsuo
Where	2	IEEE	IJITDM	IJITDM
When	2011	2010	2011	2009
What	Content	Content	Content	Content
Why	Abstract	Intermedia ry	Detailed	Abstract

Table 2 Link factor- sub matrix

Link Factors	Doc 1	Doc 2	Doc 3	Doc 4
Citation count	-	22	3	2
Title Relevancy	-	0.67	0.52	0.28
Reference Relevancy	-	0.8	0.12	0.45
Inlink Count	21	-	-	-
Quality inlink	4	-	-	-
Knowledge Path	1	-	-	-
Knowledge Path for outlink	3	-	-	-

Table 3 Weightage of Provenance Matrix values

Docs/ factrs	Doc1	Doc2	Doc3	Doc4
Who	0.2	0.5	0.3	0.2
Where	0.3	0.9	0.5	0.7
When	0.92	0.8	0.9	0.74
What	0.78	0.67	0.23	0.89
Why	0.20	0.50	0.9	.35
How	0.6	0.8	0.3	0.5

Table 4 Computed trust values

Document ID	Trust Value
Doc1	**0.452**
Doc2	**0.671**
Doc3	**0.890**
Doc4	**0.823**
Doc5	**0.035**
Doc6	**0.582**
Doc7	**0.621**
Doc8	**0.724**
Doc9	**0.192**
Doc10	**0.201**

Initially, 50 documents are taken and trust value is analyzed for each factor. Using the separate analysis of each factor its multiplicative factor is assigned. From the result, it is observed that Syntactic (What) and Semantic (Why) factor results in more relevant documents. Author (Who) and Link (How) factor results in more trustable documents. So they should be given high priority next to syntactic and semantic factors. Temporal (when) and location (where) factor moderately contribute in trustable documents. So both should be given weights lesser than Who and How factor. The following equation 5 defines trust

$$\boldsymbol{Trust} = \mathbf{0.25Pwhat} + \mathbf{0.1Pwhen} + \mathbf{0.05Pwhere} + \mathbf{0.2Pwho} + \mathbf{0.25Pwhy} + \mathbf{0.15Phow} \quad (5)$$

Pwhat is the weight for what factor and so the respective factors. Its multiplicative values are assigned such that the resultant trust value will range between 0 and 1. The trust values calculated for 10 documents is shown in Table 4.

Documents are clustered based on their similarity value and then trust values are computed for all the documents. So now near duplicate documents are detected and have to be eliminated For each cluster, If the cluster has single document, it is added to the list of original documents. Else Choose the highest trust valued document as the original one, the rest of the documents are near duplicates.For re ranking, Sort the document list based on trust values. i.e the documents with higher trust value will appear first.

4 Evaluation

Google ranking is same for all the Webpages in a particular website. For example, all the papers in an ieee website is ranked same in Google ranking, But in the proposed system, each paper or each page in a website is ranked separately based on the provenance factors.

Mean Average Precision for the collected queries is 0.9661. Calculated F-measure values are plotted in graph which is shown in figure 2. From the graph it is observed that the collected dataset for 25 domain specific queries has f-measure value above 0.85. F- measure reaches 1 at its best option. So it is observed that it gives most relevant and trusted results for domain specific queries.

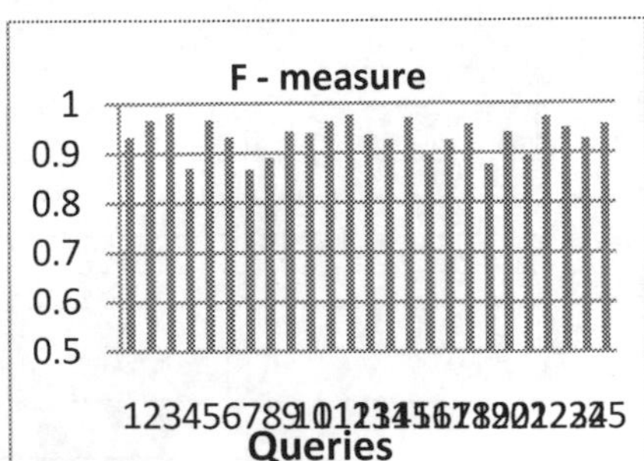

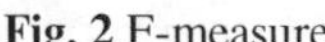

Fig. 2 F-measure

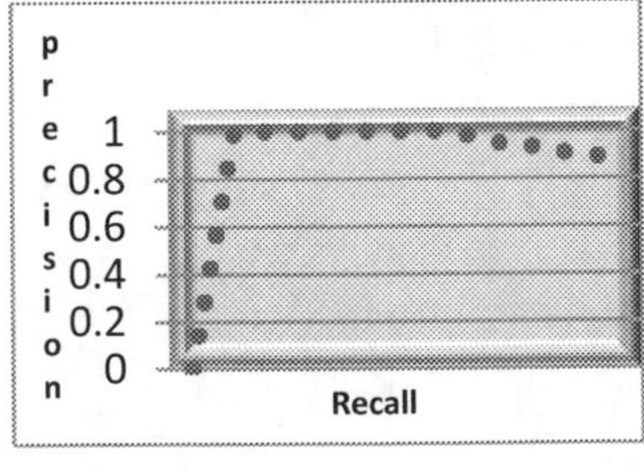

Fig. 3 ROC curve

ROC curve is drawn using Precision and Recall values which is shown in figure 3. If it is relevant, then both precision and recall increase, and the curve jags up and to the right. From figure 4 it is observed that ranked 10 results of proposed system give trusted results except in the 8th or 9th or 10th position for certain querie.

5 Conclusion and Future Work

Thus a novel method has been proposed to retrieve the trusted Web search results by removing the untrusted, duplicate and near-duplicate documents through Provenance factors. In future Work, the architecture of search engine can be effectively designed using provenance for the purpose of trusted search results.

References

1. Singh, B., Singh, H.K.: Web Data Mining Research: A Survey. In: Computational Intelligence and Computing Research (ICCIC), pp. 1–10 (2010)
2. Hartig, O.: Provenance Information in the Web of Data. In: Proceedings of the Linked Data on the Web (LDOW) Workshop at the World Wide Web Conference (WWW), Madrid, Spain, pp. 1–7 (April 2009)
3. Ma, Q., Miyamori, H., Kidawara, Y., Tanaka, K.: Content-coverage Based Trust-oriented Evaluation Method for Information Retrieval. In: Proceedings of the Second International Conference on Semantics, Knowledge, and Grid (SKG 2006), pp. 22–26 (2006)
4. Li, X., Yang, Q., Zeng, L.: Clustering Web Retrieval Results Accompanied by Removing Duplicate Documents. In: 2010 International Conference on Web Information Systems and Mining, pp. 259–261 (2010)
5. Bollegala, D., Matsuo, Y., Ishizuka, M.: A Web Search Engine-Based Approach to Measure Semantic Similarity between Words. IEEE Transactions on Knowledge and Data Engineering 23, 977–990 (2011)
6. Anderson, N.: Putting Search in Context: Using Dynamically-Weighted Information Fusion to Improve Search Results. In: 2011 Eighth International Conference on Information Technology, pp. 66–71 (2011)
7. Pandey, S.K., Mishra, R.B.: Intelligent Web Mining Model to Enhance Knowledge Discovery on the Web. In: Proceedings of the Seventh International Conference on Parallel and Distributed Computing, Applications and Technologies, pp. 339–343 (2006)
8. Taylan, D., Poyraz, M., Akyokuş, S., Ganiz, M.C.: Intelligent Focused Crawler: Learning which Links to crawl, pp. 504–508. IEEE (2011)
9. Tanaka, K.: Knowledge Search and Trust-oriented Search. In: International Conference on Informatics Education and Research for Knowledge-Circulating Society, pp. 81–86 (2008)
10. Huang, C., Chen, Y., Wang, W., Cui, Y., Wang, H., Du, N.: A novel social search model based on trust and popularity. In: Proceedings of IC-BNMT, pp. 1030–1034 (2010)
11. Vasquez, I., Gomadam, K., Patterson, S.: Data Provenance in next-gen information systems: Adding, extracting and analyzing information in the Web services domain (2008)

12. Syed Mudhasir, Y., Deepika, J., Sendhilkumar, S., Mahalakshmi, G.S.: Near-Duplicates De-tection and Elimination Based on Web Provenance for Effective Web Search. (IJIDCS) International Journal on Internet and Distributed Computing Systems 1(1), 22–32 (2011)
13. Subhashini, R., Akilandeswari, J.: A Survey On Ontology Construction Methodologies. International Journal of Enterprise Computing and Business Systems 1(1), 60–72 (2011)
14. Biryukov, M., Wang, Y.: Classification of Personal Names with Application to DBLP. In: Third International Conference on Digital Information Management (ICDIM), pp. 131–137 (2008)
15. Beel, J., Gipp, B.: Google Scholar's ranking algorithm: The impact of citation counts (An empirical study). In: Third International Conference on Research Challenges in Information Science (RCIS), pp. 439–446 (2009)
16. Poomagal, S., Hamsapriya, T.: K-Means for Search Results Clustering Using URL and Tag Contents. In: International Conference on Process Automation, Control and Computing (PACC), pp. 1–7 (2011)

A Filter Tree Approach to Protect Cloud Computing against XML DDoS and HTTP DDoS Attack

Tarun Karnwal, Sivakumar Thandapanii, and Aghila Gnanasekaran

Abstract. Cloud computing is an internet based pay as use service which provides three type of layered services (Software as a Service, Platform as a Service and Infrastructure as a Service) to its consumer on demand. These on demand service facilities is being provide by cloud to its consumers in multitenant environment but as facility increases complexity and security problems also increase. Here all the resources are at one place in data centers. Cloud uses public and private APIs (Application Programming Interface) to provide services to its consumer in multitenant environment. In this environment Distributed Denial of Service attack (DDoS), especially HTTP, XML or REST based DDoS attacks may be very dangerous and may provide very harmful effects for availability of services and all consumers may get affected at the same time. One other reason is that because the cloud computing users make their request in XML and then send this request using HTTP protocol and build their system interface with REST protocol (such as Amazon EC2 or Microsoft Azure) hence XML attack more vulnerable. So the threaten coming from distributed REST attacks are more and easy to implement by the attacker, but to security expert very difficult to resolve. So to resolve these attacks this paper introduces a comber approach for security services called filtering tree. This filtering tree has five filters to detect and resolve XML and HTTP DDoS attack.

Keywords: Economical Distributed Denial of Service (EDDoS), Militant environment, Distributed Denial of Service(DDoS) Attacks, Pay as Use, Cloud Security, SaaS, Paas, IaaS.

1 Introduction

Cloud computing is a combination of distributed system, utility computing and grid computing. Cloud Computing uses combination of all these three in

Tarun Karnwal · Sivakumar Thandapanii · Aghila Gnanasekaran
Dept. of Computer Science, Pondicherry University
Puducherry, India
e-mail: {karnwals,Aghila}@gmail.com,
tsivakumar.csc@pondiuni.edu.in

A. Abraham and S.M. Thampi (Eds.): Intelligent Informatics, AISC 182, pp. 459–469.
springerlink.com

virtualized manner. Cloud computing converts desktop computing into service based computing using server cluster and huge databases at data center. Cloud computing gives advanced facility like on demand, pay per use, dynamically scalable and efficient provisioning of resources. Cloud computing the new emerged technology of distributed computing systems changed the phase of entire business over internet and set a new trend. The dream of Software as a Service becomes true; Cloud offers Software as a Service (SaaS), Platform as a Service (PaaS) and Infrastructure as a Service (IaaS). Cloud offers these services with the help of Web Services.

Cloud computing providing services to its consumers at abstract level and take care of all the internal complex tasks. With cloud computing consumer life became easy. But "as the nature rule with increase in facility vulnerability also increases".

Similarly Cloud provides the facility to consumers in the same way it provides facility to attackers also. There are more chance of attacks in cloud computing. As cloud computing mainly provides three types of services so in each layer have some soft corners which invite attackers to attack. Some of these soft corners are (1) SaaS vulnerability as Insecure Application Programming Interface (API), Account or Service hacking, Attack on cloud firewall / Attack on public firewall, Attack on consumer browser, Integrity, Confidentiality and Availability (2) PaaS vulnerability as Insecure Application Programming Interface (API), Unknown risk profile (Heartland Data Breach), Integrity, Confidentiality and Availability (3)IaaS vulnerability as Data leakage in Virtual Machine, Shared technology issues, Integrity, Confidentiality and Availability

So among all these different vulnerabilities Availability affects all three layers and more harmful. Every Cloud has its own APIs or adapters that need to be installed or consumed if anyone wants to use that Cloud. These adapters are publically available and this paper objective is to provide security to this Open API from HTTP and XML based Denial of service attacks.

The largest DDoS attacks have now grown to 40 gigabit barrier this year and may reach to 100 gigabits soon. So if someone threatens to bring down the cloud system with DDoS attack cloud may become worrisome. XML-based DDoS and HTTP-based DDoS are more destructive than the traditional DDoS because of these protocols widely used in cloud computing and lack of the real defense against them. HTTP and XML are important elements of cloud computing so security become crucial to safeguard the healthy development of cloud platforms. But as a virtual environment, cloud poses new security threats that differ from attacks on physical system.

2 Related Work

As Cloud Computing is new research area so security in cloud computing is also a very new and open challenge. Lot of research is going in security aspects in cloud computing. There are various latest real time examples in which cloud is suffering from new attacks among them HTTP and XML DDoS attacks are more common.

Since cloud computing security follows the idea of cloud computing, there are two main areas that security experts look at security in a cloud system: These are VM (Virtual Machine) vulnerabilities and message Availability between cloud systems. IaaS layer is more vulnerable as in [5] Shared Technology issues work on IaaS layer.

In [6] Data Loss or Data Leakage is a big problem on IaaS layer. There are many ways to compromise data deletion and alteration of records without a back-up of original content is an obvious example.

In [7] insecure API is big threat in cloud computing. Cloud computing providers exploit a set of software APIs that customer use to manage and interact with cloud services.

Various solutions and techniques exist for detection and protection from HTTP flood attack and XML attack in Cloud Computing.

Chu-Hsing Lin et. al. [8] is using Semantic Web concept to find flooding attack by dividing attacks in three categories but this solution limited to identifying malicious browsing behaviors. Tuncer et. al. [9] is using fuzzy logic to find flooding attack. This solution will give more false positive results. Liming Lu et. al. [10] using Probabilistic Packet Marking for IP Traceback. This method is useful only when we already have attackers IP Address in traceback but in real time it is not possible. Suriadi et. al. using client puzzle but if each packet of client request will pass through client puzzle filter then this solution will face time bottleneck problem. Ashley Chonka et. al. [11] are using BPNN scheme to detect DDoS attack but this scheme work on expert system based approximate threshold value. Until attacker will not cross threshold value it is possible that attacker may attack. M.A. Rahaman et. al. [12] are using inline approach but the disadvantage of this method is that it is securing only some properties of SOAP message. Further N. Gruschka et. al. are introducing first real XML SOAP Message wrapping attack on Amazon EC2 services in 2008. Here attackers are changing XML tags and making vulnerability in SOAP Message request validation. By which any unauthorized user can access the services of Amazon's EC2.

3 System Architecture

The Proposed architecture is having five modules from Client to Cloud Provider. Client requests resources from Cloud Providers by using SOAP message. This

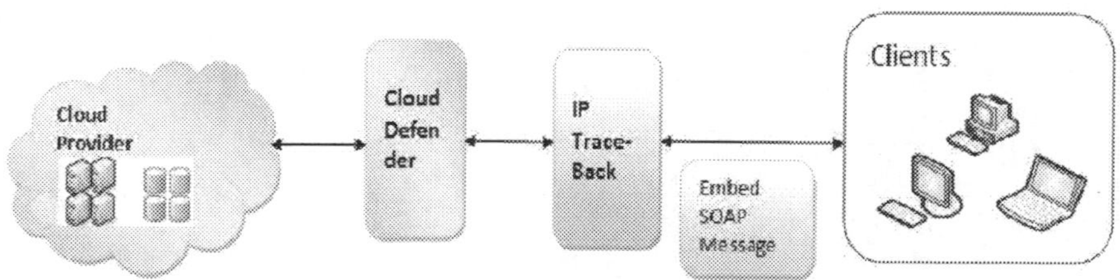

Fig. 1 Proposed Architecture Model

SOAP message has vulnerability of XML DDoS and HTTP DDoS attacks. So this paper introduces three modules which will provide security to SOAP message from these attacks. As a virtual environment, cloud poses new security threats that differ from attacks on physical system.

4 Embed Soap Message

Clients or Consumers use SOAP message to request any resource from cloud providers. SOAP message written in XML only because XML is universally acceptable language and it can run at any platform.

4.1 SOAP Signature

SOAP message is nothing but XML tags. The process of SOAP signature as: for every message part a reference element is created and the message part is hashed and cannibalized. The resulting digest added with digest value as well as the reference of signed message is added in URI field. In last this message part and digest cannibalized and put in Signed Info part and Signature element is added in security header.

4.2 Double Signature

To give the extra protection against XML rewriting attack Double Signature has been used by marking parameters (as number of children, number of header element and number of body element) in SOAP message and keeping these signed parameter in SOAP Header

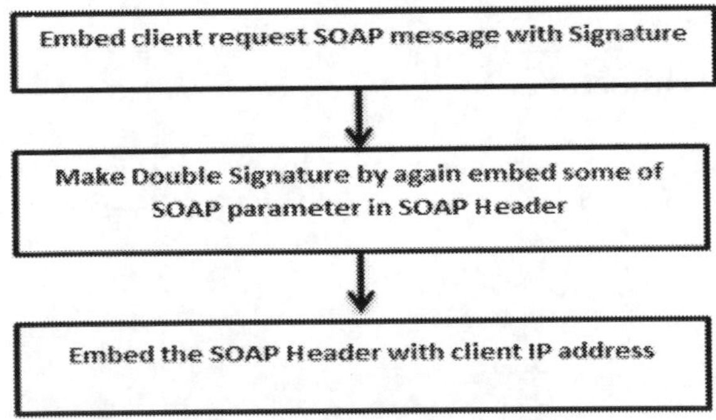

Fig. 2 Embed SOAP Message

4.3 IP Marking

SOAP message will mark at edge router using Flexible Deterministic Packet Marking scheme (FDPM). Three fields in the IP header are used for marking; they are Type of Service (TOS), Fragment ID, and Reserved Flag. A total of 25 bits (8 bit from TOS, 16 bit from TOS, 16 bit from identification field and 1 bit from offset field) are available for the storage of mark information if the protected network allows overwriting on TOS.

Algorithm 1: *Embeded SOAP Message algorithm*

```
Input: SOAP Message with XML tags, at router R, in network N
Output: client request
Requested SOAP message will embed with signature and header marked at router R
if wsse: Contained = true
computed_digest ← hash(wsse:QueryKey)
else
computed_digest ← hash(wsse:SmallerValue, ec2:GreaterValue)
for all h ← wsse:Hash ∈ wsse:HashList
computed_digest ← hash(computed_digest, h)
replace(Data1.Value, computed_digest)
for all h ← wsse: count (children, header element, body element)
computed_digest ← hash(computed_digest, h)
replace(Data2.Value, computed_digest)
return Data1, Data2
set the bit array digest and mark to 0
if N does not utilize TOS
reserved_Flag:=0
7th and 8th bit of TOS:=0
length_of_Mark:=24
else
reserved_Flag:=1
if N utilizes Differentiated Services Field
7th and 8th bit of TOS:=1
length_of_Mark:=16
else if h support Precedence but not priority
7th bit of TOS = 1 and 8th bit of TOS = 0
length_of_Mark:=19
else if N support Priority but not Precedence
7th bit of TOS = 0 and 8th bit of TOS = 1
length_of_mark:=19
decide the lengths of each part in the mark
digest:=hash(A)
for i=0 to k-1
mark[i].Digest:= Digest
mark[i].Segment_number:=i
mark[i].Address_bit:=A[i]
for each incoming p passing the encoding router
j:=random integer from 0 to k-1 // message divide in k bits
write Mark[j] into p.mark
```

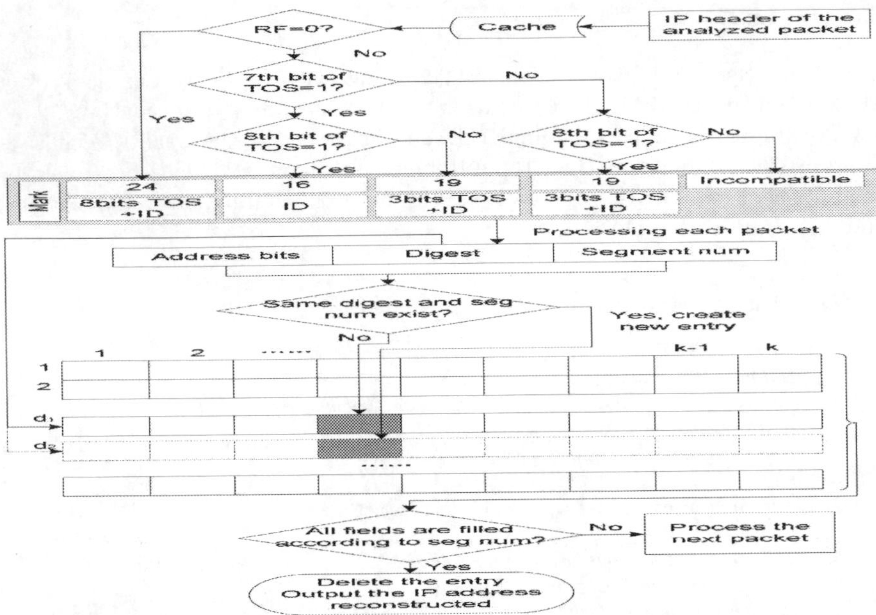

Fig. 3 Deterministic Packet marking

5 IP Trace-Back

IP Trace-Back is a logical file system. In proposed architecture IP Trace-Back stores vulnerable IP address provided by Cloud Defender. When client message request comes to IP Trace-Back, it matches coming message source IP address with already stored vulnerable IP address. If IP matched then it discard request message otherwise it send request message to Cloud Defender.

6 Cloud Defender

Cloud defender filters the attack in five stages. These five stages are

(1) Sensor Filter
(2) Hop Count Filter
(3) IP Frequency Divergence Filter
(4) Confirm legitimate user IP Filter
(5) Double Signature Filter

First four filters detect HTTP DDoS attack and fifth filter detects XML DDoS attack.

6.1 Detect Suspicious Message

6.1.1 Sensor

Sensor monitors the incoming request messages. If the sensor finds that there is hypothetical increase in the number of request messages coming from any particular consumer then it marks those messages as suspicious IP otherwise send to next filter.

6.1.2 HOP Count Filter

It will calculate the Hop Count value and compare with stored Hop Count value. If no match then it marks those messages as suspicious IP otherwise send to next filter.

6.1.3 IP Frequency Divergence

Because in DDoS attacker will not generate different request message every time so he has need to send same request messages again and again. If found same frequency of IP messages then it marks those messages as suspicious IP otherwise send to next filter.

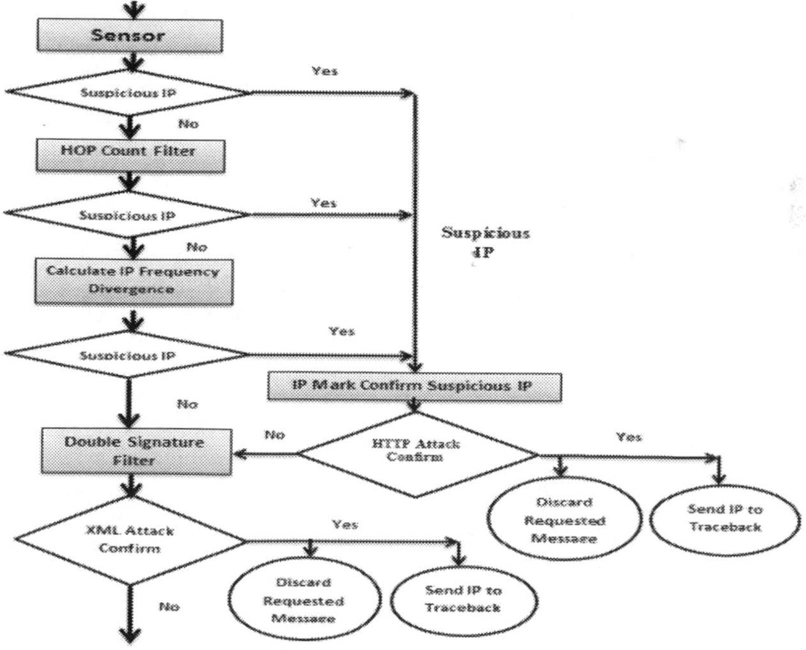

Fig. 4 Cloud Defender

6.2 Detect HTTP DDoS Attack

All suspicious packets come to the Puzzle Resolver. It resolves the SOAP header of these suspicious messages. Firstly it finds the suspicious messages IP addresses and then send the puzzles to these IP address If the suspense IP address send the correctly solved puzzle to puzzle resolver it means it is genuine client request otherwise puzzle resolver drops the request message and send suspicious IP address to IP Trace-Back otherwise it send the request message to Double signature filter.

6.3 Detect Coercive Parsing/XML DDoS Attack

Check the incoming request message for any open tag. If open tag found in incoming message then it discards that message otherwise send the request message to cloud provider to provide services to clients.

Because of cloud defender is working in multi-tenant environment. So in order to send multiple client requests to cloud defender we will limit the number of client request at a particular moment of time. Suppose coming client request are N and threshold is N1. If number of client requests are greater than N1 (here N>N1) then Cloud Defender will send these requests to different filters otherwise Cloud Defender will send the request packets directly to Double Signature Filter to check for XML DDoS attack.

In a DDoS attack, take place with zombies by a single attacker (master). These are nothing but zombie machines or attacker uses virtual machines and open thousands of tags by using these virtual machines and make DDoS attack on cloud provider. General attack traffic distribution will obey Poisson distribution approximately. The Poisson distribution function for DDoS attack traffic is shown below

$$P_k = \lambda^N * e^{-\lambda} / N! \quad (1)$$

Where λ is a positive real number, equal to the expected number of occurrences that occur during the given interval, and k is a non-negative integer, N=0, 1, 2,... In information theory, the information entropy is a measure of the uncertainty associated with a random variable.

Algorithm 2. *Cloud Defender algorithm*

1. ***Input****: N, N1, ∀N ∈[1, . . . , n], the final TTL Tf, the initial TTL Ti, stored hop-count in IP packet Hs, N is the length incoming request*
2. ***Input****: H_{fd}= frequency divergence.*
3. ***Output****: Legitimate client request*
4. ***If*** *N>N1 then*
5. ***for each*** *packet N compute the hop-count*
6. *Hc=Ti-Tf*
7. ***If*** *Hc!=Hs then*
8. *Continue*
9. ***else if***
10. *mark request packet suspicious and send to detector;*
11. $P_k = (\lambda^N * e^{-\lambda}) / N!$

12. $H_{fd} = \lambda[1\text{-}log_2\lambda]+ e^{-\lambda} \Sigma Pk$
13. ***If*** $H_{fd} > 0.5$ *then*
14. *Continue*
15. ***else***
16. *mark request packet suspicious and send to detector*
17. *reconstruction at victim V, in network N*
18. ***for each*** *coming packet p passing the reconstruction point mark recognition(length and fields)*
19. ***if*** *all fields in one entry are filled*
20. *output the source IP*
21. *delete the entry and drop the consumer request*
22. ***else if*** *same digest and segment number exist*
23. *create new entry*
24. *fill the address bits into entry and send to Double signature filter*
25. *Check for tag value*
26. ***If*** *tag value !=1*
27. *XML DDoS verify and drop the packet*
28. ***else***
29. *send the requested SOAP Message to cloud*

7 Experimental Result

7.1 Experimental Environment

Scalable simulation framework (SSFNet) is a collection of Java components used for modeling and simulation of IPs and networks. In experiment SSFNet Simulator has been used to simulate the whole process from embedding SOAP Header with signature to marking it. Darpa 1999 tcpdump (510 MB) dataset has been used as input traffic and analyzed it by using wireshark analyzer on a window7 OS, I3 Processor, 3 GB Memory and 300 GB hard disk.

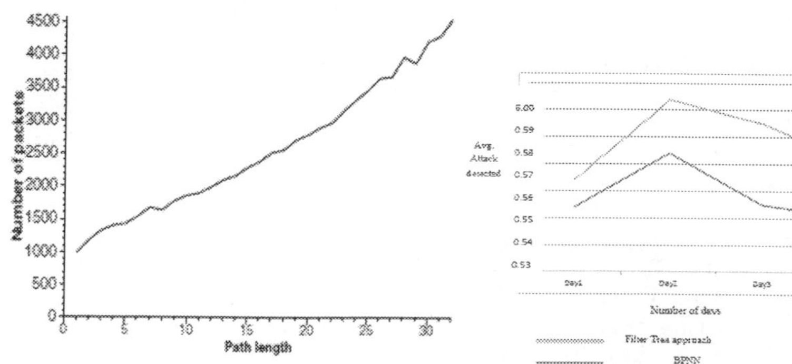

Fig. 5 Experimental results for number of packets needed to reconstruct paths of varying length

Fig. 6 Comparison between Filter Tree Approach and BCNN

7.2 Experimental Result

As in Fig. 6. simulation results, as red line showing in start BPNN takes time to make its system trained meanwhile most of the attacked packet enter in cloud so in start time less attack detected and as long as BPNN system gives knowledge to its system it detects more attacked packets but in Filter tree there is no need any previous knowledge so in start filter tree works well.

8 Conclusion

The denial-of-service attacks have become more targeted on cloud servicesand effected Client economically as eDDoS attack. This threat seems unlikely to fade away in absence of an active defense technique which pressures attacker's resources and raises the costs for delivering attack traffic. SOAP packet marking offer such a defense: An adversary cannot seize the victim's resources without committing its own resources first, which therefore limits its attack capability. Moreover Cloud Defender provides proactive defense approach. Intruder will get identify before get enter into the Cloud. Cloud defender will not check traceback for each request message, firstly it will identify the suspicious packet and will check only for those suspicious request packet. So this paper is filtering service request messages at different stages firstly matching the request client IP with previously stored suspicious IP in Trace-Back and then cloud defender is using for detecting the HTTP DDoS, Coercive parsing DDoS, XML DDoS at different stages. Cloud Defender is firstly identifying suspicious messages and then detecting attacks. This will reduce the computation cost and vurnability.

Proposed system works for HTTP interface further we can extend it to provide security for REST based APIs.

References

[1] Cloud Security Alliance (Online), https://cloudsecurityalliance.org/topthreats (viewed December 21, 2011)

[2] Europe Network and Information Security Agency (Online), http://www.enisa.europa.eu/act/rm/files/deliverables/cloud-computing-risk-assessment (viewed January 21, 2012)

[3] Microsoft Security Bulletin MS10 (Online), http://www.microsoft.com/technet/security/bulletin/ms10-070.mspx (updated October 26, 2011)

[4] Security of data (Online), http://news.cnet.com/8301-138463-20052571-62 (viewed July 02, 2011)

[5] Security labs Blog (Online), http://securitylabs.websense.com/content/Blogs/3402.asp (viewed November 21, 2011)

[6] Nurmi, D., Wolski, R., Grzegorczyk, C., Obertellli, G., Soman, S., Youseff, L., Zagorodnov, D.: The Eucalyptus Open-source Cloud computing System, http://www.eucalyptus.com/whitepapers

[7] Bhuya, R., Ranjan, R., Calheiros, R.N.: Modeling and Siulation of Scalable Cloud Computing Environments and the CloudSim Toolkit: Challenges and Opportunities. In: Proceedings of the 7th High Performance Computing and Simulation Conference, Leipzig, Germany, June 21-24 (2009)
[8] Lin, C.-H., et al.: A Group Tracing and Filtering Tree for REST DDoS in Cloud Computing. International Journal of Degital Content Technology and its Applications 4(9) (December 2010)
[9] Tuncer, T., Tatar, Y.: Detection SYN Flooding Attacks Using Fuzzy Logic. In: International Conference on Information Security and Assurance, ISA 2008, April 24-26, pp. 321–325 (2008)
[10] Lu, L., et al.: A General Model of Probabilistic Packet Marking for IP Traceback. In: ASIACCS 2008, March 18-20. ACM, Tokyo (2008)
[11] Chonka, A., Xiang, Y., Zhou, W., Bonti, A.: Cloud security defense to protect cloud computing against HTTP -DoS and XML-DoS attacks. Journal of Network and Computer Applications 34, 1097–1107 (2011)
[12] Rahaman, M.A., Schaad, A., Rits, M.: Towards secure SOAP message exchange in a SOA. In: SWS 2006: Proceedings of the 3rd ACM Workshop on Secure Web Services, pp. 77–84. ACM Press (2006)

Cloud Based Heterogeneous Distributed Framework

Anirban Kundu, Chunlin Ji, and Ruopeng Liu

Abstract. In this paper, the main target is to achieve distributed atmosphere of Cloud configuring heterogeneous structure framework with accessible common equipments, technologies, and configurable high-end servers concerning no additional expenditure to the system network. In proposed system, different categories of machines are being utilized to generate efficient diverse background. Server-side background mode of operation is to be conducted for accessing dedicated servers as well as all-purpose servers which are not only assigned for the user specific responsibilities. This is an additional challenge for this approach to make it happen using any kind of server machines. In this approach, unicast topology is going to be used for avoiding network congestion. Minimization of time and maximization of speed are the objectives of proposed system structure. Earlier version of this paper has been published in [1].

Keywords: Distributed environment, Heterogeneous Environment, Software-as-a-Service (SaaS), Cloud.

1 Introduction

The use of concurrent processes that communicate by message-passing has its roots in operating system architectures studied in the 1960s. The first widespread distributed systems were local-area networks such as Ethernet that was invented in the 1970s. E-mail became the most successful application of ARPANET, and it is probably the earliest example of a large-scale distributed application [2]. Client programs typically handle user interactions and often request data or initiate some data modification on behalf of a user [3]. The study of distributed computing became its own branch of computer science. A distributed environment is chosen

Anirban Kundu · Chunlin Ji · Ruopeng Liu
Shenzhen Key Laboratory of Artificial Microstructure Design
State Key Laboratory of Meta-RF Electromagnetic Modulation Technology
Kuang-Chi Institute of Advanced Technology
Shenzhen, Guangdong, P.R. China 518057
e-mail: {anirban.kundu,chunlin.ji,ruopeng.liu}@kuang-chi.org

A. Abraham and S.M. Thampi (Eds.): Intelligent Informatics, AISC 182, pp. 471–478.
springerlink.com

for various applications where data is generated at various geographical locations and needs to be available locally most of the time. The data and software processing are distributed to reduce the impact of any particular site or hardware failure. Location transparency eliminates the need to know the actual physical location of the data [4].

Organization of rest of the paper is as follows: Section 2 presents related works section. Proposed approach is described in Section 3. Experimental results have been shown in Section 4. Section 5 concludes the paper.

2 Related Works

Distributed performance computing [5-6] in heterogeneous systems employs the distributed objects as applications [7]. These applications are arranged in such a manner that the same type of user requests can be executed in distinct machines which are situated in different locations (in case of Wide Area Network (WAN)). Sometimes, these machines fall in the same group or cluster at same location (in case of Local Area Network (LAN)) [8]. Huge collection of heterogeneous resources offers an opportunity for delivering high performance on a range of applications. Successful scheduling of system resources achieves high performance [9]. It is being shown that scheduling can be performed automatically, efficiently, and profitably for a range of computations in this environment. Load balancing techniques can be used to balance the network based activities. Effective Scheduling is quite necessary for maintaining loosely coupled processors [10]. Software issues have been resolved by some well-known methods in effective processing systems. The success of any system network depends on its distribution. Heterogeneous multi-computers are designed to form the backbone of the network [11].

3 Proposed Approach

In this paper, a distributed computing environment has been considered. It contains typical computer resources, high performance workstations, and cluster computers, connected by one or more networks. This assembly of machines presents a large comprehensive computing resource including memory, cycles, storage space, and communication bandwidth. For this reason the proposed system has a great potential for high performance computing. An important characteristic of this system is that it exhibits heterogeneity of many types including hardware, operating system, file system, and network. Heterogeneity creates a challenge that it must be managed to enable the parts of the proposed system to work together. At the same time, it also presents an opportunity that is the variety of different resources which suggests that it is possible to select the best resources for a particular user request. The variety and amount of computing resources in the proposed system offers a great prospective for high performance computing.

The steps of the proposed approach have been shown in Fig. 1. Initially, input data and the user specific program are received from Internet to our system network. Data is received at Cloud interface. Subsequently, the network selection and corresponding suitable server selection have been carried over the network for error-free communication between interface and server using typical three-way handshaking in TCP/IP environment. Code and data are migrated as required time to time. After completion of the user based program's execution, the outputs are being saved at particular places of storage system for temporary basis until the user would not download it at the client-end.

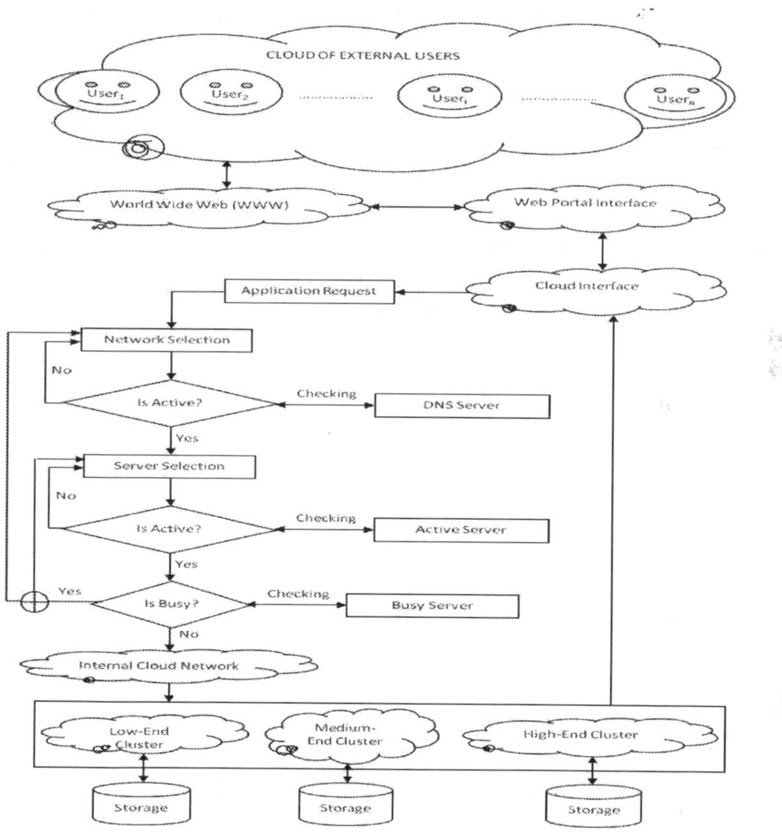

Fig. 1 Cloud System Design Internals

In proposed approach, unicast topology is followed for information propagation as prescribed data replication method is capable of selection of server machines within the network in run-time. Therefore, broadcast topology is not used in the server-side network. So, the network congestion is never happened in the proposed framework structure. The mathematical analysis has been shown to demonstrate the superiority of the proposed distributed structure compared to the typical structures in respect of network congestion.

Documents for client's Internet Protocol (IP) address has been stored by the Web portal system using TCP/IP connection at server-end and subsequently transferred to interface. User documents for input data should also be dumped in similar method. Input (I/P) sender handles these documents to prepare required materials in a proper format for further processing targeting particular sub-network within server-side system having dynamic scheduling. Scheduler checks for an active server of a particular network, and also checks whether that server is busy. Particular server is being selected for the user prescribed operation only if the server is active and free at any time instance. I/P sender also sends required materials to I/P receiver module following standardized typical communication techniques at port level. I/P receiver initiates particular application. Output files would be stored in a fixed area of each server machines. Output (O/P) sender collects data and sends it back to interface. O/P receiver module fetches output files at a particular port level and stores required information to pre-defined disk spaces related to external IP address of particular user. Web portal handles generated documents and sends them to specific destination address using Internet (refer Fig. 1).

Algorithm 1: Find keywords
Input: Executable file (i.e., *.m, *.mat, *.vb)
Output: Successful submission of user request
Step 1: Do loop – start
Step 2: Scan a line
Step 3: If ('%' at the beginning of a line)
Step 4: Then skip the right-side of that particular line
Step 5: Else
Step 6: If ("save" or "print" or "printopt" or "store" or "write")
Step 7: If ('\' or '/')
Step 8: Print ("User must not use the path")
Step 9: Else
Step 10: Go to next line of the file at Step 2
Step 11: Stop

Algorithm 1 briefly shows the methodology to find the keywords within the user submitted programs through Web portal system. User has to follow specific rules about their programs at the time of submission. Otherwise, their program would not be accepted by the Web portal. For example, if the user has mentioned some specific path(s) for saving the output(s), then it cannot be accepted by the proposed system as because the proposed system would save the outputs in some specific paths in the server-side system network.

Mapping between networks and servers within each network can be treated as the hierarchical structure of proposed design. 'n' number of networks have been considered and 's' number of servers have been considered in each network. Each server consists of some applications as predefined by the network system administrator. Proposed system framework detects the nearest network and server for a particular application at any specific time instance based on the user request.

4 Experimental Results

In this section, all stages of proposed approach have been depicted using relevant evidences. Initially, user has been registered in Web portal. User has to submit some personal and technical information to the system's database for future correspondences. After registration, user is able to login within the distributed system for allocating server-side resources for own programs. Sometimes users can also make use of system defined programs which are available in the server databases.

Server-side background mode of operation is prepared and successfully executed in all the servers of proposed system network. These servers are not like typical server systems which are always dedicated to some specific tasks. In this approach, each server can be used to different tasks. These servers are also used by the internal users for their personal usage. So, the proposed distributed system is well controlled as the collection of background tasks which are placed within the server systems of the network in a hidden mode. Therefore, the users are not aware of these processes running in their systems.

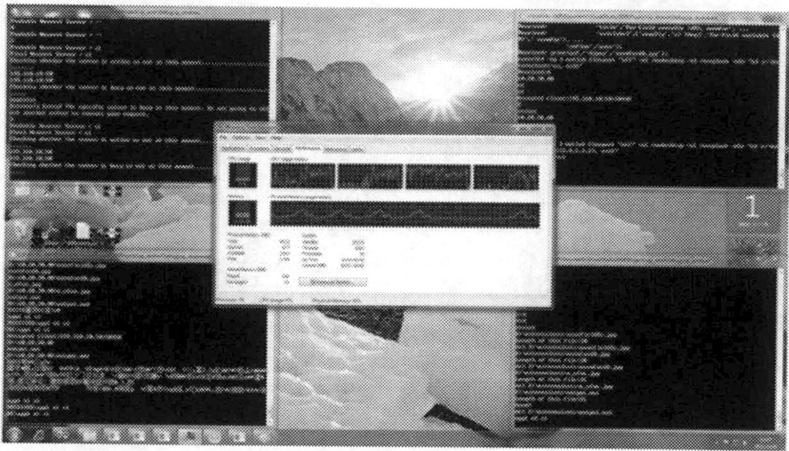

Fig. 2 0% to 9% CPU Usage of a computer when program is not running

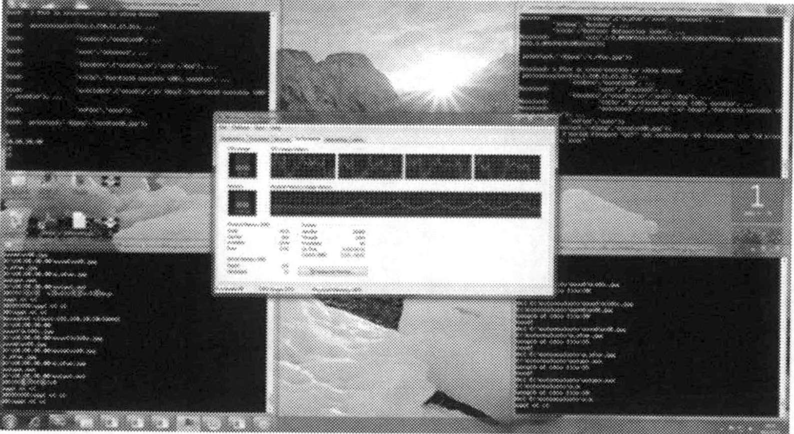

Fig. 3 3% to 27% CPU Usage of a computer when program is running

System performance of a server system is shown in Fig. 2 while program is not running. In this time instance, concerned CPU usage is about 0% to 9%. The same system shows 3% to 27% CPU usage when the program is running (refer Fig. 3). 30% to 85% CPU usage has been observed when "Comsol" and "Matlab" are running for an instance (refer Fig. 4).

It has been found through the experimentation that the best performance can be achieved in these server machines within the network using 3 to 6 number of threads at a time. Performance would be degraded if the number of threads increases over this limit.

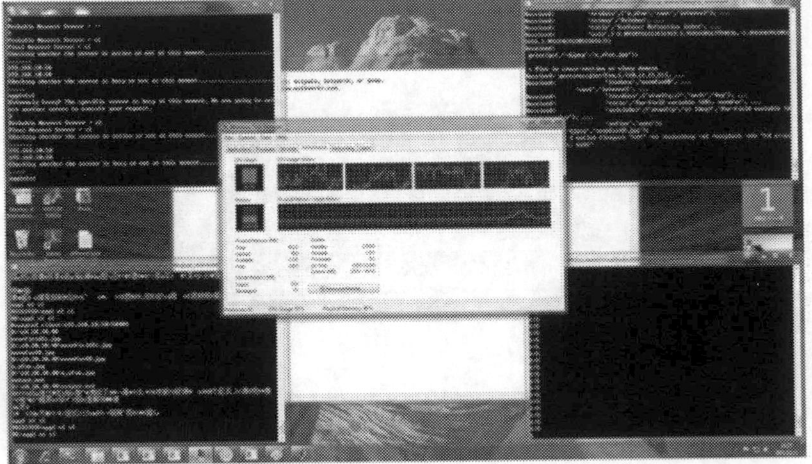

Fig. 4 30% to 85% CPU Usage of a computer when COMSOL is running

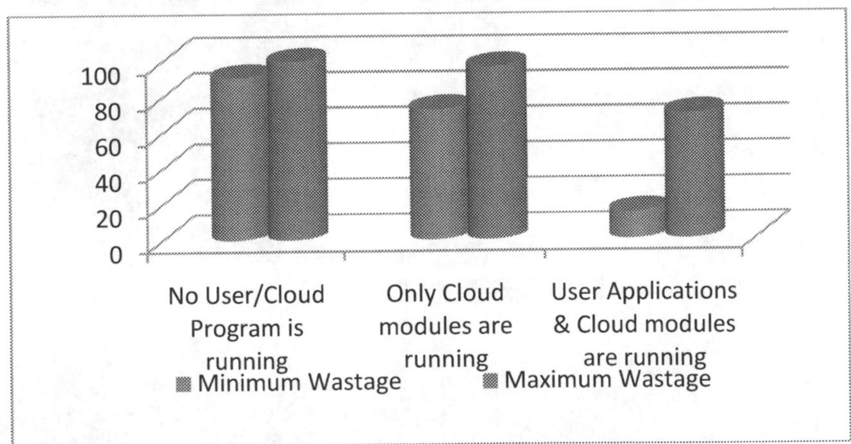

Fig. 5 CPU Wastages in different situations

Fig. 5 describes the CPU wastage situations in various conditions in Cloud compute nodes. Y-axis represents percentage (%) of CPU wastage. On horizontal axis, "No User/Cloud Program is running" represents the wastage range from 91% to 100% in the system node. "Only Cloud modules are running" represents the wastage range from 73% to 97%. Therefore, a lot of other programs can be executed at that time instance. It shows that the programs occupy less processor time and memory as applicable based on the situation. "User Applications & Cloud modules are running" represents the wastage range from 15% to 70% while a "Comsol" &"Matlab" have been executed in system nodes.

Therefore, it has been observed in different conditions that each resource is not properly utilized all the time. So, a lot of wastage happens in any scenario. Thus, dedicated as well as non-dedicated servers can be used for this purpose in Cloud environment.

5 Conclusion

In this paper, distributed environment has been accomplished for heterogeneous framework for accessing the data and program modules in specific locations at server-side of the network. Dynamic scheduling has been maintained all over the approach to determine every possible sub-network and the related servers for the specific user targeted tasks. Server-side background mode of operation has also been implemented successfully. Finally, a system framework consisting of dedicated and non-dedicated servers has been successfully developed having a target to exhibit efficient usage of machines within a Cloud server-side network.

Acknowledgments. This research was supported by grants from Project of Introducing Innovation Research Team by Guangdong Province.

References

1. Kundu, A., Ji, C.: Swarm Intelligence in Cloud Environment. In: The Third International Conference on Swarm Intelligence, Shenzhen, P. R. China, June 17-20 (2012)
2. Andrews, G.R.: Foundations of Multithreaded, Parallel, and Distributed Programming. Addison-Wesley (2000) ISBN 0-201-35752-6
3. http://en.wikipedia.org/wiki/Distributed_computing
4. http://www.answers.com/topic/distributed-computing
5. Buschmann, F., Henney, K., Schmidt, D.C.: Pattern-Oriented Software Architecture Volume 4: A Pattern Language for Distributed Computing. Wiley, Chichester (2007)
6. Spinnato, P., Albada, G.D.V., Sloot, P.M.A.: Performance Modeling of Distributed Hybrid Architectures. IEEE Transactions on Parallel and Distributed Systems 15(1) (January 2004)
7. Radojevic, I., Salcic, Z., Roop, P.S.: Design of Distributed Heterogeneous Embedded Systems in DDFCharts. IEEE Transactions on Parallel and Distributed Systems 22(2) (February 2011)

8. Liu, Y., Xiao, L., Liu, X., Ni, L.M., Zhang, X.: Location Awareness in Unstructured Peer-to-Peer Systems. IEEE Transactions on Parallel and Distributed Systems 16(2) (February 2005)
9. Xu, J., Tang, X., Lee, W.: Time-Critical On-Demand Data Broadcast: Algorithms, Analysis, and Performance Evaluation. IEEE Transactions on Parallel and Distributed Systems 17(1) (January 2006)
10. Anger, F.D., Hwang, J., Chow, Y.: Scheduling with Sufficient Loosely Coupled Processors. Journal of Parallel and Distributed Computing 9 (1990)
11. Armstrong, J.B., Watson, D.W., Siegel, H.J.: Software Issues for the PASM Parallel Processing System. In: Kowalik, J.S. (ed.) Software for Parallel Computation. Springer, Berlin (1993)

An Enhanced Load Balancing Technique for Efficient Load Distribution in Cloud-Based IT Industries

Rashmi KrishnaIyengar Srinivasan, V. Suma, and Vaidehi Nedu

Abstract. The advent of technology has led to the emergence of new technologies such as cloud computing. Evolution of IT industry has oriented towards the consumption of large scale infrastructure and development of optimal software products, thereby demanding heavy capital investment by the organizations. Cloud computing is one of the upcoming technologies that have enabled to allocate apt resources on demand in a pay-go approach. However, the existing techniques of load balancing in cloud environment are not efficient in reducing the response time required for processing the requests. Thus, one of the key challenges of the state-of- art of research in cloud is to reduce the response time, which in turn reduces starvation and job rejection rates. This paper, therefore aims to provide an efficient load balancing technique that can reduce the response time to process the job requests that arrives from various users of cloud. An enhanced Shortest Job First Scheduling algorithm, which operates with threshold (SJFST), is used to achieve the aforementioned objective. The simulation results of this algorithm shows the realization of efficient load balancing technique which has resulted in reduced response time leading to reduced starvation and henceforth lesser job rejection rate. This enhanced technique of SJFST proves to be one of the efficient techniques to accelerate the business performance in cloud atmosphere.

Keywords: Cloud Computing, Virtualization, Response Time, Starvation, Job Rejection, Load Balancing.

Rashmi KrishnaIyengar Srinivasan
Post Graduate Programme, Computer Science and Engineering, Department of Information Science and Engineering, Dayananda Sagar College of Engineering, Bangalore, India
e-mail: rashmiks.ks57@gmail.com

V. Suma · Vaidehi Nedu
Research and Industry Incubation Center, Dayananda Sagar Institutions, Bangalore, India

V. Suma · Vaidehi Nedu
Department of Information Science and Engineering,
Dayananda Sagar College of Engineering, Bangalore, India
e-mail: {sumavdsce,dm.vaidehi}@gmail.com

A. Abraham and S.M. Thampi (Eds.): Intelligent Informatics, AISC 182, pp. 479–485.
springerlink.com

1 Introduction

The advent of cloud computing has paved way for the IT organizations to have an access to a shared pool of virtualized resources. Further, cloud computing is a promising technology to provide the resources on demand and to service the received requests within a stipulated time. Henceforth, high availability of the resources is very critical in the cloud environment to service the requests received. Moreover, the management of resources is yet another challenge as it influences the business performance of the cloud service provider. Load balancing provides a solution to overcome the aforementioned issues. Load balancing strategies distributes the execution load among the virtual machines (VMs) to utilize the resources in an efficient manner.

Cloud being a pay-go-model and one of the emerging technologies to provide resources on demand, there is a requisite for the resources to be always available. To achieve the aforementioned objective, load balancing becomes an essential factor. The load balancer provides a means for allocating and de-allocating the resources automatically by either increasing or decreasing of resources to service the arrived requests on demand.

However, the existing load balancing algorithms are not efficient in successfully distributing the resources efficiently in order to process all the jobs of various users in cloud. Authors in [8] have therefore, proposed an enhanced scheduling strategy to accelerate the business performance of the cloud system for reducing the job rejections by achieving less response time. It is worth to note that reduction in job rejection rate can also be realized through efficient load balancing technique. Therefore, this paper provides an enhanced load balancing algorithm which operates on Shortest Job First Scheduling with Threshold (SJFST). Accordingly, this load balancing strategy reduces the waiting time to respond to the arrived job based on the burst time and it further reduces starvation.

The paper is organized as follows. Section 2 describes the Related Work followed by Research Design in Section 3. Section 4 discuses the enhanced load balancing algorithm. Finally, Section 5 describes the conclusion and followed by references.

2 Related Work

Cloud computing is an upcoming technology in the IT industries that has opened several challenges to the researchers working in this domain. Xiaoqiao Meng et al. 2010 have suggested an efficient resource allocation technique using VM multiplexing approach to achieve high utilization of resources. However, VM selection process suffers from overhead [1].

Sewook Wee and Huan Liu 2010 have proposed client side load balancer architecture using cloud to directly deliver static contents while allowing a client to choose a corresponding back-end web server for dynamic contents. However, load balancing for dynamic contents leads to overhead problems [2].

Shu-Ching Wang et al. 2010 have proposed a load balancing in a three –level cloud computing network using Opportunistic Load Balancing (OLB) and Load Balance Min-Min (LBMM) scheduling algorithm which has performed efficiently. However, this approach is not dynamic and suffers an overhead during selection of the node [3].

Jinhua Hu et al. 2010 have presented a scheduling strategy on load balancing of virtual machine resources in cloud computing environment based on genetic algorithm that can minimize the migration cost [4].

Wenhong Tian et al. 2011 have recommended a dynamic and integrated load balancing scheduling algorithm (DAIRS) for cloud datacenters to distribute the total and average load on each server in the cloud datacenter by sorting. Therefore, the limitation of this algorithm is that it consumes more time by sorting [5].

Vlad Nae et al. 2010 have implemented cost-efficient hosting and load balancing to reduce hosting costs and to achieve resource allocation by load distribution. But, this approach fails to optimize the distribution of load [6].

Shiyao Chen et al. 2011 have recommended secondary job scheduling with deadlines under the time constraint to address resource reutilization issue in cloud. However, this approach suffers from capacity transformation overheads [7].

3 Research Design

An efficient strategy is essential to enhance the performance of computing system in cloud environment by reducing the job rejections at the peak hour. It is evident from the progress of the research in cloud environment that currently existing load balancing strategies are inefficient in reducing job rejections and starvation. Hence, there is a need for an effective load balancing strategy that can reduce job rejections and starvation. To achieve the above said objective, we have introduced Shortest Job First Scheduling with Threshold (SJFST) algorithm. It is achieved by sequence of research activities. Accordingly, to analyze the efficiency of cloud service provider for successful processing of jobs arriving at the peak hour we have collected secondary data. Secondary data is a processed data collected from the leading IT organizations, which are operating in this domain. An analysis on this data infers that huge number of jobs that arrive to the system gets rejected due to the lack of effective load balancing strategy.

Henceforth, an effective load balancing strategy that uses Shortest Job First Scheduling with Threshold (SJFST) is introduced in this paper to overcome the aforementioned issues. Further, the simulation is carried out by configuring the simulation set up and its results indicate the efficiency of the proposed strategy.

4 An Enhanced Load Balancing Algorithm

Load balancing is an important factor in the cloud environment as the resources have to be always available to provide service to the arrived jobs. Therefore, the load balancing in cloud is achieved through Shortest Job First Scheduling (SJFS) technique where, a job with less burst time is scheduled first. However, in this technique the job in execution will not be preempted to check for the newly arrived jobs. Therefore, to preempt the job in execution, a threshold (timer) can be set, so that newly arrived jobs can be monitored at regular intervals to avoid starvation. Hence, Shortest Job First Scheduling with threshold (SJFST) algorithm, which is an integration of SJFS and threshold, has been proposed in this paper.

Fig. 1. depicts the load balancing in cloud using SJFST. The Load Balancer is an entity in the cloud service provider's datacenter, which schedules the arrived jobs according to the strategies incorporated to maintain the even distribution of load on the available resources in the datacenter. The Resource Allocator is another entity in the cloud datacenter for allocating the apt Virtual Machine (VM) to the scheduled job.

In the Load Balancer shown in Fig. 1., SJFST algorithm has been incorporated. Accordingly, the Load Balancer extracts the jobs from the queue, in which jobs are indexed according to their arrival patterns and subsequently schedules and transfers the job, which requires less burst time to the Resource Allocator. Henceforth, the job in execution routinely gets preempted after crossing the set threshold time. This preemption is carried out to reduce starvation and job rejection. On preemption of the job, the Load Balancer monitors the queue for the newly arrived jobs for modifying the scheduling decision if necessary.

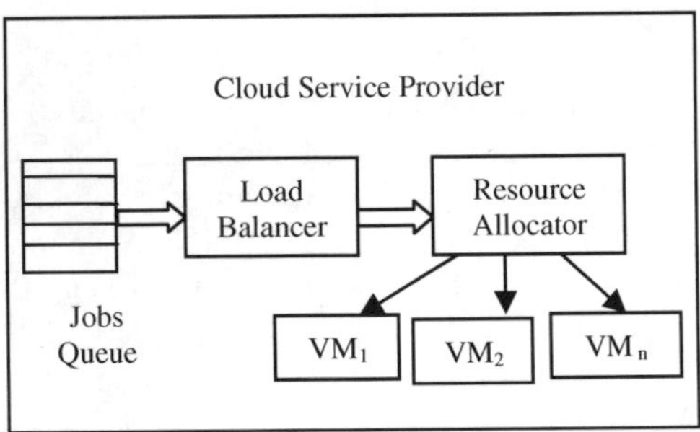

Fig. 1 Load Balancing in Cloud

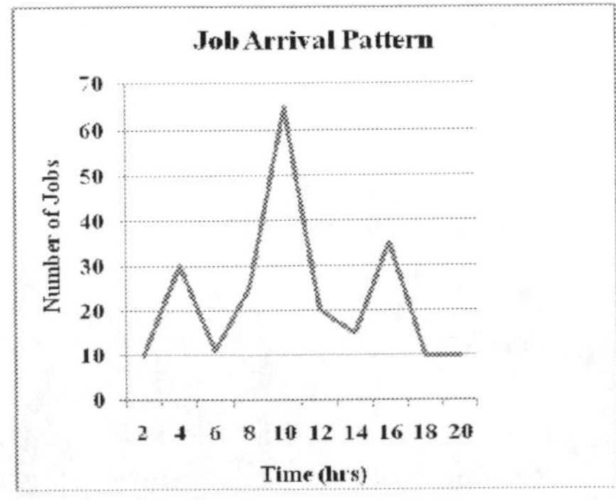

Fig. 2 Job Arrival Pattern

Table 1. Simulation Configurations

Job ID	Arrival Time (hrs)	Burst Time (hrs)
Job 1	0	7
Job 2	.5	.5
Job 3	2	4
Job 4	4	5
Job 5	4.16	1
Job 6	6	3

Fig. 2. indicates the job arrival pattern captured against the time profile. Fig. 2. infers the existence of high demand for the jobs to be processed during the peak hours, which is indicated in the form of spikes. Thus, in order to realize the efficient utilization of the resources with less starvation and job rejection an effective load balancing approach is introduced.

Henceforth, the implementation of SJFST is carried out using the following algorithm:

Step 1: Ensure the availability of all VMs.
Step 2: On arrival of the jobs, the load balancer will first schedule the job with less burst time and transfers it to the resource allocator for allocation.
Step 3: On receiving the request for allocation of the VM, the resource allocator parses the data structure to identify the feasible VM to service the request.
Step 4: The feasible VM will be assigned to the requesting job until the job completes or it gets preempted.
Step 5: The job in execution will be automatically preempted once it crosses the set threshold value.
Step 6: On preemption, the load balancer monitors the jobs queue for the newly arrived requests and the cycle repeats from step 2.

Table 1. shows a sample simulation configuration that has been set to evaluate the efficiency of SJFST algorithm used for load balancing, which includes Job ID associated uniquely with each job, Arrival time and the Burst time of each job.

Simulation results of SJFST for load balancing has yielded appreciable results in terms of response time, where response time indicates the time interval between job arrival time and the time for first dispatch. In order to have a justifiable inference to incorporate threshold concept in SJFS algorithm, a comparison table of response time of the jobs computed with SJFS and with SJFST is also presented here. Table 2. depicts the above said comparative results that are obtained using various threshold values.

Table 2. infers the job arriving with varying nature in terms of arrival time and burst time are processed using SJFS technique. Accordingly, the response time is comparatively high as against the same jobs when processed using SJFS with various threshold values. It is worth to note at this point that with decrease in threshold value, response time also decreases. However, it is also desirable not to have too less threshold value, which only leaves to increase in preemption time and its associated overheads.

Fig. 3. shows the variation of average Response Time obtained with different thresholds.

Fig. 3. infers that with increase in the threshold value the response time increases. Henceforth, it is ideal to set lower threshold to get less response time which in turn results in less job rejection and starvation. Fig. 4. depicts the comparison of SJFS and SJFST (threshold =15 minutes) in terms of response time.

Table 2 Comparison between SJFS and SJFST with respect to the response time

Job ID	Response Time (hrs)					
	SJFS	SJFST				
		T =15	T =30	T=45	T =50	T=60
Job 1	0	0	0	0	0	0
Job 2	6.5	0	0	0.25	0.33	0.5
Job 3	9.5	0	0	0	0.16	0.5
Job 4	11.5	6	6	6	6.16	11.5
Job 5	3.34	0.08	0.33	0.08	0.49	0.34
Job 6	2.5	1	1	1	1.19	1.5
Average	6.7	1.17	1.22	1.22	1.38	2.6

T - Threshold measured in minutes

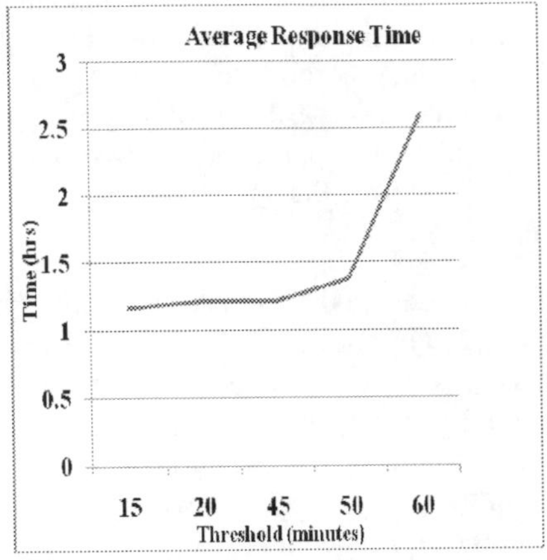

Fig. 3 Comparative graph showing variation of Average Response Time obtained with different Thresholds

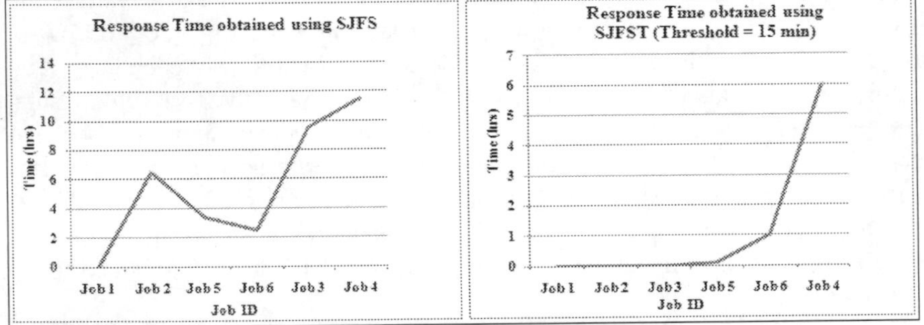

Fig. 4 Graphical comparison between SJFS and SJST in terms of Response Time

It is observed from Fig. 4. that SJFST has yielded less response time compared to SJFS, which in turn reduces starvation and job rejections.

5 Conclusions

Cloud Computing is the state-of the-art of the technology to provides on demand resources to the cloud users. Hence, it is imperative in cloud for the constant availability of resources in addition to their efficient utilization. The most popularly used Shortest Job First Scheduling (SJFS), which is a load balancing technique, is inefficient in servicing the jobs with lesser response time and starvation in addition to increased job rejection rate. This paper, presents a case study of leading IT industry where an enhanced SJFS is implemented using various threshold values. The simulation results indicate beneficial impact in the business with lesser job rejection rate. Additionally, implementation of Shortest Job First Scheduling with Threshold (SJFST) leads to lesser response time and starvation time with reduced threshold values. Further, it is also suggested to operate with desirable threshold value in order to reduce overheads. However, our future work is to provide a mathematical model to implement the SJFST for load balancing.

References

1. Meng, X., et al.: Efficient Resource Provisioning in Compute Clouds via VM Multiplexing. In: ICAC 2010, Washington, DC, USA, June 7-11 (2010), ©2010 ACM 978-1-4503-0074-2/10/06
2. Wee, S., Liu, H.: Client-side Load Balancer using Cloud. In: SAC 2010, Sierre, Switzerland, March 22-26 (2010) © 2010 ACM 978-1-60558-638-0/10/03
3. Wang, S.-C., et al.: Towards a Load Balancing in a Three-level Cloud Computing Network. In: 3rd IEEE Conference on Computer Science and Information Technology (ICCSIT), pp. 108–113 (2010)
4. Hu, J., et al.: A scheduling strategy on load balancing of virtual machine resources in cloud computing environment. In: 3rd International Symposium on Parallel Architectures, Algorithms and Programming, pp. 89–96 (2010)
5. Tian, W., et al.: A Dynamic and Integrated Load balancing Scheduling Algorithm for Cloud Datacenters. In: Proc. IEEE CCIS 2011, pp. 311–315 (2011)
6. Nae, V., et al.: Cost-Efficient Hosting and Load Balancing of Massively Multiplayer Online Games. In: 11th IEEE/ACM International Conference on Grid Computing, pp. 9–16 (2010)
7. Chen, S., et al.: Secondary Job Scheduling in the Cloud with Deadlines. In: IEEE International Symposium on Parallel and Distributed Processing Workshops and PhD Forum (2011)
8. Gopalakrishnan Nair, T.R., Vaidehi, M., Rashmi, K.S., Suma, V.: An Enhanced Scheduling Strategy to Accelerate the Business Performance of the Cloud System. In: Satapathy, S.C., Avadhani, P.S., Abraham, A. (eds.) Proceedings of the InConINDIA 2012. AISC, vol. 132, pp. 461–468. Springer, Heidelberg (2012)
9. Lee, R., Jeng, B.: Load Balancing Tactics in Cloud. In: International Conference on Cyber-Enabled Distributed Computing and Knowledge Discovery (CyberC), pp. 447–454 (2011)
10. Jin, J., et al.: BAR: An Efficient Data Locality Driven Task Scheduling Algorithm for Cloud Computing. In: 11th IEEE/ACM International Symposium on Cluster, Cloud and Grid Computing, pp. 295–304 (2011)

PASA: Privacy-Aware Security Algorithm for Cloud Computing

Ajay Jangra and Renu Bala

Abstract. Security is one of the most challenging ongoing research area in cloud computing because data owner stores their sensitive data to remote servers and users also access required data from remote cloud servers which is not controlled and managed by data owners. This paper Proposed a new algorithm PASA (Privacy-Aware Security Algorithm) for cloud environment which includes the three different security schemes to achieve the objective of maximizing the data owners control in managing the privacy mechanisms or aspects that are followed during storage, Processing and accessing of different Privacy categorized data. The Performance analysis shows that the proposed algorithm is highly Efficient, Secure and Privacy aware for cloud environment.

Keywords: privacy, cloud computing, cryptography, security, intercept detection, data security.

1 Introduction

Cloud Computing reduces the investment in an organization's computing infrastructure and brought up the major advancement to the IT-Industry by providing the capability to use computing and storage resources on pay as you go basis. According to market research and analysis firm IDC there is 27% rise in usage of cloud services from 2008 to 2012 [4]. Cloud provides "x-as-a service (xaas)" where x could be software, hardware, platform or storage etc. In spite of all the advantages provided by cloud computing, according to the recent IDC enterprise survey 74% IT companies has to be taken security and privacy as top challenges that prevents the adoption of cloud computing. [4]

In Cloud Computing there is rapid expansion in security and privacy challenges because the storage, processing and accessing of sensitive data is all done through remote machines (CSP) that are not owned or even managed by data owner themselves. As there is storage and accessing of data from cloud servers, the

Ajay Jangra · Renu Bala
Dept. of Computer Science & Engg., UIET, Kurukshetra University Kurukshetra, India
e-mail: er_jangra@yahoo.co.in, ranarenu.22@gmail.com

A. Abraham and S.M. Thampi (Eds.): Intelligent Informatics, AISC 182, pp. 487–497.
springerlink.com

concerns about data confidentiality, authentication and integrity are being increased. Besides these issues there would also be chance of using a part of data or whole by cloud server for their financial gain which results the economic losses to data owner. The main reason behind the above defined issues is that the cloud servers are existed outside from trusted boundaries of data owner.

This paper presents privacy aware security algorithm (PASA). The main objective of this algorithm is to store, process and access the data according to their sensitivity or privacy need and data owner is able to control and manage the privacy mechanisms required to maintain the security of sensitive data. The above defined objective is achieved with the help of three different security schemes where each security scheme is different from each other in the manners of privacy aspects followed to store, process and access the privacy categorized data.

The proposed algorithm PASA is a novel and innovative data security scheme which maintains the network level and storage level security in terms of confidentiality, authentication and integrity. The remainder of the paper is organized as follows. Section 2 reviews the related research. Section 3 presents the data security algorithm and assumptions. Section 4 describes the proposed PASA algorithm for cloud environment. In section 5 we further continued with performance analysis of proposed algorithm. Finally section 6 concludes the paper and presents the direction for future work.

2 Related Work

A few research efforts are dealt directly with the issues of secure and privacy aware data storage and accessing in cloud computing. Sunil Sanka et al [1] proposed capability based access control technique which is combined approach of access control and cryptography for secure data access in cloud computing. The modified D-H key exchange model is also presented for user to access the outsourced data efficiently and securely from CSP's infrastructure. DR. S.N. Panda, Gaurav Kumar [3] proposed an effective intercept detection algorithm for packet transmission which uses the Exclusive-OR operation based unique encryption and decryption technique. In this scheme the forensic database also keeps the track of invalid unauthorized access and malicious activities for analyzing the behaviour of intercepts and to avoid such attempts in future.

Data privacy research in cloud computing is still in its early stages. Wassim Itani et al [2] proposed privacy as a service (PasS) protocol which ensures the privacy and legal compliance of customer data in cloud computing. PasS supports three trust levels in CSP: first is Full Trust in which CSP is fully under trusted domain of data owner. Second is compliance based trust in this the data owner trusts on CSP to store their data in encrypted form and third is No Trust where data owner is fully responsible to maintain the data privacy. Robert Gellman [8] presents the report which discuss the various risk imposed on data privacy by the adoption of cloud computing on data privacy. Pearson [9] and S. Pearson et al [10] presents various guidelines that are considered during designing of privacy aware cloud computing services. But the detailed analyses and evaluation of fully

Privacy-Aware security schemes are still an open research topic in the field of cloud computing.

3 Data Security Algorithm and Assumptions

In PASA algorithm, we assume that the four parties are involved during the communication for data storage and accessing: Data owner, cloud service provider, user and trusted module. We also assume that each party is preloaded with public keys of other so that there is no need of any PKI for distribution of public keys of each other's. For large storage and computation capacity we assume the CSP as conglomeration of several service providers like Google, Amazon and Microsoft.

4 Proposed Algorithm

This section describes the proposed algorithm PASA (**P**rivacy **A**ware **S**ecurity **A**lgorithm) for cloud environment. In which before storing and processing the data in storage pool of CSP, the data owner classified it into three categories according to their sensitivity:

- No privacy (NP): In this category the data is not sensitive and there is no need of any form of encryption. But for network security the data can be sent via SSL.
- Privacy with trusted provider (PTP): Here the cloud provider is fully trusted by data owner. Data owner provides the sensitive data to trusted provider where the cloud provider itself is responsible for encrypting the data for maintaining its confidentiality and integrity.
- Privacy with Non-Trusted Provider (PNTP): In this category the data is highly sensitive that also needs to be concealed from cloud provider. This kind of data is encrypted on data owner side and then stored at cloud service provider.

The main focus of PASA is to maximize the data owner's control in managing all aspects of privacy mechanisms required to maintain the security of sensitive data. To achieve above defined objective PASA further includes three different security schemes for each privacy categorized data (NP, PTP and PNTP) that has different privacy aspects according to the need of sensitive data.

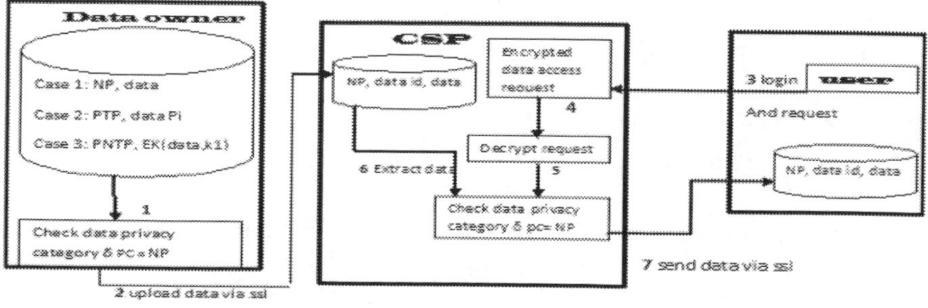

Fig. 1(a) Functional flow diagram of security scheme for NP category

The functional flow diagrams for NP and PTP are shown in figure 1(a)- 1(b) and their pseudo codes are shown in figure 2(a)-2(b).There is no encryption and decryption of data during the storage and accessing of NP privacy category data. The user only used double encryption during the request made for accessing the data from CSP. The only requirement is that each party must be authenticated before starting their communication. Whereas in PTP security scheme the CSP is responsible to make the security of data. CSP used the X-OR operation based encryption and decryption technique which automatically generates the key without any complexity. This scheme also has unique feature of intrusion detection which prevents from various malicious activities

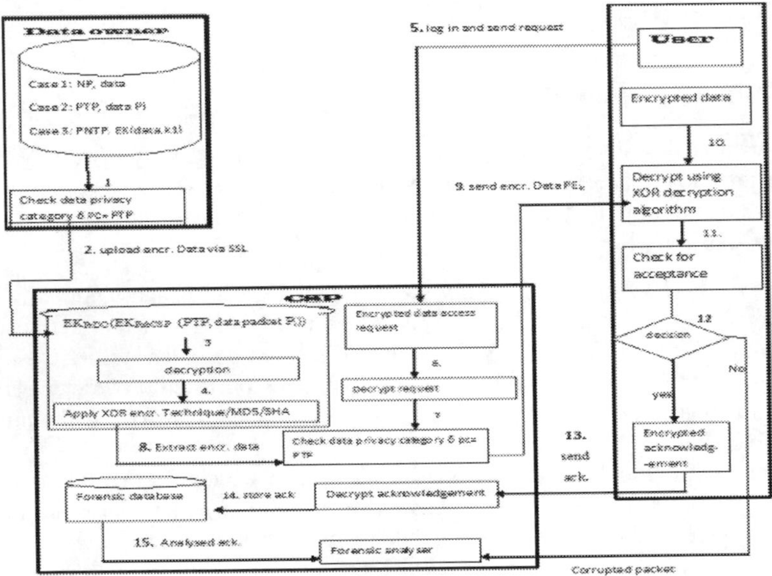

Fig. 1(b) Functional flow diagram of security scheme for PTP category

// Storage of NP privacy category data

Step 1: Data owner checks for data privacy category:

(a) δ_{pc} = NP // *here δ_{pc} represent data privacy category*

Step 2: Storage of data in storage pool of CSP:

(a) Data owner sends the data to CSP via SSL for network security where data is

$\delta_{DO} \longleftarrow$ (NP, data id, data)

// *δ_{DO} represent data send by data owner*

(b) CSP store data (δ_{DO}) in its storage pool

// secured data access between user and CSP for NP data category

Step3: User login to CSP and send request for data access

$REQ_{user} \longleftarrow EK_{PrUser}(EK_{PubCSP}$ (Uid, access control required, request))

Fig. 2(a) Algorithm for secure storage and accessing data from CSP for NP category

Step 4: CSP check for data category and send data to user:

(a) Cloud provider receive and decrypt the request with public key of user and his own private Key

(b) check to which category requested data belong

δ_{pc} = NP *//here data Privacy category is equal to No Privacy (NP)*

(c) CSP sends the data to user via SSL for network security where data is:

$\delta_{CSP} \leftarrow$ (NP, data)

Step 5: End of algorithm

Fig. 2(a) (*continued*)

// Storage of PTP privacy category data

Step 1: Data owner checks for data privacy category:

(a) δ_{pc} = PTP // *here δ_{pc} represent data privacy category*

(b) Go to step 2

Step 2: Storage of encrypted data in storage pool of CSP:

(a)Data owner send the data to CSP in encrypted form via SSL where data is

$\delta_{DO} \leftarrow EK_{PrDO}(EK_{PubCSP}$ (PTP, data packet P_i))

(b) CSP decrypt δ_{DO} and store the data in encrypted form using XOR gate based encryption technique:

1. CSP Generate a Random Key K_R by analyzing number of 1s in data PacketP_i .(a) Develop a routine to count bits in the Data Packet

(b) Set N := Count(P_i) // *Count Number of 1's in the Data Packet.*

(c) Set K_R :=N // *Store N in Random Number K_R*

2. Apply XOR (Exclusive-OR) Operation

(a) Set EK $:= P_i \oplus K_R$

(b) The Encrypted Packet E_K is generated using XOR Operation.

(c) Set PE_K : =

PF_i	E_K	PR_i

// *where PF_i and PR_i has key- id and E_K Utilize as Encr.packet*

3. Data Packet PE_K equipped for Transmission

// secured data access between user and CSP for PTP data category

Step3: User login to CSP and send request for data access

$REQ_{user} \leftarrow EK_{PrUser}(EK_{PubCSP}$(Uid, access control required, request))

Step 4: CSP check for data category and send data to user:

(a) Cloud provider receive and decrypt the request with public key of user and his own private Key

(b) check to which category requested data belong

δ_{Pc} = PTP *//here data Privacy category is equal to Privacy with trusted provider(PTP)*

(c) Go to step 5

Step5: Communication between CSP and user for accessing data of PTP category

Fig. 2(b) Algorithm for secure storage and accessing data from CSP for PTP category

(a) CSP send PE_K to user via SSL and user decrypt the data using following decrypting technique :

1: User Receive the Encrypted Packet PE_K

2: Check the Front PF_i and Rear End PR_i of Packet

if ($PF_i = PR_i$) Accept PFi and Set $K_R := PF_i$

Else goto Step 6

3: Generate the Binary Equivalent of K_R i.e. PB_i = Binary(K_R)

4: Perform XOR Operation i.e

$PB_i \oplus E_K$

Decryption the data E_k and get original packet P_i and Go to step 5

5: check whether the Packet is accepted or not

if (K_R = no of I's in each byte of decrypted data P_i)

Decryption Successful and Accept the Packet and user send ACK to csp where

ACK := $EK_{PrUser}(EK_{PubCSP}$(Uid, K_R, "successful acceptance of packet"))

CSP store ACk data to Forensic Database

Else goto step 6

6: Insert the Record of Corrupt Packet in Forensic Database of cloud

Step 6: End of algorithm

Fig. 2(b) (*continued*)

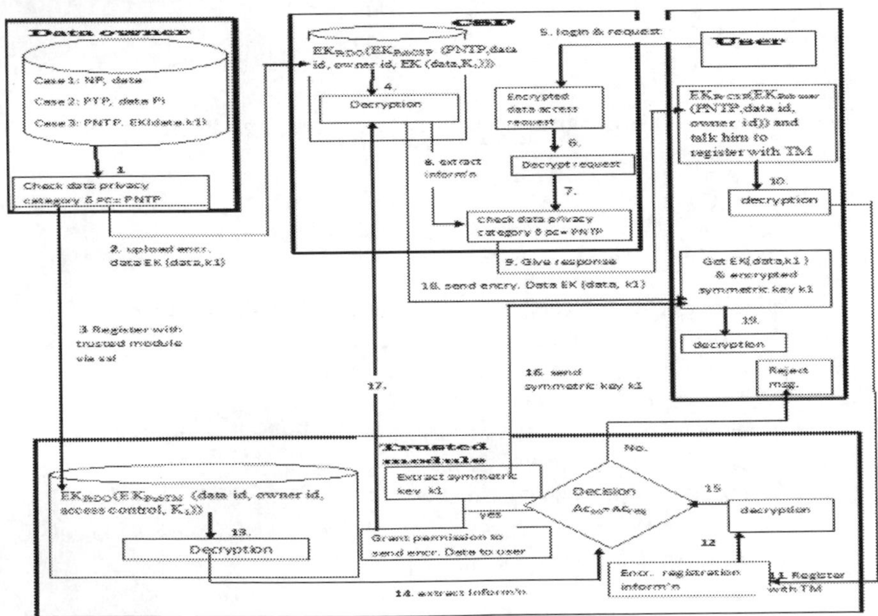

Fig. 3(a) Functional flow diagram of security scheme for PNTP category

The functional flow diagram and pseudo code for PNTP is shown in figure 3(a)- 3(b). The PNTP privacy category data is highly sensitive data which is also concealed from CSP. For this, the data owner provides the encrypted data to CSP and the symmetric key used to encrypt and decrypt the data is sent to trusted module so that both trusted module and CSP are not able to breach the security of PNTP privacy categorized data.

// Storage of PNTP privacy category data

Step 1: Data owner checks for data privacy category:

(a) δ_{pc} = PNTP // *here δ_{pc} represent data privacy category*

(b) Go to step 2

Step 2: Data owner Store encrypted data in storage pool of CSP and register with trusted module:

(a) Data owner encrypted the data with symmetric key K_1 and send data δ_{DO} to CSP via SSL:

$\delta_{DO} \longleftarrow EK_{PrDO}(EK_{PubCSP}$ (PNTP,data id, owner id, EK (data,K_1)))

(b) Data owner register with trusted module and send the following information via SSL:

$RI_{DO} \longleftarrow EK_{PrDO}(EK_{PubTM}$ (PNTP, data id, owner id, access control, K_1))

(c) TM decrypts the information and get symmetric key K_1 with all other necessary Inform'n

// secured data access between user and CSP for PNTP data category

Step3: User login to CSP and send request for data access

$REQ_{user} \longleftarrow EK_{PrUser}(EK_{PubCSP}$ (Uid, access control required, request))

Step 4: CSP check for data category and send data to user:

(a) Cloud provider receive and decrypt the request with public key of user and his own private Key

(b) check to which category requested data belong

δ_{pc} = PNTP //*here data Privacy category is equal to Privacy with non trusted provider (PNTP)*

(c) Go to step 5

Step 5: Communication between CSP and User for accessing data of PNTP category

(a) CSP decrypts the data δ_{DO} by public key of data owner and his own private key and get

Data $\longleftarrow$ (PNTP,data id, owner id,EK(data,k_1))

(b) CSP provides δ_{CSP} to user and talk him to register with trusted module where:

$\delta_{CSP} \longleftarrow EK_{Pr\ CSP}(EK_{Pub\ user}$ (PNTP,data id, owner id))

Step6: User decrypt δ_{CSP} and register with trusted module as:

$RI_{user} \longleftarrow EK_{PrUser}(EK_{PubTM}$(Uid, access control required, owner id, data id, PNTP))

Fig. 3(b) Algorithm for secure storage and accessing data from CSP for PNTP category

Step7: Trusted module receive the registration information from user and check whether the Request is valid or not

(a) check $AC_{DO} = AC_{requested}$ *// here the request is valid if the requested access control is equal to access control of data permitted by data owner*

"Request is valid"

Go to Step 8

(b) Otherwise Go to Step 11

Step 8: Trusted module communicate with CSP and user:

(a) Trusted module sends user information to CSP as:

Send (EK_{PrTM} (EK_{PubCSP} (Uid, access control, owner id, data id))) And grants permission to CSP to send the encrypted data EK(data,k_1) to user

(b) Trusted module send the symmetric key K_1 to user in encrypted form i.e.

$Enc_{key} \longleftarrow EK_{Pr\ TM}(EK_{Pub\ user}(K_1))$

// Enc_{key} represent encrypted symmetric key

Step9: CSP send the Encrypted data EK(data,k_1) to user via SSL.

Step10: User access the original data

(a) User decrypts the Enc_{key} by using public key of Trusted Module and his own private key, and get symmetric key K_1 i.e: $K_1 \longleftarrow DK_{pub\ TM}(DK_{pr\ user}(Enc_{key}))$

(b) By using K_1 user decrypt EK(data,k_1) and get original data

(c) Otherwise

Go to step 11

Step11: Send the rejection message to user:Send ("User request can't be granted")

Step 12: End of algorithm

Fig. 3(b) (*continued*)

5 Performance Analysis

This section analysed the security performance and computational efficiency of proposed algorithm (PASA).

A. Security Analysis: This subsection demonstrates the robustness of proposed security algorithm towards four factors (i) Confidentiality (ii) authentication (iii) Integrity (iv) performance against attacks.

1. Data confidentiality: The proposed scheme is mainly privacy aware and data owner centric where the confidentiality is maintained according to their sensitivity and the data owner is flexible to control and manage all privacy aspects of sensitive data. This is depicted in figure 4 which is a 3-D cube chart for performance analysis of NP, PTP and PNTP. Where X, Y, Z defines security, computational efficiency and approval of packet transmission. The thickness of triangle described the level of each performance characteristics for eg. the confidentiality is considered both by PTP and PNTP but PNTP has more thick

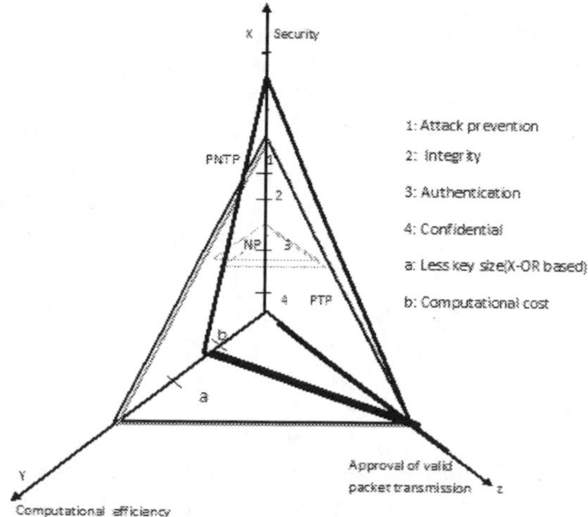

Fig. 4 3-D cube chart for performance analysis of NP, PTP and PNTP

triangle and hence more confidentiality. Both PTP and PNTP also considered the Z axis which approves the packet transmission according to acceptance condition.

2. *Authentication:* The involved parties in the proposed algorithm are authenticated by encrypting the data files every time with their own private key. In our scheme the authentication is also maintained by using SSL or login process.

3. *Integrity*: In the proposed algorithm (PASA) we ensures the integrity of data files by comparing the decimal form of key K_R with the number of 1's in each byte of decrypted packet. If they do not match the integrity violation is reported which is demonstrated in figure 2(b). The integrity of key is also maintained by forensic analyzer presents on server of cloud provider.

4. *Performance against attacks:* In PASA, double encryption scheme is used i.e. $Ek_{prsender}(Ek_{pub\ receiver}$ (data)) which has longer key length and prevents from *brute force attack* because this attack become difficult on cipher. The X-OR operation based cryptography technique used for PTP in figure 2(b) behave as intrusion detection. In which the forensic analyzer analysed the various malicious activities and prevents from various types of *intrusions.* Our scheme also prevents from *man-in-the-middle* attack by using forensic analyzer because it detects the modification of data or even key. For PTP and PNTP, the CSP stores the data in encrypted form which prevents it from *inside channel attack.*

B. computational complexity: The computational cost is increased rapidly in public key encryption when the key size increased but in proposed algorithm the X-OR operation based cryptography technique generates the key for data packet automatically without any complexity. The computational complexity of X-OR based technique as compared to public key is depicted in figure 5.

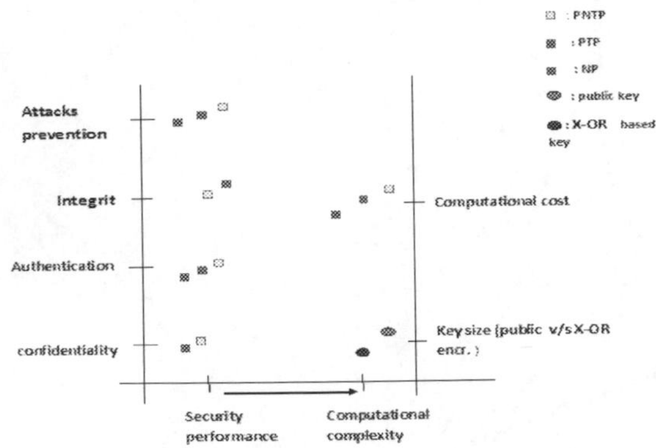

Fig. 5 Shows the performance characteristics for NP, PTP and PNTP Where PNTP is highly secured from attacks than PTP and NP. In this figure each category is defined by different colours and describes the level of their security performance and computational efficiency.

The comparison of computational cost between three security schemes i.e. for NP, PTP and PNTP is shown in table 1. Note that only dominant computation is considered i.e. encryption, decryption and authentication.

Table 1 Computational cost of NP, PTP and PNTP

	NP	PTP	PNTP
Data owner	1 DO Auth	$1EK_D$, 1 DO Auth	$2EK_D$, $1EK_{SYM}$, 2 DO Auth
CSP	$1DK_D$, 1 CSP Auth	$2DK_D$, $1EK_x$, 1 CSP Auth	$3DK_D$, $1EK_D$, 1 CSP Auth
USER	$1EK_D$, 1 user Auth	$1EK_D$, $1EK_x$	$2EK_D$, $2DK_D$, $1DK_{SYM}$, 2 user Auth
Trusted module	Not included	Not included	$2EK_D$, $2DK_D$

From this table it is clear that: Computational Cost for PNTP > Computational Cost for PTP > Computational Cost for NP where EK_D and DK_D describes the double encryption and decryption. DO Auth, User Auth and CSP Auth shows the authentication done by data owner, user and cloud provider. EKsym and DKsym describes the encryption and decryption by symmetric key, and EK_x and DK_x describes the X-OR based encryption and decryption.

6 Conclusion

This paper presented PASA (Privacy Aware Security Scheme) for cloud computing. The security solutions are mainly privacy aware and data owner centric. The paper discussed the three different privacy aware security schemes for NP, PTP and PNTP categorized data which followed the different privacy aspects according to their requirements. The paper also presented the performance analysis of proposed algorithm PASA. Future extension will: (1) consider the optimization of scheme in terms of bandwidth, memory and transmission channel consumption (2) Provide detailed evaluation of algorithm implementation.

Acknowledgement. The authors would like to thank and acknowledge the support of DR. S.N. Panda, Professor & Principal, Regional Institute of Management and Technology [RIMT], Punjab and Gaurav Kumar, Managing director, for scientific research and support providing during the designing of PASA.

References

[1] Sanka, S., Hota, C., Rajarajan, M.: Secure data access in cloud computing. In: 2010 IEEE 4th International Conference on Internet Multimedia Services Architecture and Application, 978-1-4244-7932-0/10 ©, pp. 1–6. IEEE (2010)

[2] Itani, W., Kayssi, A., Chehab, A.: Privacy as a service: privacy- aware data storage and processing in cloud computing architecture. In: 2009 Eighth IEEE International Conference on Dependable Autonomic and Secure Computing, 978-0-7695-3929-4/09 ©, pp. 711–717. IEEE (2009)

[3] Panda, S.N., Kumar, G.: IDATA – An Effective Intercept Detection Algorithm for Packet Transmission in Trust Architecture (POT-2010-0006). Selected for publication in IEEE Potentials ISSN: 0278-6648

[4] Gleeson, E.: Computing industry set for a shocking change (April 2009), MoneyWeek, http://www.moneyweek.com/investmentadvice/computing-industryset-for-a-shocking-change

[5] Jangra, A., Bala, R.: A Survey on various possible vulnerabilities and attacks in cloud computing environment. IJCBR 3(1) (January 2012) ISSN (Online): 2229-6166

[6] Wang, W., Li, Z., Owens, R., Bhargava, B.: Secure and efficient access to outsourced data. In: Proc. of ACM Cloud Computing Security Workshop 2009, pp. 55–65 (2009)

[7] di Vimercati, S.D.C., Foresti, S., Jajodia, S., Paraboschi, S., Samarati, P.: Over-encryption: Management of access control evolution on outsourced data. In: Proc. of VLDB 2007 (2007)

[8] Gellman, R.: WPF REPORT: Privacy in the Clouds: Risks to Privacy and Confidentiality from Cloud Computing, February 23 (2009)

[9] Pearson, S.: Taking Account of Privacy when Designing Cloud Computing Services. In: Proceedings of ICSE-Cloud 2009, Vancouver (2009)

[10] Pearson, S., Charlesworth, A.: Accountability as a Way Forward for Privacy Protection in the Cloud. HP Labs Technical Report, HPL-2009178 (2009), http://www.hpl.hp.com/techreports/2009/HPL-2009-178.pdf

[11] Diffie, W., van Oorschot, P.C., Wiener, M.J.: Authentication and authenticated key exchanges. Designs, Codes and Cryptography 2, 107–125 (1992)

[12] Kamara, S., Lauter, K.: Cryptographic Cloud Storage. In: Proc. of Financial Cryptography: Workshop on Real Life Cryptographic Protocols and Standardization (2010), http://research.microsoft.com/pubs/112576/crypto-cloud.pdf

A New Approach to Overcome Problem of Congestion in Wireless Networks

Umesh Kumar Lilhore, Praneet Saurabh, and Bhupendra Verma

Abstract. During past few years the wireless network has grown in leaps and bounds as it offered the end users more flexibility which enabled a huge array of services. All these services are achieved due to the network which is the backbone. The concept of the wireless network and the wireless devices also brings a lot of challenges such as energy consumption, dynamic configuration and congestion. Congestion in a network occurs when the demand on the network resources is greater than the available resources and due to increasing mismatch in link speeds caused by intermixing of heterogeneous network technologies. It is not limited to a particular point in the network but it can occur at various points in the network and it results into high dropping and queuing delay for packets, low throughput and unmaintained average queue length. It is factor that affects a network in a negative manner. Queue management provides a mechanism for protecting individual flows from congestion. One of the technique which uses Active Queue Management technique is RED. The basic idea behind RED queue management is to detect incipient congestion early and to convey congestion notification to the end-hosts, allowing them to reduce their transmission rates before queues in the network overflow and packets are dropped. Carnegie Mellon University proposed a new queue based technique for wireless network called CMUQ. The basic philosophy behind CMU queue is to prevent congestion. This paper introduces a new range variable and priority queue for existing CMU queue and for RED algorithm.

Keywords: Active queue management, Drop Tail, RED, Wireless Networks, CMUQ.

1 Introduction

Wireless network is a network set up by using radio signal frequency to communicate among computers and other network devices [1, 3].A wireless mobile ad-hoc network (MANET) is a network consisting of two or more mobile

e-mail: {umeshlilhore,praneetsaurabh}@gmail.com, bk_verma3@rediffmail.com

A. Abraham and S.M. Thampi (Eds.): Intelligent Informatics, AISC 182, pp. 499–506.
springerlink.com

nodes equipped with wireless communication and networking capabilities, but lacking any pre-existing network infrastructure. Each node in the network acts both as a mobile host and a router, offering to forward traffic on behalf of other nodes within the network [6].

1.1 Major Challenges in Wireless Network

Wireless Networks have following issues.

I. **Packet delay and drop -** A poor network performance can be offered due to congestion, e.g. high dropping and queuing delay for packets, low throughput and unmaintained average queue length which may not prevent the router buffers from building up, then dropping packets [2].

II. **Degradation of the throughput** - Degradation of throughput is an important issue in Wireless networks, due to congestion throughput degraded [3]. It is the ratio between the numbers of sent packets vs. received packets [7].

III. **Routing -** The concern of routing packets between any pair of nodes becomes a challenging task since the topology of the network is frequently changing[12]. Due to the random movement of nodes within the network the multicast tree is no longer static so multicast routing is another challenge. Routing is becoming more complex and challenging because routes between nodes may potentially contain multiple hops, than the single hop communication[1,8].

IV. **Internetworking -** Harmonious mobility management is a challenge in mobile device due to coexistence of routing protocols [8].

V. **Security and Reliability -** Wireless network has its particular security problems due to e.g. nasty neighbor relaying packets in spite of accumulation to the frequent vulnerabilities of wireless connection[13].

VI. **Quality of Service (QoS) -**It will be a challenge on pro-viding various qualities of service levels in a persistently varying environment[2].

VII. **Power Consumption -** Power-aware routing and Maintenance of power must be taken into consideration [6].

2 Congestion in Wireless Network

Congestion can be defined, "If, for any time interval, the total sum of demands on a source is more than its available capacity, the source is said to be congested for that Interval". Mathematically congestion can be defines by following equation.

$$\sum \text{Demand} > \text{Available Resources}$$

2.1 Congestion Control

When the aggregate demand for resources (e.g., bandwidth) exceeds the capacity of the link, congestion results. Congestion is characterized by delay and loss of packets in delivery. In congestion Control [4], system controls the network parameters after realizing congestion (reactive); whereas, in congestion avoidance; system controls the network parameters before congestion (proactive).

2.1.1 Congestion Control in Wired Network

Most of the traffic (around80%) in the Internet are TCP traffic [4]. TCP's congestion control in wired network is based on Adaptive Window Management technique. In this technique, congestion window (cwnd) increases or decreases based on packet drops and dupacks.

2.1.2 Congestion Control in Wireless Network

The adaptive window based congestion control mechanism used by TCP for wired network may not be appropriate for wireless network [9]. This is due to the time varying nature of a wireless channel and interference due to other nodes causing packet loss, which is different from packet loss due to congestion. But, TCP's congestion control mechanism does not discriminate packet loss due to congestion and that due to bad channel or interference; rather apply the same congestion control mechanism for both.

3 Literature Survey

The Wireless communication technology is playing an increasingly important role in data networks. Gianluigi Pibiri proposed an AQM algorithm for wireless network. In particular, a new priority queue algorithm is introduced which helps to manage both control and routing packets in wireless network. This priority queue algorithm offers an improvement in QoS for some services; specifically for TCP traffic & real time services that are sensitive to the loss of control and routing packets. Key AQM schemes (RED, REM, AVQ, Blue and RIO) are extended to incorporate the priority queue algorithm [1].G.G Ali Ahammed and Reshma Banu, analyzed several AQM algorithms with respect to their abilities of maintaining high resource utilization, identifying and restricting disproportionate bandwidth usage, and their deployment complexity [2]. Reshma Banu compared the performance of FRED, BLUE, SFB, and Choke based on simulation results, using RED and Drop Tail as the evaluation baseline [2].C V Hollot and Vishal Mishra use a previously developed nonlinear dynamic model of TCP to analyze and design Active Queue Management (AQM) control systems using RED [3].

Victor Firoiu, Marty Borden investigated a new mechanism for Internet congestion control in general, and Random Early Detection (RED). M. Alshanyour and Uthman presented Bypass-AODV, a local recovery protocol, to enhance the performance of AODV routing protocol by overcoming several inherited problems[5]. Bypass-AODV uses cross-layer MAC-notification to identify mobility-related link break, and then setup a bypass between the broken-link end nodes via an alternative node while keeps on the rest of the route[5].Alex Hills,David B.Johnson presented an approach to support mobile computing research, including the development of software which will allow seamless access to multiple wireless data networks, they presented a approach for building a wireless data network infrastructure at Carnegie Mellon University. This infrastructure will allow researchers and other members of the campus community to use mobile computers to gain access to data networks while they are on-campus or while they are off-campus in the greater Pittsburgh area [6].Thomas D Dyer, examine the performance of the TCP protocol for bulk data transfers in mobile ad hoc networks (MANETs). The number of TCP connections and compare the performances of three recently proposed on-demand (AODV and DSR) and adaptive proactive (ADV) routing algorithms. A simple heuristic, called fixed RTO, is developed to distinguish between route loss and network congestion and thereby improve the performance of the routing algorithms [7].

4 Problem Identification

Congestion leads to both waste of communication and energy resources of the nodes and also hampers the event detection reliability because of packet losses [1, 8]. The following problems were identified, based on Literature survey.

1) Packet loss rate at router and buffer level.
2) Delay of packet at transmission time between source and destination.
3) Poor packet delivery ratio %.

5 Proposed Algorithm

This section describes proposed algorithm, based RED, CMU and priority queue. The proposed algorithm CMUPQ is as follows.

Step 1: *Initialization of variable critical zone-CZ and radio zone –RZ with threshold value Th,*
If (CZ > Th)
Communication between nodes not possible
else If (RZ<= Th)
allow to enter in network range
Step 2:- *Calculate average queue size (avg) to determine whether or not to drop the packet.*
*avg =(1-wQ)avg+wQ*Q*
Step 3*-//The router first checks whether a node is in range of the router*
If Dis <=Rg
Then allow source to send packet to the router
Else if Dis > Rg
then display unreachable
Step 4*-//Now if node comes under the range of router. avg <- 0 and count <- -1*
//Initialization of variable (For each packet arrival)
if the queue is non
$avg \leftarrow (1-\omega q)\times avg+\omega q\times q$
else
$m \leftarrow f(time-q\_time)$
$avg \leftarrow (1-\omega q)m\times avg$
If $minth \leq avg < maxth$

```
Increment count with probability pa and mark the arriving packet
count <- 0
      else if maxth < avg
                mark the arriving packets
      count <- 0
      else count <- -1
When queue become empty
      q_time<-time
//Packet delivered successfully
```

6 Simulation and Result Analysis

This algorithm is implemented in NS-2 simulator.

6.1 Simulation Environment

For simulation following environments were used.

Terrain	1000*1000
Connection	CBR
Transport protocol	TCP
Node placement	Random way point
No of Nodes	10 -60
Simulation Time	200 Seconds
Seed	1

6.2 Performance Determining Parameters

For comparison the performance of proposed CMUQ algorithm with existing RED algorithm, following determining parameters were calculated.

I-Packet Delivery Ratio- Packet delivery ratio is calculated by dividing the number of packets received by the destination through the number of packets originated by the source.

II- Average End to End Delay -Average End to End Delay signifies the average time taken by packets to reach one end to another end (Source to Destination).

6.3 Simulation 1

In the first simulation parameter no. of nodes is used as a variable and rest of the parameters like as pause time, mobility and terrain were constant. Protocol for simulation is AODV. The outputs are given below.

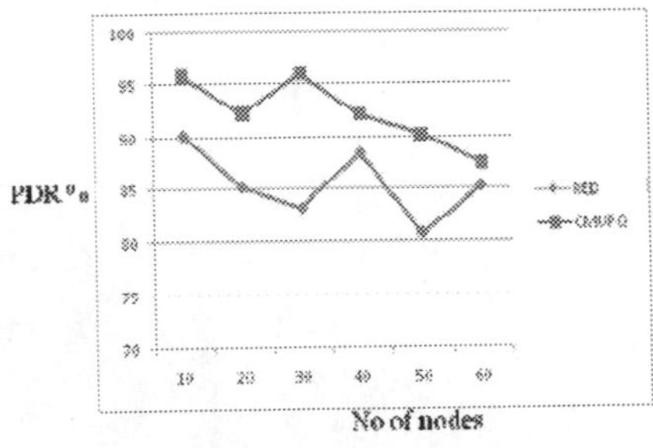

Fig. 1 Graph between packet delivery Ratio (PDR %) Vs No. of nodes

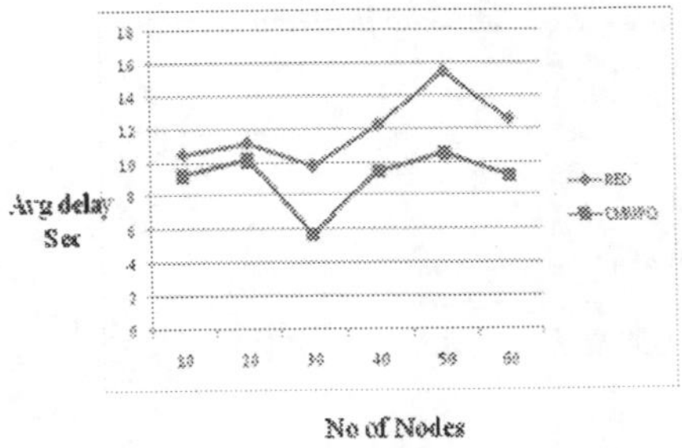

Fig. 2 Graph between Avg end to end delay Vs No. of nodes

The above graph 1 shows PDR (packet delivery ratio %). Here CMUPQ have higher packet delivery ratio % as compared to RED. Since higher packet delivery ratio % signifies the better performance so CMPQ scores and fairs better as compared to RED. Graph 2 shows results for Avg end to end delay in seconds. It indicates that packets delivered by using CMUPQ have lesser avg end to end delay as compared to RED. Lesser end to end is desired for better performance.

6.4 Simulation 2

In the Second simulation parameter Pause time is used as a variable; it increases from 10 to 50 seconds and rest of the parameters like as no. of nodes, mobility and terrain were constant. The following results are drawn.

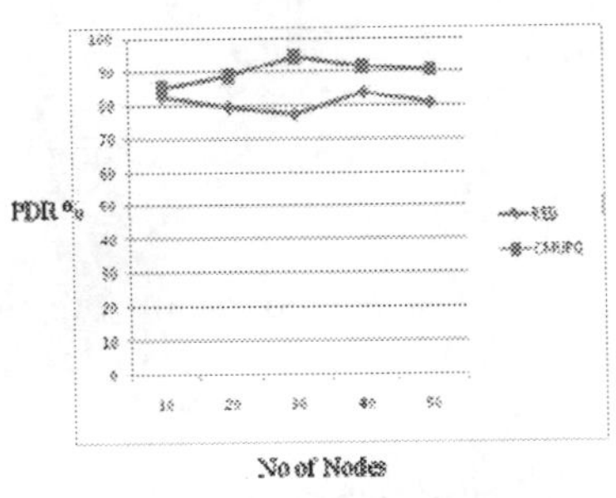

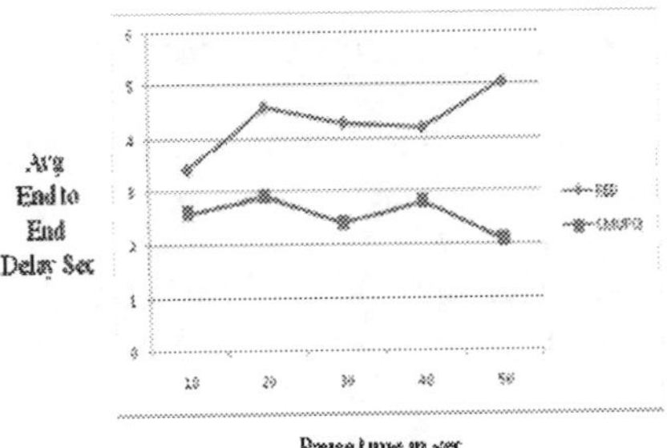

Fig. 3 Graph between packet delivery Ratio (PDR %) Vs No. of nodes

Fig. 4 Graph between Avg end to end delay Vs No. of nodes

The above graph 3 and 4 shows the performance of CMPQ is better against all the performance driven parameters that are such as Packet delivery ratio and average end to end delay. CMPQ proves to be more reliable and effective in countering the menace of congestion.

7 Conclusion and Future Work

This whole new approach counters the problem of congestion in a much more prudent manner as compared to its predecessors and gives very encouraging and significant results. A CMU Priority queue based AQM scheme, which uses the packet arrival rate and queue size to decide the probability of dropping/marking a packet to minimize the effect of net-work congestion. Comparison with the prior AQM schemes RED indicates that our algorithm not only outperforms the other schemes by achieving lower packet loss rate and higher throughput, but also is more resilient to dynamic workloads in maintaining a stable queue. Stability of the queue is a desirable feature of an AQM policy since it helps in lowering the packet loss rate, and the developed policy CMUPQ has this distinct feature.

Future work will see these schemes evaluated within the wider family of IEEE 802.11 protocols such as TORA, DSDV and number of dropped packet practically near zero. The scheme needs to be developed further and optimized for scenarios where the wireless stations are mobile. A more thorough analysis should be completed for more complex traffic mixes and topologies.

References

[1] Pibiri, G., Goldrick, C.M., Huggard, M.: Using Active Queue Management to Enhance Performance in IEEE802.11. Published by ACM 978-1-60558-621-2/09/10
[2] Ali Ahammed, G.G., Banu, R.: Analyzing the Performance of Active Queue Management Algorithms. In: IJCNC 2010, vol. 2(2) (2010)
[3] Hollot, C.V., Misra, V., Towsley, D., Gong, W.-B.: A Control Theoretic Analysis of RED. In: IEEE INFOCOM 2001 (2001)
[4] Gupta, P., Kumar, P.R.: The Capacity of Wireless Networks. IEEE Transactions on Information Theory 46(2) (2000)
[5] Dyer, T.D., Boppana, R.V.: A Comparison of TCP Performance over Three Routing Protocols for Mobile Ad Hoc Networks. ACM (2001) ISBN:1-58113-428-2
[6] Guo, S., Dang, C., Liao, X.: Distributed algorithms for resource allocation of physical and transport layers in wireless cognitive ad hoc networks. Springer Science Business Media, LLC (2010)
[7] Alshanyour, A.M., Baroudi, U.: Bypass AODV Improving Performance of Ad Hoc On-Demand Distance Vector (AODV) Routing Protocol in Wireless Ad Hoc Networks. ICST (2008)
[8] Marbach, P., Lu, Y.: Active Queue Management and Scheduling for Wireless Networks: The Single-Cell Case, vol. 3. IEEE (2008)
[9] Fengy, W.-C., Kandlurz, D.D., Sahaz, D., Shiny, K.G.: A Self-Configuring RED Gateway. UM-CSE-TR-349-2004 (November 2004)
[10] Firoiu, V., Borden, M.: A Study of Active Queue Management for Congestion Control. In: IEEE INFOCOM 2000 (2000)
[11] Wang, C., Daneshmand, M., Li, B., Sohraby, K.: A Survey of Transport Protocols for Wireless Networks. IEEE Network Magazine Special Issue on Wireless Networking
[12] Kulkarni, V., Juny, M., Falb, P.: Active queue management for tcp governed wireless networks. In: IEEE INFOCOM, vol. 7 (2002)

[13] Biaz, S., Vaidya De-randomizing, N.: Congestion Losses to Improve TCP Performance over Wired-Wireless Networks. In: Proc. of IEEE Global Telecommun. Conf.

[14] Balakrishnan, H., Padmanabhan, V., Seshan, S., Katz, R.: Eectiveness of loss labeling in improving TCP performance in wired/wireless networks. In: Proceedings of ICNP 2002: The 10th IEEE International Conference on Network Protocols, Paris, France (November 2002)

[15] Brakmo, L., O'Malley, S.: TCP-Vegas: New techniques for congestion detection and avoidance. In: ACM SIGCOMM 1994, pp. 24–35 (October 1994)

[16] Yavatkar, R., Bhagwat, N.: Improving End-to-End Performance of TCP over Mobile Internetworks. In: Proc of Workshop on Mobile Computing Systems and Applications (December 1994)

[17] Balakrishnan, H., Seshan, S., Amir, E., Katz, R.H.: Improving TCP/IP Performance over Wireless Networks. In: Proc. 1st ACM Conf. on Mobile Computing and Networking (November 1995)

[18] Floyd, S., McCanne, S.: Network Simulator, LBNL public domain software, Available via ftp from, `http://frp.ee.lbl.gov`

[19] Floyd, S.: TCP and Explicit Congestion Notification. ACM Computer Communication Review 24(5) (October 1994)

[20] Bakshi, B., Krishna, P., Vaida3, N., Pradhan, D.: Improving performance of TCP over wireless networks. In: Proceedings of 17th Int. Conf. on Distributed Computing Systems, pp. 693–708 (May 1997)

[21] Jacobson, V.: Modified TCP congestion avoidance algorithm (April 1990), Mailing list, end2endinterst@isi.edu

[22] Bhaskar Reddy, T., Ahammed, A., Banu, R.: Performance Comparison of Active Queue Management Techniques. IJCSNS 9(2), 405–408 (2009)

[23] Biaz, S., Vaidya, N.: "De-randomizing" Congestion Losses to Improve TCP Performance over Wired-Wireless Networks. In: Proc. of IEEE Global Telecommun. Conf. (2009)

[24] Jacobson, V.: Modified TCP congestion avoidance algorithm (April 1990), Mailing list, end2endinterst@isi.edu

[25] Bhaskar Reddy, T., Ahammed, A., Banu, R.: Performance Comparison of Active Queue Management Techniques. IJCSNS 9(2), 405–408 (2009)

[26] Biaz, S., Vaidya, N.: "De-randomizing" Congestion Losses to Improve TCP Performance over Wired-Wireless Networks. In: Proc. of IEEE Global Telecommun. Conf. (2009)

[27] Agrawal, D.P., Zing, Q.A.: Introduction to Wireless and Mobile Systems. Thomson Publication (2003)

CoreIIScheduler: Scheduling Tasks in a Multi-core-Based Grid Using NSGA-II Technique

Javad Mohebbi Najm Abad, S. Kazem Shekofteh, Hamid Tabatabaee, and Maryam Mehrnejad

Abstract. Load balancing has been known as one of the most challenging problems in computer sciences especially in the field of distributed systems and grid environments; hence, many different algorithms have been developed to solve this problem. Considering the revolution occurred in the modern processing units, using mutli-core processors can be an appropriate solution. one of the most important challenges in multi-core-based grids is scheduling. Specific computational intelligence methods are capable of dealing with complex problems for which there is no efficient classic method-based solution. One of these approaches is multi-objective genetic algorithm which can solve the problems in which multiple objectives are to be optimized at the same time. CoreIIScheduler, the proposed approach uses NSGA-II method which is successful in solving most of the multi-objective problems. Experimental results over lots of different grid environments show that the average utilization ratio is over 90% whilst for FCFS algorithm, it is only about 70%. Furthermore, CoreIIScheduler has an improvement ratio of 60% and 80% in wait time and makespan, respectively which is relative to FCFS.

Keywords: Grid Computing, Multi-objective Genetic Algorithm, Load Balancing, Multi-core processor.

1 Introduction

During the recent years the advancement in network technology and other communication infrastructures results in more complex systems such as cluster

Javad Mohebbi Najm Abad · Hamid Tabatabaee
Department of Computer Engineering, Quchan Branch, Islamic Azad University, Quchan, Iran
e-mail: {javad.mohebi,hamid.tabatabaee}@gmail.com

S. Kazem Shekofteh · Maryam Mehrnejad
Department of Computer Engineering, Ferdowsi University of Mashhad, Mashhad, Iran
e-mail: kazem.shekofteh@stu-mail.um.ac.ir, maryam.mehrnejad@gmail.com

A. Abraham and S.M. Thampi (Eds.): Intelligent Informatics, AISC 182, pp. 507–518.
springerlink.com

and grid systems. In addition, the need for more computational power leads us to use some kinds of high-speed systems such as grid systems in which internet connections are considered as the underlying communication platform.

Due to uneven task arrival patterns and unequal computing capabilities, some resources of grid may be overloaded while others may be underutilized. As a result, to take full advantage of such grid systems, task scheduling and resource management are essential functionalities provided at the service level of the grid software infrastructure.

Load balancing mechanism targets to equally spread the load on each computing node, maximizing their utilization and minimizing the total task execution time[1]. Generally, in terms of the location where the load balancing decisions are made, load balancing algorithms can be roughly categorized as centralized or decentralized. In centralized methods, there is an independent unit which makes decision about how to distribute the tasks among existing computational resources. On the other hand, in decentralized mode, almost all of the resources take part in making this decision.

Load balancing and task scheduling problems are generally NP-complete and there is no polynomial solution to solve such problems[2]. Hence intelligent methods such as evolutionary approaches are involved to overcome these difficulties. In order to solve a problem using these methods, first of all the objectives should be defined. In a grid environment there are multiple objectives to be considered such as processing node utilization, number of executed tasks per some periods of time and also the average execution time or makespan for each task. Therefore, any successful intelligent method is faced with a multi-objective problem.

One of the most successful and popular categories of intelligent approaches are evolutionary methods such as genetic algorithm, ant colony, etc. Genetic algorithms are inspirations from nature which generally use population of solutions in each generation and try to use genetic operators on some individuals in order to produce new populations and finally find the best solution in a timely fashion. There are several types of GAs proposed up to now. One of the most efficient classes of them which we used in this work is called Non-dominated Sorting Genetic Algorithm II (NSGA-II) which is a pareto-based approach which sorts individuals based on non-domination in different levels [3].

The rest of the paper is organized as follows. In the second section some works in the field of load balancing in grid environments are discussed. Section 3 introduces the major idea of the proposed approach and its components. The simulation of the proposed algorithm and also the results are discussed in the fourth section. Finally, we conclude the work and express potential future works in Section 5.

2 Related Work

There are several different approaches for solving the problem of load balancing proposed since about more than 50 years ago. Some of them use soft computing algorithms –especially GA – to overcome the difficulties of the problem. Since there are lots of works using GA-based methodologies we categorized these approaches into two classes. The methods in the first class use single-objective GA, which are named single objective-based methods, whereas the approaches in second category use multi-objective GA, called multi-objective approaches.

2.1 Single Objective-Based Methods

There are several kinds of load balancing methods using GA in which most of them include simple GA and hybrid multi-objective issues combined with other soft computing methods such as agents. As an example Cao *et al.* [4] proposed a method which involves both intelligent agents and multi-agent solutions in scheduling the tasks in a grid to balance the load among different resources. In [5] a two-step method is introduced in which an FCFS (First Come First Served) algorithm is first used when there are few tasks for running and uses GA when number of ready tasks increases. This method uses makespan as fitness function for each individual representing a possible schedule. In [6] Singh and Bawa proposed a combination of SexualGA and simulated annealing which tries to minimize makespan and cumulative delay in meeting user specified deadline time. Since simple GA is an insufficiently a powerful technique, in [7] a new class of GA based on roulette cycle is proposed which is called Rank-based Roulette Wheel Selection Genetic Algorithm (RRWSGA). Zhu *et al.* in [8] focused on the speed of convergence in their proposed GA. They overcome the low speed of convergence by using a Hybrid Adaptive Genetic Algorithm (HAGA) which is capable of performing a local search by appending the environmental configurations to the problem. On the other hand, in order to alleviate the drawbacks of rapid convergence, this method is always changing the ratio of crossover and mutation in a non-linear adaptive way.

2.2 Multi Objective-Based Methods

Most of the works based on single objective GA focus on optimizing just one parameter of the problem such as makespan or the cost of communication. Hence considering multiple parameters for optimizing is sometimes challenging since some of these objectives are in contrast with each other.

In [9] a pareto-based technique is proposed which does not convert the objectives into one, which is the way that most of the works in this field perform to simplify the problem. The authors in [8] formulate a novel Evolutionary Multi-Objective (EMO) approach by using the Pareto dominance and the objectives are formulated independently. Talukder *et al.*[10] used a Multi-objective Differential Evolution method in order to satisfy the user quality of service factors i.e. cost and time. A multi-objective resource load balancing scheme is proposed in [11] which deals with dependent relationships of jobs which regards multi-dimensional QoS metrics, completion time and execution cost of jobs, as multi-objective. In this work based on pareto sorting and niched sharing methods, optimal solutions are determined.

3 Proposed Method

Since the load balancing –and also task scheduling– problem involves multiple objectives in the environment such as processor utilization, task waiting time, it must be solved by a multi-objective technique. In the following we introduce the CoreIIScheduler and also the components of the grid environment. First of all the overall architecture of CoreIIScheduler is illustrated in Fig. 1.

In this architecture, tasks from different nodes are submitted to a central node called Task Manager which is responsible for scheduling and assigning them to the processing nodes.

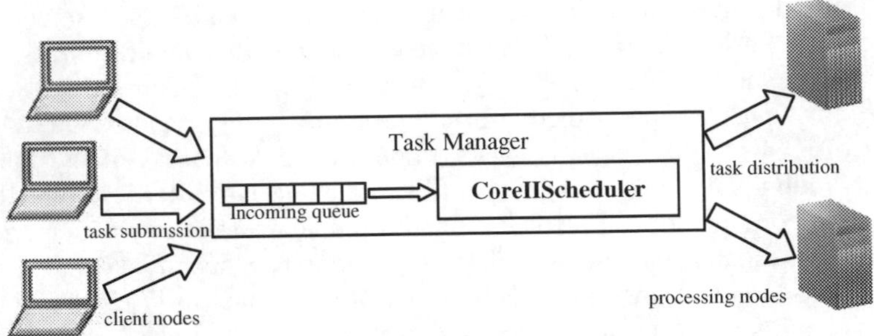

Fig. 1 The overall architecture of CoreIIScheduler

3.1 The Underlying Grid Environment

Since in this work it is assumed that the computational nodes are of multi-core type, they are capable of performing multiple tasks at a time. Multi-core processors have been considered as the new generation of processors and have been used in most general and scientific applications. This can cause the scheduler to take more than physical number of resources into account. The scheduler is a central unit and all the tasks are first submitted to this unit and after the scheduler performs the scheduling algorithm, the tasks are transferred to suitable resources.

The tasks are independent of each other. Each task has an arrival time which is the time that the task was input to the Task Manager. Another property of each task is the number of instructions the task is composed of. The number of instructions of the task is expressed by IC(i).

Computational power of the processors is expressed in terms of number of instructions completed per second. Computational requirement of tasks is evaluated in terms of number of instructions. Therefore the estimated time for completing a task is calculated as equation (1):

$$CT(i,j) = IC(i) / IPS(j) \quad (1)$$

Where CT(i,j) represents the estimated time required for task i to be completed on processor j and IPS(j) stands for the number of instructions that the processor j can execute per second. First of all, CoreIIScheduler will inspect the list of ready tasks in some time slots, then a multi-objective GA starts running, and it will return the optimal –or almost optimal– assignment of tasks to processors.

3.2 *The Proposed Genetic Algorithm*

As any GA consists of several components which require being adapted to the problem and its specifications, in the following we are introducing the components of the proposed GA.

3.2.1 Chromosome Coding

In most of methods of scheduling *n* tasks on *m* resources based on genetic approaches the well-known coding is a chromosome consists of *m* genotypes each represents a processor. This is illustrated in Fig. 2 for 6 processors.

The coding used in the proposed method is similar to the mentioned coding in Fig. 2. But the difference is that multi-core processors are capable of running multiple tasks simultaneously. This property should be embedded into the coding. So each genotype has multiple inner parts meaning that each genotype has a capacity of more than one task. An instance of such method is shown in Fig. 3.

In Fig. 3, the first processor is a quad-core processor and the second and third processors are dual-core. In order to formulate the representation of the chromosome we will show it as equation (2).

T3	T4	T7	T2	T10	T5

Fig. 2 A sample chromosome

T3 T8 T1	T6	T5 T9	T2 T7	T4

Fig. 3 A sample chromosome

$$S = \{\langle T3;T8;T1\rangle , \langle T6\rangle , \langle T5;T9\rangle , \langle T2;T7\rangle , \langle T4\rangle\} \quad (2)$$

The set of tasks assigned to processor *i* is called $Tasks_i$.

Each individual must be valid. The validity conditions are:

- Each task must be included in the list of assigned tasks of one and only one processor.
- The assigned tasks to each processor should be chosen from the list of ready tasks.

3.2.2 Fitness Function

Since CoreIIScheduler uses a multi-objective genetic algorithm, first two fitness functions are introduced in the following, and then the way they are combined will be discussed in the final proposed fitness function.

3.2.2.1 First Fitness Function: Makespan

The first fitness function which is used in most of the scheduling methods is mainly based on makespan metric. Makespan of a schedule is the finish time of the last task in the list of scheduled tasks. The smaller the value of the makespan is, the better the scheduling pattern is. Equation (3) shows how to calculate the makespan of each individual representing the scheduling pattern S.

$$makespan(S) = \max(makespan(Tasks_i)) \, , i \in P \tag{3}$$

P is the set of processors and makespan($Tasks_i$) represents the finish time of the list of assigned tasks to processor i. As one of the objectives, CoreIIScheduler targets to minimize this value.

3.2.2.2 Second Fitness Functions: Load Balance Level

One of the most important targets of the algorithm is to balance the distribution of the load among the existing processors. In order to evaluate the load balance of the scheduling pattern we use the method introduced in [5]. In this method the load balance of the scheduling pattern is calculated as equation (4).

$$\overline{loadbalance} = \left(\sum_{i=1}^{m} utilization_{P_i} \right) \Big/ m \tag{4}$$

In this equation m is the number of processors and $utilization_{Pi}$ represents the utilization level of processor i. In other word, the load balance of a scheduling pattern is calculated by dividing the summation of utilization of all processors by number of them. Utilization of each processor is calculated as equation (5). As another objective CoreIIScheduler tries to maximize this value.

$$utilization_{P_i} = makespan_{P_i} \Big/ makespan(S) \tag{5}$$

3.2.3 Selection Operator

The selection operator tries to choose a suitable region of search space for the next generation. There exist different selection operators such as roulette wheels, ranking, tournament etc.[12]. The selection method strongly affects the speed of the convergence. In the proposed approach a tournament selection is used with the selection parameter of 10. Using this approach, 10 individuals are selected randomly and the best individual among them is transferred to the next generation.

3.2.4 Crossover Operator

The crossover operator is responsible for producing a new individual from two existing individuals. The new individual must meet the validity conditions mentioned in 0. In a common one point cross over operator, a crossover point is selected randomly for two typical chromosomes and the left part of one individual is concatenated to the right part of another individual and vice versa. In the proposed method using the simple concatenation procedure after selecting a crossover point may result in an individual which does not essentially meet the validity conditions. This individual is named a malformed chromosome. This situation is illustrated in Fig. 4, i.e. the tasks which lead to malformed chromosome are colored in both right and left part of the individuals with the same color.

In order to overcome to this problem the following technique will be used: All of the tasks in the right part of the first parent which exist in the left part of the second parent should be removed from the set of candidate tasks for performing crossover. It means that those tasks will not participate in crossover operation and are fixed in their location and are directly transferred to new offspring with no change in their location. This is also done for the right part of the second parent.

3.2.5 Mutation Operator

The second effective operator is mutation which allows the algorithm to explore randomly far points in the search space different from the points in the current gradient. To this end, two gens of that individual and also one of their tasks are chosen randomly and exchanged with each other. The mutation over a sample individual is illustrated in Fig. 5. In this figure, two selected gens are highlighted and the selected tasks are colored red.

Since the individual holds the validity conditions, unlike the crossover operator, the resulting individual also holds those two conditions.

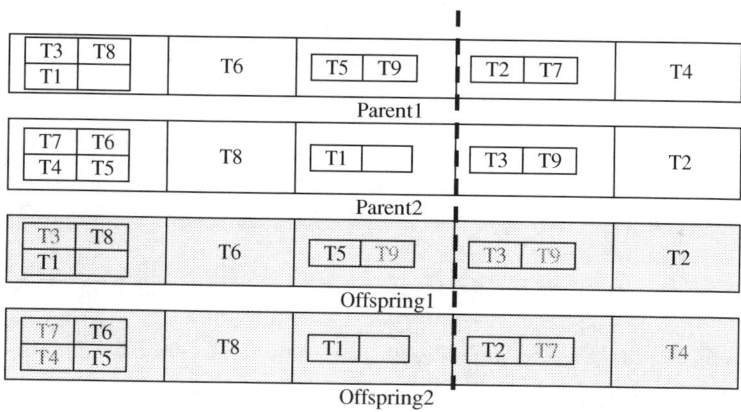

Fig. 4 A simple crossover over two parents which leads to two malformed offsprings

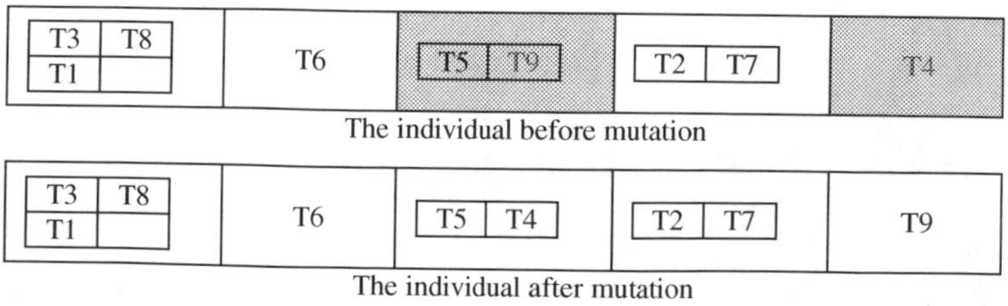

T3 T8 T1
T6
T5 T9
T2 T7
T4

The individual before mutation

T3 T8 T1
T6
T5 T4
T2 T7
T9

The individual after mutation

Fig. 5 The mutation operator

3.2.6 Optimization Algorithm – NSGA-II

Unlike most of the state-of-the-art works which perform a linear combination of objectives and convert it to a single-objective problem, in the proposed method we use NSGA-II, a multi-objective GA, which is mainly based on the pareto-front. This method is first introduced by K. Deb [3] in 2002 and suggests a non-dominated sorting-based multi-objective evolutionary algorithm. Simulation results on difficult test problems show that NSGA-II is able, for most problems, to find a much better spread of solutions and better convergence near the true Pareto-optimal front compared to the Pareto-archived evolution strategy and the strength-Pareto evolutionary algorithm - two other elitist MOEAs that pay special attention to creating a diverse Pareto-optimal front [3]. We use the implementation of NSGA-II located in the website of Dr. Deb's laboratory [13]. The crossover and mutation probability is set to 0.8 and 0.1 respectively. The population includes 50 individuals and the main loop is iterated 1000 times.

4 Experimental Results

CoreIIScheduler and also GRID environment are simulated and implemented in Matlab software. Generating the tasks is done using P-Method [14]. Each experiment is run for 50 times and the average of results is reported. The experiments are also performed using FCFS method and the results are compared to each other. The processors are considered to be homogenous.

4.1 Comparison Metrics

The experiments are designed such that in each of them an important metric of a scheduling method is compared between CoreIIScheduler method and FCFS. The metrics are as follows:

Average Processor Utilization: This metric shows how much the scheduling method averagely makes use of the processing power in unit time slot, i.e. how much percent of the power of a processor is used by the scheduling method.

Improvement Percentage for Wait Time: This ratio expresses the relative goodness of CoreIIScheduler compared to FCFS form the wait time point of view. This is illustrated in equation (6).

Improvement Percentage for Makespan: The relative goodness of CoreIIScheduler compared to FCFS form the makespan point of view.

4.2 Experiment No. 1: Comparing the Processor Utilization

In the first class of experiments, utilization of the processors is captured during the processing phases and is compared to FCFS method. Processor utilization is calculated using equation (5). Fig. 6 illustrates the comparison.

The parameters of these experiments are listed in Table 1 in which, μ_{IPT} represents average number of instructions per task, σ_{IPT} is the standard deviation of the number of instructions per task, CPP stands for the number of cores per processor, μ_{PS} and σ_{PS} are the average and standard deviation for the speed of processors in instructions per cycle, respectively.

The advantage of CoreIIScheduler over FCFS is obvious especially when number of tasks increases. When there are 400 tasks, utilization of processors is less than 75% while for CoreIIScheduler, this value is near 95%.

4.3 Experiment No. 2: Comparing the Average Wait Time

In this class of experiments, the decision factor is the average waiting time for the tasks. Wait time for a task is the time that task is entered to the system until the time that task starts to run on a processor.

Table 1 Parameters of the first experiments

# Processors	μ_{IPT}	σ_{IPT}	CPP	μ_{PS}	σ_{PS}	Network Sparsity
8	1000	50	variable in range [2-16]	30	10	0.5

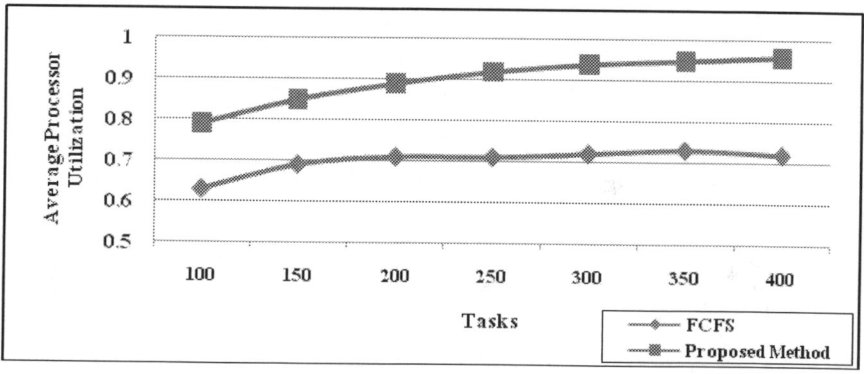

Fig. 6 Comparing CoreIIScheduler with FCFS over average processor utilization

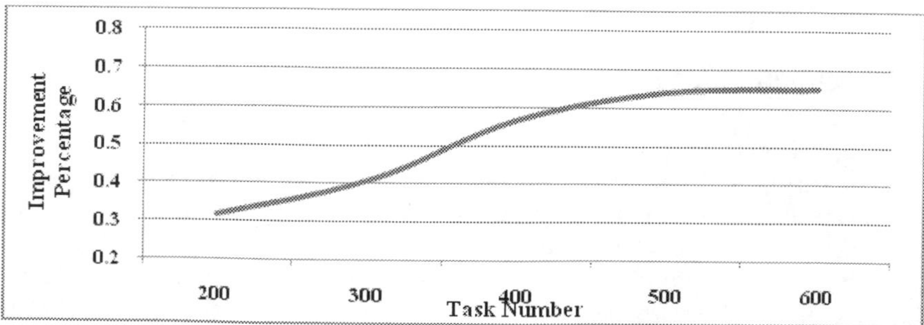

Fig. 7 Improvement percentage for waiting time of CoreIIScheduler relative to FCFS

The parameters of these experiments are listed in Table 1. The values of average waiting time are normalized relative to ones for FCFS. The relative improvement percentage of CoreIIScheduler over FCFS is drawn in Fig. 7. Improvement percentage for wait time is calculated using equation (6).

Increasing the number of tasks will result in more improvement ratio. The reason is that the task selection process of FCFC is performed with no intelligence.

4.4 Experiment No. 3: Comparing the Average Execution Time

In these experiments, the decision factor is makespan. The parameters of these experiments are identical to ones in Table 1, the only difference is the first column which is replaced with number of tasks which equals to 200.

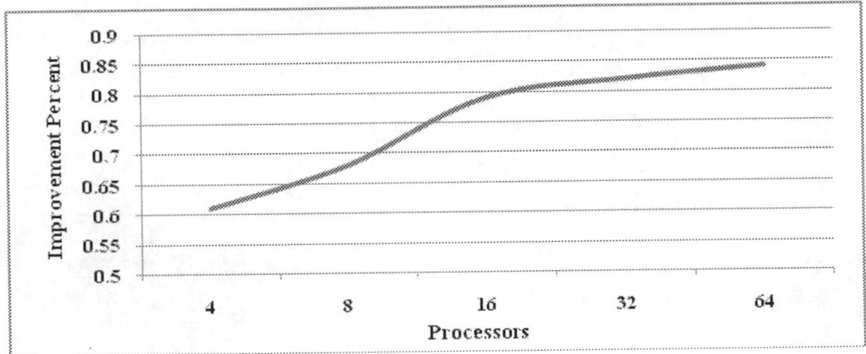

Fig. 8 Improvement percentage for makespan of CoreIIScheduler relative to FCFS

$$Improvement\ percentage = \frac{WaitTime_{FCFS} - WaitTime_{CoreIIScheduler}}{WaitTime_{FCFS}} \tag{6}$$

Equation (6) shows the normalization of makespan relative to ones for FCFS. The relative improvement of CoreIIScheduler over FCFS is illustrated in Fig. 8.

5 Conclusion and Future Works

Load balancing is one of the most important problems in computer sciences especially in the field of distributed systems and GRID environments and up to now there are lots of different works in this area. Computational intelligence techniques can overcome the difficulty of complex problems for which there is no effective classic solution. Multi-objective GA is one of these techniques. Load balancing in GRID environment is a suitable sample of a multi criterion problem which could be coded in multi-objective GA easily. The proposed algorithm uses

NSGA-II optimization method that is one of the most successful approaches in solving multi-objective problems.

Experimental results show that the average utilization level of the processors are more than 90% while this value for FCFS is about 70%. Furthermore, CoreIIScheduler has an improvement ratio of 60% in waiting time and 80% in makespan of tasks relative to FCFS.

Since the processors in this work are regarded as multi-core ones, we should take some considerations such as requiring a common memory while running tasks assigned to different cores of a processor. This consideration may lead the algorithm to be more precise and practical. This is our future work to make the scheduler involves the common memory among multiple cores to achieve a better solution.

Acknowledgments. The authors would like to thank officials in Azad University of Quchan for their financial support of the scientific project.

References

1. Zomaya, A.Y., Teh, Y.-H.: Observations on using genetic algorithms for dynamic load-balancing. IEEE Transactions on Parallel and Distributed Systems 12, 899–912 (2001)
2. Kwok, Y.K., Ahmad, I.: Dynamic Critical-Path Scheduling: An Effective Technique for Allocating Task Graphs to Multiprocessors. IEEE Trans. Parallel and Distributed Systems 7, 506–521 (1996)
3. Deb, K.: A fast and elitist multiobjective genetic algorithm: NSGA-II. IEEE Transactions on Evolutionary Computation 6, 182–197 (2002)
4. Cao, J., Spooner, D.P., Jarvis, S.A., Nudd, G.R.: Grid load balancing using intelligent agents. Future Generation Computer Systems 21, 135–149 (2005)
5. Li, Y., Yang, Y., Ma, M., Zhou, L.: A hybrid load balancing strategy of sequential tasks for grid computing environments. Future Generation Computer Systems 25, 819–828 (2009)
6. Singh, B., Bawa, S.: HybridSGSA: Sexual GA and Simulated Annealing based Hybrid Algorithm for Grid Scheduling. Global Journal of Computer Science and Technology 10, 78–81 (2010)
7. Abdulal, W., AlJadaan, O., Jabas, A., Ramchandraram, S.: Rank-based Genetic Algorithm with Limited Iteration for Grid Scheduling. In: First International Conference on Computational Intelligence, Communication Systems and Networks, India (2009)
8. Zhu, Y., Guo, X.: Grid Dependent Tasks Scheduling Based on Hybrid Adaptive Genetic Algorithm. In: Global Congress on Intelligent Systems (2009)
9. Grosan, C., Abraham, A., Helvik, B.: Multi-objective Evolutionary Algorithms for Scheduling Jobs on Computational Grids. In: International Conference on Applied Computing, Salamanca, Spain, pp. 459–463 (2007)
10. Talukder, A.K.M.K.A., Kirley, M., Buyya, R.: Multiobjective differential evolution for scheduling workflow applications on global Grids. Concurrency and Computation: Practice & Experience - Special Issue: Advanced Strategies in Grid Environments 21 (2009)

11. Ye, G., Rao, R., Li, M.: A Multiobjective Resources Scheduling Approach Based on Genetic Algorithms in Grid Environment. In: Fifth International Conference on Grid and Cooperative Computing Workshops (GCCW 2006), Hunan, China, pp. 504–509 (2006)
12. Gog, A., Dumitrescu, D., Hirsbrunner, B.: New Selection Operators based on Genetic Relatedness for Evolutionary Algorithms in Congress on Evolutionary Computation (2007)
13. Deb, K. (2009), Kanpur Genetic Algorithms Laboratory, `http://www.iitk.ac.in/kangal/`
14. Al-Sharaeh, S., Wells, B.E.: A Comparison of Heuristics for List Schedules using The Box-method and P-method for Random Digraph Generation. In: Proceedings of the 28th Southeastern Symposium on System Theory, pp. 467–471 (1996)

Author Index